Genome Engineering
for Crop Improvement

Genome Engineering for Crop Improvement

Edited by

Santosh Kumar Upadhyay
Department of Botany
Panjab University
Chandigarh, India

WILEY

This edition first published 2021

Registered Office(s)
John Wiley & Sons, Inc., 111 River Street, Hoboken, NJ 07030, USA
John Wiley & Sons Ltd, The Atrium, Southern Gate, Chichester, West Sussex, PO19 8SQ, UK

Editorial Office
The Atrium, Southern Gate, Chichester, West Sussex, PO19 8SQ, UK

For details of our global editorial offices, customer services, and more information about Wiley products visit us at www.wiley.com.

Wiley also publishes its books in a variety of electronic formats and by print-on-demand. Some content that appears in standard print versions of this book may not be available in other formats.

Library of Congress Cataloging-in-Publication Data

Name: Upadhyay, Santosh Kumar, editor.
Title: Genome engineering for crop improvement / edited by Santosh Kumar Upadhyay, Department of Botany, Panjab University, Chandigarh, India.
Description: First edition. | Hoboken, NJ : Wiley, 2021. | Includes bibliographical references and index.
Identifiers: LCCN 2020029680 (print) | LCCN 2020029681 (ebook) | ISBN 9781119672364 (hardback) | ISBN 9781119672388 (adobe pdf) | ISBN 9781119672401 (epub)
Subjects: LCSH: Crops–Genetic engineering. | Plant biotechnology.
Classification: LCC SB123.57 .G4835 2021 (print) | LCC SB123.57 (ebook) | DDC 631.5/233–dc23
LC record available at https://lccn.loc.gov/2020029680
LC ebook record available at https://lccn.loc.gov/2020029681

Cover Design: Wiley
Cover Image: © vchal / Getty Images

Set in 9.5/12.5pt STIXTwoText by SPi Global, Pondicherry, India
Printed and bound by CPI Group (UK) Ltd, Croydon, CR0 4YY

10 9 8 7 6 5 4 3 2 1

This book is dedicated to all the Corona warriors for their endless efforts to save the human life.

Contents

List of Contributors

Sipla Aggarwal
Department of Biotechnology
National Institute of Plant Genome Research
Jawaharlal Nehru University Campus
New Delhi, India

Sibtain Ahmad
University of Agriculture Faisalabad
Sub-Campus Depalpur
Okara
Punjab, Pakistan
Animal Breeding and Genetics
Faculty of Animal husbandry
University of Agriculture Faisalabad
Punjab, Pakistan

Habib Ali
Department of Agricultural Engineering
Khawaja Fareed University of Engineering and Information Technology
Rahim Yar Khan
Punjab, Pakistan

Kazim Ali
National Institute for Genomics and Advanced Biotechnology
National Agricultural Research Centre
Park Road, Islamabad 45500
Pakistan

Mohsin Ali
University of Agriculture Faisalabad
Sub-Campus Depalpur
Okara
Punjab, Pakistan

Imran Amin
Agricultural Biotechnology Division
National Institute for Biotechnology and Genetic Engineering (NIBGE)
Constituent College of Pakistan Institute of Engineering and Applied Sciences
Faisalabad, Pakistan

Galal Bakr Anis
Rice Research and Training Center (RRTC)
Rice Research Department
Field Crops Research Institute
Agricultural Research Center
Sakha
Kafr El-Sheikh, Egypt

Naveed Anjum
Agricultural Biotechnology Division
National Institute for Biotechnology and Genetic Engineering (NIBGE)
Constituent College of Pakistan Institute of Engineering and Applied Sciences
Faisalabad, Pakistan
Department of Biotechnology
University of Sialkot
Sialkot
Punjab, Pakistan

Bilal Ahmad Asad
University of Agriculture Faisalabad
Sub-Campus Depalpur
Okara
Punjab, Pakistan

Muhammad Qasim Aslam
Agricultural Biotechnology Division
National Institute for Biotechnology and Genetic Engineering (NIBGE)
Constituent College of Pakistan Institute of Engineering and Applied Sciences
Faisalabad, Pakistan

Muhammad Jawad Akbar Awan
Agricultural Biotechnology Division
National Institute for Biotechnology and Genetic Engineering (NIBGE)
Constituent College of Pakistan Institute of Engineering and Applied Sciences
Faisalabad, Pakistan

Uzma Badar
Cell and Systems Biology
University of Toronto
Toronto
Ontario, Canada

Poorwa Kamal Badola
CSIR-National Botanical Research Institute
Council of Scientific and Industrial Research (CSIR-NBRI)
Lucknow
Uttar Pradesh, India
Academy of Scientific and Innovative Research (AcSIR)
Ghaziabad,
Uttar Pradesh, India

Deepak Singh Bagri
Laboratory of Plant Molecular Biology
Department of Biotechnology
Dr Harisingh Gour Central University
Sagar
Madhya Pradesh, India
Department of Botany
Government Degree College
Amanganj, Panna
Madhya Pradesh, India

Kaushal Kumar Bhati
Louvain Institute of Biomolecular Science
Catholic University of Louvain
Louvain-la-Neuve, Belgium

Siddhant Chaturvedi
National Agri-Food Biotechnology Institute (NABI)
Department of Biotechnology
Ministry of Science and Technology (Government of India)
Mohali
Punjab, India
Department of Biotechnology
Panjab University
Chandigarh, India

Roni Chaudhary
National Agri-Food Biotechnology Institute (NABI)
Department of Biotechnology
Ministry of Science and Technology (Government of India)
Mohali
Punjab, India

Sameer Dixit
Department of Biology
University of Western Ontario
London
Ontario, Canada

Conrado Dueñas, Jr.
Trait and Genome Engineering Cluster
Strategic Innovation Platform
International Rice Research Institute
Manila, Philippines

Sajid Fiaz
Department of Plant Breeding and Genetics
The University of Haripur 22620
Haripur
Khyber Pakhtunkhwa, Pakistan

Mahmoud Mohamed Gaballah
Rice Research and Training Center (RRTC)
Rice Research Department
Field Crops Research Institute
Agricultural Research Center
Sakha
Kafr El-Sheikh, Egypt

Vikrant Gautam
ICAR-NBPGR
New Delhi, 110012, India

Ghyda Murad Hashim
Cell and Systems Biology
University of Toronto
Toronto
Ontario, Canada

Kathleen Hefferon
Cell and Systems Biology
University of Toronto
Toronto
Ontario, Canada

Boštjan Jenčič
Jozef Stefan Institute
Ljubljana, Slovenia

Ajinder Kaur
School of Agricultural Biotechnology
Punjab Agricultural University
Ludhiana
Punjab, India

Gurminder Kaur
Institute of Bioscience and Technology
Shri Ramswaroop Memorial University
Barabanki
Uttar Pradesh, India

Sher Aslam Khan
Department of Plant Breeding and Genetics
The University of Haripur 22620
Haripur
Khyber Pakhtunkhwa, Pakistan

Muhammad Zuhaib Khan
Agricultural Biotechnology Division
National Institute for Biotechnology and Genetic Engineering (NIBGE)
Constituent College of Pakistan Institute of Engineering and Applied Sciences
Faisalabad, Pakistan

Ivan Kreft
Nutrition Institute
Ljubljana, Slovenia

Ajay Kumar
Molecular Biology and Biotechnology
Council of Scientific and Industrial Research
National Botanical Research Institute (CSIR-NBRI)
Lucknow
Uttar Pradesh, India

Anil Kumar
Department of Biotechnology
National Agri-Food Biotechnology Institute
Mohali
Punjab, India

Pankaj Kumar
ICAR-CPRI
Shimla, 171001, India

Peter Kump
Jozef Stefan Institute
Ljubljana, Slovenia

Nirbhay Kumar Kushwaha
Department of Plant Biology
Swedish University of Agricultural Sciences
Uppsala, Sweden

Shahid Mansoor
Agricultural Biotechnology Division
National Institute for Biotechnology and Genetic Engineering (NIBGE)
Constituent College of Pakistan Institute of Engineering and Applied Sciences
Faisalabad, Pakistan

Varsha Meena
Department of Biotechnology
National Agri-Food Biotechnology Institute
Mohali
Punjab, India

Gela Myan B. Mota
Trait and Genome Engineering Cluster
Strategic Innovation Platform
International Rice Research Institute
Manila, Philippines

Mehmood Ali Noor
Institute of Crop Sciences
Chinese Academy of Agricultural Sciences
Key Laboratory of Crop Physiology and Ecology
Ministry of Agriculture
Beijing 100081, China

Azka Noureen
Agricultural Biotechnology Division
National Institute for Biotechnology and Genetic Engineering (NIBGE)
Constituent College of Pakistan Institute of Engineering and Applied Sciences
Faisalabad, Pakistan
University Institute of Biochemistry and Biotechnology (UIBB)
Pir Mehar Ali Shah Arid Agriculture University
Rawalpindi, Pakistan

Norman Oliva
Trait and Genome Engineering Cluster
Strategic Innovation Platform
International Rice Research Institute
Manila, Philippines

Ajay K. Pandey
Department of Biotechnology
National Agri-Food Biotechnology Institute
Mohali
Punjab, India

Ashish Kumar Pathak
Department of Botany
Charles University
Praha 2
Czech Republic

Primož Pelicon
Jozef Stefan Institute
Ljubljana, Slovenia

Paula Pongrac
Biotechnical Faculty
University of Ljubljana
Ljubljana, Slovenia
Jozef Stefan Institute
Ljubljana, Slovenia

Pravin Prakash
Molecular Biology and Biotechnology
Council of Scientific and Industrial Research
National Botanical Research Institute (CSIR-NBRI)
Lucknow
Uttar Pradesh, India

Krishan Mohan Rai
Department of Plant and Microbial Biology
University of Minnesota
Saint Paul
MN, USA

Hasthi Ram
Department of Agri-Biotechnology
National Agri-Food Biotechnology Institute
Mohali
Punjab, India

Gurwinder Singh Rana
AFC
New Delhi, India

Marjana Regvar
Biotechnical Faculty
University of Ljubljana
Ljubljana, Slovenia

Riyazuddin Riyazuddin
Department of Plant Biology and Doctoral School in Biology
Faculty of Science and Informatics
University of Szeged
Szeged, Hungary

Joy K. Roy
National Agri-Food Biotechnology Institute (NABI)
Mohali
Punjab, India

Shweta Roy
Department of Cell and Molecular Biology
Uppsala University
Uppsala, Sweden

Suchismita Roy
Skaggs School of Pharmacy and Pharmaceutical Sciences
University of California
San Diego
CA, USA

Ashish Sharma
CSIR-National Botanical Research Institute
Council of Scientific and Industrial Research (CSIR-NBRI)
Lucknow
Uttar Pradesh, India
Academy of Scientific and Innovative Research (AcSIR)
Ghaziabad
Uttar Pradesh, India

Chhaya Sharma
School of Agricultural Biotechnology
Punjab Agricultural University
Ludhiana
Punjab, India

Erum Shoeb
Cell and Systems Biology
University of Toronto
Toronto
Ontario, Canada

Akanchha Shukla
Department of Biology
University of Western Ontario
London
Ontario, Canada

Anuradha Singh
French Associates Institute for Agriculture and Biotechnology of Drylands (FAAB)
Ben-Gurion University of the Negev
Sede Boqer Campus
Sede Boqer, Israel

Harpal Singh
Department of Nematology
University of California
Riverside
CA, USA

Sudhir P. Singh
Center of Innovative and Applied Bioprocessing (DBT-CIAB)
Mohali
Punjab, India

Vinayak Singh
Plant Molecular Biology and Genetic Engineering Division
CSIR-National Botanical Research Institute
Lucknow
Uttar Pradesh, India

Inez H. Slamet-Loedin
Trait and Genome Engineering Cluster
Strategic Innovation Platform
International Rice Research Institute
Manila, Philippines

Praveen Soni
Department of Botany
University of Rajasthan
Jaipur
Rajasthan, India

Sushmita
Molecular Biology and Biotechnology
Council of Scientific and Industrial Research
National Botanical Research Institute
(CSIR-NBRI)
Lucknow
Uttar Pradesh, India
Academy of Scientific and Innovative
Research (AcSIR)
Ghaziabad
Uttar Pradesh, India

Rakesh Srivastava
Molecular Biology and Biotechnology
Council of Scientific and Industrial Research
National Botanical Research Institute
(CSIR-NBRI)
Lucknow
Uttar Pradesh, India

Siddharth Tiwari
National Agri-Food Biotechnology
Institute (NABI)
Department of Biotechnology
Ministry of Science and Technology
(Government of India)
Mohali
Punjab, India

Prabodh Kumar Trivedi
CSIR-National Botanical Research Institute
Council of Scientific and Industrial
Research (CSIR-NBRI)
Lucknow
Uttar Pradesh, India
Academy of Scientific and Innovative
Research (AcSIR)
Ghaziabad
Uttar Pradesh, India
CSIR-Central Institute of Medicinal and
Aromatic Plants (CSIR-CIMAP)
Lucknow
Uttar Pradesh, India

Johannes Tuen van Elteren
National Institute of Chemistry
Ljubljana, Slovenia

Santosh Kumar Upadhyay
Department of Botany
Panjab University
Chandigarh, India

Chandrama Prakash Upadhyaya
Laboratory of Plant Molecular Biology
Department of Biotechnology
Dr Harisingh Gour Central University
Sagar
Madhya Pradesh, India

Chirag Uppal
School of Agricultural Biotechnology
Punjab Agricultural University
Ludhiana
Punjab, India

Primož Vavpetič
Jozef Stefan Institute
Ljubljana, Slovenia

Srividhya Venkataraman
Cell and Systems Biology
University of Toronto
Toronto
Ontario, Canada

Praveen Chandra Verma
Molecular Biology and Biotechnology
Council of Scientific and Industrial Research
National Botanical Research Institute
(CSIR-NBRI)
Lucknow
Uttar Pradesh, India
Academy of Scientific and Innovative
Research (AcSIR)
Ghaziabad
Uttar Pradesh, India

Katarina Vogel-Mikuš
Biotechnical Faculty
University of Ljubljana
Ljubljana, Slovenia
Jozef Stefan Institute
Ljubljana, Slovenia

Tian Wang
Key Laboratory of Food Nutrition and Safety of Shandong Normal University
College of Life Science
Shandong Normal University
Jinan, China

Santosh Watpade
ICAR-IARI Regional Station (CHC)
Shimla, 171004, India

Amit Yadav
Zuckerberg Institute for Water Research (ZIWR)
Ben-Gurion University of the Negev
Israel

Afifa Younas
Department of Botany
Lahore College for Women University
Lahore, Pakistan

Kashaf Zafar
Agricultural Biotechnology Division
National Institute for Biotechnology and Genetic Engineering (NIBGE)
Constituent College of Pakistan Institute of Engineering and Applied Sciences
Faisalabad, Pakistan
Department of Biotechnology
Balochistan University of Information Technology
Engineering and Management Sciences (BUITEMS)
Quetta, Pakistan

Hongyan Zhang
Key Laboratory of Food Nutrition and Safety of Shandong Normal University
College of Life Science
Shandong Normal University
Jinan, China

Hongliang Zhu
College of Food Science and Nutritional Engineering
China Agricultural University
Beijing, China

Preface

About half of the world population is suffering from nutritional deficiency, which is due to either improper diet uptake, or non-availability and unaffordability of balanced nutritive food. Therefore, nutritional enrichment of common food items has been of great interest to fulfill the demands of nutritional deficiency. For instance, development of carotenoids-rich banana and tomato, and iron-rich rice and wheat has been proposed to solve major issues of vitamin A deficiency and anemia. Further, the distribution of important mineral elements in grains is also a major challenge for their bioavailability. For example, the majority of elements like iron are enriched in the aleurone layer of wheat grain and, therefore, it is not in flour for consumers. These are the major challenges which can be addressed by utilizing genome-engineering mechanisms.

Genome engineering has been evolved as an efficient method for targeted mutation and modification in the genome of various organisms. During the last decade, genome engineering has been utilized for numerous trait improvement programs in several important crop plants. Genome-engineering tools like ZFNs, TALENs, and CRISPR-Cas system have been used for *in-vivo* gene editing, as well as much biological pathway engineering for the nutritional improvement of various agricultural and horticultural crop plants. They have been of greater interest due to their precise editing mechanisms and very high specificity. Over the past few years, they have been utilized in numerous agricultural and horticultural crop plants including rice, wheat, legumes, tomato, potato, banana, grapes, etc. They are found to be highly specific and precise in their function. They have also been used for the engineering of numerous metabolic pathways to develop nutrient-rich produce. They are used to decrease the anti-nutrients in crop plants to improve bioavailability of minerals and vitamins, development of zero calorie/ sugar-free potato, indigestible starch-rich grains, allergenic gluten-free wheat and various other processes. Further, they can also be used for the mobilization of minerals from unavailable locations to the bioavailable location for grains.

Researchers have been working hard to increase the nutritional value of various important crop species by increasing the nutraceuticals and minerals such as carotenoids and flavonoid-rich cereals, tomato and fruit crops, iron-enriched rice, maize, and wheat, etc. On the other hand, people are also working to decrease the anti-nutrients such as phytic acids and other chelating elements in crops to increase the bioavailability of the available micronutrients. Further, development of sugar-free potato and gluten-free wheat, resistant starch rich rice and wheat is in great demand by people suffering from various diseases or as preventive measures.

The development of new strategies to cope up with the various challenges has always been a top priority over recent years. Agronomical practices, chemical applications, biofortifications and transgenic expression of protein-coding genes have been explored to alleviate these problems to some extent. However, new and specific technology is always in demand. Genome engineering can be the best alternative for nutritional improvement of agricultural and horticultural crops. Further, it will be highly specific to a particular gene and plant, therefore, it has minimal bio-safety risk. Taking this, together with progress made in crop improvement using genome engineering methods, this book has covered all the recent knowledge and development in the area related to the nutritional value, mineral and nutrient localization, and engineering for stress resistance as well as nutritional improvement of various crop plants. None of the currently available books have covered this topic at such a broad scale. Further, this area is of tremendous importance, not only in the present scenario but also for future decades.

Much effort has been made by all those involved to provide a well-informed and knowledgeable book. I hope that this stimulates discussion and inspires future research and discussion on such an important topic.

About the Editor

Dr. Santosh Kumar Upadhyay is currently working as an Assistant Professor at the Department of Botany, Panjab University, Chandigarh, India. Prior to this, Dr. Upadhyay was DST-INSPIRE faculty at the National Agri-Food Biotechnology Institute, Mohali, Punjab, India. He did his doctoral work at the CSIR-National Botanical Research Institute, Lucknow and received his PhD in Biotechnology from UP Technical University, Lucknow, India. He has been working in the field of Plant Biotechnology for more than 14 years. His present research focuses in the area of functional genomics. He is involved in the characterization of various insect toxic proteins from plant biodiversity, and defense and stress-signaling genes in bread wheat. His research group at PU has characterized numerous important gene families and long non-coding RNAs related to the abiotic and biotic stress tolerance and signaling in bread wheat. He has also established the method for genome editing in bread wheat using CRISPR-Cas system and developed a tool, SSinder, for CRISPR target-site prediction. His research contribution led to the publication of more than 55 research papers in leading journals of international repute. Further, there are more than five national and international patents,17 book chapters and four books to his credit.

In recognition of his strong research record he has been awarded NAAS Young scientist award (2017–2018) and NAAS-Associate (2018) from the National Academy of Agricultural Sciences, India, INSA Medal for Young Scientist (2013) from the Indian National Science Academy, India, NASI- Young Scientist Platinum Jubilee Award (2012) from the National Academy of Sciences, India, and Altech Young Scientist Award (2011). He has also been the recipient of the prestigious DST-INSPIRE Faculty Fellowship (2012), and SERB-Early Career Research Award, (2016) from the Ministry of Science and Technology, Government of India. Dr. Upadhyay also serves as a member of the editorial board and reviewer of a number of peer-reviewed international journals.

Acknowledgments

I am thankful to the Panjab University, Chandigarh, India for providing the facility to complete this book. I am grateful to all the esteemed authors for their exceptional contributions and reviewers for their critical evaluation and suggestions to improve the quality of the text.

I would like to thank Miss Rebecca Ralf (Commissioning Editor), Miss Kerry Powell (Managing Editor) and Nora Naughton (Copy Editor) from John Wiley & Sons, Ltd for their excellent management of this project, and anonymous reviewers for their positive recommendations about the book.

I also appreciate the support of my research students whose discussion and comments were very useful in shaping this book. I thank Dr. Prabodh K. Trivedi, Dr. Praveen C. Verma, Dr. Krishan Mohan Rai, Dr. Sameer Dixit, Dr. Sudhir P. Singh and Dr. Prashant Misra for direct or indirect help with this project. I wish to express my gratitude to my parents and my beloved wife for her endless support, patience, and inspiration. I thank my daughter, who missed me during this project. I would like to warmly thank the faculties and staff of the department and university for providing a great working environment. Last, but not least, my sincere thanks to Lord Krishna for endowing me to live with joy and success in the form of this book.

1

An Overview of Genome-Engineering Methods

Sushmita[1,3#], Gurminder Kaur[2#], Santosh Kumar Upadhyay[4], and Praveen Chandra Verma[1,3]

[1] *Molecular Biology and Biotechnology, Council of Scientific and Industrial Research, National Botanical Research Institute (CSIR-NBRI), Lucknow, Uttar Pradesh, India*
[2] *Institute of Bioscience and Technology, Shri Ramswaroop Memorial University, Barabanki, Uttar Pradesh, India*
[3] *Academy of Scientific and Innovative Research (AcSIR), Ghaziabad, Uttar Pradesh, India*
[4] *Department of Botany, Panjab University, Chandigarh, India*
 # Equal Contribution

CHAPTER MENU

1.1 Introduction

Agricultural practices, combined with advanced plant breeding and modern technologies, provided food security to millions of people. However, increasing global population demands significant increase in world food production (Parry and Hawkesford 2012). Nevertheless, climate change, depletion of natural resources, increased pollution, and political instabilities are a threat to the food and nutritional security for future generations in the twenty-first century. Unfortunately, the amount of remaining arable land is limited, necessitating an increase in food production on currently-used land. Compounding these challenges are the predicted crop losses due to extreme temperatures, pest attacks, and pathogen outbreaks. A powerful approach that may help overcome these challenges is to modify DNA sequences within plant chromosomes for trait improvement (Sedeek et al. 2019). Further, plants can be engineered to have increased tolerance to environmental stresses and pathogens (Han and Kim 2019; Ji et al. 2015; Makarova et al. 2011). In addition to improving the genetic makeup of the crops to meet increasing food demands and control crop loss, genome engineering can also be used to produce valuable plants or products for non-agricultural purposes (Chen et al. 2019). For example, there is great

Genome Engineering for Crop Improvement, First Edition. Edited by Santosh Kumar Upadhyay.

potential for plants to be used as bioreactors for pharmaceutical proteins. Genetic engineering for increasing the secondary metabolite production in plants would be another use of this technology which would help the perfumery, cosmetic and medical industries, as the secondary metabolites produced from plants have a number of uses (El-Mounadi et al. 2020). However, to realize the potential benefits of these applications, we must generate effective tools and approaches for editing plant DNA (Miroshnichenko et al. 2019; Tang and Tang 2017).

Introduction of programmed sequence-specific nucleases (SSNs) and their applications in precise genome editing unfurled a new dimension in genome engineering (Kim and Kim 2014; Voytas 2013). Over the last few decades, researchers reported a few important SSNs, which could be easily engineered and reprogrammed to create double-stranded breaks (DSBs) at the desired location inside the chromosome. There are three major genome engineering methods, ZFNs, TALENs, and CRISPR-Cas system (Figure 1.1A) (Jang and Joung 2019; Mahfouz et al. 2014), that have been utilized so far for a variety of purposes, and these have been discussed in detail in the coming sections. Further, we have also described the recently added CRISPR-Cpf1 system of genome engineering.

1.2 ZFNs

Zinc-finger nucleases are chimeric fusion proteins consisting of a DNA-binding domain and a DNA-cleavage domain. The DNA-binding domain is composed of a set of Cys2His2 zinc fingers (usually three to six). Each zinc finger primarily contacts 3 bp of DNA and a set of three to six fingers recognize 9–18 bp, respectively. The DNA-cleavage domain is derived from the cleavage domain of the FokI restriction enzyme. FokI activity requires dimerization; therefore, to site-specifically cleave DNA, two zinc-finger nucleases are designed in a tail-to-tail orientation (Kim et al. 1996).

Zinc-finger nucleases can be remodified to recognize different DNA sequences. However, one limitation with redirecting targeting is that it depends on the context of the host. For example, a zinc finger that recognizes GGG may not recognize this sequence when fused to other zinc fingers. As a result, the modular assembly of zinc fingers has had limited success (Ramirez et al. 2008). One of the more successful methods for redirecting targeting involves generating a library of three zinc-finger variants from a pre-selected pool of zinc-finger monomers (Maeder et al. 2008). The resulting library of zinc-finger arrays can then be interrogated using a bacterial two- hybrid screen, where binding of the zinc-finger array to a pre-determined sequence results in the expression of a selectable marker gene. This method has generated highly-active zinc-finger nuclease (ZFN) pairs for sites within animal and plant genomes. Since the development of ZFN technology, several studies have been done to engineer specific zinc-finger modules for each of the 64 codon triplets (Bae et al. 2003; Dreier et al. 2001; Pabo et al. 2001). Until now, several ZFNs have been designed and used in numerous species. The developments for more specific and efficient technologies also gave rise to fewer off-target effects. There are three most commonly available tools for engineering the ZF domains: context-dependent Assembly (CoDA), Oligomerized Pool Engineering (OPEN), and Modular Assembly (MA). Several softwares are available for designing engineered ZFs (ZiFiT), containing the database of ZFs (ZiFDB) and

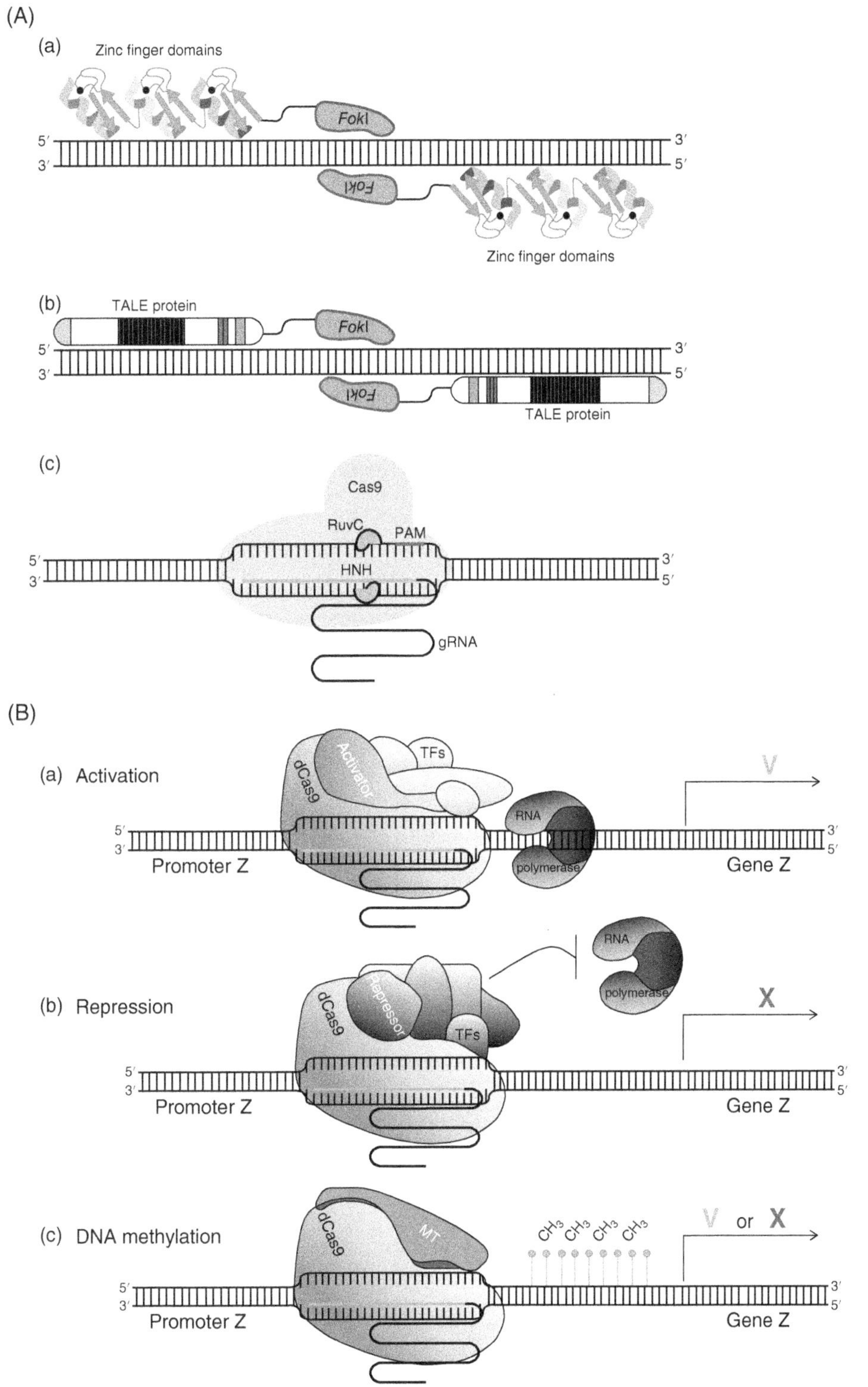

Figure 1.1 (A) Diagrammatic representation of (a) Zinc-finger nucleases (ZFNs), (b) Transcription activator-like effector nucleases (TALENs) and (c) Clustered regularly interspaced short palindromic repeats (CRISPR)/Cas9 mediates DSBs formation. (B) dCas9-based targeted genome regulation by (a) activation of gene expression, (b) repression of gene expression and (c) DNA methylation. *Source:* Adapted from Mahfouz et al. (2014) © 2014. Reproduced with the permission of John Wiley & Sons.

identification of potential targets for ZFNs in several model organisms (ZFNGenome) (Kim et al. 2009; Mandell and Barbas 2006; Sander et al. 2007).

Zinc-finger nucleases have been widely used for plant genome engineering. Plant species that have been modified using zinc-finger nucleases include, Arabidopsis, maize, soybean, tobacco, etc. (Ainley et al. 2013; Cai et al. 2009; Curtin et al. 2011; Lloyd et al. 2005; Marton et al. 2010; Osakabe et al. 2010; Shukla et al. 2009; Townsend et al. 2009; Wright et al. 2005; Zhang et al. 2010). With their relatively small size (~300 amino acids per zinc-finger nuclease monomer), and the further advancements in methods for redirecting targeting (Sander et al. 2011a), zinc-finger nucleases should continue to be an effective technology for editing plant.

1.3 TALENs

Transcription activator-like effectors nucleases (TALENs) are fusion proteins, consisting of a DNA-binding domain and a DNA-cleavage domain. Whereas the DNA-cleavage domain is the same between zinc-finger nucleases and TALENs (the catalytic portion of FokI), the DNA binding domains are different. The TALEN DNA-binding domain is derived from TALE proteins found in the plant pathogen Xanthomonas. These proteins are composed of direct repeats of 33–35 amino acids, and nearly all arrays found in Xanthomonas contain a final, half repeat, consisting of the first 20 amino acids from the normal repeat. Two amino acids within these repeats (positions 12 and 13) are responsible for recognizing a single nucleotide base (these amino acids are referred to as repeat-variable diresidues; RVDs). When the TALE effector code was broken (i.e. the relationship between the RVD and corresponding target base) (Boch et al. 2009; Moscou and Bogdanove 2009), the ability to redirect targeting, and their use as a genome engineering tool was realized (Christian et al. 2010; Li et al. 2011; Mahfouz et al. 2011). To make TALENs useful in gene targeting, the basic requirement is the modular assembly of repeat sequences containing the appropriate RVD corresponding to the nucleotide target. The most widely used RVDs and their nucleotide targets are HD, cytosine; NG, thymine; NI, adenine; NN, guanine, and adenine; NS, adenine, cytosine, and guanine; N*, all four nucleotides. This one-to-one correspondence of a single RVD to a single DNA base has eliminated construction challenges due to context-dependency seen with zinc-fingers and meganucleases. However, one limitation when using TALENs is that the target sequence must have thymine at the −1 position (Boch et al. 2009). Further, the long and repetitive nature of TALENs puts a strain on delivery methods where cargo capacity or stability is a limitation.

The assembly of engineered TALE repeat arrays can be challenging from nearly similar repeat sequences; therefore, a number of platforms have been designed to facilitate this assembly. These can be classified into three categories: standard restriction enzyme and ligation-based cloning methods (Huang et al. 2011; Sander et al. 2011); Golden Gate assembly methods (Briggs et al. 2012; Cermak et al. 2011; Engler et al. 2008) and solid-phase assembly methods (Heigwer et al. 2013; Wang et al. 2012).

Several online tools are available for designing TALE effectors to target specific gene sequence and off-target analysis. For example- E-TALEN (Lin et al. 2014), Scoring Algorithm for Predicting TALEN Activity (SAPTA) (Neff et al. 2013), Mojo-hand (Coordinators 2013),

TAL Effector-Nucleotide Targeter (TALE-NT), etc. TALE-NT is a collection of versatile web-based tools like-TALEN Targeter, TAL Effector Targeter, Target finder, Paired Target Finder, and TALEN Targeter Off-Target Counter (Christian et al. 2013).

Several studies have demonstrated the usefulness of TALENs in different plant species, including Arabidopsis (Zhang et al. 2013), tobacco (Wang et al. 2012; Wendt et al. 2013), barley (Li et al. 2012), rice (Shan et al. 2013a) and Brachypodium (Reyon et al. 2011). Taken together, the modular nature of TALE repeats, along with efficient methods for assembling repetitive DNA sequences (Garneau et al. 2010; Wang et al. 2012), have enabled TALENs to become one of the premier tools for plant genome engineering.

1.4 CRISPR-Cas System

The most recent addition to the SSN family is the CRISPR/Cas system that is normally present within bacteria and archaea, and provides an adaptive immunity against invading plasmids or viruses. CRISPR/Cas system functions to destroy invading nucleic acids by introducing targeted DNA breaks (Garneau et al. 2010).

There are three major types of CRISPR/Cas system: Types I – III (Makarova et al. 2011). The Type II system was adopted for genome engineering a few years ago (Cong et al. 2013; Zhang et al. 2011). In this system, two components enable targeted DNA cleavage: a Cas9 protein and an RNA complex consisting of a CRISPR RNA (crRNA; contains 20 nucleotides of RNA that are homologous to the target site) and a trans-activating CRISPR RNA (tracrRNA). Cas9 protein causes double-stranded DNA break at the sequences homologous to the crRNA sequence and upstream of a protospacer-adjacent motif (PAM) (PAM; e.g. NGG for *Streptococcus pyogenes* Cas9). For genome engineering purposes, the complexity of the system was reduced by fusing the crRNA and tracrRNA to generate a single-guide RNA (gRNA). Moreover, off-target cleavage is a limitation of the CRISPR/Cas system (Cho et al. 2014; Fu et al. 2013).

The target site recognition in CRISPR-Cas system is facilitated through RNA: DNA interaction (as opposed to a protein: DNA interaction used by meganucleases, zinc-finger nucleases, and TALENs). Redirecting of Cas9 targets involves modification of 20 nucleotides within the crRNA or gRNA. These 20 nucleotides are used to direct Cas9 binding and cleavage, the system has been shown to tolerate mismatches, with a higher tolerance closer to the 5′ end of the target sequence (Fu et al. 2013). Results from recent studies suggest the first 8–12 nucleotides, in addition to the PAM sequence, are most critical for target site recognition (Sternberg et al. 2014; Wu et al. 2014). To reduce off-targeting, several methods have been developed, including dual-nicking of DNA (Mali et al. 2013; Ran et al. 2013), a fusion of catalytically-dead Cas9 to FokI (Guilinger et al. 2014; Tsai et al. 2014) and shortening of gRNA sequence (Fu et al. 2014). Several softwares and programs have been developed in recent years for the identification of target sequences in the genome and the design of specific gRNA, which are listed in Table 1.1.

The Cas9 is an endonuclease consisting of two discrete nuclease domains: the HNH domain which is responsible for the cleavage of the DNA strand complementary to the guide RNA sequence (target strand) and the RuvC-like domain that cleaves the DNA strand opposite the complementary strand (Chen et al. 2014; Gasiunas et al. 2012; Jinek et al. 2012).

Table 1.1 List of available softwares and programs for designing gRNA.

Software	Features	Link	References
Cas-OFFinder	Identifies gRNA target sequence from an input sequence and checks off-target binding site	http://www.rgenome.net/cas-offinder	Bae et al. (2014)
Cas-Designer	Identifies gRNA target sequence from an input with low probability of off-target effect	http://www.rgenome.net/cas-designer/	Park et al. (2015)
Cas9 Design	Designs gRNA	http://cas9.cbi.pku.edu.cn/database.jsp	Ma et al. (2013)
E-CRISP	Designs gRNA	http://www.e-crisp.org/E-CRISP/designcrispr.html	Heigwer et al. (2014)
CRISPR-P	Designs gRNA	http://cbi.hzau.edu.cn/crispr2/	Lei et al. (2014)
CHOP	Identifies target site	https://chopchop.rc.fas.harvard.edu/	Montague et al. (2014)
CRISPR-PLANT	Designs gRNA	http://www.genome.arizona.edu/crispr/	Xie et al. (2014)
CCTop	Identifies candidate gRNA target sites with reduced off-target quality	http://crispr.cos.uni-heidelberg.de/	Stemmer et al. (2015)
CRISPRdirect	Identifics candidate gRNA target sequences	http://crispr.dbcls.jp/	Naito et al. (2015)
COSMID	Identifies target sites	https://crispr.bme.gatech.edu	Cradick et al. (2014)
CRISPR Finder	Identifies CRISPR	http://crispr.u-psud.fr/Server	Grissa et al. (2007)
CrisprGE	Identifies target sites	http://crdd.osdd.net/servers/crisprge	Kaur et al. (2015)
CRISPR Multitargeter	Identifies target sites	http://www.multicrispr.net	Prykhozhij et al. (2015)
CRISPRseek	Identifies target specific guide RNAs	http://www.bioconductor.org/packages/release/bioc/html/CRISPRseek.html	Zhu et al. (2014)
flyCRISPR	Identifies target sites and evaluate its specificity	http://flycrispr.molbio.wisc.edu	Gratz et al. (2014)
GT-SCAN	Identifies target sites and ranking them with their potential off target sites	http://flycrispr.molbio.wisc.edu	O'Brien and Bailey (2014)
sgRNAcas9	Identifies target sites with their potential off target sites	www.biootools.com	Xie et al. (2014)
SSFinder	Identifies target sites	https://code.google.com/p/ssfinder	Upadhyay and Sharma (2014)
ZiFiT	Identifies target sites	http://zifit.partners.org/ZiFiT	Mandell and Barbas (2006)
sgRNA Designer	Guide RNA design based on efficiency score	http://broadinstitute.org/rnai/public/analysis-tools/sgrna-design	Doench et al. (2014)

The double-strand breaks (DSBs) are repaired through Non-Homologous End Joining or Homology directed Repair in the presence of a template. Mutations in both nuclease domains (Asp10 → Ala, His840 → Ala) result in an RNA-guided DNA-binding protein without endonuclease activity that is called dCas9 (Jinek et al. 2012; Qi et al. 2013). This dCas9 is then supplemented with effector domains for the execution of distinct functions in the genome (Figure 1.1B). Fusion of a transcriptional activator VP64 with dCas9 exhibited targeted gene activation by altering the flowering time regulation in Arabidopsis (Xu et al. 2019). Similarly, dCas9-VP64 regulated transcriptional activation of endogenous genes and dCas9-SRDX-regulated transcriptional repression in *Arabidopsis* and tobacco (Lowder et al. 2015, 2018). These regulatory domains can also perform multiplex gene targeting using multiple sgRNAs. As a new dimension to CRISPR/Cas technology, there are the base editing enzymes, for example, cytidine deaminase fused with the dCas9, which can replace specific bases in the targeted region of DNA and RNA.

Several plant species have been edited using CRISPR/Cas system, including rice, wheat (Shan et al. 2013b; Upadhyay et al. 2013), sorghum (Jiang et al. 2013), tobacco (Li et al. 2013), Arabidopsis (Fauser et al. 2014; Feng et al. 2014; Li et al. 2013), *Brassica napus* (Kang et al. 2018), watermelon (Tian et al. 2018), etc. (Table 1.2). Moreover, dCas9 can be

Table 1.2 List of examples of genes edited by CRISPR Cas system in various plant species.

Plant system	Gene	Description of Experiment	References
Arabidopsis thaliana	GSS21/2	Host adaptation against *P. xylostella*	Chen et al. (2020)
Rice	EPFL9	a positive regulator of stomatal development	Yin et al. (2017)
	OsDEP1 OsROC5 OsPDS	Carotenoid biosynthesis, leaf morphology	Tang and Tang (2017)
	OsDL and *OsALS*	loss of midrib in the leaf blade	Endo et al. (2016)
	OsPDS, OsBEL	Herbicide resistant and Nutritional improvement	Xu et al. (2017)
	OsRLK, OsBEL	receptor-like kinases	Wang et al. (2017)
	OsPDS OsDL	Herbicide resistant and loss of midrib in the leaf blade	Tang et al. (2019)
	OsERF922	Enhanced resistance to blast disease	Wang et al. (2016)
	GW2, GW5, and TGW6	Improvement of grain weight	Xu et al. (2016)
	ALS	Enhanced herbicide resistance	Sun et al. (2016)
	SBEIIb and SBEI	Generation of high amylose rice	Sun et al. (2017)
	Hd 2, Hd 4, and Hd 5	Early maturity of rice varieties	Li et al. (2017)
	OsMATL	Induction of haploid plants	Yao et al. (2018)

(*Continued*)

Table 1.2 (Continued)

Plant system	Gene	Description of Experiment	References
	ALS	Herbicide resistance	Butt et al. (2017)
	EPSPS	Herbicide resistance	Li et al. (2016)
	ALS	Herbicide resistance	Endo et al. (2016)
	Gn1a, GS3, DEP1	Enhanced yield, dense erect panicles	Li et al. (2016)
	LAZY1	Tiller-spreading	Miao et al. (2013)
	OsSWEET13	Bacterial blight resistance	Zhou et al. (2015)
	OsDEP1 OsROC5	Herbicide resistant	Yao et al. (2018)
Soybean	FAD2-1A, FAD2-1B	Biosynthesis of lipids	Kim et al. (2017)
	ALS	Herbicide resistance	Li et al. (2015)
	GmPDS11&18	Carotenoid Biosynthesis	Du et al. (2016)
Tobacco	FAD2-1A, FAD2-1B	Lipid biosynthesis	Kim et al. (2017)
	NtPDS and NtPDR6	etiolated leaves for the psd mutant and more branches for the pdr6 mutant	Gao et al. (2015)
Cotton	Cloroplastos alterados (GhCLA)	Photosynthesis	Li et al. (2019)
	CABs, replication associated protein (Rep) and non-coding intergenic regions (IR), a-Satellite Rep and b-Sat IR.	CLCuD associated Begomoviruses (CABs) and Helper begomoviruses a and b satellites.	Iqbal et al. (2016), Uniyal et al. (2019)
Ashbya gossypii	HIS3, ADE2, TRP1, LEU2 and URA3	auxotrophic markers	Jiménez et al. (2020)
Maize	Maize glossy2 gene	Cuticular wax deposition	Lee et al. (2019)
	ARGOS8	Novel variants of ARGOS8 for drought-tolerance	Shi et al. (2017)
	ALS	Herbicide resistance	Svitashev et al. (2015)
	ZmIPK	Reduction of anti-nutritional compound phytic acid	Liang et al. (2014)
	TMS5	Thermosensitive male-sterile	Li et al. (2017)
Wheat	MLO	Resistance to powdery mildew	Wang et al. (2014)
	GW2	Enhanced yield	Zhang et al. (2018)
	EDR1	Powdery mildew resistance	Zhang et al. (2017)

Table 1.2 (Continued)

Plant system	Gene	Description of Experiment	References
Barley	HvPM19	Positive regulation of grain dormancy	Lawrenson et al. (2015)
	HvCKX1/3	Cytokinin metabolism and root morphology	Gasparis et al. (2019)
Tomato	SlMlo1	Resistant to powdery mildew	Nekrasov et al. (2017)
	SlWUS	Increased fruit size	Rodríguez-Leal et al. (2017)
	SlAGL6	Facultative parthenocarpy	Klap et al. (2017)
	SP5G	Day neutrality and early flowering	Soyk et al. (2016)
	SP, SP5G, CLV3, WUS, GGP1	Tomato domestication	Li et al. (2018)
	SlAN2	Anthocyanin biosynthesis	Zhi et al. (2020)
	SlJAZ2	Bacterial speck resistance	Ortigosa et al. (2019)
Potato	VInv	Reduction of sugar accumulation	Clasen et al. (2016)
	ALS	Herbicide resistance	Butler et al. (2016)
	Wx1	High amylopectin content	Andersson et al. (2017)
	StPPO	Reduced Enzymatic Browning in Tubers	González et al. (2020)
Cassava	MePDS	Carotenoid Biosynthesis	Odipio et al. (2017)
Citrus paradise	CsLOB1	Citrus canker resistance	Jia et al. (2017)
Citrus sinensis	CsLOB1 promoter	Citrus canker resistance	Peng et al. (2017)
Citrus paradise	CsLOB1 promoter	Alleviated citrus canker	Jia et al. (2016)
Manihot esculenta	EPSPS	Herbicide resistance	Hummel et al. (2018)
Cucumas sativus	eIF4E	Virus resistance	Chandrasekaran et al. (2016)
Camelina sativa	FAD2	Low polyunsaturated fatty acids	Jiang et al. (2017)
Linumusitatissimum	EPSPS	Herbicide resistance	Sauer et al. (2016)
Carrot	DcPDS and DcMYB113-like genes	depigmented carrot plants	Xu et al. (2019)
Strawberry	PDS	Carotenoid Biosynthesis	Wilson et al. (2019)

(*Continued*)

Table 1.2 (Continued)

Plant system	Gene	Description of Experiment	References
Grapes	VvPDS	Carotenoid Biosynthesis	Nakajima et al. (2017)
	CsLOB1	Increased resistance to citrus canker	Jia et al. (2017)
	MLO-1	Negative regulator of resistance to Powdery mildew	Malnoy et al. (2016)
Pear	MdPDS and TFL1.1	Carotenoid Biosynthesis & Floral repression	Charrier et al. (2019)
Apple	PDS	Carotenoid Biosynthesis	Charrier et al. (2019, Nishitani et al. (2016)
	DIMP-1/2/3	Negative regulator of resistance to blight disease	Malnoy et al. (2016)
Banana	eBSV	Resistance against Banana streak virus (eBSV)	Tripathi et al. (2019)
Chicory	phytoene desaturase gene (CiPDS)	Fruit ripening	Jansing et al. (2019)
Rubber	HbFT HbTFL1	Delayed-flowering and early-flowering	Fan et al. (2020)
Brassica napus L.)	BnLPAT2 and BnLPAT5	size of the oil bodies increased	Zhang et al. (2019)
Populus	phytoene desaturase gene 8	Carotenoid Biosynthesis	Fan et al. (2015)
Flax	EPSPS	Aromatic amino acid biosynthesis	Sauer et al. (2016)

fused with various epigenetic regulatory factors which can modulate DNA acetylation/methylation, post- translational histone modification, ubiquitination and protein sumoylation and phosphorylation to carry out epigenetic modifications (Shrestha et al. 2018; Yamamuro et al. 2016). This has been more recently explored in Arabidopsis for demethylation (Gallego-Bartolomé et al. 2018).

1.5 CRISPR-Cpf1

Introduction of CRISPR-Cpf1 also known as CRISPR-Cas12a further diversified the genome engineering methods (Figure 1.2). Cpf1 is an endonuclease which belongs to the class II CRISPR family (Alok et al. 2020; Zaidi et al. 2017). It was identified from *Prevotella* and *Francisella1*, therefore named as named as Cpf1. This system become popular as it was able to fill the gaps of previous genome editing tools and can substitute the CRISPR-Cas9

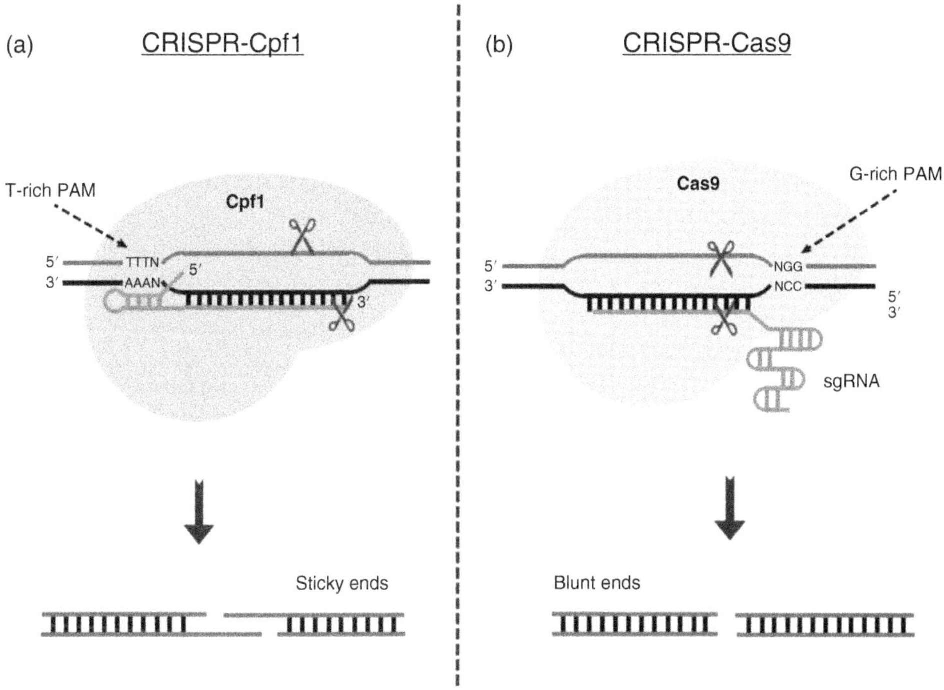

Figure 1.2 Comparison of various features of CRISPR-Cpf1 (a) and CRISPR-Cas9. *Source:* Adapted from Zaidi et al. (2017) © 2017. Reproduced with the permission of Elsevier.

in an efficient way (Moon et al. 2018). It is smaller in size than Cas9 and required a shorter CRISPR RNA for proper functioning (Liu et al. 2017). Unlike the CRISPR/Cas9, tracrRNA is no longer necessary to process Cpf1 associated mature CRISPR RNAs (Zetsche et al. 2015). Further, in contrast to the G rich PAM at 3′ end in CRISPR Cas9, it requires a T-rich PAM sequence at the 5′-end to perform cleavage efficiently, which enables the targeting of AT-rich regions in the genome. Further, Cpf1 creates staggered cuts where insertion of a DNA fragment is easily possible via HDR (Gao et al. 2017). The off-target binding of CRIPSR-Cpf1 is comparatively less than CRISPR-Cas9 which could be an additional benefit (Kim et al. 2016; Kleinstiver et al. 2016; Yan et al. 2017). A number of studies showed the application of CRISPR-Cpf1 for targeted genome editing in various eukaryotes including plants (Kim et al. 2017; Zetsche et al. 2015). It has also been used for multiple targets in the genome (Wang et al. 2017).

1.6 Conclusions

Genome-editing technologies enable us to make precise changes in the genome of any living organism. These changes may be diverse-insertion, deletion or even replacement of a particular stretch of DNA from the genome. Targeting these changes in accordance with our needs has always been the ambition of the scientific community. It was a challenging job until the development of recent genome-editing technologies. Among these, the most popular and successful ones till now have been ZFNs, TALENs, and CRISPR-Cas9. The breakthrough in the field of genome editing came after the discovery of CRISPR/Cas9 system as it is an RNA- guided and easy-to-design system. This system is fascinatingly repurposed as a genome editing tool and is till now the most efficient, cost effective and least demanding genome editing technique. Moreover, the recent addition of CRISPR-Cpf1 that is a variant of the CRISPR-Cas system further diversified the application of genome engineering tools by overcoming the various shortcomings of earlier systems.

Acknowledgements

Ms. Sushmita is grateful to DST-INSPIRE fellowship Program, DST, New Delhi, India for providing financial support. Authors are also thankful to CSIR, New Delhi for financial support in the form of "FBR Genome Editing Network Project" (MLP-007).

Institute's Manuscript Number is 'CSIR-NBRI_MS/2020/06/24.

References

Ainley, W.M., Sastry-Dent, L., Welter, M.E. et al. (2013). Trait stacking via targeted genome editing. *Plant Biotechnology Journal* 11: 1126–1134.

Alok, A., Sandhya, D., Jogam, P. et al. (2020). The rise of the CRISPR/Cpf1 system for efficient genome editing in plants. *Frontiers in Plant Science* 11: 264.

Andersson, M., Turesson, H., Nicolia, A. et al. (2017). Efficient targeted multiallelic mutagenesis in tetraploid potato (Solanum tuberosum) by transient CRISPR-Cas9 expression in protoplasts. *Plant Cell Reports* 36 (1): 117–128.

Bae, K.H., Do Kwon, Y., Shin, H.C. et al. (2003). Human zinc fingers as building blocks in the construction of artificial transcription factors. *Nature Biotechnology* 21 (3): 275–280.

Bae, S., Park, J., and Kim, J.S. (2014). Cas-OFFinder: a fast and versatile algorithm that searches for potential off-target sites of Cas9 RNA-guided endonucleases. *Bioinformatics* 30 (10): 1473–1475.

Boch, J., Scholze, H., Schornack, S. et al. (2009). Breaking the code of DNA binding specificity of TAL-type III effectors. *Science* 326: 1509–1512.

Brazelton, V.A. Jr., Zarecor, S., Wright, D.A. et al. (2015). A quick guide to CRISPR sgRNA design tools. *GM Crops and Food* 6 (4): 266–276.

Briggs, A.W., Rios, X., Chari, R. et al. (2012). Iterative capped assembly: rapid and scalable synthesis of repeat-module DNA such as TAL effectors from individual monomers. *Nucleic Acids Research* 40 (15): e117–e117.

Butler, N.M., Baltes, N.J., Voytas, D.F., and Douches, D.S. (2016). Geminivirus-mediated genome editing in potato (Solanum tuberosum L.) using sequence-specific nucleases. *Frontiers in Plant Science* 7: 1045.

Butt, H., Eid, A., Ali, Z. et al. (2017). Efficient CRISPR/Cas9-mediated genome editing using a chimeric single-guide RNA molecule. *Frontiers in Plant Science* 8: 1441.

Cai, C.Q., Doyon, Y., Ainley, W.M. et al. (2009). Targeted transgene integration in plant cells using designed zinc finger nucleases. *Plant Molecular Biology* 69: 699–709.

Cermak, T., Doyle, E.L., Christian, M. et al. (2011). Efficient design and assembly of custom TALEN and other TAL effector-based constructs for DNA targeting. *Nucleic Acids Research* 39 (12): e82–e82.

Chandrasekaran, J., Brumin, M., Wolf, D. et al. (2016). Development of broad virus resistance in non-transgenic cucumber using CRISPR/Cas9 technology. *Molecular Plant Pathology* 17 (7): 1140–1153.

Charrier, A., Vergne, E., Dousset, N. et al. (2019). Efficient targeted mutagenesis in apple and first time edition of pear using the CRISPR-Cas9 system. *Frontiers in Plant Science* 10: 40.

Chen, H., Choi, J., and Bailey, S. (2014). Cut site selection by the two nuclease domains of the Cas9 RNA-guided endonuclease. *Journal of Biological Chemistry* 289 (19): 13284–13294.

Chen, K., Wang, Y., Zhang, R. et al. (2019). CRISPR/Cas genome editing and precision plant breeding in agriculture. *Annual Review of Plant Biology* 70: 667–697.

Chen, W., Dong, Y., Saqib, H.S.A. et al. (2020). Functions of duplicated glucosinolate sulfatases in the development and host adaptation of Plutella xylostella. *Insect Biochemistry and Molecular Biology* 119: 103316.

Cho, S.W., Kim, S., Kim, Y. et al. (2014). Analysis of off-target effects of CRISPR/Cas-derived RNA-guided endonucleases and nickases. *Genome Research* 24: 132–141.

Christian, M., Cermak, T., Doyle, E.L. et al. (2010). Targeting DNA double-strand breaks with TAL effector nucleases. *Genetics* 186: 757–761.

Christian, M., Qi, Y., Zhang, Y., and Voytas, D.F. (2013). Targeted mutagenesis of Arabidopsis thaliana using engineered TAL effector nucleases. *G3: Genes, Genomes, Genetics* 3 (10): 1697–1705.

Clasen, B.M., Stoddard, T.J., Luo, S. et al. (2016). Improving cold storage and processing traits in potato through targeted gene knockout. *Plant Biotechnology Journal* 14 (1): 169–176.

Cong, L., Ran, F.A., Cox, D. et al. (2013). Multiplex genome engineering using CRISPR/Cas systems. *Science* 339: 819–823.

Coordinators, N.R. (2013). Database resources of the national center for biotechnology information. *Nucleic Acids Research* 41 (Database issue): D8.

Cradick, T.J., Qiu, P., Lee, C.M. et al. (2014). COSMID: a web-based tool for identifying and validating CRISPR/Cas off-target sites. *Molecular Therapy – Nucleic Acids* 3: e214.

Curtin, S.J., Zhang, F., Sander, J.D. et al. (2011). Targeted mutagenesis of duplicated genes in soybean with zinc-finger nucleases. *Plant Physiology* 156 (2): 466–473.

Doench, J.G., Hartenian, E., Graham, D.B. et al. (2014). Rational design of highly active sgRNAs for CRISPR-Cas9–mediated gene inactivation. *Nature Biotechnology* 32 (12): 1262.

Dreier, B., Beerli, R.R., Segal, D.J. et al. (2001). Development of zinc finger domains for recognition of the 5′-ANN-3′ family of DNA sequences and their use in the construction of artificial transcription factors. *Journal of Biological Chemistry* 276 (31): 29466–29478.

Du, H., Zeng, X., Zhao, M. et al. (2016). Efficient targeted mutagenesis in soybean by TALENs and CRISPR/Cas9. *Journal of Biotechnology* 217: 90–97.

El-Mounadi, K., Morales-Floriano, M.L., and Garcia-Ruiz, H. (2020). Principles, applications, and biosafety of plant genome editing using CRISPR-Cas9. *Frontiers in Plant Science* 11: 56.

Endo, A., Masafumi, M., Kaya, H., and Toki, S. (2016). Efficient targeted mutagenesis of rice and tobacco genomes using Cpf1 from Francisella novicida. *Scientific Reports* 6: 38169.

Endo, M., Mikami, M., and Toki, S. (2016). Biallelic gene targeting in rice. *Plant Physiology* 170 (2): 667–677.

Engler, C., Kandzia, R., and Marillonnet, S. (2008). A one pot, one step, precision cloning method with high throughput capability. *PLoS One* 3 (11): e3647.
Fan, D., Liu, T., Li, C. et al. (2015). Efficient CRISPR/Cas9-mediated targeted mutagenesis in Populus in the first generation. *Scientific Reports* 5: 12217.
Fan, Y., Xin, S., Dai, X. et al. (2020). Efficient genome editing of rubber tree (hevea brasiliensis) protoplasts using CRISPR/Cas9 ribonucleoproteins. *Industrial Crops and Products* 146: 112146.
Fauser, F., Schiml, S., and Puchta, H. (2014). Both CRISPR/Cas-based nucleases and nickases can be used efficiently for genome engineering in *Arabidopsis thaliana*. *The Plant Journal* 79: 348–359.
Feng, Z., Mao, Y., Xu, N. et al. (2014). Multigeneration analysis reveals the inheritance, specificity, and patterns of CRISPR/Casinduced gene modifications in *Arabidopsis*. *Proceedings of the National Academy of Sciences of the United States of America* 111: 4632–4637.
Fu, Y., Foden, J.A., Khayter, C. et al. (2013). High-frequency off-target mutagenesis induced by CRISPR-Cas nucleases in human cells. *Nature Biotechnology* 31 (9): 822–826.
Fu, Y., Sander, J.D., Reyon, D. et al. (2014). Improving CRISPR-Cas nuclease specificity using truncated guide RNAs. *Nature Biotechnology* 32 (3): 279.
Gallego-Bartolomé, J., Gardiner, J., Liu, W. et al. (2018). Targeted DNA demethylation of the Arabidopsis genome using the human TET1 catalytic domain. *Proceedings of the National Academy of Sciences* 115 (9): E2125–E2134.
Gao, J., Wang, G., Ma, S. et al. (2015). CRISPR/Cas9-mediated targeted mutagenesis in Nicotiana tabacum. *Plant Molecular Biology* 87 (1–2): 99–110.
Gao, L., Cox, D.B., Yan, W.X. et al. (2017). Engineered Cpf1 variants with altered PAM specificities. *Nature Biotechnology* 35 (8): 789.
Garneau, J.E., Dupuis, M.È., Villion, M. et al. (2010). The CRISPR/Cas bacterial immune system cleaves bacteriophage and plasmid DNA. *Nature* 468 (7320): 67–71.
Gasiunas, G., Barrangou, R., Horvath, P., and Siksnys, V. (2012). Cas9–crRNA ribonucleoprotein complex mediates specific DNA cleavage for adaptive immunity in bacteria. *Proceedings of the National Academy of Sciences* 109 (39): E2579–E2586.
Gasparis, S., Przyborowski, M., Kała, M., and Nadolska-Orczyk, A. (2019). Knockout of the HvCKX1 or HvCKX3 gene in barley (Hordeum vulgare L.) by RNA-guided Cas9 nuclease affects the regulation of cytokinin metabolism and root morphology. *Cell* 8 (8): 782.
González, M.N., Massa, G.A., Andersson, M. et al. (2020). Reduced enzymatic Browning in potato tubers by specific editing of a polyphenol oxidase gene via ribonucleoprotein complexes delivery of the CRISPR/Cas9 system. *Frontiers in Plant Science* 10: 1649.
Gratz, S.J., Ukken, F.P., Rubinstein, C.D. et al. (2014). Highly specific and efficient CRISPR/Cas9-catalyzed homology-directed repair in drosophila. *Genetics* 196 (4): 961–971.
Grissa, I., Vergnaud, G., and Pourcel, C. (2007). CRISPRFinder: a web tool to identify clustered regularly interspaced short palindromic repeats. *Nucleic Acids Research* 35 (suppl_2): W52–W57.
Guilinger, J.P., Thompson, D.B., and Liu, D.R. (2014). Fusion of catalytically inactive Cas9 to FokI nuclease improves the specificity of genome modification. *Nature Biotechnology* 32: 577–582.
Han, Y.J. and Kim, J.I. (2019). Application of CRISPR/Cas9-mediated gene editing for the development of herbicide-resistant plants. *Plant Biotechnology Reports* 13: 447–457.
Heigwer, F., Kerr, G., Walther, N. et al. (2013). E-TALEN: a web tool to design TALENs for genome engineering. *Nucleic Acids Research* 41 (20): e190–e190.

Heigwer, F., Kerr, G., and Boutros, M. (2014). E-CRISP: fast CRISPR target site identification. *Nature Methods* 11 (2): 122.
Huang, P., Xiao, A., Zhou, M. et al. (2011). Heritable gene targeting in zebrafish using customized TALENs. *Nature Biotechnology* 29 (8): 699–700.
Hummel, A.W., Chauhan, R.D., Cermak, T. et al. (2018). Allele exchange at the EPSPS locus confers glyphosate tolerance in cassava. *Plant Biotechnology Journal* 16 (7): 1275–1282.
Iqbal, Z., Sattar, M.N., and Shafiq, M. (2016). CRISPR/Cas9: a tool to circumscribe cotton leaf curl disease. *Frontiers in Plant Science* 7: 475.
Jang, G. and Joung, Y.H. (2019). CRISPR/Cas-mediated genome editing for crop improvement: current applications and future prospects. *Plant Biotechnology Reports* 13 (1): 1–10.
Jansing, J., Schiermeyer, A., Schillberg, S. et al. (2019). Genome editing in agriculture: technical and practical considerations. *International Journal of Molecular Sciences* 20 (12): 2888.
Ji, X., Wang, D., and Gao, C. (2015). CRISPR editing-mediated antiviral immunity: a versatile source of resistance to combat plant virus infections. *Science Bulletin* 60: 1332.
Jia, H., Orbovic, V., Jones, J.B., and Wang, N. (2016). Modification of the PthA4 effector binding elements in type I Cs LOB 1 promoter using Cas9/sg RNA to produce transgenic Duncan grapefruit alleviating XccΔpthA4: dCs LOB 1.3 infection. *Plant Biotechnology Journal* 14 (5): 1291–1301.
Jia, H., Zhang, Y., Orbović, V. et al. (2017). Genome editing of the disease susceptibility gene Cs LOB 1 in citrus confers resistance to citrus canker. *Plant Biotechnology Journal* 15 (7): 817–823.
Jiang, W., Zhou, H., Bi, H. et al. (2013). Demonstration of CRISPR/Cas9/sgRNA-mediated targeted gene modification in Arabidopsis, tobacco, sorghum and rice. *Nucleic Acids Research* 41 (20): e188–e188.
Jiang, W.Z., Henry, I.M., Lynagh, P.G. et al. (2017). Significant enhancement of fatty acid composition in seeds of the allohexaploid, Camelina sativa, using CRISPR/Cas9 gene editing. *Plant Biotechnology Journal* 15 (5): 648–657.
Jiménez, A., Hoff, B., and Revuelta, J.L. (2020). Multiplex genome editing in Ashbya gossypii using CRISPR-Cpf1. *New Biotechnology* 57: 29–33.
Jinek, M., Chylinski, K., Fonfara, I. et al. (2012). A programmable dual-RNA–guided DNA endonuclease in adaptive bacterial immunity. *Science* 337 (6096): 816–821.
Kang, B.C., Yun, J.Y., Kim, S.T. et al. (2018). Precision genome engineering through adenine base editing in plants. *Nature Plants* 4 (7): 427–431.
Kaur, K., Tandon, H., Gupta, A.K., and Kumar, M. (2015). CrisprGE: a central hub of CRISPR/Cas-based genome editing. *Database* 2015: bav055.
Kim, H. and Kim, J.S. (2014). A guide to genome engineering with programmable nucleases. *Nature Reviews Genetics* 15 (5): 321–334.
Kim, Y.G., Cha, J., and Chandrasegaran, S. (1996). Hybrid restriction enzymes: zinc finger fusions to Fok I cleavage domain. *Proceedings of the National Academy of Sciences* 93 (3): 1156–1160.
Kim, H.J., Lee, H.J., Kim, H. et al. (2009). Targeted genome editing in human cells with zinc finger nucleases constructed via modular assembly. *Genome Research* 19 (7): 1279–1288.
Kim, D., Kim, J., Hur, J.K. et al. (2016). Genome-wide analysis reveals specificities of Cpf1 endonucleases in human cells. *Nature Biotechnology* 34 (8): 863.

Kim, H.K., Song, M., Lee, J. et al. (2017). In vivo high-throughput profiling of CRISPR–Cpf1 activity. *Nature Methods* 14 (2): 153.

Kim, H., Kim, S.T., Ryu, J. et al. (2017). CRISPR/Cpf1-mediated DNA-free plant genome editing. *Nature Communications* 8 (1): 1–7.

Klap, C., Yeshayahou, E., Bolger, A.M. et al. (2017). Tomato facultative parthenocarpy results from Sl AGAMOUS-LIKE 6 loss of function. *Plant Biotechnology Journal* 15 (5): 634–647.

Kleinstiver, B.P., Tsai, S.Q., Prew, M.S. et al. (2016). Genome-wide specificities of CRISPR-Cas Cpf1 nucleases in human cells. *Nature Biotechnology* 34 (8): 869.

Lawrenson, T., Shorinola, O., Stacey, N. et al. (2015). Induction of targeted, heritable mutations in barley and Brassica oleracea using RNA-guided Cas9 nuclease. *Genome Biology* 16 (1): 258.

Lee, K., Zhang, Y., Kleinstiver, B.P. et al. (2019). Activities and specificities of CRISPR/Cas9 and Cas12a nucleases for targeted mutagenesis in maize. *Plant Biotechnology Journal* 17 (2): 362–372.

Lei, Y., Lu, L., Liu, H.Y. et al. (2014). CRISPR-P: a web tool for synthetic single-guide RNA design of CRISPR-system in plants. *Molecular Plant* 7 (9): 1494–1496.

Li, T., Huang, S., Jiang, W.Z. et al. (2011). TAL nucleases (TALNs), hybrid proteins composed of TAL effectors and FokI DNAcleavage domain. *Nucleic Acids Research* 39: 359–372.

Li, T., Liu, B., Spalding, M.H. et al. (2012). High-efficiency TALEN-based gene editing produces disease-resistant rice. *Nature Biotechnology* 30 (5): 390.

Li, J.F., Norville, J.E., Aach, J. et al. (2013). Multiplex and homologous recombination-mediated genome editing in *Arabidopsis* and *Nicotiana benthamiana* using guide RNA and Cas9. *Nature Biotechnology* 31: 688–691.

Li, Z., Liu, Z.B., Xing, A. et al. (2015). Cas9-guide RNA directed genome editing in soybean. *Plant Physiology* 169 (2): 960–970.

Li, J., Meng, X., Zong, Y. et al. (2016). Gene replacements and insertions in rice by intron targeting using CRISPR–Cas9. *Nature Plants* 2 (10): 1–6.

Li, M., Li, X., Zhou, Z. et al. (2016). Reassessment of the four yield-related genes Gn1a, DEP1, GS3, and IPA1 in rice using a CRISPR/Cas9 system. *Frontiers in Plant Science* 7: 377.

Li, X., Zhou, W., Ren, Y. et al. (2017). High-efficiency breeding of early-maturing rice cultivars via CRISPR/Cas9-mediated genome editing. *Journal of Genetics and Genomics= Yi chuan xue bao* 44 (3): 175.

Li, J., Zhang, H., Si, X. et al. (2017). Generation of thermosensitive male-sterile maize by targeted knockout of the ZmTMS5 gene. *Journal of Genetics and Genomics= Yi chuan xue bao* 44 (9): 465.

Li, T., Yang, X., Yu, Y. et al. (2018). Domestication of wild tomato is accelerated by genome editing. *Nature Biotechnology* 36 (12): 1160–1163.

Li, B., Rui, H., Li, Y. et al. (2019). Robust CRISPR/Cpf1 (Cas12a)-mediated genome editing in allotetraploid cotton (Gossypium hirsutum). *Plant Biotechnology Journal* 17 (10): 1862–1864.

Liang, Z., Zhang, K., Chen, K., and Gao, C. (2014). Targeted mutagenesis in Zea mays using TALENs and the CRISPR/Cas system. *Journal of Genetics and Genomics* 41 (2): 63–68.

Lin, Y., Fine, E.J., Zheng, Z. et al. (2014). SAPTA: a new design tool for improving TALE nuclease activity. *Nucleic Acids Research* 42 (6): e47–e47.

Liu, Y., Han, J., Chen, Z. et al. (2017). Engineering cell signaling using tunable CRISPR–Cpf1-based transcription factors. *Nature Communications* 8 (1): 1–8.

Lloyd, A., Plaisier, C.L., Carroll, D., and Drews, G.N. (2005). Targeted mutagenesis using zinc-finger nucleases in Arabidopsis. *Proceedings of the National Academy of Sciences* 102 (6): 2232–2237.

Lowder, L.G., Zhang, D., Baltes, N.J. et al. (2015). A CRISPR/ Cas9 toolbox for multiplexed plant genome editing and transcriptional regulation. *Plant Physiology* 169: 971–985.

Lowder, L.G., Zhou, J., Zhang, Y. et al. (2018). Robust transcriptional activation in plants using multiplexed CRISPR-Act2.0 and mTALE-act systems. *Molecular Plant* 11: 245–256.

Ma, M., Ye, A.Y., Zheng, W., and Kong, L. (2013). A guide RNA sequence design platform for the CRISPR/Cas9 system for model organism genomes. *BioMed Research International* 2013: 270805.

Maeder, M.L., Thibodeau-Beganny, S., Osiak, A. et al. (2008). Rapid "open-source" engineering of customized zinc-finger nucleases for highly efficient gene modification. *Molecular Cell* 31 (2): 294–301.

Mahfouz, M.M., Li, L., Shamimuzzaman, M. et al. (2011). *De novo*-engineered transcription activator-like effector (TALE) hybrid nuclease with novel DNA binding specificity creates double-strand breaks. *Proceedings of the National Academy of Sciences of the United States of America* 108: 2623–2628.

Mahfouz, M.M., Piatek, A., and Stewart, C.N. Jr. (2014). Genome engineering via TALENs and CRISPR/Cas9 systems: challenges and perspectives. *Plant Biotechnology Journal* 12 (8): 1006–1014.

Makarova, K.S., Haft, D.H., Barrangou, R. et al. (2011). Evolution and classification of the CRISPR–Cas systems. *Nature Reviews Microbiology* 9 (6): 467–477.

Mali, P., Aach, J., Stranges, P.B. et al. (2013). CAS9 transcriptional activators for target specificity screening and paired nickases for cooperative genome engineering. *Nature Biotechnology* 31: 833–838.

Malnoy, M., Viola, R., Jung, M.H. et al. (2016). DNA-free genetically edited grapevine and apple protoplast using CRISPR/Cas9 ribonucleoproteins. *Frontiers in Plant Science* 7 (1904).

Mandell, J.G. and Barbas, C.F. (2006). Zinc finger tools: custom DNA-binding domains for transcription factors and nucleases. *Nucleic Acids Research* 34 (suppl_2): W516–W523.

Marton, I., Zuker, A., Shklarman, E. et al. (2010). Nontransgenic genome modification in plant cells. *Plant Physiology* 154 (3): 1079–1087.

Miao, J., Guo, D., Zhang, J. et al. (2013). Targeted mutagenesis in rice using CRISPR-Cas system. *Cell Research* 23 (10): 1233–1236.

Miroshnichenko, D.N., Shulga, O.A., Timerbaev, V.R., and Dolgov, S.V. (2019). Achievements, challenges, and prospects in the production of nontransgenic, genome-edited plants. *Applied Biochemistry and Microbiology* 55 (9): 825–845.

Montague, T.G., Cruz, J.M., Gagnon, J.A. et al. (2014). CHOPCHOP: a CRISPR/Cas9 and TALEN web tool for genome editing. *Nucleic Acids Research* 42 (W1): W401–W407.

Moon, S.B., Lee, J.M., Kang, J.G. et al. (2018). Highly efficient genome editing by CRISPR-Cpf1 using CRISPR RNA with a uridinylate-rich 3′-overhang. *Nature Communications* 9 (1): 1–11.

Moscou, M.J. and Bogdanove, A.J. (2009). A simple cipher governs DNA recognition by TALeffectors. *Science* 326: 1501.

Naito, Y., Hino, K., Bono, H., and Ui-Tei, K. (2015). CRISPRdirect: software for designing CRISPR/Cas guide RNA with reduced off-target sites. *Bioinformatics* 31 (7): 1120–1123.

Nakajima, I., Ban, Y., Azuma, A. et al. (2017). CRISPR/Cas9-mediated targeted mutagenesis in grape. *PLoS One* 12 (5): e0177966.

Neff, K.L., Argue, D.P., Ma, A.C. et al. (2013). Mojo hand, a TALEN design tool for genome editing applications. *BMC Bioinformatics* 14 (1): 1.

Nekrasov, V., Wang, C., Win, J. et al. (2017). Rapid generation of a transgene-free powdery mildew resistant tomato by genome deletion. *Scientific Reports* 7 (1): 1–6.

Nishitani, C., Hirai, N., Komori, S. et al. (2016). Efficient genome editing in apple using a CRISPR/Cas9 system. *Scientific Reports* 6 (1): 1–8.

O'Brien, A. and Bailey, T.L. (2014). GT-Scan: identifying unique genomic targets. *Bioinformatics* 30 (18): 2673–2675.

Odipio, J., Alicai, T., Ingelbrecht, I. et al. (2017). Efficient CRISPR/Cas9 genome editing of phytoene desaturase in cassava. *Frontiers in Plant Science* 8: 1780.

Ortigosa, A., Gimenez-Ibanez, S., Leonhardt, N., and Solano, R. (2019). Design of a bacterial speck resistant tomato by CRISPR/Cas9-mediated editing of Sl JAZ 2. *Plant Biotechnology Journal* 17 (3): 665–673.

Osakabe, K., Osakabe, Y., and Toki, S. (2010). Site-directed mutagenesis in Arabidopsis using custom-designed zinc finger nucleases. *Proceedings of the National Academy of Sciences* 107 (26): 12034–12039.

Pabo, C.O., Peisach, E., and Grant, R.A. (2001). Design and selection of novel Cys2His2 zinc finger proteins. *Annual Review of Biochemistry* 70: 313–340.

Park, J., Bae, S., and Kim, J.S. (2015). Cas-designer: a web-based tool for choice of CRISPR-Cas9 target sites. *Bioinformatics* 31 (24): 4014–4016.

Parry, M.A.J. and Hawkesford, M.J. (2012). An integrated approach to crop genetic improvement F. *Journal of Integrative Plant Biology* 54 (4): 250–259.

Peng, A., Chen, S., Lei, T. et al. (2017). Engineering canker-resistant plants through CRISPR/Cas9-targeted editing of the susceptibility gene Cs LOB 1 promoter in citrus. *Plant Biotechnology Journal* 15 (12): 1509–1519.

Prykhozhij, S.V., Vinothkumar Rajan, D.G., and Berman, J.N. (2015). CRISPR multitargeter: a web tool to find common and unique CRISPR single guide RNA targets in a set of similar sequences. *PLoS One* 10 (3): e0119372.

Qi, L.S., Larson, M.H., Gilbert, L.A. et al. (2013). Repurposing CRISPR as an RNA-guided platform for sequence-specific control of gene expression. *Cell* 152 (5): 1173–1183.

Ramirez, C.L., Foley, J.E., Wright, D.A. et al. (2008). Unexpected failure rates for modular assembly of engineered zinc fingers. *Nature Methods* 5 (5): 374–375.

Ran, F.A., Hsu, P.D., Lin, C.Y. et al. (2013). Double nicking by RNA-guided CRISPR Cas9 for enhanced genome editing specificity. *Cell* 154: 1380–1389.

Reyon, D., Kirkpatrick, J.R., Sander, J.D. et al. (2011). ZFNGenome: a comprehensive resource for locating zinc finger nuclease target sites in model organisms. *BMC Genomics* 12 (1): 83.

Rodríguez-Leal, D., Lemmon, Z.H., Man, J. et al. (2017). Engineering quantitative trait variation for crop improvement by genome editing. *Cell* 171 (2): 470–480.

Sander, J.D., Zaback, P., Joung, J.K. et al. (2007). Zinc finger Targeter (ZiFiT): an engineered zinc finger/target site design tool. *Nucleic Acids Research* 35 (suppl_2): W599–W605.

Sander, J.D., Dahlborg, E.J., Goodwin, M.J. et al. (2011a). Selection-free zinc-finger-nuclease engineering by context-dependent assembly (CoDA). *Nature Methods* 8: 67–69.

Sander, J.D., Cade, C., Khayter, C. et al. (2011b). Targeted gene disruption in somatic zebrafish cells using engineered TALENs. *Nature Biotechnology* 29: 697–698.

Sauer, N.J., Narváez-Vásquez, J., Mozoruk, J. et al. (2016). Oligonucleotide-mediated genome editing provides precision and function to engineered nucleases and antibiotics in plants. *Plant Physiology* 170 (4): 1917–1928.

Sedeek, K.E., Mahas, A., and Mahfouz, M. (2019). Plant genome engineering for targeted improvement of crop traits. *Frontiers in Plant Science* 10: 114.

Shan, Q., Wang, Y., Chen, K. et al. (2013a). Rapid and efficient gene modification in rice and Brachypodium using TALENs. *Molecular Plant* 6 (4): 1365–1368.

Shan, Q., Wang, Y., Li, J. et al. (2013b). Targeted genome modification of crop plants using the CRISPR-Cas system. *Nature Biotechnology* 31: 686–688.

Shi, J., Gao, H., Wang, H. et al. (2017). ARGOS 8 variants generated by CRISPR-Cas9 improve maize grain yield under field drought stress conditions. *Plant Biotechnology Journal* 15 (2): 207–216.

Shrestha, A., Khan, A., and Dey, N. (2018). Cis–trans engineering: advances and perspectives on customized transcriptional regulation in plants. *Molecular Plant* 11 (7): 886–898.

Shukla, V.K., Doyon, Y., Miller, J.C. et al. (2009). Precise genome modification in the crop species *Zea mays* using zinc-finger nucleases. *Nature* 459: 437–441.

Singh, R., Kuscu, C., Quinlan, A. et al. (2015). Cas9-chromatin binding information enables more accurate CRISPR off-target prediction. *Nucleic Acids Research* 43 (18): e118–e118.

Soyk, S., Müller, N.A., Park, S.J. et al. (2016). Variation in the flowering gene SELF PRUNING 5G promotes day-neutrality and early yield in tomato. *Nature Genetics* 49: 162–168.

Stemmer, M., Thumberger, T., del Sol Keyer, M. et al. (2015). CCTop: an intuitive, flexible and reliable CRISPR/Cas9 target prediction tool. *PLoS One* 10 (4): e0124633.

Sternberg, S.H., Redding, S., Jinek, M. et al. (2014). DNA interrogation by the CRISPR RNA-guided endonuclease Cas9. *Nature* 507: 62–67.

Sun, Y., Zhang, X., Wu, C. et al. (2016). Engineering herbicide-resistant rice plants through CRISPR/Cas9-mediated homologous recombination of acetolactate synthase. *Molecular Plant* 9 (4): 628–631.

Sun, Y., Jiao, G., Liu, Z. et al. (2017). Generation of high-amylose rice through CRISPR/Cas9-mediated targeted mutagenesis of starch branching enzymes. *Frontiers in Plant Science* 8: 298.

Svitashev, S., Young, J.K., Schwartz, C. et al. (2015). Targeted mutagenesis, precise gene editing, and site-specific gene insertion in maize using Cas9 and guide RNA. *Plant Physiology* 169 (2): 931–945.

Tang, W. and Tang, A.Y. (2017). Applications and roles of the CRISPR system in genome editing of plants. *Journal of Forestry Research* 28 (1): 15–28.

Tang, X., Ren, Q., Yang, L. et al. (2019). Single transcript unit CRISPR 2.0 systems for robust Cas9 and Cas12a mediated plant genome editing. *Plant Biotechnology Journal* 17 (7): 1431–1445.

Tian, S., Jiang, L., Cui, X. et al. (2018). Engineering herbicide-resistant watermelon variety through CRISPR/Cas9-mediated base-editing. *Plant Cell Reports* 37 (9): 1353–1356.

Townsend, J.A., Wright, D.A., Winfrey, R.J. et al. (2009). High-frequency modification of plant genes using engineered zinc-finger nucleases. *Nature* 459 (7245): 442–445.

Tripathi, J.N., Ntui, V.O., Ron, M. et al. (2019). CRISPR/Cas9 editing of endogenous banana streak virus in the B genome of Musa spp. overcomes a major challenge in banana breeding. *Communications Biology* 2 (1): 1–11.

Tsai, S.Q., Wyvekens, N., Khayter, C. et al. (2014). Dimeric CRISPR RNA-guided FokI nucleases for highly specific genome editing. *Nature Biotechnology* 32: 569–576.

Uniyal, A.P., Yadav, S.K., and Kumar, V. (2019). The CRISPR–Cas9, genome editing approach: a promising tool for drafting defense strategy against begomoviruses including cotton leaf curl viruses. *Journal of Plant Biochemistry and Biotechnology* 28: 121–132.

Upadhyay, S.K. and Sharma, S. (2014). SSFinder: high throughput CRISPR-Cas target sites prediction tool. *BioMed Research International* 2014: 742482.

Upadhyay, S.K., Kumar, J., Alok, A., and Tuli, R. (2013). RNA-guided genome editing for target gene mutations in wheat. *G3: Genes, Genomes, Genetics* 3 (12): 2233–2238.

Voytas, D.F. (2013). Plant genome engineering with sequence-specific nucleases. *Annual Review of Plant Biology* 64: 327–350.

Wang, Z., Li, J., Huang, H. et al. (2012). An integrated chip for the high-throughput synthesis of transcription activator-like effectors. *Angewandte Chemie International Edition* 51 (34): 8505–8508.

Wang, Y., Cheng, X., Shan, Q. et al. (2014). Simultaneous editing of three homoeoalleles in hexaploid bread wheat confers heritable resistance to powdery mildew. *Nature Biotechnology* 32: 947–951.

Wang, F., Wang, C., Liu, P. et al. (2016). Enhanced rice blast resistance by CRISPR/Cas9-targeted mutagenesis of the ERF transcription factor gene OsERF922. *PLoS One* 11 (4): e0154027.

Wang, M., Mao, Y., Lu, Y. et al. (2017). Multiplex gene editing in rice using the CRISPR-Cpf1 system. *Molecular Plant* 10 (7): 1011–1013.

Wendt, T., Holm, P.B., Starker, C.G. et al. (2013). TAL effector nucleases induce mutations at a pre-selected location in the genome of primary barley transformants. *Plant Molecular Biology* 83 (3): 279–285.

Wilson, F.M., Harrison, K., Armitage, A.D. et al. (2019). CRISPR/Cas9-mediated mutagenesis of phytoene desaturase in diploid and octoploid strawberry. *Plant Methods* 15 (1): 45.

Wright, D.A., Townsend, J.A., Winfrey, R.J. Jr. et al. (2005). High-frequency homologous recombination in plants mediated by zinc-finger nucleases. *The Plant Journal* 44 (4): 693–705.

Wu, X., Scott, D.A., Kriz, A.J. et al. (2014). Genome- wide binding of the CRISPR endonuclease Cas9 in mammalian cells. *Nature Biotechnology* 32: 670–676.

Xie, S., Shen, B., Zhang, C. et al. (2014). sgRNAcas9: a software package for designing CRISPR sgRNA and evaluating potential off-target cleavage sites. *PLoS One* 9 (6): e100448.

Xu, R., Yang, Y., Qin, R. et al. (2016). Rapid improvement of grain weight via highly efficient CRISPR/Cas9-mediated multiplex genome editing in rice. *Journal of Genetics and Genomics= Yi chuan xue bao* 43 (8): 529.

Xu, R., Qin, R., Li, H. et al. (2017). Generation of targeted mutant rice using a CRISPR-Cpf1 system. *Plant Biotechnology Journal* 15 (6): 713–717.

Xu, X., Gao, J., Dai, W. et al. (2019). Gene activation by a CRISPR-assisted trans enhancer. *eLife* 8: e45973. Published 2019 Apr 11. doi:https://doi.org/10.7554/eLife.45973.

Xu, Z.S., Feng, K., and Xiong, A.S. (2019). CRISPR/Cas9-mediated multiply targeted mutagenesis in orange and purple carrot plants. *Molecular Biotechnology* 61 (3): 191–199.

Yamamuro, C., Zhu, J.K., and Yang, Z. (2016). Epigenetic modifications and plant hormone action. *Molecular Plant* 9 (1): 57–70.

Yan, W.X., Mirzazadeh, R., Garnerone, S. et al. (2017). BLISS is a versatile and quantitative method for genome-wide profiling of DNA double-strand breaks. *Nature Communications* 8 (1): 1–9.

Yao, L., Zhang, Y., Liu, C. et al. (2018). OsMATL mutation induces haploid seed formation in indica rice. *Nature Plants* 4 (8): 530–533.

Yin, X., Biswal, A.K., Dionora, J. et al. (2017). CRISPR-Cas9 and CRISPR-Cpf1 mediated targeting of a stomatal developmental gene EPFL9 in rice. *Plant Cell Reports* 36 (5): 745–757.

Zaidi, S.S.E.A., Mahfouz, M.M., and Mansoor, S. (2017). CRISPR-Cpf1: a new tool for plant genome editing. *Trends in Plant Science* 22 (7): 550–553.

Zetsche, B., Gootenberg, J.S., Abudayyeh, O.O. et al. (2015). Cpf1 is a single RNA-guided endonuclease of a class 2 CRISPR-Cas system. *Cell* 163 (3): 759–771.

Zhang, F., Maeder, M.L., Unger-Wallace, E. et al. (2010). High frequency targeted mutagenesis in Arabidopsis thaliana using zinc finger nucleases. *Proceedings of the National Academy of Sciences* 107 (26): 12028–12033.

Zhang, F., Cong, L., Lodato, S. et al. (2011). Efficient construction of sequence-specific TAL effectors for modulating mammalian transcription. *Nature Biotechnology* 29 (2): 149.

Zhang, Y., Zhang, F., Li, X. et al. (2013). Transcription activator-like effector nucleases enable efficient plant genome engineering. *Plant Physiology* 161: 20–27.

Zhang, Y., Bai, Y., Wu, G. et al. (2017). Simultaneous modification of three homoeologs of ta EDR 1 by genome editing enhances powdery mildew resistance in wheat. *The Plant Journal* 91 (4): 714–724.

Zhang, Y., Li, D., Zhang, D. et al. (2018). Analysis of the functions of ta GW 2 homoeologs in wheat grain weight and protein content traits. *The Plant Journal* 94 (5): 857–866.

Zhang, K., Nie, L., Cheng, Q. et al. (2019). Effective editing for lysophosphatidic acid acyltransferase 2/5 in allotetraploid rapeseed (Brassica napus L.) using CRISPR-Cas9 system. *Biotechnology for Biofuels* 12 (1): 1–18.

Zhi, J., Liu, X., Li, D. et al. (2020). CRISPR/Cas9-mediated SlAN2 mutants reveal various regulatory models of anthocyanin biosynthesis in tomato plant. *Plant Cell Reports* 39 (6): 799–809.

Zhou, J., Peng, Z., Long, J. et al. (2015). Gene targeting by the TAL effector PthXo2 reveals cryptic resistance gene for bacterial blight of rice. *The Plant Journal* 82 (4): 632–643.

Zhu, L.J., Holmes, B.R., Aronin, N., and Brodsky, M.H. (2014). CRISPRseek: a bioconductor package to identify target-specific guide RNAs for CRISPR-Cas9 genome-editing systems. *PLoS One* 9 (9): e108424.

2

Distribution of Nutritional and Mineral Components in Important Crop Plants

Katarina Vogel-Mikuš[1,2], Paula Pongrac[1,2], Ivan Kreft[3], Primož Pelicon[2], Primož Vavpetič[2], Boštjan Jenčič[2], Johannes Teun van Elteren[4], Peter Kump[2], Sudhir P. Singh[5], and Marjana Regvar[1]

[1]*Biotechnical Faculty, University of Ljubljana, Ljubljana, Slovenia*
[2]*Jozef Stefan Institute, Ljubljana, Slovenia*
[3]*Nutrition Institute, Ljubljana, Slovenia*
[4]*National Institute of Chemistry, Ljubljana, Slovenia*
[5]*Center of Innovative and Applied Bioprocessing (DBT-CIAB), Mohali, Punjab, India*

CHAPTER MENU

2.1 Introduction

Plants provide us with essential energy-rich compounds, minerals, vitamins, amino and fatty acids, and a wealth of health-promoting compounds (Eckardt 2011). The cultivation of new, high-yielding varieties and the ever-increasing use of mineral fertilizers and pesticides has led to a significant increase in crop yields. However, the quality of edible plants related to human nutrition did not receive proper attention until the 1990s, when an increasing number of reports started to appear on inadequate diet and unbalanced nutrition in vulnerable, low-income populations in many countries, leading to poor health, low productivity and an increase in chronic diseases (Eckardt 2011).

The quality of crops and their produce is a highly complex concept as it depends on the large number of plant traits which, in turn, are controlled by numerous underlying genetic

Genome Engineering for Crop Improvement, First Edition. Edited by Santosh Kumar Upadhyay.

and exogenous factors. In addition, the global increase in temperature, UV radiation, CO_2 and ozone pollution, and drought and field salinization, place an even greater strain on crop performance. The need to adapt to constantly changing abiotic conditions leads to unpredictable changes in the biochemical composition of crops and, consequently, in the quality of food. In addition, biotic stress such as bacterial, fungal, and viral infections induce changes in the metabolic pathways with the aim of synthesizing self-protecting biochemicals, but often at the expense of energy-rich and health-promoting compounds such as anthocyanins.

This chapter will focus on cereals and pseudo-cereal plants that produce starchy grains since they are part of almost every meal we eat. Grains consists of the plant embryo – a miniature plant – and the seed coat that protects the embryo from abiotic and biotic stress. Storage compounds (starch, proteins, and lipids) that provide energy and building blocks for the young, developing plant after germination, are stored in the endosperm or in the cotyledons, while minerals are mainly stored in the aleurone layer or in the embryo. Nutrients and minerals are not evenly distributed within the grain (Pongrac et al. 2011; Singh et al. 2013, 2014; Vogel-Mikuš et al. 2014; Mantouvalou et al. 2017). This is of high relevance for human nutrition, as cereals are usually processed before consumption, and certain nutritious compounds stored in the seed coat or embryonic tissue may be lost. There is, therefore, an urgent need to develop research tools that will provide better insights into the 2D and 3D spatial distribution of nutrients and minerals in cereals and the mechanisms behind these distribution patterns.

2.2 Exploring Nutrient Distribution in Grain

The revolution in *omics* technology, particularly metabolomics, has been closely linked to technological developments in mass spectrometry (MS) (Boughton et al. 2016). MS is currently the most efficient technology for the characterization of biomolecular composition and has made significant progress in providing a comprehensive understanding of biological functions. Unfortunately, in MS-based studies, where the analysis is performed on a tissue homogenate (e.g. liquid or gas chromatography), spatial information is lost (Dong et al. 2016a). Spatial information in mass spectrometric analysis can be preserved if the extraction process is limited to very small areas of the sample, which is then analyzed using new ionization techniques, such as Matrix Assisted Laser/Desorption/Ionization (MALDI) (Dong et al. 2016a). These techniques account for the majority of applications in biomolecular imaging. Techniques that can also be used for spatially resolved studies of molecular profiles are also Laser Desorption Ionization (LDI) (Lee et al. 2012), Secondary Ion Mass Spectroscopy (SIMS) or Desorption Electrospray Ionization (DESI) (Lee et al. 2012). In the aforementioned techniques, the analytes are raster-scanned and ionized and desorbed directly from the surface, thus avoiding traditional liquid extraction. By automating the process, images can be generated that show the spatial distributions of all detected compounds (Dong et al. 2016a; Jenčič et al. 2016).

2.2.1 Matrix Assisted Laser/Desorption/Ionization

MALDI is mainly used in biomedicine to identify biomarkers associated with various diseases (Grassl et al. 2011), while applications in plant sciences remain limited (Boughton et al. 2016; Dong et al. 2016a). With this technique, large molecules such as proteins can be detected with a lateral resolution of 0 micrometers in novel, state-of-the-art instruments (Römpp and Spengler 2013). The working principle of MALDI is based on the co-crystallization of the analyte with a chemical matrix which absorbs the laser energy and releases the analyte into the gas phase in a process that leads to ionization (Bjarnholt et al. 2014). The addition of the matrix has a number of advantages: in particular, it enables the ionization of compounds that do not themselves absorb light. Moreover, since the energy of the laser light is absorbed by the matrix and not directly by the analyte, MALDI is a soft ionization technique that exhibits little fragmentation in the mass spectra (Bjarnholt et al. 2014). Unfortunately, when performing MALDI imaging with higher resolution (laser spot sizes smaller than 1 μm), the ion yield (and thus the sensitivity) decreases significantly. Further, the choice of the type of matrix becomes extremely critical, since the size of the matrix crystals incorporated with the analytes should be smaller than the required lateral resolution of the imaging experiment. Also, MALDI is unable to detect metabolites in the low mass range (m/z 1000) due to interference from the high matrix ion background (Bjarnholt et al. 2014).

Recently, workflows were developed for MALDI imaging of plant carbohydrate, lipid, and protein as well as secondary metabolite localization (Dong et al. 2016b; Peukert et al. 2016; Gupta et al. 2019; Lim et al. 2020).

Oligofructans, for example, represent one of the most important groups of sucrose-derived water–soluble carbohydrates in the plant kingdom. In cereals, oligofructans accumulate in aboveground parts of the plants (stems, leaves, seeds), and their biosynthesis leads to the formation of both types of glycosidic bonds (ß (2,1); ß(2,6)-fructans) or to mixed patterns. In recent studies, MALDI-MS imaging has revealed tissue- and development-specific distribution patterns of the different types of oligofructans in barley grains, which may be related to the different phases of grain development, such as cellular differentiation of grain tissue and accumulation of storage products (Peukert et al. 2016).

In the cereal grain, sucrose is converted into storage carbohydrates; mainly starch, fructan, and mixed-linkage (1,3;1,4)-β-glucan(MLG) are formed in ratios, that seem to be tissue-dependent. Previously, endosperm specific overexpression of the *HvCslF6* gene in hull-less barley led to high MLG and low starch content in mature grains (Lim et al. 2020). Morphological changes included inwardly elongated aleurone cells, irregular cell shapes of the peripheral endosperm and smaller starch granules of the starch-containing endosperm. Lim et al. (2020) also investigated the physiological basis for these defects by investigating how changes in the carbohydrate composition of the developing grain affect the morphology of mature grains. Visualization of oligosaccharides was performed by MALDI-MS and complemented by histological studies pointing out the importance of the regulation of carbohydrate distribution for the maintenance of grain-cell formation and filling processes. Augmented MLG coincided with increased levels of soluble carbohydrates in the cavity and endosperm during the storage phase. Transcript levels of genes related to cell wall, starch, sucrose, and fructan metabolism were disturbed in all tissues. The cell walls of endosperm transfer cells (ETC) in transgenic cereals were thinner and showed reduced mannan labeling compared to the wild type. In the early storage phase, a rupture of the unevenly developed ETC and disorganized neighboring endosperm cells was observed. Soluble sugars accumulated in the developing cereal cave, suggesting a disruption of carbohydrate flow from the cave toward the endosperm, resulted in a shrunken phenotype of the mature grain.

MALDI-MS imaging was also performed to localize metabolites during the first 7 days of barley germination. Up to 100 mass signals were detected, of which, 85 signals were identified as 48 different metabolites with highly tissue-specific localizations. Oligosaccharides were observed in the endosperm and in parts of the embryo. Lipids in the endosperm co-localized, depending on their fatty acid composition, with changes in the distribution of diacyl phosphatidylcholines during germination. Also, 26 potentially antifungal hordatines were detected in the embryo with tissue-specific localizations of their glycosylated, hydroxylated and O-methylated derivatives (Gorzolka et al. 2016).

MALDI-MS also proved to be the technique of choice for studies in plant lipidomics. While profiles of lipids in seed extracts provide detailed measurements of the amounts and types of lipids present, the information about where these lipids are located in the seed is lost with sample preparation and solvent extraction. At present, mass spectrometric imaging allows the detailed identification and localization of lipids and other metabolites in situ, e.g. directly on tissue samples. The application of this technology to oil seeds has shown that lipid metabolites are unevenly distributed in the different tissues of the seed, indicating a spatial complexity of lipid metabolism that was previously unknown (Sturtevant et al. 2018).

2.2.2 Secondary Ion Mass Spectroscopy

Although SIMS was the first of the imaging MS techniques presented (introduced for imaging in the1980s by Briggs 1983), its use for imaging of plant material remains limited. SIMS operates under high vacuum and uses high-energetic (5–40 keV or MeV) primary ions for ionization (Jeromel et al. 2014). Its primary advantage over all other MS imaging techniques is very high spatial resolution (less than 100 nm) (Bjarnholt et al. 2014), while the major disadvantage lies in a high degree of molecular fragmentation. It is expected that high-throughput analysis and construction of a large database of molecular fragments may overcome this drawback in the near future.

With the introduction of the new cluster and swift ion beams (MeVSIMS) (Nakata et al. 2008), the ion yield in the higher mass range has been significantly improved (Jeromel et al. 2014), allowing the detection of intact lipids and secondary metabolites (Jenčič et al. 2016, 2017). However, ion intensities in them/z ~ 800 range are still orders of magnitude lower than intensities around m/z ~ 100, and sensitivity to intact metabolites is thus significantly lower than with other MS imaging techniques. Due to the high degree of fragmentation in SIMS experiments, characteristic fragments of analytes are often imaged rather than the intact analytes. SIMS does not require much sample preparation (no matrix is needed as with MALDI), but the fact that the analysis takes place in a vacuum means that the samples need to be freeze-dried before being introduced into the vacuum or measured under cryogenic conditions (Dickinson et al. 2006).

An example of MeV-SIMS imaging of organic compounds is presented in Figure 2.1 obtained as described previously (Jeromel et al. 2014).

2.2.3 Fourier Transform Infrared Spectroscopy

Fourier Transform Infrared (FTIR) imaging represents an important addition to the spatially resolved study of the molecular composition of plants. FTIR spectroscopy with a focused beam provides reliable information on biochemical composition (structural molecules) in a spatially resolved manner, without chemical fixation or staining of the sample, but plant tissue samples still need to be cryofixed and cutinto10 μm thin sections to perform micro-FTIR measurements

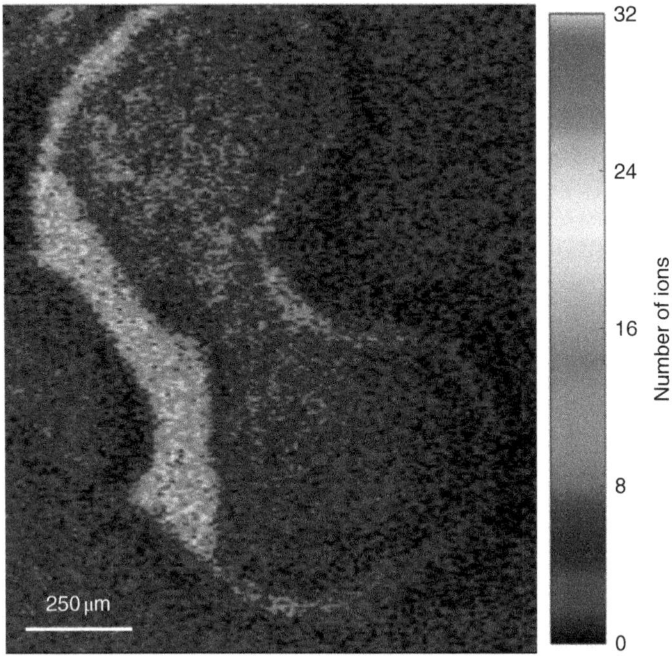

Figure 2.1 Spatial distribution of organic compound fitting the peak of m/z = 182, presumably belonging to phenolics in a wheat grain cross-section imaged by MeV-SIMS. Color bar presents the total ion yield of the selected peak. (For detailed information please see the colour figure in online version.)

in the most common transmission mode (Regvar et al. 2013). An important advantage of FTIR over other imaging techniques is that infrared radiation does not cause radiation damage to biological material, so that the chemical nature of the sample constituents can be studied without breaking the chemical bonds (Miller and Dumas 2006). This enables subsequent complementary studies using other techniques on the same samples and areas of interest.

FTIR imaging can discriminate between individual cells in plant samples. Cell walls and large structures within cells such as starch granules and protein bodies, in intact hydrothermally processed and digested wheat (*Triticum aestivum*) kernels, were observed. Gelatinized and native starch within cells could be distinguished, and also the loss of starch during wheat digestion was observed (Warren et al. 2015).

In barley grain, FTIR imaging was used to visualize the spatial distribution of sucrose (Suc). The intake of Sucin barley caryopsis occurs via the main vascular bundle, which is located in the ventral pericarp. As a result, concentration gradients are formed along the allocation path, which are nicely represented in the Suc distribution maps obtained by FTIR. The concentration of Suc in the post-phloem regions of the pericarp drops slightly and increases again in the region of the ETC. Such gradients could be an indication of passive unloading at the vein, but an active loading against the concentration gradient at the ETC via a set of sugar transporters. From the ETC, most of the Suc moves passively into the storage tissue, resulting in the formation of a steep negative gradient from the ETC toward the peripheral endosperm. Besides Suc, FTIR can also analyze the distribution of other metabolites such as starch. The quantitative data identified a characteristic pattern of starch distribution, which has already been described, and is largely explained by different local conditions that favor starch biosynthesis under hypoxic conditions in the endosperm (Guendel et al. 2018).

Of the edible grains, buckwheat is a pseudo-cereal and it is one of the main traditional ancient food sources in some of the populations around the Himalayan region (southwest China, Bhutan, Nepal, northern India and northeast Pakistan). It is also a popular food in several other countries such as Slovenia, Poland, the Czech Republic, Ukraine, Korea, and Japan (Bonafaccia et al. 2003). In common (*Fagopyrum esculentum*) and Tartary (*Fagopyrum tataricum*) buckwheat grains, which produce gluten-free starch-rich grains, chemical profiling using a combination of ATR and synchrotron radiation FTIR revealed the distinct composition of each of the three main tissues (husk, cotyledons and endosperm) and illustrated their difference from the whole grain (Figure 2.2).

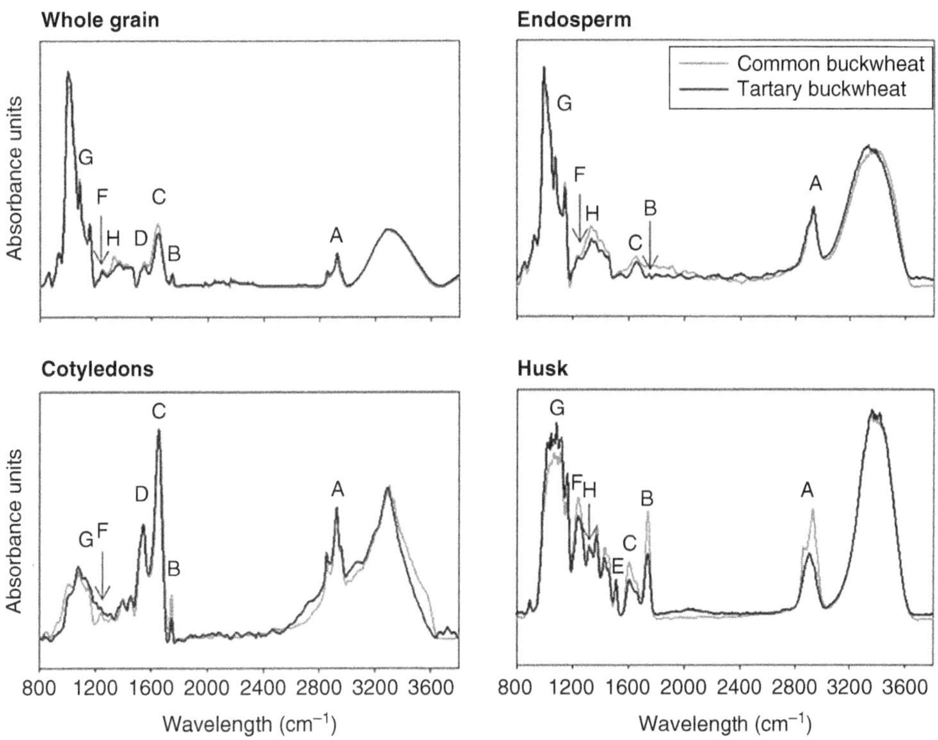

Figure 2.2 Comparison of the infrared spectra of common buckwheat (pink) and Tartary buckwheat (blue) whole grain and grain tissues endosperm, cotyledons and husk. The determination of the main molecular groups present in the entire grain was firstly assessed via the Attenuated Total Reflection (ATR) mode. Tissue-specific spectra were collected from cryo-sectioned and freeze-dried grain and data was processed as described previously (Regvar et al. 2013). The recorded spectra were corrected for atmospheric changes, truncated to a range of 3800–800 cm^{-1}, baseline corrected and normalized to permit an accurate comparison of the different spectral domains and their intensities. Peak A, lipids (2990–2813 cm^{-1}); peak B, ester band (1767–1714 cm^{-1}); peak C, amide I (1700–1586 cm^{-1}); peak D, amide II (1573–1490 cm^{-1}); peak E, lignin (1530–1490 cm^{-1}); peak F, cellulose (1278–1219 cm^{-1}); peak G, total carbohydrates (1182–886 cm^{-1}); and peak H, carboxyl groups from ligands, proteins, various polysaccharides (1349–1308 cm^{-1}). (For detailed information please see the colour figure in online version.)

Chemical mapping of a cross-section of the Tartary buckwheat grain clearly illustrates two distinct grain tissues, the cotyledons and the endosperm (Figure 2.3). Likewise, in wheat grain, husk, aleurone and endosperm also exhibit distinct chemical profiles (Figure 2.4).

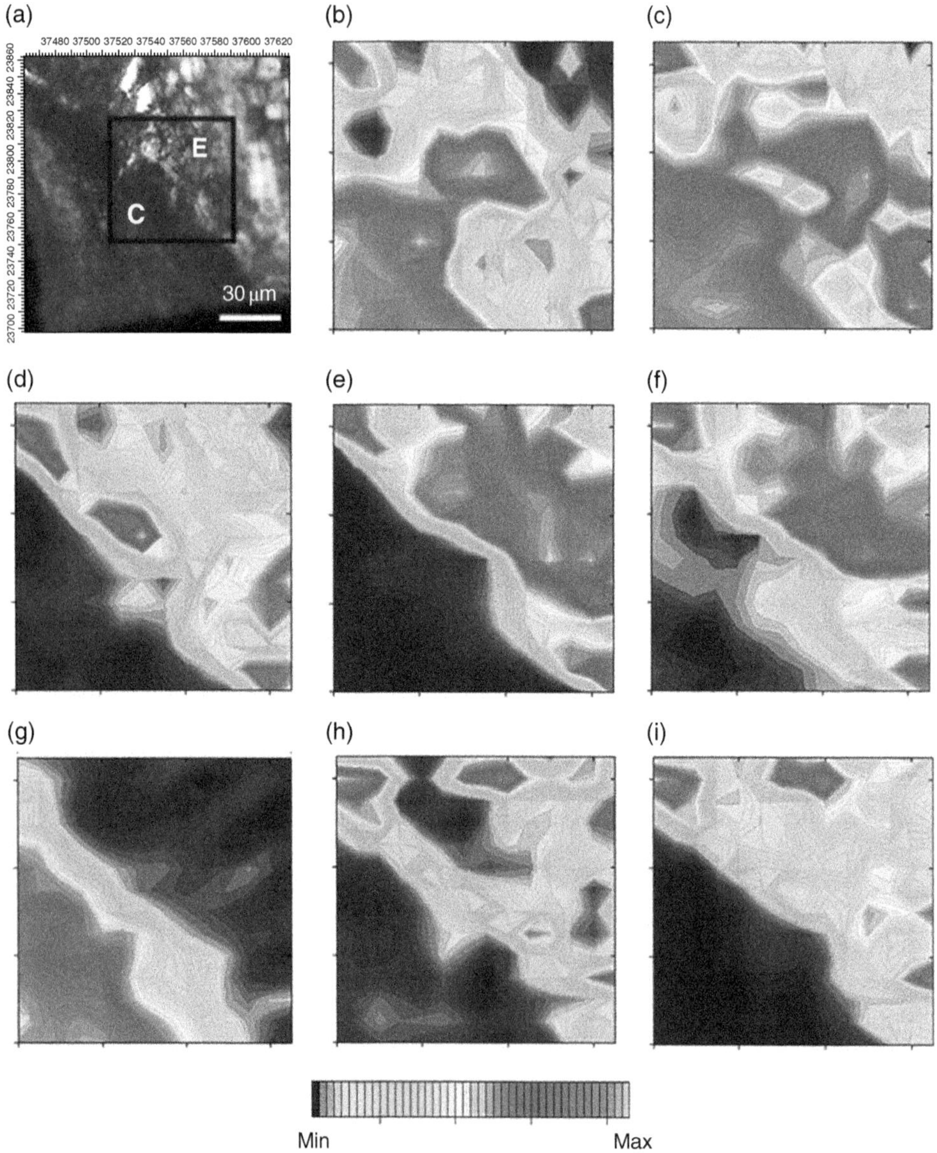

Figure 2.3 Representative SR-FTIR chemical mapping of the Tartary buckwheat grain cross-section comprising the cotyledon (c) and endosperm (e) as shown in the microscopic image (a) with the red rectangle indicating the area of mapping and the corresponding maps related to lipids (CH_2, b), lipids (CH_3, c), esters (d), amide I (e), amide II (f), total carbohydrates (g), cellulose (h) and lignin (i) SR-FTIR spectra were recorded on the SISSI beamline, Elettra, Trieste, Italy as described previously (Regvar et al. 2013). (For detailed information please see the colour figure in online version.)

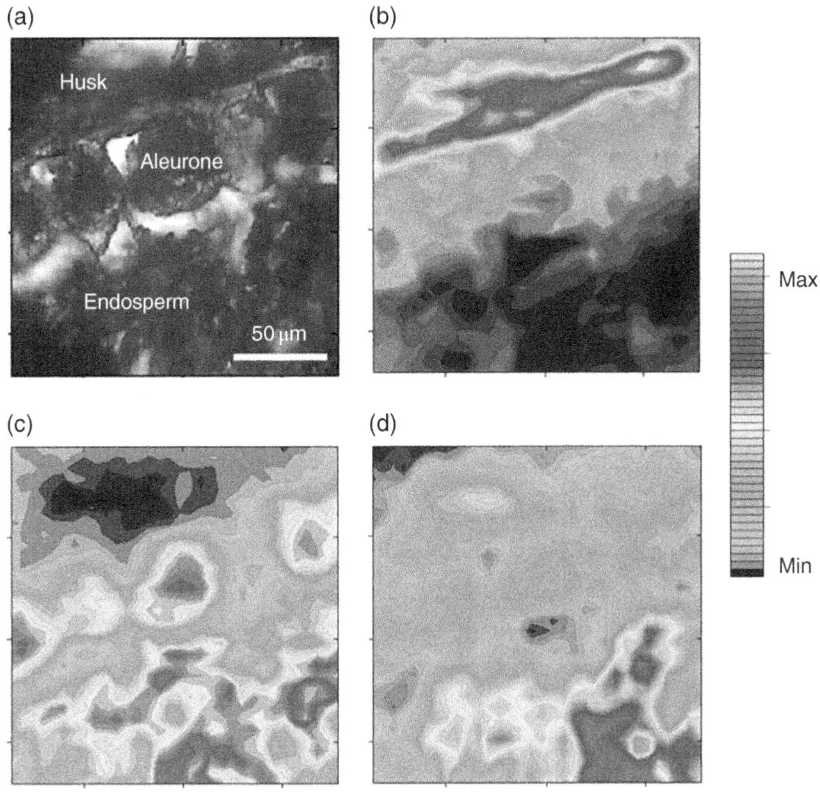

Figure 2.4 Representative SR-FTIR chemical mapping of the wheat grain cross-section comprising the husk, aleurone layer and endosperm as shown in the microscopic image (a) and corresponding maps related to esters, representing fats (b), amide II representing proteins (c) and coupling of the CO and CC stretching representing starch (d). SR-FTIR spectra were recorded on the SISSI beamline, Elettra, Trieste, Italy using a global source and a focal plane array detector. The maps represent 64 × 64 pixels with each pixel including 128 scans. Spectra were processed and maps were generated in Opus software (Bruker, USA). (For detailed information please see the colour figure in online version.)

2.3 Exploring the Mineral Distribution in Grain

Development of accelerator-based nuclear microprobes (micro-Particle Induced X-ray Emission; PIXE) (Moretto 1996), the third and fourth generation of synchrotron facilities with micro-and nano-focused beams (SR-micro-X-ray Fluorescence (XRF) (Kaulich et al. 2009; Tolrà et al. 2011; Martínez-Criado et al. 2016; Cotte et al. 2017) and Laser Ablation-Inductively Coupled Mass Spectrometry (LA-ICPMS) (Limbeck et al. 2015) have advanced the imaging of the elemental distribution in plants down to the subcellular level (Koren et al. 2013). Each of these techniques has its advantages and limitations, but provide a means to complement biomolecular imaging with elemental distributions.

2.3.1 Nuclear Microprobe

With the nuclear microprobe or micro-PIXE it is possible to image a broad spectrum of elements (from NatoU) with lateral resolution down to the submicron range (Vavpetič et al. 2017), but the method is less sensitive for light and heavy elements, with a detection limit in the $\mu g\, g^{-1}$ range for middle Z elements (Nečemer et al. 2008). Microprobes are normally operated in vacuum, therefore samples must be vacuum compatible – dehydrated or measured under cryogenic conditions (Simičič et al. 2002; Vogel-Mikuš et al. 2014).

The localization of minerals in whole grain cross sections of wheat was first performed in 1981 and 1985 with micro-PIXE by Mazzolini, Pallaghy and Legge (Mazzolini et al. 1981, 1985), with special emphasis on the Zn and Mn distribution. It was found that both elements are mainly distributed in the aleurone layer and in the embryo. However, in the embryo, the highest Zn concentration was measured in the scutellum and the highest Mn concentration in the coleoptyle. Almost 30 years later, the same question of Zn allocation in the wheat grain, after Zn fertilization, was addressed and confirmed that Zn is preferentially localized in scutellum and aleurone (Pongrac et al. 2013a). With Zn fertilization (Zn added as $ZnNO_3$), the Zn concentrations in the aforementioned parts increased. In addition, almost twice as high Zn concentrations were also found in the endosperm, the part used for flour production. These results underline agronomic biofortification, i.e. increasing the mineral element composition of staple foods through the application of fertilizers, as a means to successfully increase the Zn concentration in the edible parts of the grain, and demonstrate the applicability of imaging techniques such as micro-PIXE to evaluate the success of biofortification. Recently, grain cross-section micro-PIXE analysis of contrasting lines of barley (*Hordeum vulgare* L.) that accumulate different total amounts of Zn, additionally confirmed that differences in grain Zn accumulation apply to all parts of the grain, including the endosperm (Detterbeck et al. 2016), while also highlighting the possibility of selecting high Zn content lines for biofortification purposes.

According to the World Health Organization, WHO, iron (Fe) deficiency affects almost two thirds of the world population, especially women in their fertility period and children (Collings et al. 2013). In wheat grains, Fe is found in the outer layers of bran, most of which is lost during milling and processing (Zhang et al. 2010; Regvar et al. 2011). The flour is almost free of Fe. Therefore, the development of wheat varieties with Fe-enriched endosperm offers an important strategy to improve the nutritional availability of Fe in wheat flour. In a study conducted by Singh et al. (Singh et al. 2013, 2014), the variability of the mineral distribution in wheat was investigated by comparing three wheat genotypes and a wild wheat relative, *Aegilops kotschyi* (A.kot), which differ in the content of Fe in whole grain. In the genotypes with high Fe content (IITR26, A.kot), an increased concentration of Fe was observed in the aleurone layer, but also in the endosperm (Singh et al. 2014).

Beans are staple foods in many countries worldwide, especially in South and Central America. Thirteen genotypes of *Phraseoulus vulgaris*, *P. coccineus* and *P. lunatus* were investigated for Fe distribution (Cvitanich et al. 2010). It was shown that in *P. vulgaris* and *P. coccineus* high Fe concentrations are distributed in the cells surrounding the provascular tissue. Using the Pearls staining method, Fe was detected in the cytoplasm of epidermal cells of the cotyledons, in cells near the epidermis and in cells surrounding the provascular tissue. In contrast, the protein ferritin, which has been proposed as the most important Fe storage protein in legumes, was only detected in the amyloplasts of the seed cotyledons. With micro-PIXE it was shown that the tissue near the provascular bundles contained up to 500 $\mu g g^{-1}$ Fe, depending on the genotype. In contrast to *P. vulgaris* and *P. coccineus*, *P. lunatus* did not show Fe accumulation in the cells surrounding the provascular tissues of the cotyledons. The results presented emphasize the importance of complementing research on model organisms with analysis in crops and suggest that Fe distribution criteria should be integrated into selection strategies for the Fe biofortification of beans (Cvitanich et al. 2010).

The application of micro-PIXE has enabled us to generate quantitative element maps of the nutritionally most important elements in the grain tissue of common and Tartary buckwheat (Vogel-Mikuš et al. 2009; Pongrac et al. 2011, 2013b, 2020). About half of the buckwheat grain (57% of common and 43% of Tatary buckwheat) consists of starchy endosperm, which is only a moderate source of minerals. This could be due to dilution effects, with starch granules being formed in the last stages of assimilation and filling of the endosperm cells, which are not as rich in elements as other grain components. However, the majority of elements, which are found in both buckwheat varieties, are localized in the cotyledons, while Ca is mainly localized in the seed coat (Vogel-Mikuš et al. 2009; Pongrac et al. 2011, 2013b). Though the localization patterns are quite similar in both buckwheat species, higher contractions of Mg, P, Mn, Fe, Cu and Zn are observed in the cotyledons of Tartary buckwheat grain (Figure 2.5), making the latter a better source of minerals compared to common buckwheat grain. During the processing of buckwheat grains, the outer parts of the grains are removed (husk, seed coat and aleurone layer), while the inner parts with endosperm and cotyledons are preserved – the de-husked product is called groats. In wheat, however, the flour-milling fraction contains only parts of the endosperm, which is rather poor in minerals. A comparison between the edible grain fractions of buckwheat (endosperm and cotyledons) and three different wheat genotypes (endosperm) shows that buckwheat groats are a much richer source of minerals than wheat flour. However, the bioavailability of Fe, Zn, Mn, and Mg in buckwheat groats may be affected negatively because of the high P concentrations found in the groats (more precisely, in cotyledons). In grain, the majority of P is present in the form of phytate, which limits bioavailability of phytate-bound essential elements such as Mg, Mn, Fe and Zn found in buckwheat cotyledons.

2.3.2 Synchrotron Radiation X-Ray Florescence Spectrometry

SR-micro-XRF is a state-of-the-art imaging technique that enables 2D and 3D imaging. Separate beamlines are usually dedicated to low or high energy X-ray fluorescence. Low SR energy micro-XRF is operated in vacuum (e.g. TwinMic, Synchrotron Elettra Trieste (Gianoncelli et al. 2016), ID21, ESRF, Grenoble (Cotte et al. 2017), while hard X-ray microprobes are operated in air with limited sensitivity for low Z elements (Martínez-Criado et al. 2016). The detection limits range down to below $\mu g g^{-1}$ and the lateral resolution to a few tens of nanometers (Martínez-Criado et al. 2016). The highest chemical sensitivity can be achieved by tuning the photon beam energy above the absorption edge of the element of interest; however, this limits the number of elements that can be examined during the fingerprinting of the samples. In addition to SR-micro-XRF, a bench-top laboratory XRF

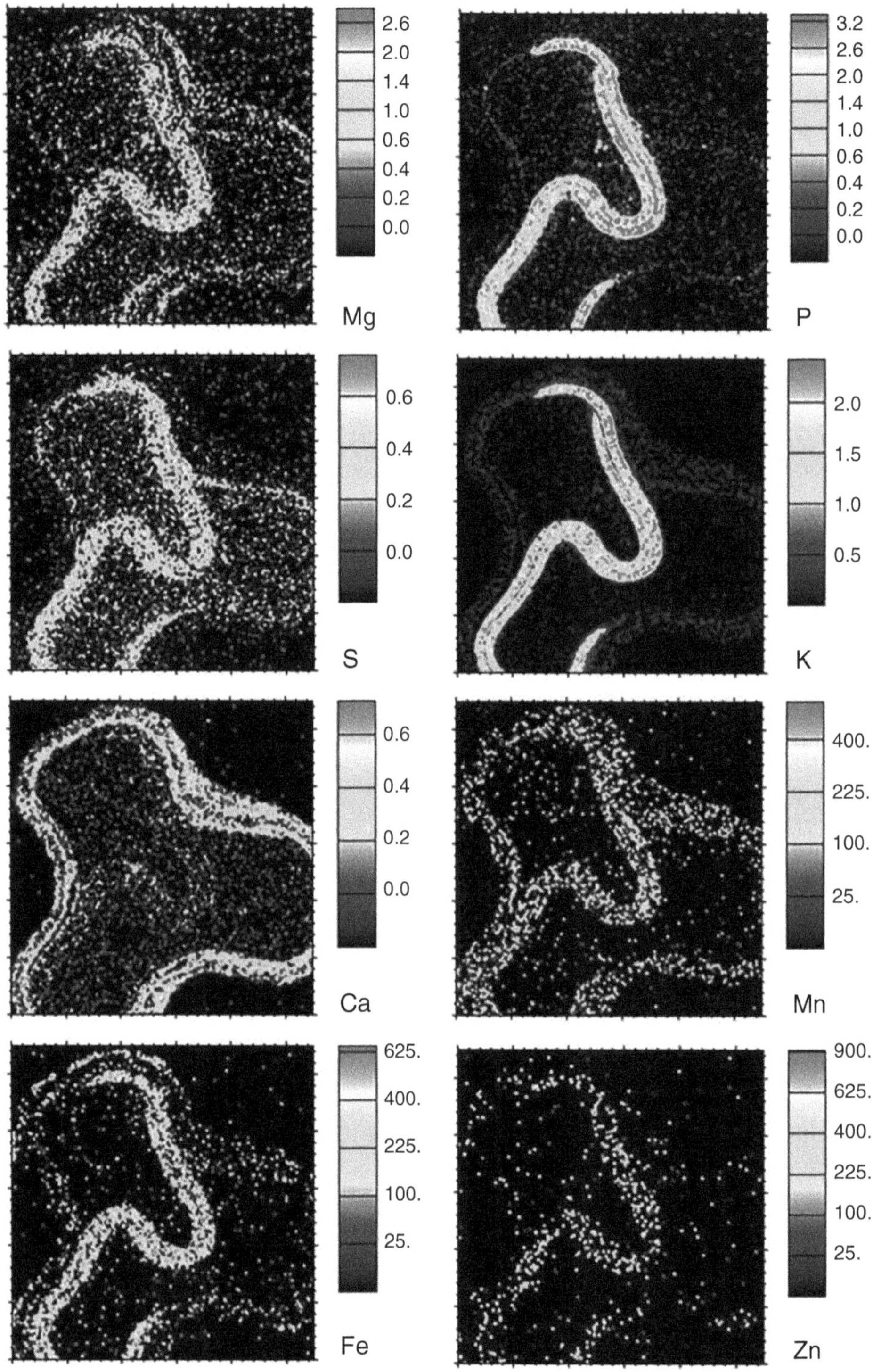

Figure 2.5 Quantitative mineral-element distribution maps of a Tartary buckwheat grain cross-section, comprising centrally-positioned cotyledons surrounding the endosperm and the pericarp as obtained by micro-PIXE and described previously (Pongrac et al. 2013b). The color scales for Mg, P, S, K, and Ca are in weight %, while for Mn, Fe and Zn they are in mg kg^{-1} dry weight. (For detailed information please see the colour figure in online version.)

instrument with a focused beam and polychromatic excitation (e.g. Tornado, Bruker) can be used, but with two orders of magnitude (or more) lower sensitivity (a few $\mu g g^{-1}$) and lower lateral resolution (a few tens of micrometers) (Mantouvalou et al. 2017; Rodrigues et al. 2018).

Synchrotron micro-X-ray fluorescence was, for example, used to determine the *in vivo* mineral distribution patterns in rice (*Oryza sativa*) grains and shifts in these distribution patterns during progressive germination stages. The results of bulk analyses of hulled, brown and polished rice showed that half of the total Zn, two thirds of the total Fe and most of the total K, Ca, and Mn were removed by the milling process when the hull and bran were thoroughly polished. The concentrations of all elements were high in the regions of the embryo, though local distributions within the embryo varied between the elements. The mobilization of minerals from certain seed locations during germination was also element specific. A high mobilization of K and Ca from the grains to the growing roots and leaf primordia was observed; the flow of Zn to these expanding tissues was slightly lower than that of K and Ca; the mobilization of Mn or Fe was relatively low, at least during the first days of germination (Lu et al. 2013).

Understanding the spatial distribution of inorganic nutrients within edible parts of plant products helps biofortification efforts to identify and focus on specific uptake pathways and storage mechanisms. Thus, the distribution of inorganic nutrients was studied in maize and sweetcorn. The results show that localization of elements is largely similar between maize and sweetcorn, but defer markedly depending upon the maturity stage after further embryonic development. The micronutrients Zn, Fe, and Mn accumulated primarily in the scutellum of the embryo during early kernel development, while trace amounts of these were found in the aleurone layer at the mature stage. Though P accumulated in the scutellum, there was no direct relationship between the concentrations of P and those of the micronutrients, compared to the linear trend between Zn and Fe concentrations (Cheah et al. 2019).

Localization of elements in Khorasan wheat (*Triticum turgidum ssp. turanicum*) with SR-micro-XRF showed an increased Fe accumulation in scutellum together with Zn, Mn and Ca, while in *T. aestivum* aleurone was the most Fe dense tissue (Figure 2.6).

In a more detailed analysis of wheat aleurone, synchrotron radiation soft X-ray full-field imaging mode (FFIM) provided detailed images of globoids covered by oleosomes. Low-energy X-ray fluorescence (LEXRF) spectro-microscopy showed that these structural features were connected to subcellular distribution of elements (Zn, Fe, Na, Mg, Al, Si, and P) (Regvar et al. 2011). This evidence suggest that membranous globular structures provide a basic structural scaffold for deposition on mineral elements within the aleurone cells and most likely affect their bioavailability.

2.3.3 Laser Ablation-Inductively Coupled Plasma Mass Spectrometry

LA-ICPMS is characterized by its superior chemical sensitivity down to the $\mu g kg^{-1}$ level, and is, therefore, suitable for imaging trace element distributions. Further, it is a benchtop laboratory technique with significantly lower operating costs than accelerator- or synchrotron-based techniques. The lateral resolution reaches down to 1 μm thanks to laser ablation cells with a fast washout and sensitive ICPMS detectors (Gundlach-Graham and Günther 2016;

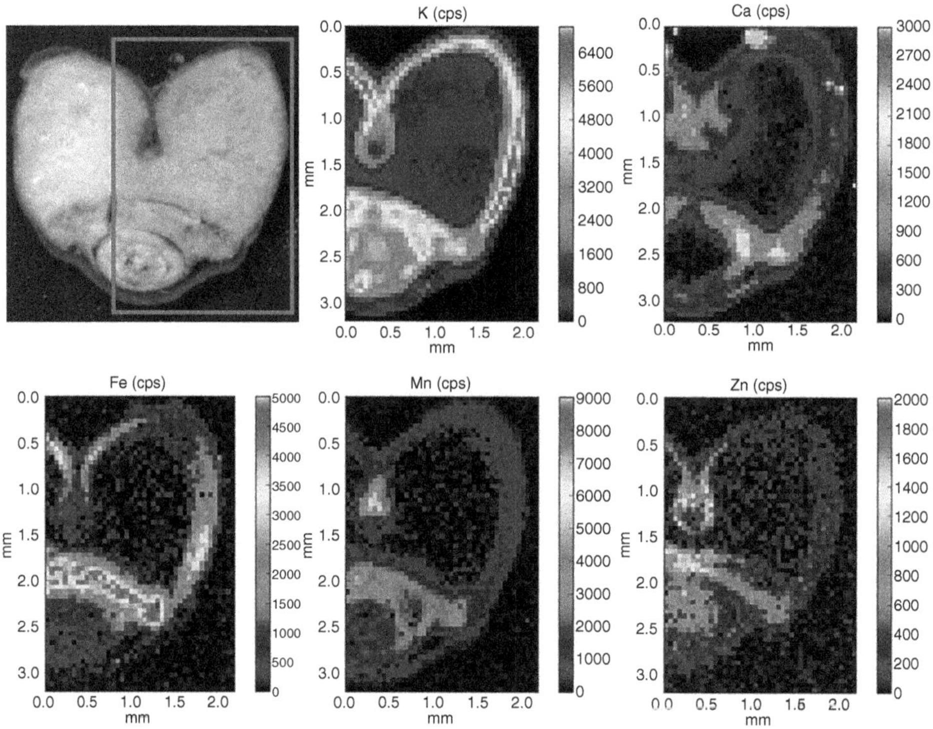

Figure 2.6 Element (K, Ca, Fe, Mn, and Zn) localization in Khorasan wheat (*Triticum turgidum ssp. turanicum*) by micro-XRF, Synchrotron Light Research Institute, Thailand, lateral resolution 50 μm, step size 50 μm, polychromatic excitation, data fitted by PyMCA (Solé et al. 2007). Cps, counts per second. (For detailed information please see the colour figure in online version.)

Van Elteren et al. 2016; Van Malderen et al. 2016). In contrast to micro-PIXE and micro-XRF techniques, LA-ICPMS belongs to the destructive methods, because biological samples are usually completely ablated during the measurement (Wu and Becker 2012) and after LA-ICPMS no other analytical technique can be applied.

LA-ICPMS was used to map the Zn localization in durum wheat at different nitrogen supply (Persson et al. 2016). Elemental distribution imaging was combined with the analysis of Zn-binding proteins by liquid chromatography ICPMS and orbitrap MS. The increase in Zn and N supply had a major impact on the Zn concentration in the endosperm and reached concentrations higher than current breeding targets. The sulfur concentration also increased, but S was only partially co-localized with Zn. The mutual accumulation of Zn and S was reflected in much more Zn bound to small cysteine-rich proteins (apparent size, 10–30 kDa), while the response of larger proteins (apparent size, 50 kDa) was only moderate. Most Zn-reactive proteins were associated with redox and stress processes.

In assessing quality parameters in grain, it is also important to track potentially hazardous metals like Cd. Cd is one of the most mobile hazardous elements in plants, therefore, it readily accumulates in grains of rice, wheat and other cereals, posing threats to consumers. Besides essential macro in micronutrients, LA-ICPMS was used to image the distribution of hazardous Cd in durum wheat (Yan et al. 2020). Wheat grains accumulate extremely low

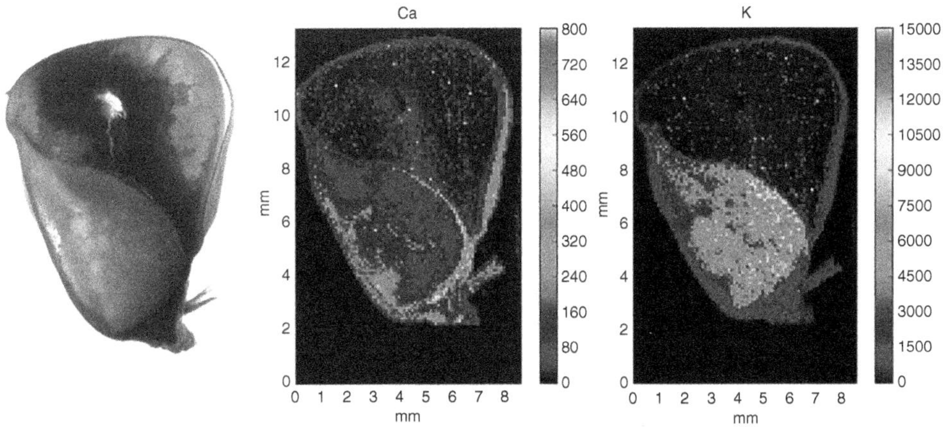

Figure 2.7 LA-ICPMS images of the Ca and K distribution in the maize (*Zea mays* L.) kernel. The lateral resolution is 50 μm. (For detailed information please see the colour figure in online version.)

concentrations of Cd ($<1\,\mu g\,g^{-1}$), so it is impossible to probe the distribution using X-ray based techniques. Cd is a high energy element with a Cd—K line of 23.17 keV and a Cd-L_3 line of 3.13 keV. The K line is hard to excite and the L_3 line is hidden in the tail of the macronutrient potassium K-line, raising the limit of detection to tens of $\mu g\,g^{-1}$ (Vogel-Mikus et al. 2008). In durum wheat, Cd accumulated mostly in the crease region, and inside the crease, in the pigment strand. In the vascular bundle, Cd was mainly associated to Sulfur ligands, and a relatively high concentration was therefore found in endosperm, the Sulfur-containing protein (gluten) fraction.

Localization of the major elements Ca and K in maize (*Zea mays*) (Figure 2.7) shows similar patterns as in other monocotyledonous species. Ca is mainly localized in the seed coat, while K is found in high concentrations in embryonic tissues.

2.3.4 Nano Secondary Ion Mass Spectrometry

Nano Secondary Ion MS (nano SIMS) works based on a coaxial optical design of the ion gun and secondary ion extraction, and on an original magnetic sector mass spectrometer with multicollection. Nano SIMS fills a unique niche as an MS tool for biological analysis by providing unmatched lateral resolution of elemental and isotopic distributions in samples of interest (Nuñez et al. 2018).

Cereals are an important source of Se to humans; many people suffer from Se deficiency in their diet causing malfunctioning of Se-enzymes, thyroid hormones and the immune system. Conversely, As is toxic and may accumulate in grain at levels that pose a health risk. Due to good sensitivity, it is especially suitable to image distribution of trace mineral nutrients like selenium (Se) or hazardous elements like arsenic (As) (Moore et al. 2010). Therefore, Nano SIMS was used to determine the localization of Se in wheat and As in rice. Combined synchrotron X-ray fluorescence (S-XRF) and Nano SIMS analysis utilized the strengths of both techniques. Selenium was concentrated in the protein surrounding the starch granules in the starchy endosperm cells and more homogeneously distributed in the aleurone cells but with Se-rich hotspots. Arsenic was concentrated in the sub-aleurone endosperm cells associated with the protein matrix, rather than in the aleurone cells. Nano SIMS indicated that the high intensity of As, identified in the S-XRF image was localized in micrometer-sized hotspots near the ovular vascular trace and nucellar projection.

In addition, Nano SIMS was also used to localize Ag in rice grains. The images revealed that the silver is concentrated in the aleuronic layer of the rice bran. Its concentration decreases in the sub-aleurone and becomes negligible in the endosperm. Accumulation of silver does not alter the grain morphology and chemical characteristics. The metal may be extracted from the bran after milling of the rice, thereby causing no hazards in associated foodstuffs (Sen Gupta et al. 2017).

2.3.5 Sample Preparation

Sample preparation is a highly demanding but crucial step that is directly related to the quality and accuracy of imaging results. Though state-of-the-art imaging tools are available today, the process is still severely limited by slow and demanding sample preparation protocols. In order to keep the anatomy and biochemistry of the cellular components as undisturbed as possible, it is essential that all biochemical and proteolytic processes are inactivated and the structures are immobilized and fixed in space and time by a "fixation step." Two approaches are normally used to "fix" biological samples: a chemical and a physical one (Dong et al. 2016b). Though chemical fixation is the most common approach for specimen preservation in light and electron microscopy, it is not suitable for spectroscopic analysis where it is essential to obtain the anatomy to be able to retrieve quantitative spatial biochemical information (Perrin et al. 2015). Since chemical fixation can lead to artifacts such as contamination, leaching of mobile elements and modification of biomolecules, cryofixation is accepted as the only reliable fixation method for spectroscopic analysis (Schneider et al. 2002; Vogel-Mikuš et al. 2014; Perrin et al. 2015; Dong et al. 2016b). The samples must be frozen in such a way that the growth of ice crystals is prevented in a process called vitrification (Scheloske et al. 2004). The current cryofixation workflows for the preparation of plant samples for imaging the elemental distribution (e.g. micro-PIXE) (Vogel-Mikuš et al. 2014) include manual cutting of plant material into small pieces, embedding of the pieces in a freezing medium (OCT), rapid freezing by immersing the sample in a cryogen cooled with liquid nitrogen (e.g. isopentane, propane), cryotome sectioning and freeze drying. Due to their hardness, grains can be imbibed (soaked for two to four hours, or more for very hard seeds, at 4 °C) and hand cut with a sharp razor blade to 100–200 μm thick slices, especially when analyzing the samples with MS-based and FTIR techniques, where OCT could contaminate sections or can interact with the sample composition. Carboxymethylcellulose (CMC), gelatine and their combinations have been successfully used as media compatible with biomolecular imaging (Dong et al. 2016b).

After cutting, freeze-drying is performed to obtain vacuum-compatible samples (micro-PIXE, micro-MeV-SIMS) and samples with low water content, which are required in micro-FTIR. During dehydration, sample shrinkage is often observed, which usually results in an uneven surface, affecting surface spectroscopic techniques such as micro-FTIR, micro-Raman and micro-MeV-SIMS. Shrinkage can also lead to a mismatch between biochemical and anatomical features (e.g. when plant protoplasts shrink and adhere to the cell wall, it is impossible to distinguish whether the biochemical information originates from the cell wall or from the inner parts of the cell), making biological interpretation extremely difficult (Dong et al. 2016b).

2.4 Prospect

The revolution in *omics* technology, particularly in metabolomics, has been closely linked to technological developments in mass spectrometry, currently the most efficient technology for characterizing molecular structures, and has led to significant progress in providing a comprehensive understanding of biological functions. Spatial information in mass spectrometric analysis can be preserved if the extraction process is avoided or limited to very small areas of the sample, which is then analyzed using new ionization techniques, such as MALDI and MeV or keV SIMS. In these techniques, the analytes are ionized and desorbed directly from the surface, thus avoiding traditional liquid extractions. They allow the sample to be scanned with the ion beam and the mass spectrum to be recorded at any point, resulting in the exact location in the sample where a particular compound was detected. By automating the process, images can be generated that show the spatial distributions of all compounds detected.

Development of accelerator-based nuclear microprobes (micro-PIXE), the third and fourth generation of synchro tron facilities with micro-and nano-focused beams (SR-micro-XRF), LA-ICPMS and Nano SIMShaveadvanced2D imaging of the elemental distribution in plants down to the subcellular level. Each of these techniques has its advantages and limitations, but they all provide a means to complement biomolecular imaging with elemental distributions. Inevitably, suitable sample preparation is a crucial step in the application of the existing up-to-date imaging techniques.

Cereals and pseudo-cereal plants that produce starchy grains are part of almost every meal we eat. Cereals are usually processed before consumption and certain nutritious compounds stored in the husk or embryonic tissue may be lost. Therefore, there is an urgent need to further explore these research tools that will provide better insights into the 2D and 3D spatial distribution of nutrients and minerals in cereals, and the mechanisms behind these distribution patterns.

Acknowledgement

The work was supported by Slovenian Research Agency (ARRS) project no. J7-9418, N1-0105, N1-0090, L4-9305, J7-9398 and N7-0077, and program group Plant Biology (P1-0212). Synchrotron Light Research Institute, Thailand is acknowledged for provision of beamtime (proposal number 3457) and Jitrin Chaiprapa for assistance with the measurements. Synchrotron Elettra, Italy and EU are acknowledged for provision of beamtime (proposal number 2008223 and 2008203) and Diane Eichert and Lisa Vaccari for assistance with the measurements.

References

Bjarnholt, N., Li, B., D'Alvise, J., and Janfelt, C. (2014). Mass spectrometry imaging of plant metabolites – principles and possibilities. *Nat. Prod. Rep.* 31: 818–837. https://doi.org/10.1039/c3np70100j.

Bonafaccia, G., Marocchini, M., and Kreft, I. (2003). Composition and technological properties of the flour and bran from common and tartary buckwheat. *Food Chem.* 80: 9–15. https://doi.org/10.1016/S0308-8146(02)00228-5.

Boughton, B.A., Thinagaran, D., Sarabia, D. et al. (2016). Mass spectrometry imaging for plant biology: a review. *Phytochem. Rev.* 15: 445–488. https://doi.org/10.1007/s11101-015-9440-2.

Briggs, D. (1983). Analysis of polymer surfaces by SIMS, 3—preliminary results from molecular imaging and microanalysis experiments. *Surf. Interface Anal.* 5: 113–118. https://doi.org/10.1002/sia.740050307.

Cheah, Z.X., Kopittke, P.M., Harper, S.M. et al. (2019). In situ analyses of inorganic nutrient distribution in sweetcorn and maize kernels using synchrotron-based x-ray fluorescence microscopy. *Ann. Bot.* 123: 543–556. https://doi.org/10.1093/aob/mcy189.

Collings, R., Harvey, L.J., Hooper, L. et al. (2013). The absorption of iron from whole diets: a systematic review. *Am. J. Clin. Nutr.* 98: 65–81. https://doi.org/10.3945/ajcn.112.050609.

Cotte, M., Pouyet, E., Salomé, M. et al. (2017). The ID21 X-ray and infrared microscopy beamline at the ESRF: status and recent applications to artistic materials. *J. Anal. At. Spectrom.* 32: 477–493. https://doi.org/10.1039/C6JA00356G.

Cvitanich, C., Przybyłowicz, W.J., Urbanski, D.F. et al. (2010). Iron and ferritin accumulate in separate cellular locations in Phaseolus seeds. *BMC Plant Biol.* 10: 26. https://doi.org/10.1186/1471-2229-10-26.

Detterbeck, A., Pongrac, P., Rensch, S. et al. (2016). Spatially resolved analysis of variation in barley (Hordeum vulgare) grain micronutrient accumulation. *New Phytol.* 211: 1241–1254. https://doi.org/10.1111/nph.13987.

Dickinson, M., Heard, P.J., Barker, J.H.A. et al. (2006). Dynamic SIMS analysis of cryo-prepared biological and geological specimens. *Appl. Surf. Sci.* 252: 6793–6796. https://doi.org/10.1016/j.apsusc.2006.02.236.

Dong, Y., Li, B., and Aharoni, A. (2016a). More than pictures: when MS imaging meets histology. *Trends Plant Sci.* 21: 686–698. https://doi.org/10.1016/j.tplants.2016.04.007.

Dong, Y., Li, B., Malitsky, S. et al. (2016b). Sample preparation for mass spectrometry imaging of plant tissues: areview. *Front. Plant Sci.* 7: 60. https://doi.org/10.3389/fpls.2016.00060.

Eckardt, N.A. (2011). Plant science in the Service of Human Health and Nutrition. *Plant Cell* 23: 2476–2476. https://doi.org/10.1105/tpc.111.230715.

Gianoncelli, A., Kourousias, G., Merolle, L. et al. (2016). Current status of the TwinMic beamline at Elettra: a soft X-ray transmission and emission microscopy station. *J. Synchrotron Radiat.* 23: 1526–1537. https://doi.org/10.1107/S1600577516014405.

Gorzolka, K., Kölling, J., Nattkemper, T.W., and Niehaus, K. (2016). Spatio-temporal metabolite profiling of the barley germination process by MALDI MS imaging. *PLoS One* 11: e0150208. https://doi.org/10.1371/journal.pone.0150208.

Grassl, J., Taylor, N.L., and Millar, H. (2011). Matrix-assisted laser desorption/ionisation mass spectrometry imaging and its development for plant protein imaging. *Plant Methods* 7: 21. https://doi.org/10.1186/1746-4811-7-21.

Guendel, A., Rolletschek, H., Wagner, S. et al. (2018). Micro imaging displays the sucrose landscape within and along its allocation pathways. *Plant Physiol.* 178: 1448–1460. https://doi.org/10.1104/pp.18.00947.

Gundlach-Graham, A. and Günther, D. (2016). Toward faster and higher resolution LA-ICPMS imaging: on the co-evolution of la cell design and ICPMS instrumentationyoung investigators in analytical and bioanalytical science. *Anal. Bioanal. Chem.* 408: 2687–2695. https://doi.org/10.1007/s00216-015-9251-8.

Gupta, S., Rupasinghe, T., Callahan, D.L. et al. (2019). Spatio-temporal metabolite and elemental profiling of salt stressed barley seeds during initial stages of germination by MALDI-MSI and μ-XRF spectrometry. *Front. Plant Sci.* 10: 1139. https://doi.org/10.3389/fpls.2019.01139.

Jenčič, B., Jeromel, L., Ogrinc Potočnik, N. et al. (2016). Molecular imaging of cannabis leaf tissue with MeV-SIMS method. *Nucl. Instruments Methods Phys. Res. Sect. B Beam Interact. with Mater. Atoms* 371: 205–210. https://doi.org/10.1016/j.nimb.2015.10.047.

Jenčič, B., Jeromel, L., Ogrinc Potočnik, N. et al. (2017). Molecular imaging of alkaloids in khat (Catha edulis) leaves with MeV-SIMS. *Nucl. Instruments Methods Phys. Res. Sect. B Beam Interact. with Mater. Atoms* 404: 140–145. https://doi.org/10.1016/j.nimb.2017.01.063.

Jeromel, L., Siketić, Z., Ogrinc Potočnik, N. et al. (2014). Development of mass spectrometry by high energy focused heavy ion beam: MeV SIMS with 8 MeV Cl7+ beam. *Nucl. Instruments Methods Phys. Res. Sect. B Beam Interact. with Mater. Atoms* 332: 22–27. https://doi.org/10.1016/j.nimb.2014.02.022.

Kaulich, B., Gianoncelli, A., Beran, A. et al. (2009). Low-energy X-ray fluorescence microscopy opening new opportunities for bio-related research. *J. R. Soc. Interface* 6 (Suppl 5): S641–S647. https://doi.org/10.1098/rsif.2009.0157.focus.

Koren, Š., Arčon, I., Kump, P. et al. (2013). Influence of CdCl2 and CdSO4 supplementation on cd distribution and ligand environment in leaves of the cd hyperaccumulator Noccaea (Thlaspi) praecox. *Plant Soil* 370: 125–148. https://doi.org/10.1007/s11104-013-1617-0.

Lee, Y.J., Perdian, D.C., Song, Z. et al. (2012). Use of mass spectrometry for imaging metabolites in plants. *Plant J.* 70: 81–95. https://doi.org/10.1111/j.1365-313X.2012.04899.x.

Lim, W.L., Collins, H.M., Byrt, C.S. et al. (2020). Overexpression of HvCslF6 in barley grain alters carbohydrate partitioning plus transfer tissue and endosperm development. *J. Exp. Bot.* 71: 138–153. https://doi.org/10.1093/jxb/erz407.

Limbeck, A., Galler, P., Bonta, M. et al. (2015). Recent advances in quantitative LA-ICP-MS analysis: challenges and solutions in the life sciences and environmental chemistry. *Anal. Bioanal. Chem.* 407: 6593–6617. https://doi.org/10.1007/s00216-015-8858-0.

Lu, L., Tian, S., Liao, H. et al. (2013). Analysis of metal element distributions in Rice (Oryza sativa L.) seeds and relocation during germination based on X-ray fluorescence imaging of Zn, Fe, K, Ca, and Mn. *PLoS One* 8: e57360. https://doi.org/10.1371/journal.pone.0057360.

Mantouvalou, I., Lachmann, T., Singh, S.P.S.P. et al. (2017). Advanced absorption correction for 3D elemental images applied to the analysis of pearl millet seeds obtained with a laboratory confocal micro X-ray fluorescence spectrometer. *Anal. Chem.* 89: 5453–5460. https://doi.org/10.1021/acs.analchem.7b00373.

Martínez-Criado, G., Villanova, J., Tucoulou, R. et al. (2016). ID16B: a hard X-ray nanoprobe beamline at the ESRF for nano-analysis. *J. Synchrotron Radiat.* 23: 344–352. https://doi.org/10.1107/S1600577515019839.

Mazzolini, A.P., Legge, G.J.F., and Pallaghy, C.K. (1981). The distribution of trace elements in mature wheat seed using the Melbourne proton microprobe. *Nucl. Inst. Methods Phys. Res. A* 191: 583–589. https://doi.org/10.1016/0029-554X(81)91066-1.

Mazzolini, A., Pallaghy, C., and Legge, G. (1985). Quantitative microanalysis of Mn, Zn and other elements in mature wheat seed. *New Phytol.* 100: 1985.

Miller, L.M. and Dumas, P. (2006). Chemical imaging of biological tissue with synchrotron infrared light. *Biochim. Biophys. Acta Biomembr.* 1758: 846–857. https://doi.org/10.1016/j.bbamem.2006.04.010.

Moore, K.L., Schröder, M., Lombi, E. et al. (2010). NanoSIMS analysis of arsenic and selenium in cereal grain. *New Phytol.* 185: 434–445. https://doi.org/10.1111/j.1469-8137.2009.03071.x.

Moretto, P. (1996). Nuclear microprobe: a microanalytical technique in biology. *Cell. Mol. Biol. (Noisy-le-grand)* 42: 1–16.

Nakata, Y., Yamada, H., Honda, Y. et al. (2008). Imaging mass spectrometry with swift heavy ions. *J. Mass Spectrom. Soc. Jpn.* 56: 201–208. https://doi.org/10.5702/massspec.56.201.

Nečemer, M., Kump, P., Ščančar, J. et al. (2008). Application of X-ray fluorescence analytical techniques in phytoremediation and plant biology studies. *Spectrochim. Acta Part B At. Spectrosc.* 63: 1240–1247. https://doi.org/10.1016/j.sab.2008.07.006.

Nuñez, J., Renslow, R., Cliff, J.B., and Anderton, C.R. (2018). NanoSIMS for biological applications: current practices and analyses. *Biointerphases* 13: 03B301. https://doi.org/10.1116/1.4993628.

Perrin, L., Carmona, A., Roudeau, S., and Ortega, R. (2015). Evaluation of sample preparation methods for single cell quantitative elemental imaging using proton or synchrotron radiation focused beams. *J. Anal. At. Spectrom.* 30: 2525–2532. https://doi.org/10.1039/C5JA00303B.

Persson, D.P., de Bang, T.C., Pedas, P.R. et al. (2016). Molecular speciation and tissue compartmentation of zinc in durum wheat grains with contrasting nutritional status. *New Phytol.* 211: 1255–1265. https://doi.org/10.1111/nph.13989.

Peukert, M., Thiel, J., Mock, H.P. et al. (2016). Spatiotemporal dynamics of oligofructan metabolism and suggested functions in developing cereal grains. *Front. Plant Sci.* 6 https://doi.org/10.3389/fpls.2015.01245.

Pongrac, P., Vogel-Mikuš, K., Regvar, M. et al. (2011). Improved lateral discrimination in screening the elemental composition of buckwheat grain by micro-PIXE. *J. Agric. Food Chem.* 59: 1275–1280. https://doi.org/10.1021/jf103150d.

Pongrac, P., Kreft, I., Vogel-Mikus, K. et al. (2013a). Relevance for food sciences of quantitative spatially resolved element profile investigations in wheat (Triticum aestivum) grain. *J. R. Soc. Interface* 10: 1742–5662. https://doi.org/10.1098/rsif.2013.0296.

Pongrac, P., Vogel-Mikuš, K., Jeromel, L. et al. (2013b). Spatially resolved distributions of the mineral elements in the grain of tartary buckwheat (Fagopyrum tataricum). *Food Res. Int.* 54: 125–131. https://doi.org/10.1016/j.foodres.2013.06.020.

Pongrac, P., Kelemen, M., Vavpetič, P. et al. (2020). Application of micro-PIXE (particle induced X-ray emission) to study buckwheat grain structure and composition. *Fagopyrum* 37: 5–10.

Regvar, M., Eichert, D., Kaulich, B. et al. (2011). New insights into globoids of protein storage vacuoles in wheat aleurone using synchrotron soft X-ray microscopy. *J. Exp. Bot.* 62: 3929–3939. https://doi.org/10.1093/jxb/err090.

Regvar, M., Eichert, D., Kaulich, B. et al. (2013). Biochemical characterization of cell types within leaves of metal-hyperaccumulating Noccaea praecox (Brassicaceae). *Plant Soil* 373: 157–171. https://doi.org/10.1007/s11104-013-1768-z.

Rodrigues, E.S., Gomes, M.H.F., Duran, N.M. et al. (2018). Laboratory microprobe X-ray fluorescence in plant science: emerging applications and case studies. *Front. Plant Sci.* 871: 1588. https://doi.org/10.3389/fpls.2018.01588.

Römpp, A. and Spengler, B. (2013). Mass spectrometry imaging with high resolution in mass and space. *Histochem. Cell Biol.* 139: 759–783. https://doi.org/10.1007/s00418-013-1097-6.

Scheloske, S., Maetz, M., Schneider, T. et al. (2004). Element distribution in mycorrhizal and nonmycorrhizal roots of the halophyte Aster tripolium determined by proton induced X-ray emission. *Protoplasma* 223: 183–189. https://doi.org/10.1007/s00709-003-0027-1.

Schneider, T., Scheloske, S., Povh, B., and Traxel, K. (2002). A method for cryosectioning of plant roots for proton microprobe analysis. *Int. J. PIXE* 12: 101–107. https://doi.org/10.1142/S0129083502000196.

Sen Gupta, S., Baksi, A., Roy, P. et al. (2017). Unusual accumulation of silver in the Aleurone layer of an Indian Rice (Oryza sativa) landrace and sustainable extraction of the metal. *ACS Sustain. Chem. Eng.* 5: 8310–8315. https://doi.org/10.1021/acssuschemeng.7b02058.

Simičič, J., Pelicon, P., Budnar, M., and Šmit, Ž. (2002). The performance of the Ljubljana ion microprobe. *Nucl. Instruments Methods Phys. Res. Sect. B Beam Interact. with Mater. Atoms* 190: 283–286. https://doi.org/10.1016/S0168-583X(01)01258-7.

Singh, S.P., Vogel-Mikus, K., Arcon, I. et al. (2013). Pattern of iron distribution in maternal and filial tissues in wheat grains with contrasting levels of iron. *J. Exp. Bot.* 64: 3249–3260. https://doi.org/10.1093/jxb/ert160.

Singh, S.P., Vogel-Mikuš, K., Vavpetič, P. et al. (2014). Spatial X-ray fluorescence micro-imaging of minerals in grain tissues of wheat and related genotypes. *Planta* 240: 277–289. https://doi.org/10.1007/s00425-014-2084-4.

Solé, V.A.A., Papillon, E., Cotte, M. et al. (2007). A multiplatform code for the analysis of energy-dispersive X-ray fluorescence spectra. *Spectrochim. Acta B At. Spectrosc.* 62: 63–68. https://doi.org/10.1016/j.sab.2006.12.002.

Sturtevant, D., Aziz, M., and Chapman, K.D. (2018). Visualizing the oilseed lipidome. *Int. News Fats, Oils Relat. Mater.* 29: 21–24. https://doi.org/10.21748/inform.04.2018.21.

Tolrà, R., Vogel-Mikuš, K., Hajiboland, R. et al. (2011). Localization of aluminium in tea (Camellia sinensis) leaves using low energy X-ray fluorescence spectro-microscopy. *J. Plant Res.* 124: 165–172. https://doi.org/10.1007/s10265-010-0344-3.

Van Elteren, J.T., Izmer, A., Šelih, V.S., and Vanhaecke, F. (2016). Novel image metrics for retrieval of the lateral resolution in line scan-based 2D LA-ICPMS imaging via an experimental-modeling approach. *Anal. Chem.* 88: 7413–7420. https://doi.org/10.1021/acs.analchem.6b02052.

Van Malderen, S.J.M., Managh, A.J., Sharp, B.L., and Vanhaecke, F. (2016). Recent developments in the design of rapid response cells for laser ablation-inductively coupled plasma-mass spectrometry and their impact on bioimaging applications. *J. Anal. At. Spectrom.* 31: 423–439. https://doi.org/10.1039/c5ja00430f.

Vavpetič, P., Kelemen, M., Jenčič, B., and Pelicon, P. (2017). Nuclear microprobe performance in high-current proton beam mode for micro-PIXE. *Nucl. Instruments Methods Phys. Res. Sect. B Beam Interact. with Mater. Atoms* 404: 69–73. https://doi.org/10.1016/J.NIMB.2017.01.023.

Vogel-Mikus, K., Regvar, M., Mesjasz-Przybyłowicz, J. et al. (2008). Spatial distribution of cadmium in leaves of metal hyperaccumulating Thlaspi praecox using micro-PIXE. *New Phytol.* 179: 712–721. https://doi.org/10.1111/j.1469-8137.2008.02519.x.

Vogel-Mikuš, K., Pelicon, P., Vavpetič, P. et al. (2009). Elemental analysis of edible grains by micro-PIXE: common buckwheat case study. *Nucl. Instruments Methods Phys. Res. Sect. B Beam Interact. with Mater. Atoms* 267: 2884–2889. https://doi.org/10.1016/j.nimb.2009.06.104.

Vogel-Mikuš, K., Pongrac, P., and Pelicon, P. (2014). Micro-PIXE elemental mapping for ionome studies of crop plants. *Int. J. PIXE* https://doi.org/10.1142/S0129083514400142.

Warren, F.J., Perston, B.B., Galindez-Najera, S.P. et al. (2015). Infrared microspectroscopic imaging of plant tissues: spectral visualization of Triticum aestivum kernel and Arabidopsis leaf microstructure. *Plant J.* 84: 634–646. https://doi.org/10.1111/tpj.13031.

Wu, B. and Becker, J.S. (2012). Imaging techniques for elements and element species in plant science. *Metallomics* 4: 403. https://doi.org/10.1039/c2mt00002d.

Yan, B., Isaure, M.P., Mounicou, S. et al. (2020). Cadmium distribution in mature durum wheat grains using dissection, laser ablation-ICP-MS and synchrotron techniques. *Environ. Pollut.* 260: 113987. https://doi.org/10.1016/j.envpol.2020.113987.

Zhang, Y., Shi, R., Rezaul, K.M. et al. (2010). Iron and zinc concentrations in grain and flour of winter wheat as affected by foliar application. *J. Agric. Food Chem.* https://doi.org/10.1021/jf103039k.

3

Application of Genome Engineering Methods for Quality Improvement in Important Crops

Sajid Fiaz[1], Sher Aslam Khan[1], Galal Bakr Anis[2], Habib Ali[3], Mohsin Ali[4], Kazim Ali[5], Mehmood Ali Noor[6], Sibtain Ahmad[4,7], and Bilal Ahmad Asad[4]

[1]*Department of Plant Breeding and Genetics, The University of Haripur 22620, Haripur, Khyber Pakhtunkhwa, Pakistan*
[2]*Rice Research and Training Center (RRTC), Rice Research Department, Field Crops Research Institute, Agricultural Research Center, Sakha, Kafr El-Sheikh, Egypt*
[3]*Department of Agricultural Engineering, Khawaja Fareed University of Engineering and Information Technology, Rahim Yar Khan, Punjab, Pakistan*
[4]*University of Agriculture Faisalabad, Sub-Campus Depalpur, Okara, Punjab, Pakistan*
[5]*National Institute for Genomics and Advanced Biotechnology, National Agricultural Research Centre, Park Road, Islamabad 45500, Pakistan*
[6]*Institute of Crop Sciences, Chinese Academy of Agricultural Sciences, Key Laboratory of Crop Physiology and Ecology, Ministry of Agriculture, Beijing 100081, China*
[7]*Animal Breeding and Genetics, Faculty of Animal husbandry, University of Agriculture Faisalabad, Punjab, Pakistan*

CHAPTER MENU

3.1 Introduction

The adequate supply of food providing calories and nutrients is essential for human survival. The inadequate supply of food leads toward food insecurity, threatening people around the globe. Moreover, with the rapid upsurge of population, the global population is expected to reach 8.3 billion by 2030 (United Nations, Department of Economic and Social Affairs, Population Division 2017). In response, the demand for food, fuel and shelter will

Genome Engineering for Crop Improvement, First Edition. Edited by Santosh Kumar Upadhyay.

also increase which will further intensify the pressure on available resource to grow more (Sundström et al. 2014). The growing population has a multi-dimensional impact from the emergence of new threats, i.e. abiotic stress due to climate change, reduction in arable land, salinization and biotic stress. To ensure a constant food supply, there is a need to exploit available resource wisely in spite of natural threats, including climate change (Godfray et al. 2010; Jones et al. 2014). Plant breeders have utilized both natural and induced mutation techniques along with heterosis breeding to ensure food security. However, more efforts are needed to meet with current and future challenges. The current breeding approaches are focusing to maintain or increase the production per unit area, and to reduce risk of crop failure. To increase production, the breeder mostly focused on agronomic attributes, i.e. grain yield per unit area, plant population per unit area and the grain size, etc. To minimize production there is a need to understand yield stability, the plants tolerant to several biotic and abiotic stresses are prerequisite. The biotic and abiotic stresses are now considered a major threat to global food security and thus it is much more important to understand the genomics of the plant to withstand against these challenges (Butt et al. 2018). There are great efforts required to understand the loci related to disease resistance and introgression of those loci to elite germplasm. Further, diseases also cause negative impact on grain quality of major cereals whereas, balancing the plant energy required to resist against diseases without penalties of yield and quality is challenging.

To enhance the nutritional value of crops, current breeding efforts emphasized the provision of divergent and balanced diet with optimum level of vitamins and minerals to boost human health. The quality requirements vary around the globe as consumer preferences are variable depending on region and cultural values. The nutritional aspect is of unique importance as the nutrients are required in less quantity and almost two billion people across globe suffer from malnutrition, especially Zinc (Zn), Iron (Fe), Iodine (I) and vitamin A (Bhullar and Gruissem 2013). Analysis of grain quality attributes required much time and effort along with a great number of samples, mostly taken at the early stage of breeding and standard protocols must be followed (Cruz and Khush 2000).The latest accomplishments in the field of crop biotechnology enable to breakdown of information to understand the key enzymes playing their vital role in metabolic pathways, increasing the contents of required vitamins and reducing compounds, i.e. phytic acids and acrylamide-forming amino acids. There are some cereals, i.e. Wheat, Maize and Rice which are biofortified to enhance food quality through fixing nutrition deficiencies. (Gil-Humanes et al. 2014; Ye et al. 2000). A documented example is of Golden Rice, which is genetically engineered to markedly increase the β-carotene to solve the problem of vitamin A deficiency (Ye et al. 2000). In this chapter, we have highlighted the application of targeted genome engineering approaches to understand and enhance the food quality with special emphasis on major crops. Based on previous findings, we have proposed the genome engineering techniques are an ultimate choice to enhance the food quality and ensure food security to the global human population.

3.2 Evolution and Historical Perspective of Genome Engineering

The utility of biological organism, procedures and techniques to develop products for improving human life is coined as genetic engineering/biotechnology. The natural variation has been taking place since the inception of plant kingdom, with natural selection mechanisms providing the means for plants survival under adverse situations. Moreover, humans have been utilizing the phenomena of artificial selection for around 10 000 years. Modern day crops are the outcome of drastic genetic alternations caused by artificial selection procedure. These genetic variations caused by natural mutation or artificial selection are important for any crop improvement program. The hypothesis gained more ground when it became evident that phenotypic changes can occur via induced mutations. Therefore, researchers utilized reagents, radiations and chemical mutagenic agents to generate mutation in targeted genome resulting in phenotypic changes (Shu et al. 2012). The mutation breeding concept conceived in the 1940s, revolutionized plant breeding ultimately leading toward the 1970s green revolution. A major breakthrough took place with the discovery of *Agrobacterium tumefaciens* utilized for carrying the gene of interest in a plant's genome producing transgenic plants (Nester 2014; Schubert and Williams 2006). The following technique carries several drawbacks, i.e. disruption of endogenous genes, random mutations, foreign DNA residues called genetically modified organisms (GMOs) and failure to utilize the native plant genetic repertoire. Keeping all these disadvantages in mind, there was a dire need to precisely and efficiently edit the genome of interest for the designated objectives of plant breeding.

In the 1980s, the first genome targeting system was introduced by Mario Capecchi along with the idea of double-strand breaks (DSBs) for genome manipulations (Capecchi 1980), and later the ability to engineer genome by site-specific DSBs was developed (Jasin 1996). After DSBs are generated, the cell's own repair machinery can be harvested to dictate the genetic outcome through the imprecise repair process of non-homologous end joining (NHEJ) or the precise repair process of homology-directed repair (HDR) (Schaart et al. 2016).The NHEJ mechanism possesses the ability to make frame shift mutation through the knockout of genes (Lloyd et al. 2012). With the ability to develop multiple DSB, it became possible to make further changes in the genome, i.e. chromosomal deletion, DSBs on different genome, chromosomal translocation and gene inversion (Ferguson and Alt 2001). In comparison to NHEJ, HDR methods are more efficient, producing the more precise repairs and enabling the sequence to be assembled as per user requirement (Puchta 2005). The HDR method can be efficiently utilized for precise genomic modifications allowing a variety of templates ranging from short oligonucleotides to generate DSBs for genome editing. Proteins that can been engineered and reprogrammed to bind and cleave DNA do not exist in nature. However, it is possible to program a DNA-binding domain to bind to any user-defined site-specific sequence. This domain can be fused with another domain that can cleave the DNA specifically where it binds. These bi-modular fusion proteins are the key to precise genome engineering because they can be programmed to bind to any user-selected sequence and generate a DSB. Such programmable site-specific binding proteins can carry other functional domains capable of effecting other genetic and genomic changes, including transcriptional regulation, epigenetic regulation, and even base editing without generation of DSBs (Komor et al. 2016). Complete genetic engineering events are well described. A genome-engineering tool box has three major platforms: zinc-finger nucleases (ZFNs), transcription activator-like effector nucleases (TALENs), and CRISPR/Cas systems. ZFNs and TALENs are protein-based and require protein engineering for every user-defined sequence. However, CRISPR/Cas is an RNA-guided system and can easily be engineered to bind to the DNA target (Quétier 2016). In the current genetic engineering era, the most widely utilized is CRISPR/Cas system which will be discussed in detail to achieve crops quality improvement.

3.3 CRISPR/Cas Genome Editing Systems

In the current scenario, CRISPR-Cas9 is a widely adopted genome edited system (GET) due to simplicity, efficiency and versatility. In 1987, CRISPR array was identified in *Escherichia coli* (*E. coli*) genome (Ishino et al. 1987) with unknown biological properties. During 2005, several studies were successful that revealed a CRISPR array role in adaptive immunity based on the availability of homologs spacers to viral and plasmid sequence (Pourcel et al. 2005). Jinek et al. (Jinek et al. 2012) first reported an RNA-guided DNA cleavage system with high required target efficiency. Deltcheva et al. (Deltcheva et al. 2011), uncovered CRISPR array provides protection against foreign DNA when coupled with Cas9 protein whereas, the immune system was based on RNA mediated DNA targeting. CRISPR/Cas systems are part of the adaptive immune system of bacteria and archaea, protecting them against invading nucleic acids (spacers) such as viruses by cleaving foreign DNA in a sequence dependent manner. The immunity is obtained via integration of spacers between two adjacent repeats at the proximal end of a CRISPR locus. The spacers are transcribed into CRISPR RNAs (crRNAs) approximately 40 nt in length by successive encounters with foreign DNA, and combines with trans activating CRISPR RNA (tracrRNA) to activate and guide Cas9 nuclease (Barrangou et al. 2007). The Cas9 cleaves the homologs double-stranded DNA sequence (protospacer) in the foreign DNA. The availability of PAM downstream of the target DNA is a key for the successful cleavage with frequent 5′-NGG-3′ and less frequent 5′-NAG-3′ (Hsu et al. 2013). Seed sequence approximately 12 bp upstream of the PAM are integral for pairing between RNA and target DNA (Bortesi and Fischer 2015). Multiplex editing by the introduction of double-stranded breaks (DSBs) on various sites and ability to edit several genes at the same time is a unique feature of CRISPR/Cas system (Zhou et al. 2014). It only requires monomeric Cas9 protein with sequence-specific gRNA meanwhile, ZFNs and TALENs requires dimeric proteins with reference to target site. In the CRISPR/Cas system, the recognition of target site is based on the Watson and Crick model enabling the off-target detection through sequencing analysis along with off-target mutation can be fixed by careful design of gRNA. On the contrary, ZFNs and TALENs target specificity depends on protein-DNA interaction with unpredictable and function-specific properties (Cho et al. 2014). Moreover, the scientific community equipped with CRISPR expertise had contributed a lot to disseminate the fresh information to newcomers which is opposite to ZFN proprietary nature. Several online platforms are now available and are assisting researchers for all concerns related to CRISPR (Table 3.1; Figure 3.1).

3.4 Application of CRISPR/Cas System for Crops Quality Improvement

The improvement in quality is as important as yield and plant breeding programs are focusing on the improvement of grain quality. The grain quality is multigenic in nature and controlled by a number of factors. The demand for better quality grain is increasing owing

Table 3.1 Online platform for tools of CRISPR/Cas system.

sgRNA design platform	Webpage link	References
CRISPR Design	http://crispr.mit.edu	Hsu et al. (2013)
E-CRISP	http://www.e-crisp.org/E-CRISP	Heigwer et al. (2014)
CRISPR Multi Targeter	http://www.multicrispr.net/basic_input.html	Prykhozhij et al. (2015)
sgRNA designer: CRISPRko	http://portals.broadinstitute.org/gpp/public/analysis-tools/sgrna-design	Doench et al. (2016)
Off-Spoter	https://cm.jefferson.edu/Off-Spotter	Pliatsika and Rigoutsos (2015)
CCTop	http://crispr.cos.uni-heidelberg.de/index.html	Stemmer et al. (2015)
CHOPCHOP	http://chopchop.cbu.uib.no/index.php	Montague et al. (2014)
Plasmid free access		
Add gene	https://www.addgene.org/search/advanced	Bortesi and Fischer (2015)
Expert discussion forum		
OXfCRISPR	https://www.dpag.ox.ac.uk/research/liu-group/liu-group news/oxfcrispr/	Bortesi and Fischer (2015)

Source: Extracted from (Bortesi and Fischer 2015; Chira et al. 2017).

to improvement in grain-processing techniques, environmental variations and consumers' preferences. The breeding to develop semi-dwarf cultivars and hybrids of various crops have markedly increased production with a compromise on quality. The advancement in genomics studies for grain quality has further improved the understanding of genes and biosynthesis pathways controlling essential features of grain quality. The complex network of genes controlling quality attributes can be resolved by understanding functional and comparative genomics of major genes and their faithful introgression across cultivars. The targeted mutagenesis through the CRISPR/Cas system has been successfully adopted in several crops (Fiaz et al. 2019) for traits including biotic and abiotic stresses, yield and quality improvement. The series of published articles on the following system is considered as proof of concept describing the application of CRISPR/Cas system for the knockout of genes controlling desirable phenotypes. A generalized protocol for the application of CRISPR/Cas system in several crop species has been described with the potential to be employed for improvement of quality of important crops (Figure 3.2).

3.4.1 Rice

Rice (*Oryzasativa* L.) is a major cereal and a source of calories and protein to the vast population around the globe. It has a diploid genome (2n = 24); the small genome size makes it a model crop among monocots to investigate for research objectives (Fiaz et al. 2019).

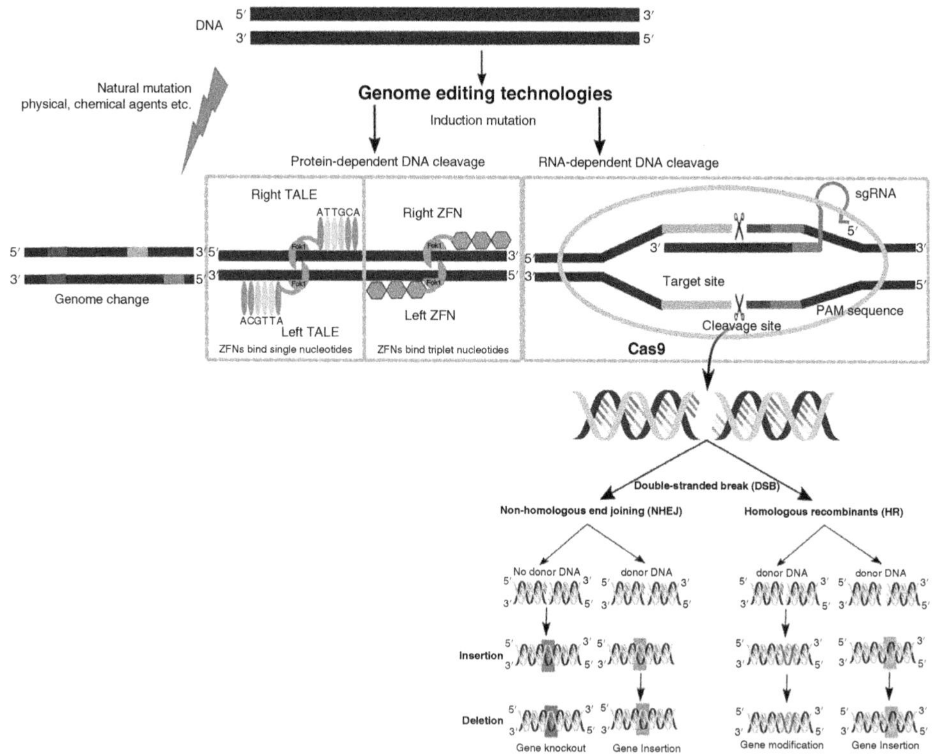

Figure 3.1 Comparison of Gets. Classical methods include natural mutation via hybridization, induced mutation via physical agents (ultra violet; x-ray and light), benzene analogs and chemical methods (nitrous acid). Site-specific genome targeting technologies; (a) protein-dependent DNA cleavage system TALE and ZFN (b) RNA-dependent DNA cleavage, e.g. CRISPR-Cas system (c) random mutation via error-prone NHEJ or targeted mutation via error-free HR. These approaches achieve genomic modification by inserting, deleting, or replacing targeted DNA sequence. *Source:* Modified from Zhang et al. (Zhang et al. 2017).

It has been reported that approximately, 499.37 million tons of rice was produced around the globe during year 2019 (FAO (Food and Agriculture Organization of The United Nations) Statistics 2017–18). The prime attributes influencing quality include cooking and eating, phytochemicals and micronutrients-related characters (Lau et al. 2015). People from different ethnic groups and/or geographical regions have different grain-quality preferences (Sabouri et al. 2012). Rice of premium quality like Indian basmati and Thai Jasmine are highly valued due to their typical aroma during cooking (Ferrero 2004). Rice grain quality parameters are very complex and easily affected by the environment. The key components which have impact on grain quality are (i) appearance quality (ii) milling quality (iii) eating-cooking quality and (iv) nutritional quality which are further divided into many other categories (Lou et al. 2009). From the last few years, much research has demonstrated the successful application of CRISPR-based targeted mutagenesis approach in rice to make efficient and reliable crop quality improvement.

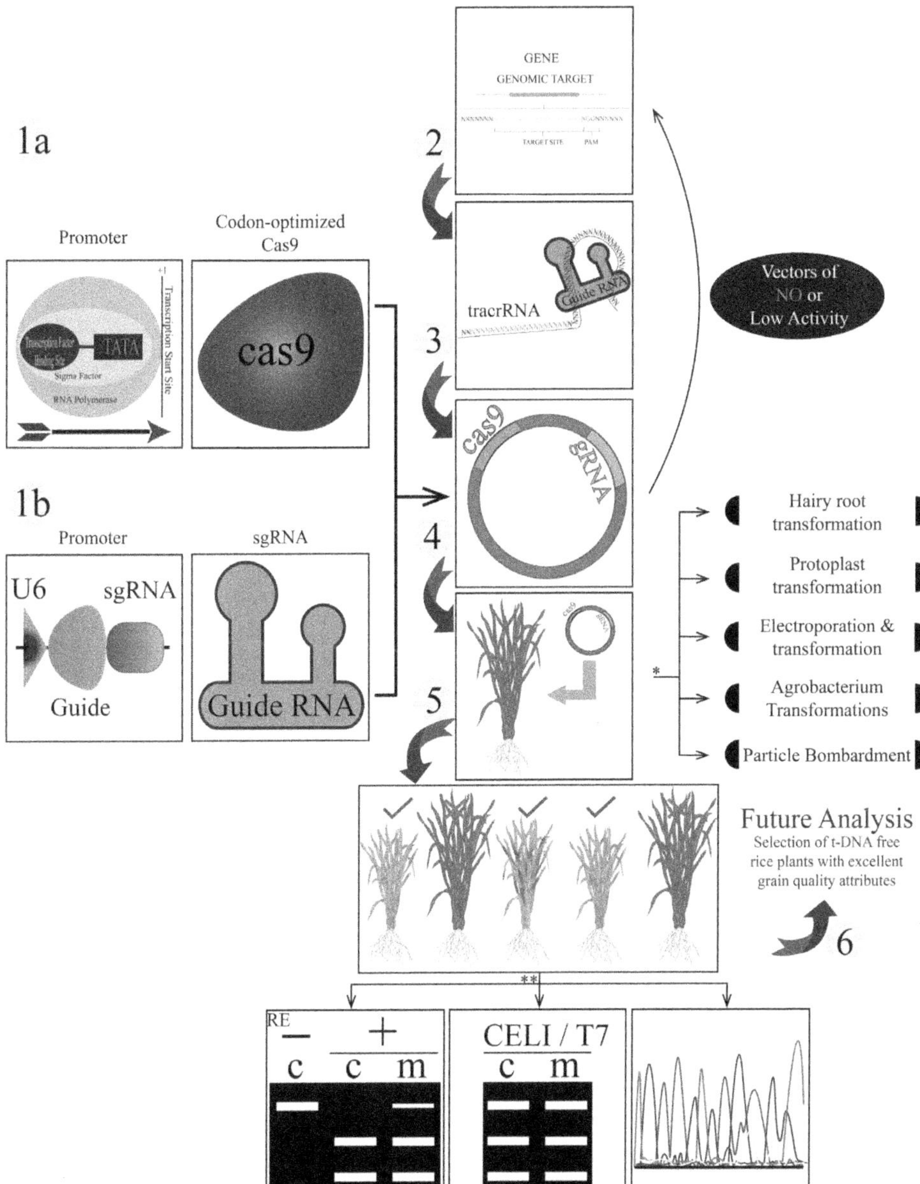

Figure 3.2 The basic scheme of CRISPR/Cas system (Fiaz et al. 2019). (1a) Promoter fused with cas9 (1b) Promoter fused with sgRNA (2) Selection of target site from gene under investigation (3) tracrRNA hybridize and join the Cas9 (4) Cas9 vector containing gRNA attached with target sequence (5) Vector transformation; Detection of mutation and selection of desirable mutant plants (6) The mutant screening to get t-DNA free plants and further analysis of phenotype in mutant plants. *Source:* Modified from Fiaz et al. (Fiaz et al. 2019).

3.4.1.1 Application of CRISPR/Cas9 for Rice Quality Improvement

Several genes controlling grain appearance attributes have been identified through functional genomics studies. To understand the underlying mechanisms, reverse genetics tools especially the CRISPR genome-editing system have been utilized. Grain shape, a yield component, plays a key role in determining the quality of rice grains. *GS3* a major gene controlling grain length, and *Gn1a* controlling the number of grains were successfully genome edited in four different cultivars of rice (Shen et al. 2017). The transgene free T_1 plants have shown longer grain length and increased thousand grain weight in comparison to wild type. There are multiple genes controlling these attributes further, the interaction and underlying function has also been characterized. The genome interaction of *S3* with *GW5* also determines the regulation of head rice recovery. The cloning and characterization of *OsSPL13* revealed that it has a significant positive impact on cell size in the grain hull, resulting in enhanced rice grain length and yield (Si et al. 2016). *Chalk5* is a major QTL and is highly expressed in the endosperm and increases H^+ concentration, resulting in water loss from the vacuoles, which ultimately leads to chalky endosperm, an undesirable attribute (Li et al. 2004). *OsSPL16/GW8* and *GW7* boost grain length and reduce grain width and chalkiness (Wang et al. 2015). Generally, three important physico-chemical indices, i.e. amylose content (AC), Gel consistency (GC), and gelatinization temperature (GT) determine the cooking and eating quality of rice grain. AC is regulated by the *Waxy* (*Wx*) gene in the endosperm (Fiaz et al. 2019), and *ALK/SSIIa* and *RSR1* control GT (Zhang et al. 2011). Several studies have been conducted to unearth the role of different genes and/or enzymes involved in the regulation and expression of the *Wx* gene. Thc *Indica* hybrid rice grown in mainland China has high AC which makes it hard and dry during the cooking process. In another effort, Ma et al. (Ma et al. 2015) knock out the *Wx* gene in a *Japonica* rice leads toward reduction in AC. Moreover, transgenic Taichung 65 rice lines with antisense *Wx* construct showed lower AC, and hybrids developed by utilizing these transgenic lines also demonstrated lower AC (Terada et al. 2000). In comparison, Zhang et al. (Zhang et al. 2018) introduced a loss of function mutation via CRISPR/Cas9 to the *Wx* gene in two widely grown *japonica* cultivars "Xiushui134" and "Wuyunjing 7." The *Wx* gene mutation caused reduced AC without affecting other desirable agronomic traits. To understand the fine structure and physical properties of starch *SBEI* and *SBEII*, they were mutated via targeted mutagenesis. The results showed that *SBEIIb* plays an important role in producing high amylose rice. The effective editing of several genes proved the reliability of CRISPR/Cas system to understand the functional genomics of genes and transcriptional factors controlling rice grain cooking and eating quality. The improvement in nutritional quality is worthy of consideration, especially those countries where rice is the staple food. It has been estimated, every day approximately 24,000 people die because of poverty based lack of food with adequate content of vitamins, minerals and essential amino-acids. (Nordin and Lantbruksakademien 2008). The food intake by much of the human population lacks in protein, iron, zinc, vitamin A and iodine (Wenefrida et al. 2009).The application of CRISPR/Cas system can help to improve the nutritional quality of rice grain. The nutritional quality is influenced by the amount of seed storage proteins (SSPs), amino acids contents, fats, vitamins and other essential micronutrients (Peng et al. 2014), but rice has the lowest protein content among the cereal grains (Shewry and Halford 2002). Some amino acids, i.e. lysine (Lys) and tryptophan (Trp) have been reported to be missing from rice grain (Liu et al. 2016). Therefore, improving nutrition in humans is intrinsically associated with improving nutrition in SSPs, especially for people of the regions where rice is being consumed as staple food. Rice contains six 5-methylcytosine (5mC) DNA methylase genes including *OsROS1*, *OsROS1b*, *OsROS1c*, *OsROS1d*, *OsDML3a*, and *OsDML3b* which can play an important role in enhancing nutritional grain quality. The number of aleurone cell layers was increased by *ta2–1* which is a weak mutant allele of *OsROS1* (Liu et al. 2018). SSPs include four different categories in rice, albumin, globulin, prolamin, and glutelin, separated by their solubility (Chen et al. 2018). However, the synthesis and accumulation of these four fractions and starch are accelerated by *qPC1* which encodes a putative amino acid transporter OsAAP6. Thus, *qPC1* significantly increases grain protein content (Peng et al. 2014). Golden rice rich in β-carotene was developed, by introgressing of four genes including *phytoene synthase*, *phytoenedesaturase*, *β-carotene desaturase*, and *lycopene β-cyclase* to overcome the night blindness caused by vitamin A deficiency in many developing countries (Ye et al. 2000). In conclusion, the genes discussed above have been manipulated through different plant breeding efforts, however, these techniques are time consuming and less efficient. So, to overcome these barriers, using the CRISPR/Cas9 system, i.e. knock-out/in and multiplex editing, could be an efficient approach to enhance rice grain nutritional value.

3.4.1.2 Wheat

Wheat (*Triticum aestivum* L.) also being a major staple food around the globe, having hexaploid genome (2n = 6x = 42, AABBDD), which is more than five times greater than the human genome (Borisjuk et al. 2019). Wheat production covers more than 240 million hectares (ha) globally and its gross world trade is greater than all other crops combined. Approximately, 21% of the world's food depends on annual wheat crop harvests, which often have relatively low stocks (FAO (Food and Agriculture Organization of The United Nations) Statistics 2014–15). However, there is a gap between current productivity gains as compared to those required to satisfy expected global demands over the coming decades. To reduce this gap, there is a dire need to explore new ways to improve wheat productivity. Genetic improvement has been a major contributor to enhance productivity in all crops important from the food security standpoint. Most of the wheat produced globally is consumed directly by humans as food. The requirement that wheat grain must be suitable to produce breads and other end products with complex and specific properties limit the rate of genetic gains in wheat production as the low grain quality varieties with high yield production are least preferred by consumers due to rises in their living standard. A clear understanding of genetic, molecular and biochemical bases of wheat grain quality might allow these requirements to be more easily satisfied and help to accelerate wheat productivity along with quality. Advances in next generation sequencing have been applied in many different ways that are relevant to wheat quality. The analysis of specific groups of genes allows identified genes influencing grain quality to be studied, while new genetic and molecular influences on wheat quality can be revealed by considering variations in the whole genome and relating all the differences to explain the basis in fluctuations in wheat grain quality.

3.4.1.3 Application of CRISPR/Cas9 for Wheat Quality Improvement

Most of the studies provided insights on starch composition and its biosynthesis pathways based on the knowledge gained, several biotechnologies approaches have been utilized to enhance the quality of starch related properties. Weichert et al. (Weichert et al. 2010) over expressed the barley sucrose transporter gene (*HvSUT1*) that uplifts sucrose uptake along with protein content of wheat grains with less compromise on starch content. In another attempt, expression of an optimized maize ADP-glucose pyrophosphorylase (*ZmAGPase*) resulted in elevated yield and enhanced photosynthetic rates in the transgenic wheat lines (Smidansky et al. 2007).

Subsequently, down regulation of the transcription factor *TaRSR1* (a wheat homolog of Rice Starch Regulator; *OsRSR1*), showed a negative regulation of some starch synthesis-related enzymes in wheat grains (Kang et al. 2013),resulting in a significant 30% increase in starch content, and also a ~ 20% increase in yield in terms of 1000-kernel weight (Liu et al. 2016). The amylose to amylopectin ratio influences the starch quality. The higher amylose content contributes more toward resistant starch (RS) in food with potential benefits to human health. There is evidence that RS can provide protection from several health conditions such as diabetes, obesity, and cardiovascular diseases (Meenu and Xu 2018). Many experiments have been conducted on down regulation of starch-branching enzymes, *SBEIIa*, and *SBEIIb*, which led to substantially increased amylose levels in wheat (Regina et al. 2006). The starch with higher concentration of AC demonstrated a health impact in rats (Regina et al. 2006), and a similar study showed obesity in humans (Sestili et al. 2010). Glutelin makes the major portion of Wheat SSPs, determining the dough viscoelastic properties (Anjum et al. 2007). Glutelin contribute 70–80% of crude protein and are composed of gliadins and glutenins. Therefore, it is essential to understand and manipulate the genes controlling/improving nutritional and dough- related properties. The application of transgenic approaches enhances the understanding for underlying attributes and their genomic architecture controlling quality traits. The genes controlling high molecular glutelin subunits (HMW-GS) were introgressed in Wheat to improve dough-related properties (Altpeter et al. 1996). The transformation of subunits 1Ax1 and 1Dx5 in common wheat cultivars modify dough properties to various extent (Alvarez et al. 2001). The genes influencing HMW-GS were introduced into Wheat cultivars Bobwhite through a transgenic approach. Further, the transgenic lines were crossed with elite varieties to improve grain qualities which demonstrated the feasibility of transgenic wheat breeding (Li et al. 2007). In comparison to HMW-GS, the gliadins contribute to dough viscosity and extensibility (Payne 1987). The gliadins are further grouped into α-, γ-, and ω-gliadins, multigenic in nature. The gliadins possess research interest owing to its contribution to influence dough quality, immunogenic epitopes linked with immune condition, e.g. wheat-dependent exercise-induced anaphylaxis (WDEIA) and celiac disease (Scherf et al. 2016).

Iron and zinc are essential micronutrients contributing to nutrition. As per the WHO reports, approximately one billion in the human population are suffering from several diseases directly linked with nutrition deficiency. Moreover, the annual death rate is approximately half a million below the age of five. The cultivable Wheat lacks in micronutrients (Cakmak et al. 2000), and one possible reason is that the micronutrients accumulate in the outer husk, aleurone and embryo, lost during the milling and polishing process (Welch and Graham 1999). The anti-nutritional factor, i.e. phytic acid also deposited in the aleurone storage vacuoles with impact on iron and zinc in the human digestive tract. The micronutrient deficiency can easily be manipulated through biofortification mechanism. The transgenic approaches have been successfully carried out to make micronutrients especially iron contents in Wheat. The plants store iron in ferritin structures deposited into no-green plastids, etioplasts and amyloplasts. The expression of gene controlling *Aspergillus niger* phytase, phytic acid degrading enzyme in Wheat aleurone and endogenous (Borg et al. 2012) or Soybean (Sui et al. 2012) ferritin genes in wheat endosperm, were the first successful attempts to fortify wheat grains with iron through transgenic approaches. Recently, Connorton et al. (Connorton et al. 2017) have isolated, characterized, and overexpressed two wheat *Vacuolar Iron Transporter (TaVIT)* genes under the control of an endosperm-specific promoter in wheat and barley. The introduction of *TaVIT2* gene through the transgenic approach enhances, almost two-fold of iron content in Wheat grains.

In wheat, the principal applicability of CRISPR/Cas9 was demonstrated in protoplasts and suspension cultures, where multiple genes were successfully targeted and achieved in the year following the publication of the original CRISPR/Cas9 principle (Upadhyay et al. 2013). Many agriculturally important traits of wheat have been targeted by genome editing among which include; (i) resistance/tolerance to biotic and abiotic stresses, (ii) yield and grain quality, and (iii) male sterility. The first experiment was conducted by Shan et al. (Shan et al. 2014) and successfully revealed the application of CRISPR/Cas9 system in wheat protoplasts for *TaMLO* gene (Mildew resistance locus O). The CRISPR *TaMLO* knockout has shown to confer resistance to powdery mildew disease caused by *Blumeriagraminis* f. sp. *Tritici* (Btg). The lipoxygenase genes, *TaLpx1* and *TaLox2*, attracted attention as potential subjects for gene editing in relation to resistance to *Fusarium*, one of the most devastating fungal diseases in wheat. Lipoxygenases hydrolyze polyunsaturated fatty acids and initiate biosynthesis of oxylipins, playing a role in the activation of jasmonic acid-mediated defense responses in plants. Silencing of the *TaLpx-1* gene has resulted in resistance to *Fusarium graminearum* in wheat (Nalam et al. 2015). *TaLpx1* and *TaLox2* genes were edited in protoplasts with a mutation frequency of 9 and 45%, respectively (Shan et al. 2014; Wang et al. 2018). Wheat plants with mutated *TaLOX2* were obtained with a frequency of 9.5%, of which homozygous mutants accounted for 44.7% (Zhang et al. 2016). With the aim of enhancing grain size and yield, several genes have been edited by the CRISPR/Cas9 system: *TaGASR7* (Liang et al. 2017), *TaGW2* (Li et al. 2017), and *TaDEP1* (Zhang et al. 2016). *TaGASR7*, a member of the Snakin/GASA gene family, has been associated with grain length in wheat. A CRISPR/Cas9 vector designed to target *TaGASR7* was delivered by particle bombardment into shoot apical meristems. Eleven (5.2%) of the 210 bombarded plants carried mutant alleles, and the mutations of three (1.4%) of these were inherited in the next generation (Wang et al. 2018). Transiently expressing the CRISPR/Cas9 DNA and using the CRISPR/Cas9 RNP mediated method were also highly effective for *TaGASR7* editing (Liang et al. 2017). The *TaGW2* gene encodes a previously unknown RING-type E3 ubiquitin ligase that was reported to be a negative regulator of grain size and thousand grain weight in wheat (Li et al. 2017). Recent studies, detailing the functionality of the allelic *TaGW2* genes through genome specific knockouts, showed that the *TaGW2* gene in wheat acts by regulating the gibberellin hormone biosynthesis pathway (Li et al. 2017), principally confirming the parallel functions of these genes in rice and wheat. T1 plants carrying knock-out mutations in all three copies of the *TaGW2* gene (genotype *aabbdd*) showed significantly increased thousand grain weight (27.7%), grain area (17.0%), grain width (10.9%), and grain length (6.1%) compared to the wild-type cultivar (Wang et al. 2018). CRISPR/Cas9 technology was also successful in obtaining low immunogenic wheat. Sanchez-Le'on et al. (Sánchez-León et al. 2018) have shown that CRISPR/Cas9 technology can be used to efficiently reduce the amount of alpha-gliadins in the seeds, providing bread and durum wheat lines with reduced immune reactivity for consumers with celiac disease. Twenty-one mutant lines were generated (15 bread wheat and 6 durum wheat), all showing a strong reduction in alpha-gliadins. Up to 35 of the 45 different genes identified in the wild-type were mutated in one of the lines, and immunoreactivity, as measured by competitive ELISA assays using two monoclonal antibodies, was reduced by 85%. These examples provide an insight into the many ways in which modern genome modification technologies are being used to mine the core research findings from model plant transgenesis, and finally harness that understanding to drive essential crop. The ability to enact targeted changes to the genome has revolutionized genetic modification for polyploidy crop species such as wheat.

3.4.1.4 Maize

Maize (*Zea mays* L.) is a widely grown C4 crop with a high rate of photosynthetic activity leading to high grain and biomass yield potential and most produced grain crop globally. The maize genome is genetically diploid and consists of 10 chromosomes with an estimate size ranging from 2.3 to 2.7 Gb (Schnable et al. 2009). The maize genome consists mostly of a non-genic, repetitive fraction punctuated by islands of unique, or low-copy DNA that harbor single genes or small groups of genes. The repetitive elements contribute significantly to the wide range of diversity within the species and include transposable elements, ribosomal DNA, and high-copy short-tandem repeats mostly present at the telomeres, centromeres, and heterochromatin knobs (Morgante 2006). It has been reported that approximately, 62 million tonnes maize was produced around the globe during the year 2019 (FAO (Food and Agriculture Organization of The United Nations) Statistics 2019–20). Its myriad end uses and the ease of cultivation over varied environmental and soil conditions has made it a desirable crop worldwide. In addition to human consumption, it is used as feed for livestock, raw materials for chemical and food industries and as biofuel (Pegoraro et al. 2011).

3.4.1.5 Application of CRISPR/Cas9 for Maize Quality Improvement

In Maize (*Zea mays*) phytic acid constitutes more than 70% of the maize seed. It is believed to be anti-nutritional as it is not digested by monogastric animals and is also an environmental pollutant. Liang et al. (Liang et al. 2014) have reported targeted knock out of genes involved in phytic acid synthesis (*ZmIPK1A*, *ZmIPK*, and *ZmMRP4*) in *Z. mays*. Similarly, Zhu et al. (Zhu et al. 2016) demonstrated gene editing of phytoene synthase gene (*PSY1*) using maize U6 snRNA promoter. *PSY1* is involved in carotenoid biosynthesis and its mutant (*psy1*) results in white kernels and albino seedlings. Among 52 T_0 lines obtained by Agrobacterium-mediated transformation, seven lines were reported to carry the *psy1* knockout furthermore, the sequencing analysis undertaken to evaluate the variations generated and mutation efficiency. The results showed that no off-target sites were edited and stable *psy1* mutants were obtained. Feng et al. (Feng et al. 2016) have demonstrated the utility of the CRISPR/Cas9 system in maize by targeting the albino marker gene, *Zmzb7* through protoplast system. Knockout of *Zmzb7* results in albino plant, with the sgRNA designed to target a region in the eighth exon of *Zmzb7* and maize U3 promoter was used for expression. Following Agrobacterium-mediated transformation of maize embryos, T_0 lines were found to show a 31% mutation efficiency. Gene editing tools that can affect multiple gene knockouts are of immense importance to accelerate and achieve efficient crop breeding. For the first time, multiplex genome editing in maize was demonstrated by Qi et al. (Qi et al. 2016) using a tRNA-RNA processing system. A multiplex editing vector can incorporate a cluster of gRNAs separated by spacers in a polycistron, producing multiple gRNAs from one primary transcript. The study targeted three transcription factor genes (*MADS*, *MYBR*, and *AP2*) for simplex editing and three other genes (RPL, PPR, and IncRNA) for multiplex editing. Increased editing efficiency (up to 100%) was observed for t-RNA processing based multiplex editing. Current high yielding maize varieties are the result of hybrid maize seed production and the production of hybrid maize requires sterilization to avoid self-fertilization. Maize thermosensitive genic male-sterile 5 (*ZmTMS5*), known to cause male sterility was targeted for genome editing by CRISPR/Cas9 approach (Li et al. 2017). Three gRNAs were used to knockout the gene, with one sgRNA targeting the first exon and the other two sgRNAs targeting the second exon. Mutation efficiency was examined in maize protoplasts using PCR/restriction enzyme assays. Analysis of mutational efficiency revealed that the sgRNA targeting the first exon had no off-targets whereas the other two sgRNAs had off-targets in the maize genome. *The AUXIN REGULATED GENE INVOLVED IN ORGAN SIZE* (*ARGOS*) gene family are negative regulators of the ethylene response and modulate ethylene signal transduction. Overexpression of *ARGOS* genes in transgenic maize plants enhances drought tolerance and identification of new allelic variants would be of immense importance in maize breeding programs. Shi et al. (Shi et al. 2017)utilized CRISPR/Cas9 genome editing to create new allelic variants of *ARGOS8*. Two genome-edited variants (*ARGOS8-v1* and *ARGOS8-v2*) were used for the production of hybrids and evaluated in the field in multi-location trials. Improved yield under stress condition was observed for the mutant than wild-type. This study demonstrated the use of CRISPR/Cas9 genome editing method for creating new variants and their application in maize crop improvement.

3.4.1.6 Cotton

Cotton is an important crop for the production of fiber, oil and biofuel. In addition, cotton serves as a cash crop for more than 20 million farmers in Asia and Africa. Despite the availability of synthetic alternatives, cotton remains an important source of fiber because of the advantages related to cost of production and unique features offered by cotton lint. Consumption of cotton products in the world is increasing day by day in a lot of places, but world cotton production is stagnant because of biotic and abiotic stresses. To meet the demands of the masses, production of cotton needs to be very high, with good quality. Cotton is also affected by diseases, causing significant losses to industry. Therefore, it became evident to utilize plant-breeding approaches to tackle threats caused by both biotic and abiotic factors ultimately reducing fiber quality.

The genus *Gossypium* includes approximately 45 diploid ($2n = 2x = 26$) and five tetraploid ($2n = 4x = 52$) species, all exhibiting disomic patterns of inheritance. Diploid species ($2n = 26$) fall into eight genomic groups (A–G, and K). The African clade, comprising the A, B, E, and F genomes (Wendel and Cronn 2003), occurs naturally in Africa and Asia, while the D genome clade is indigenous to the Americas. A third diploid clade, including

C, G, and K, is found in Australia. All 52 chromosome species, including *Gossypium hirsutum* and *Gossypium barbadense*, are classic natural allotetraploids that arose in the New World from interspecific hybridization between an A genome-like ancestral African species and a D genome-like American species. More than 95% of the annual cotton crop worldwide is *G. hirsutum*, Upland or American cotton, and the extra-long staple or Pima cotton (*G. barbadense*) accounts for less than 2% (National Cotton Council, http://www.cotton.org, 2006).The genome of both sea-island and upland cotton were sequenced in 2015, paving the way for the use of tools such as genome editing in genetic improvement programs (Zhang et al. 2015). Modern biotechnology and molecular approaches have been applied to plant breeding to promote plant-genome manipulation and enhance selection of desired traits and performance of crops. The recent availability of genome-editing technologies provides a vast opportunity to introduce targeted modifications in the genome efficiently to study the functional aspects of various components of plants.

3.4.1.7 Application of CRISPR/Cas9 for Cotton Quality Improvement

Cotton fiber quality is directly related to boost the economy determining the income of almost 100 million families from more than 100 countries (Guan et al. 2014). Tetraploid cotton retains special features such as larger fiber length and fiber strength to facilitate more spinnable cotton. Cotton fibers contain single-celled trichomes originating from outer integument cells of the ovular surface. Fiber developmental mechanisms comprising four levels, fiber-cell initiation, elongation, secondary cell wall biosynthesis and maturation (Manik and Ravikesavan 2009). The overlapping developmental stages have some special features differentiating cellular and physiology. This is due to complexity of cotton fiber transcriptome involving ~18 000 and 36 000 genes in diploid and allotetraploid cotton genomes, respectively (Arpat et al. 2004). Several key fiber-related genes have been identified creating interest to study their functions and subsequent improvement of enhance fiber quality. Some key genes including *E6* (John and Crow 1992), *GhExp1* (Harmer et al. 2002), *GhSusA1* (Jiang et al. 2012), *PIP2s* (Li et al. 2013) and *GA20ox* (Bai et al. 2014) were reported predominantly expressed during fiber initiation, secondary cell wall biosynthesis (Brill et al. 2011), and fiber elongation (Yang et al. 2014). Taking an example, protodermal factor1 gene (*GbPDF1*) specifically regulates fiber initiation by the *HDZIP2-ATATHB2* core cis element (Deng et al. 2012). Similarly, alpha-expansins (*GhExp1*) overexpression regulate fiber elongation encoding cell wall loosening proteins (Harmer et al. 2002). In addition, several related genes are highly expressed during fiber elongation. Earlier, antisense suppression of sucrose synthase (*SuSy*) was revealed to suppress fiber elongation due to change in osmosis (Ruan et al.2007). In contrast, proline-rich protein-coding (*GhPRP5*) was found as a negative regulator of fiber development (Xu et al. 2013).

Cellulose synthesis is a principal event in fiber cells during the secondary cell wall biosynthesis. Previously, many efforts have been made to explore how the cotton fiber regulates and supports the strong irreversible carbon sink characterized by secondary wall cellulose synthesis (Brill et al. 2011). It has been shown that suppression of *Sus* gene expression affects the cellulose deposition (Brill et al. 2011), emphasizing the importance of this enzyme in cellulose synthesis. Subsequently, Brill et al. (Brill et al. 2011) identified and characterized a novel *Sus* isoform (*SusC*), which was up-regulated during secondary wall cellulose synthesis in cotton fiber. Besides secondary wall cellulose synthesis, the maturation stage of fiber development begins. During fiber maturation, the majority of the expressed genes belong to cellular respiration (Kim et al. 2013). Many genes encoding transcription factors, that are *MYB*, *C2H2*, *bHLH*, *WRKY*, and *HD-ZIP* families, were also expressed during cotton fiber development. Previously, various studies indicated that MYB-related genes have high expression during fiber development in *G. hirsutum* (Machado et al. 2009). For example, expression studies of six *MYB*-related genes in *G. hirsutum* indicated that *GhMYB6* has high expression in cotton fiber (Loguercio et al. 1999), while *R2R3 MYB*-like transcription factor-encoding gene "*GhMYB109*" is particularly expressed in fiber initiation and elongation (Suo et al. 2003). The RAD-like *GbRL1* was also highly expressed in cotton ovules during fiber initiation (Zhang et al. 2011). *TCP* transcription factor has played a significant role in fiber and root hair development by controlling the jasmonic acid biosynthesis, ethylene signaling, calcium channel and reactive oxygen species (Hao et al. 2012). Though, *GhHOX3* from class IV homeo-domain-leucine zipper (*HD-ZIP*) family showed strong expression during early fiber elongation (Shan et al. 2014). Besides transcription factors, phytohormones such as ethylene, auxins and brassinosteroids (BR) also play a critical role during fiber development. Ethylene plays a vital function in fiber elongation by stimulating the pectin biosynthesis network (Qin and Zhu 2011), while gibberellins (GA) and indole-3-acetic acid (IAA) are required for fiber initiation and elongation in cotton (Xiao et al. 2010). In contrast, the persistent high concentration of jasmonic acid (JA) inhibits fiber elongation (Tan et al. 2012). Though several gene expression studies have been reported on cotton fiber development, some issues are illustrated here. First, most of the differentially expressed genes identified by the comparative analyses are associated with variations between species rather than related to fiber traits. Second, in some cases, the use of the protein-coding gene sequences from *Gossypium raimondii* and *Gossypium arboreum* may not be accurate enough for gene annotation in tetraploid cotton. Third, it is unknown whether any of the expressed genes recognized from earlier reports had sequence variations between a cotton fiber mutant and its wild type because only the differentially expressed genes having sequence differences and co-localization with target fiber traits are possible candidates for advanced cotton studies. These excellent contributions extremely facilitate vector construction for functional genes analyses and screening via the "genotype-to-phenotype" approach. Thus, the arsenal of cotton genomic manipulation urgently needs to be updated to meet the demand for rapid and precise dissecting gene functional analyses. Based on the presented facts and the well-documented functional genomics of fiber quality traits, along with the availability of genetic resources and the high transformation efficiency, the employment of the CRISPR/Cas system is a better choice for cotton fiber quality improvement.

3.4.1.8 Soybean

Soybean (*Glycine max* L.) is an old polyploidy crop with immense economic benefits. The soybean genome is complex and probably arisen by two rounds of genome-wide duplications or polyploidy events (Innes et al. 2008). Though several QTLs were mapped on the soybean genome before the publication of soybean genome sequence, the identification of underlying genes responsible for the quantitative trait of interest was rare (Watanabe et al. 2009). The availability of genome sequence of soybean has accelerated mapping and identification of genes responsible for various qualitative and quantitative traits

(Schmutz et al. 2010).Soybean is an important crop owing to its potential for supplying protein, oil and animal forages. The world soybean production in the year 2019 was approximately 358.65 million tons. The US was the largest producer with 120.52million metric tons. Other major countries such as Brazil, Argentina, and China contributed 117, 55.30, and 15.90 million metric tons, respectively (http://www.sopa.org/statistics/world-soybean-production/).The classical breeding approaches are not affected due to limited genetic variability available in the germplasm. However, there are several ways through which the classical breeding approaches deals and is able to incorporate the traits of interest in elite cultivars. However, if cultivars show low genetic diversity, their existing genes/alleles may not be able to cope with new environmental stresses. The novel genes/alleles can be incorporated in existing germplasm from wild relatives and/or created by random mutagenesis. However, the incorporation of large genome from wild relatives may cause severe damage to the elite germplasm, i.e. low production making long breeding cycles essential to purify the background while reserving the introduced attribute. The utilization of Marker-assisted breeding (MAB) which is marker-assisted selection (MAS) and genomic selection (GS) can speed up the process and has been well documented to achieve the objective (Crossa et al. 2017; Zhang et al. 2018). In modern times, the targeted mutagenesis has been successfully exploited in many crops for several desirable phenotypes with no exception to soybean.

3.4.1.9 Application of CRISPR/Cas9 for Soybean Quality Improvement

The genetic transformation methods in plants have made tremendous progress especially in soybean breeding programs, to produce novel, and genetically variable, plant material. These transformation techniques have been utilized previously to study the functional aspects of soybean (Stewart et al. 1996). These transgenic plants represented a priceless tool for molecular, genetic, biochemical and physiological studies by gene overexpression or silencing, transposon-based mutagenesis, protein sub-cellular localization and/or promoter characterization. However, there are some disadvantages to this traditional method. The introduced genes may cause non-targeted mutagenesis ultimately interrupting the endogenous or exogenous genes which may have negative results. The RNAi technique possesses the ability to silence a whole gene family hough, the silencing of only targeted gene is required. The genome modifications of soybean are complex due to the fact that genes are highly duplicated in nature. Thus, precise, efficient and straightforward methods for researching gene functions and genome engineering are required. Recently, CRISPR/Cas9 has emerged as a robust and effective technology for editing each member of a gene family without influencing other genes or simultaneously editing multiple genes of interest, thereby overcoming the shortcomings of the traditional plant-breeding methods.

Genome editing has emerged as a better choice for molecular breeding than transgenesis and it is well accepted by society. Cai et al. (Cai et al. 2015) first successfully realized CRISPR/Cas9 mediated genome editing in soybean using a single sgRNA for a trans gene (bar) and six sgRNAs that targeted different sites of two endogenous soybean genes (*GmFEI2* and *GmSHR*) and understood deficiency of the sgRNAs in a hairy root system. The targeted mutagenesis of two genomic sites in soybean chromosome 4 (DD20 and DD43) caused frame shift mutation (Li et al. 2015). The targeted gene integrations through HDR were detected by border-specific polymerase chain reaction analysis at callus stage.

Soybean GmU6–16-1 promoter was found to be more efficient in simultaneous editing of multiple homoeoallelesrelativetotheArabidopsisAtU6-26 promoter (Du et al. 2016). The role of a dominant nodulation restriction gene in soybean, *Rj4*, that inhibits nodulation by many strains of *Bradyrhizobium elkanii* was shown through both complementation and CRISPR/Cas9-mediated gene knockout experiments (Tang et al. 2016). CRISPR was used to disrupt the pathogen virulence gene (*Avr4/6*) in *Phytophtho rasojae* (Fang and Tyler 2016). Homologous genes placement of *Avr4/6* by a marker gene (*NPT II*) stimulated by the CRISPR/Cas9 system emphasized the contribution made by the virulence gene in recognition of the pathogen by plants containing the soybean R gene loci, *Rps4* and *Rps6*. CRISPR knockout of the soybean flowering time gene, *GmFT2*, was stably heritable in the subsequent T_2 generation, with homozygous *GmFT2*a mutant inhibiting late flowering under both long-day and short-day conditions (Cai et al. 2018). The first soybean product, produced by genome editing, was announced in early 2019 and will be available on the market soon. By using the transcription activator-like effector nuclease (TALEN) technique, a new high-oleic-acid soybean cultivar has been made without introducing foreign DNA into the soybean genome (Calyxt Inc2019). Vegetable oil, produced with this new soybean cultivar, contains 80% oleic acid and has 20% less saturated fat, which makes the oil healthier for human consumption, and the shelf life of the oil is also extended (Splitter 2019). It is expected that the high-oleic-acid soybean oil will be the first gene-edited plant product to be commercialized (Calyxt Inc 2019; Splitter 2019). Most of the genome editing studies were focused on disease resistance and photoperiod sensitivity in soybean, however, the nutrition aspects are less studied which need further attention. So far, the genomics of soybean nutrition is a concern and several studies have been undertaken to highlight the genomic architecture soybean nutrition.

To improve the oxidative stability and quality of soybean oil, breeding programs have mainly focused on reducing the saturated fatty acid and linolenic acid contents and increasing oleic acid in the oil. Hence, delta-12 fatty-acid desaturase 2 (FAD2), which converts oleic acid (18 : 1) to linolenic acid (18 : 2), becomes the target for modification via molecular breeding. There are two FAD2 loci, *FAD2–1A* (*Glyma10g42470*) and *FAD2–1B* (*Glyma20g24530*), in the soybean genome. To develop high-oleic-acid lines (> 80% of total oil content vs. an average of 20–50% among existing soybean cultivars), efforts were made to combine a recessive mutant allele at the *FAD2–1A* locus (*Glyma10g42470*) and a recessive mutant allele at the *FAD2–1B* locus (*Glyma20g24530*) using molecular marker assays (Pham et al. 2011). In a later study, a non-synonymous SNP on *FAD2–1A* (S117N) from 17D, an EMS mutant of Wm82, and another one on *FAD2–1B* (P137R), a mutant allele from PI 283327, were identified to be highly associated with the high-oleic-acid phenotype (Kulkarni et al. 2018). Similarly, an EMS mutant, *PE1690*, derived from the cultivar, Pungsannamul, was found to be low in linolenic acid. The phenotype was the result of a single-base mutation (W128*) in the *GmFAD3A* gene on the locus *Glyma14g37350*, which rendered the desaturase enzyme non-functional (Chahal and Gosal 2002). A derived cleaved amplified polymorphic sequence (dCAPs) assay specific for this artificial polymorphism was thus developed for future breeding activities (Kim et al. 2015). These studies have provided insights and, based on these observations, future studies can be designed utilizing modern technologies to improve soybean quality to fulfill the consumer requirements.

3.5 Regulatory Measures for Genome Engineering Crops

The classical breeding methods largely depend on selection and crossing to develop homologous recombinant. These techniques are utilized from decades ago to make selection of novel traits and introgression into elite cultivars to obtain desirable phenotype. The traditional breeding methods have helped to develop many improved cultivars in several crops important for food security, however, it requires much time and successive generations for selection (Chahal and Gosal 2002). Moreover, the unavailability of genetic variability in major crops have further hindered the classical breeding approach (Pacher and Puchta 2017). The modern breeding technologies have potential to overcome these challenges and are successfully utilized to a variety of crops. Mutation breeding for the development of transgenic and non-transgenic plants are worthy strategies utilized for crop improvement programs (Scheben et al. 2017). The traditional mutagenesis approaches have helped to develop genetic variations in traits of interest to improve yield and quality. However, the traditional mutagenesis approaches require, labor oriented screening of a large number of mutant population (Sikora et al. 2011), less precise and leaves breeding programs, that unable to meet the requirements (Scheben et al. 2017). The transgenic breeding efforts have been utilized to introduce important genes in cultivars with promising attributes. The transgenic breeding approach has potential to overcome crossing barriers to enhance genetic variability. The major concern with transgenic approaches is the public concern regarding the impact on human health and the environment, therefore, commercialization of genetically modified crops (GMOs) is under strict scrutiny (Hartung and Schiemann 2014). In European countries, the GM crops are still not allowed for commercial purposes and are prohibited in the consumer market along with, the GE non-transgenic crops also going through strict regulatory measures. The new breeding techniques, i.e. ZFNs, TALENs, and CRISPR do not fall under the category of GMOs due to no residues of foreign particles being in targeted genome. The US Department of Agriculture had exempted CRISPR edit crops from GMOs and are allowed to be cultivated and sold to consumers (Waltz 2018). The CRISPR technique will help to reduce the cost required for field trail and data collection of GMOs along with overcoming public concerns. However, there is an urgent need for intensive uniform of regulatory policy for GE tools around the globe.

3.6 Conclusion

The utilization of genome engineering for quality improvement in crops essential for food security requires substantial expression of genes without influencing other regulatory pathways essential for plant growth and development. The lack of understanding of biosynthesis pathways and their complex interaction for quality attributes in the majority of crops is a limiting factor. The availability of genomic data of several important crops helped in identification, characterization and regulation of genes controlling quality traits but, requires further research. The modern GETs have revolutionized the crops quality improvement programs but still requires further investigation. These breeding techniques possess the ability to make precise and quick targeted modifications in genes of interest compared to conventional breeding techniques. CRISPR/Cas proved to be a fundamental breakthrough in genome editing which is successfully utilized for yield enhancement, quality improvement, disease resistance and other traits of interest. In the past few years, targeted mutagenesis has been vigorously utilized in plant-based research for functional studies. Moreover, the CRISPR/Cas genome editing system has potential to be adopted as a viable approach to achieve zero hunger and nutritional imbalance for growing human population.

Acknowledgement

The authors are thankful to the scientific insights of Dr. Aamir Riaz from China National Rice Research Institute, Hangzhou, Zhejinag, P.R. China.

References

Altpeter, F., Vasil, V., Srivastava, V., and Vasil, I.K. (1996). Integration and expression of the high-molecular-eight gluten in sub unit 1Ax1 gene in to wheat. *Nature Biotechnology* 14: 1155–1159.

Alvarez, M.L., Omez, M.G., Marıa, J. et al. (2001). Analysis of dough functionality of flours from transgenic wheat. *Molecular Breeding* 8: 103–108.

Anjum, F.M., Khan, M.R., Din, A. et al. (2007). Wheat gluten:high molecular weight glutenin subunits, structure, genetics, and relation to dough elasticity. *Journal of Food Science* 72: 56–63.

Arpat, A., Waugh, M., Sullivan, J.P. et al. (2004). Functional genomics of cell elongation in developing cotton fibers. *Plant Molecular Biology* 54: 911–929.

Bai, W.Q., Xiao, Y.H., Zhao, J. et al. (2014). Gibberellin overproduction promotes sucrose synthase expression and secondary cell wall deposition in cotton fibers. *PLoS One* 9: 96537.

Barrangou, R., Fremaux, C., Deveau, H. et al. (2007). CRISPR provides acquired resistance against viruses in prokaryotes. *Science* 315: 1709–1712.

Bhullar, N.K. and Gruissem, W. (2013). Nutritional enhancement of rice for human health: the contribution of biotechnology. *Biotechnology Advances* 31: 50–57.

Borg, S., Brinch-Pedersen, H., Tauris, B. et al. (2012). Wheat ferritins: improving the iron content of the wheat grain. *Journal of Cereal Science* 56: 204–213.

Borisjuk, N., Kishchenko, O., Eliby, S. et al. (2019). Genetic Modification for Wheat Improvement: From Transgenesis to Genome Editing. *BioMed Research International* 2019: 6216304.

Bortesi, L. and Fischer, R. (2015). The CRISPR/Cas9 system for plant genome editing and beyond. *Biotechnology Advances* 33: 41–52.

Brill, E., Van, T.M., White, R.G. et al. (2011). A novel isoform of sucrose synthase is targeted to the cell wall during secondary cell wall synthesis in cotton fiber. *Plant Physiology* 157: 40–54.

Butt, H., Jamil, M., Wang, J.Y. et al. (2018). Engineering plant architecture via CRISPR/Cas9-mediated alteration of strigolactone biosynthesis. *BMC Plant Biology* 18: 1–9.

Cai, Y., Chen, L., Liu, X. et al. (2015). CRISPR/Cas9-mediated genome editing in soybean hairy roots. *PLoS One* 10: e0136064.

Cai, Y., Chen, L., Liu, X. et al. (2018). CRISPR/Cas9-mediated targeted mutagenesis of GmFT2a delays flowering time in soya bean. *Plant Biotechnology Journal* 16: 176–185.

Cakmak, I., Ozkan, H., Braun, H.J. et al. (2000). Zinc and iron concentrations in seeds of wild, primitive, and modern wheat. *Food and Nutrition Bulletin* 21: 401–403.

Capecchi, M.R. (1980). High effciency transformation by direct microinjection of DNA into cultured mammalian cells. *Cell* 22: 479–488.

Chahal, G.S. and Gosal, S.S. (2002). *Principles and Procedures of Plant Breeding: Biotechnological and Conventional Approaches*. Oxford, UK: Alpha Science Int'l Ltd.

Chen, P., Shen, Z., Ming, L. et al. (2018). Genetic basis of variation in Rice seed storage protein (albumin, globulin, Prolamin, and Glutelin) content revealed by genome-wide association analysis. *Frontiers in Plant Science* 9: 612.

Chira, S., Gulei, D., Hajitou, A. et al. (2017). CRISPR/Cas9: transcending the reality of genome editing. *Molecular Therapy – Nucleic Acids* 7: 211–222.

Cho, S.W., Kim, S., Kim, Y. et al. (2014). Analysis of off-target effects of CRISPR/Cas-derived RNA-guided endonucleases and nickases. *Genome Research* 24: 132–141.

Connorton, J.M., Jones, E.R., Rodríguez-Ramiro, I. et al. (2017). Wheat vacuolar iron transporter TaVIT2 transports Fe and Mn and is effective for biofortification. *Plant Physiology* 174: 2434–2444.

Crossa, J., Pérez-Rodríguez, P., Cuevas, J. et al. (2017). Genomic selection in plant breeding: methods, models, and perspectives. *Trends in Plant Science* 22: 961–975.

Cruz, D.N. and Khush, G.S. (2000). Rice grain quality evaluation procedures. In: *Aromatic Rices* (eds. R.K. Singh, U.S. Singh and G.S. Khush), 292. New Delhi: Mohan Primlani for Oxford & IBH Publishing Co. Pvt. Ltd.

Deltcheva, E., Chylinski, K., Sharma, C.M. et al. (2011). CRISPR RNA maturation by trans-encoded small RNA and host factor RNase III. *Nature* 471: 602.

Deng, F., Tu, L., Tan, J. et al. (2012). GbPDF1 is involved in cotton fiber initiation via the core cis-element HDZIP2ATATHB2. *Plant Physiology* 158: 890–904.

Doench, J.G., Fusi, N., Sullender, M. et al. (2016). Optimized sgRNA design to maximize activity and minimize off-target effects of CRISPR-Cas9. *Nature Biotechnology* 34: 184.

Du, H., Zeng, X., Zhao, M. et al. (2016). Efficient targeted mutagenesis in soybean by TALENs and CRISPR/Cas9. *Journal of Biotechnology* 217: 90–97.

Fang, Y. and Tyler, B.M. (2016). Effcient disruption and replacement of an effector gene in the OomycetePhytophthorasojae using CRISPR/Cas9. *Molecular Plant Pathology* 17: 127–139.

FAO (Food And Agriculture Organization of The United Nations) Statistics (2014-15) (available at http://www.fao.org/faostat/en/#data/QC).

FAO (Food And Agriculture Organization of The United Nations) (2019-20).Commodity markets: Rice. http://www.fao.org/economic/est/est-commodities/rice/en/ (accessed 17 July 2020.

FAO (Food And Agriculture Organization of The United Nations) Statistics (2017-18) (available at http://www.fao.org/faostat/en/#data/QC).

Feng, C., Yuan, J., Wang, R. et al. (2016). Efficient targeted genome modification in maize using CRISPR/Cas9 system. *Journal of Genetics and Genomics* 43: 37–43.

Ferguson, D.O. and Alt, F.W. (2001). DNA double strand break repair and chromosomal translocation: lessons from animal models. *Oncogene* 20: 5572.

Ferrero, A. (2004). Constraints and opportunities for the sustainable development of rice-based production systems in Europe.The International Conference on Sustainable Rice Systems, Rome.

Fiaz, S., Ahmad, S., Noor, M.A. et al. (2019). Applications of the CRISPR/Cas9 system for Rice grain quality improvement: perspectives and opportunities. *International Journal of Molecular Sciences* 20: 888.

Fiaz, S., Jiao, G., Sheng, Z. et al. (2019). Analysis of genomic regions governing cooking and eating quality traits using a recombinant inbred population in Rice (Oryza sativa L.). *International Journal of Agriculture and Biology* 22: 611–619.

Gil-Humanes, J., Pistón, F., Barro, F., and Rosell, C.M. (2014). The shutdown of celiac disease-related gliadin epitopes in bread wheat by RNAi provides flours with increased stability and better tolerance to over-mixing. *PLoS One* 9 (3): e91931.

Godfray, H.C.J., Beddington, J.R., Crute, I.R. et al. (2010). Food security: the challenge of feeding 9 billion people. *Science* 327: 812–818.

Guan, X., Song, Q., and Chen, Z.J. (2014). Polyploidy and small RNA regulation of cotton fiber development. *Trends in Plant Science* 19: 516–528.

Hao, J., Tu, L., Hu, H. et al. (2012). GbTCP, a cotton TCP transcription factor, confers fibre elongation and root hair development by a complex regulating system. *Journal of Experimental Botany* 63: 6267–6281.

Harmer, S., Orford, S., and Timmis, J. (2002). Characterisation of six a-expansin genes in Gossypium hirsutum (upland cotton). *Molecular Genetics and Genomics* 268: 1–9.

Hartung, F. and Schiemann, J. (2014). Precise plant breeding using new genome editing techniques: opportunities, safety and regulation in the EU. *Plant Journal* 78: 742–752.

Heigwer, F., Kerr, G., and Boutros, M. (2014). E-CRISP: fast CRISPR target site identification. *Nature Methods* 11: 122.

Hsu, P.D., Scott, D.A., Weinstein, J.A. et al. (2013). DNA targeting specificity of RNA-guided Cas9 nucleases. *Nature Biotechnology* 31: 827.

Calyxt Inc (2019). *First Commercial Sale of Calyxt High Oleic Soybean Oil.* Minneapolis-St. Paul: Calyxt Inc.

Innes, R.W., Ameline-Torregrosa, C., Ashfield, T. et al. (2008). Differential accumulation of retroelements and diversification of NB-LRR disease resistance genes in duplicated regions following polyploidy in the ancestor of soybean. *Plant Physiology* 148: 1740–1759.

Ishino, Y., Shinagawa, H., Makino, K. et al. (1987). Nucleotide sequence of the iap gene, responsible for alkaline phosphatase isozyme conversion in Escherichia coli, and identification of the gene product. *Journal of Bacteriology* 169: 5429–5433.

Jasin, M. (1996). Genetic manipulation of genomes with rare-cutting donucleases. *Trends in Genetics* 12: 224–228.

Jiang, Y., Guo, W., Zhu, H. et al. (2012). Overexpression of GhSusA1 increases plant biomass and improves cotton fiber yield and quality. *Plant Biotechnology Journal* 10: 301–312.

Jinek, M., Chylinski, K., Fonfara, I. et al. (2012). A programmable dual-RNA–guided DNA endonuclease in adaptive bacterial immunity. *Science* 337: 816–821.

John, M.E. and Crow, L.J. (1992). Gene expression in cotton (Gossypium hirsutum L.) fiber: cloning of the mRNAs. *Proceedings of the National Academy of Sciences* 89: 5769–5773.

Jones, J.D., Witek, K., Verweij, W. et al. (2014). Elevating crop disease resistance with cloned genes. *Philosophical Transactions of the Royal Society: Biological Sciences* 369: 20130087.

Kang, G., Xu, W., Liu, G. et al. (2013). Comprehensive analysis of the transcription of starch synthesis genes and the transcription factor RSR1 in wheat (*Triticum aestivum* L.) endosperm. *Genome* 56: 115–122.

Kim, H.J., Tang, Y., Moon, H.S. et al. (2013). Functional analyses of cotton (Gossypium hirsutum L.) immature fiber (im) mutant infer that fiber cell wall development is associated with stress responses. *BMC Genomics* 14: 889.

Kim, M., Song, J.T., Bilyeu, K.D., and Lee, J.D. (2015). A new low linolenic acid allele of GmFAD3A gene in soybean PE1690. *Molecular Breeding* 35: 155.

Komor, A.C., Kim, Y.B., Packer, M.S. et al. (2016). Programmable editing of a target base in genomic DNA without doubles tranded DNA cleavage. *Nature* 533: 420.

Kulkarni, K.P., Patil, G., Valliyodan, B. et al. (2018). Comparative genome analysis to identify SNPs associated with high oleic acid and elevated protein content in soybean. *Genome* 61: 217–222.

Lau, W.C., Rafii, M.Y., Ismail, M.R. et al. (2015). Review of functional markers for improving cooking, eating, and the nutritional qualities of rice. *Frontiers in Plant Science* 6: 832.

Li, J., Xiao, J., Grandillo, S. et al. (2004). QTL detection for rice grain quality traits using an interspecific backcross population derived from cultivated Asian (Oryza sativa L.) and African (Oryzaglaberrima S.) rice. *Genome* 47: 697–704.

Li, S., Li, J., Wang, N. et al. (2007). Inheritance and expression of copies of transgenes 1Dx5 and 1Ax1 in elite wheat (Triticumaestivum L.) varieties transferred from transgenic wheat through conventional crossing. *ActaBiochimicaetBiophysicaSinica* 39: 377–383.

Li, D.D., Ruan, X.M., Zhang, J. et al. (2013). Cotton plasma membrane intrinsic protein 2s (PIP2s) selectively interact to regulate their water channel activities and are required for fiber development. *New Phytology* 199: 695–707.

Li, Z., Liu, Z.B., Xing, A. et al. (2015). Cas9-guide RNA directed genome editing in soybean. *Plant Physiology* 169: 960–970.

Li, Q., Li, L., Liu, Y. et al. (2017). Influence of TaGW2-6A on seed development in wheat by negatively regulating gibberellin synthesis. *Plant Science* 263: 226–235.

Li, J., Zhang, H., Si, X. et al. (2017). Generation of thermosensitive male-sterile maize by targeted knockout of the ZmTMS5 gene. *Journal of Genetics and Genomics* 44: 465.

Liang, Z., Zhang, K., Chen, K., and Gao, C. (2014). Targeted mutagenesis in Zea mays using TALENs and the CRISPR/Cas system. *Journal of Genetics and Genomics* 41: 63–68.

Liang, Z., Chen, K., Li, T. et al. (2017). Efficient DNA-free genome editing of bread wheat using CRISPR/Cas9 ribonucleoprotein complexes. *Nature Communications* 8: 1–5.

Liu, G., Wu, Y., Xu, M. et al. (2016). Virus-induced gene silencing identifies an important role of the TaRSR1 transcription factor in starch synthesis in bread wheat. *International Journal of Molecular Sciences* 17: 1557.

Liu, J., Wu, X., Yao, X. et al. (2018). Mutations in the DNA demethylase OsROS1 result in a thickened aleurone and improved nutritional value in rice grains. *Proceedings of the National Academy of Sciences* 115: 11327–11332.

Lloyd, A.H., Wang, D., and Timmis, J.N. (2012). Single molecule PCR reveals similar patterns of non-homologous DSB repair in tobacco and Arabidopsis. *PLoSOne* 7 (2): e32255.

Loguercio, L.L., Zhang, J.Q., and Wilkins, T.A. (1999). Differential regulation of six novel MYB-domain genes defines two distinct expression patterns in allotetraploid cotton (Gossypium hirsutum L.). *Molecular Genomics and Genetics* 261: 660–671.

Lou, J., Chen, L., Yue, G. et al. (2009). QTL mapping of grain quality traits in rice. *Journal of Cereal Science* 50: 145–151.

Ma, X., Zhang, Q., Zhu, Q. et al. (2015). A robust CRISPR/Cas9 system for convenient, high-efficiency multiplex genome editing in monocot and dicot plants. *Molecular Plant* 8: 1274–1284.

Machado, A., Wu, Y., Yang, Y. et al. (2009). The MYB transcription factor GhMYB25 regulates early fibre and trichome development. *Plant Journal* 59: 52–62.

Manik, N. and Ravikesavan, R. (2009). Emerging trends in enhancement of cotton fiber productivity and quality using functional genomics tools. *Biotechnology and Molecular Biology Reviews* 4: 11–28.

Meenu, M. and Xu, B. (2018). A critical review on anti-diabetic and anti-obesity effects of dietary resistant starch. *Critical Reviews in Food Science and Nutrition* 59 (18): 3019–3031.

Montague, T.G., Cruz, J.M., Gagnon, J.A. et al. (2014). CHOPCHOP: a CRISPR/Cas9 and TALEN web tool for genome editing. *Nucleic Acids Research* 42: 401–407.

Morgante, M. (2006). Plant genome organisation and diversity: the year of the junk. *Current Opinion in Biotechnology* 17: 168–173.

Nalam, V.J., Alam, S., Keereetaweep, J. et al. (2015). Facilitation of Fusariumgraminearum infection by 9-lipoxygenases in Arabidopsis and wheat. *Molecular Plant-Microbe Interactions* 28: 1142–1152.

Nester, E.W. (2014). Agrobacterium: nature's genetic engineer. *Frontiers in Plant Science* 5: 730.

Nordin, Y. and Lantbruksakademien, K.S.O. (2008). *Golden Rice and Other Biofortified Food Crops for Developing Countries: Challenges and Potential.* Rome, Italy: FAO.

Pacher, M. and Puchta, H. (2017). From classical mutagenesis to nuclease based breeding directing natural DNA repair for a natural end-product. *Plant Journal* 90: 819–833.

Payne, P.I. (1987). Genetics of wheat storage proteins and the effect of allelic variation on bread-making quality. *Annual Review of Plant Physiology and Plant Molecular Biology* 38: 141–153.

Pegoraro, C., da Mertz, L.M., Maia, L.C. et al. (2011). Importance of heat shock proteins in maize. *Journal of Crop Science and Biotechnology* 14: 85–95.

Peng, B., Kong, H., Li, Y. et al. (2014). OsAAP6 functions as an important regulator of grain protein content and nutritional quality in rice. *Nature Communications* 5: 4847.

Pham, A.T., Lee, J.D., Shannon, J.G., and Bilyeu, K.D. (2011). A novel FAD2-1 a allele in a soybean plant introduction offers an alternate means to produce soybean seed oil with 85% oleic acid content. *Theoretical and Applied Genetics* 123: 793–802.

Pliatsika, V. and Rigoutsos, I. (2015). Off-spotter: very fast and exhaustive enumeration of genomic lookalikes for designing CRISPR/Cas guide RNAs. *Biology Direct* 10: 4.

Pourcel, C., Salvignol, G., and Vergnaud, G. (2005). CRISPR elements in Yersinia pestis acquire new repeats by preferential uptake of bacteriophage DNA, and provide additional tools for evolutionary studies. *Microbiology* 151: 653–663.

Prykhozhij, S.V., Rajan, V., Gaston, D., and Berman, J.N. (2015). CRISPR multitargeter: a web tool to find common and unique CRISPR single guide RNA targets in a set of similar sequences. *PLoS One* 10: e0119372.

Puchta, H. (2005). The repair of double-strand breaks in plants: mechanisms and consequences for genome evolution. *Journal of Experimental Botany* 56: 1–14.

Qi, W., Zhu, T., Tian, Z. et al. (2016). High-efficiency CRISPR/Cas9 multiplex gene editing using the glycine tRNA processing system-based strategy in maize. *BMC Biotechnology* 16: 58.

Qin, Y.M. and Zhu, Y.X. (2011). How cotton fibers elongate: a tale of linear cell growth mode. *Current Opinion in Plant Biology* 14: 106–111.

Quétier, F. (2016). The CRISPR-Cas9 technology: closer to the ultimate toolkit for targeted genome editing. *Plant Science* 242: 65–76.

Regina, A., Bird, A., Topping, D. et al. (2006). High-amylose wheat generated by RNA interference improves indices of large-bowel health in rats. *Proceedings of the National Academy of Sciences* 103: 3546–3551.

Ruan, Y. (2007). Rapid cell expansion and cellulose synthesis regulated by plasmodesmata and sugar: insights from the single-celled cotton fibre. *Functional Plant Biology* 34: 1–10.

Sabouri, A., Rabiei, B., Toorchi, M. et al. (2012). Mapping quantitative trait loci (QTL) associated with cooking quality in rice (Oryza sativa. L). *Australian Journal of Crop Science* 6: 808.

Sánchez-León, S., Gil-Humanes, J., Ozuna, C.V. et al. (2018). Low-gluten, nontransgenic wheat engineered with CRISPR/Cas9. *Plant Biotechnology Journal* 16: 902–910.

Schaart, J.G., Van De Wiel, C.C.M., Lotz, L.A.P., and Smulders, M.J.M. (2016). Opportunities for products of new plant breeding techniques. *Trends in Plant Science* 21: 438–449.

Scheben, A., Wolter, F., Batley, J. et al. (2017). Towards CRISPR/Cas crops–bringing together and genome editing. *New Phytology* 216: 682–698.

Scherf, K.A., Koehler, P., and Wieser, H. (2016). Gluten and wheat sensitivities an overview. *Journal of Cereal Science* 67: 2–11.

Schmutz, J., Cannon, S.B., Schlueter, J. et al. (2010). Genome sequence of the palaeopolyploid soybean. *Nature* 463: 178–183.

Schnable, P.S., Ware, D., Fulton, R.S. et al. (2009). The B73 maize genome: complexity, diversity, and dynamics. *Science* 326: 1112–1115.

Schubert, D. and Williams, D. (2006). Cisgenic; as a product designation. *Nature Biotechnology* 24: 1327.

Sestili, F., Janni, M., Doherty, A. et al. (2010). Increasing the amylose content of durum wheat through silencing of the SBEIIagenes. *BMC Plant Biology* 10: 144.

Shan, Q., Wang, Y., Li, J., and Gao, C. (2014). Genome editing in rice and wheat using the CRISPR/Cas system. *Nature Protocols* 9: 2395–2410.

Shan, Q., Wang, Y., Li, J., and Gao, C. (2014). Genome editing in rice and wheat using the CRISPR/Cas system. *Nature Protocols* 9: 2395.

Shen, L., Li, J., Fu, Y. et al. (2017). Orientation improvement of grain length and grain number in rice by using CRISPR/Cas9 system. *Chinese Journal of Rice Science* 31: 223–231.

Shewry, P.R. and Halford, N.G. (2002). Cereal seed storage proteins: structures, properties and role in grain utilization. *Journal of Experimental Botany* 53: 947–958.

Shi, J., Gao, H., Wang, H. et al. (2017). ARGOS8 variants generated by CRISPR–Cas9 improve maize grain yield under field drought stress conditions. *Plant Biotechnology Journal* 15: 207–216.

Shu, Q.Y., Forster, B., and Nakagawa, H. (2012). *Principles and Applications of Plant Mutation Breeding, in Plant Mutation Breeding and Biotechnology* (eds. Q.Y. Shu, B. Forster and H. Nakagawa), 301–326. Wallingford: CABI.

Si, L., Chen, J., Huang, X. et al. (2016). OsSPL13 controls grain size in cultivated rice. *Nature Genetics* 48: 447–456.

Sikora, P., Chawade, A., Larsson, M. et al. (2011). Mutagenesis as a tool in plant genetics, functional genomics, and breeding. *International Journal of Plant Genomics* 2011: 314829.

Smidansky, E.D., Meyer, F.D., Blakeslee, B. et al. (2007). Expression of a modified ADP-glucose pyrophosphorylase large subunit in wheat seeds stimulates photosynthesis and carbon metabolism. *Planta* 225: 965–976.

Splitter, J. (2019). *The Latest Gene-Edited Food Is a Soybean Oil that Comes with Zero Trans Fats*. New York: Forbes.

Stemmer, M., Thumberger, T., Del Sol Keyer, M. et al. (2015). CCTop: an intuitive, flexible and reliable CRISPR/Cas9 target prediction tool. *PLoS One* 10: e0124633.

Stewart, C.N., Adang, M.J., All, J.N. et al. (1996). Genetic transformation, recovery, and characterization of fertile soybean transgenic for a synthetic bacillus thuringiensiscryIAc gene. *Plant Physiology* 112: 121–129.

Sui, X., Zhao, Y., Wang, S. et al. (2012). Improvement Fe content of wheat (*Triticum aestivum*) grain by soybean ferritin expression cassette without vector backbone sequence. *Journal of Agriculture and Biotechnology* 20: 766–773.

Sundström, J.F., Albihn, A., Boqvist, S. et al. (2014). Future threats to agricultural food production posed by environmental degradation, climate change, and animal and plant diseases–a risk analysis in three economic and climate settings. *Food Security* 6: 201–215.

Suo, J., Liang, X., Pu, L. et al. (2003). Identification of GhMYB109 encoding a R2R3 MYB transcription factor that expressed specifically in fiber initials and elongating fibers of cotton (Gossypium hirsutum L.). *Biochimtry and Biophysics Acta* 1630: 25–34.

Tan, J., Tu, L., Deng, F. et al. (2012). Exogenous jasmonic acid inhibits cotton fiber elongation. *Journal of Plant Growth Regulation* 31: 599–605.

Tang, F., Yang, S., Liu, J., and Zhu, H. (2016). Rj4, a gene controlling nodulation specificity in soybeans, encodes a thaumatin-like protein but not the one previously reported. *Plant Physiology* 170: 26–32.

Terada, R., Nakajima, M., Isshiki, M. et al. (2000). Antisense waxy genes with highly active promoters effectively suppress waxy gene expression in transgenic rice. *Plant Cell Physiology* 41: 881–888.

United Nations, Department of Economic and Social Affairs, Population Division (2017). The Impact of Population Momentum on Future Population Growth. Population Facts. https://www.un.org/en/development/desa/population/publications/factsheets/index.shtml (accessed 17 July 2020.

Upadhyay, S.K., Kumar, J., Alok, A., and Tuli, R. (2013). RNA-guided genome editing for target gene mutations in wheat. *G3: Genes, Genomes, Genetics* 3: 2233–2238.

Waltz, E. (2018). With a free pass, CRISPR-edited plants reach market in record time. *Nature Biotechnology* 36: 6–7.

Wang, S., Li, S., Liu, Q. et al. (2015). The OsSPL16-GW7 regulatory module determines grain shape and simultaneously improves rice yield and grain quality. *Nature Genetics* 47: 949–954.

Wang, W., Pan, Q., He, F. et al. (2018). Transgenerational CRISPR-Cas9 activity facilitates multiplex gene editing in allopolyploid wheat. *The CRISPR Journal* 1: 65–74.

Wang, W., Simmonds, J., Pan, Q. et al. (2018). Gene editing and mutagenesis reveal inter-cultivar differences and additivity in the contribution of *TaGW2*homoeologues to grain size and weight in wheat. *Theoretical and Applied Genetics* 131: 2463–2475.

Watanabe, S., Hideshima, R., Xia, Z. et al. (2009). Map-based cloning of the gene associated with the soybean maturity locus E3. *Genetics* 182: 1251–1262.

Weichert, N., Saalbach, I., Weichert, H. et al. (2010). Increasing sucrose uptake capacity of wheat grains stimulates storage protein synthesis. *Plant Physiology* 152: 698–710.

Welch, R.M. and Graham, R.D. (1999). A new paradigm for world agriculture: meeting human needs. Productive, sustainable, nutritious. *Field Crops Research* 60: 1–10.

Wendel, J.F. and Cronn, R.C. (2003). Polyploidy and the evolutionary history of cotton. *Advancement in Agronomy* 78: 139–186.

Wenefrida, I., Utomo, H.S., Blanche, S.B., and Linscombe, S.D. (2009). Enhancing essential amino acids and health benefit components in grain crops for improved nutritional values. *Recent Advances in DNA and Gene Sequences* 3: 219–225.

Xiao, Y.H., Li, D.M., Yin, M.H. et al. (2010). Gibberellin 20-oxidase promotes initiation and elongation of cotton fibers by regulating gibberellin synthesis. *Journal of Plant Physiology* 167: 829–837.

Xu, W.L., Zhang, D.J., Wu, Y.F. et al. (2013). Cotton PRP5 gene encoding a proline-rich protein is involved in fiber development. *Plant Molecular Biology* 82: 353–365.

Yang, Z., Zhang, C., Yang, X. et al. (2014). PAG1, a cotton brassinosteroid catabolism gene, modulates fiber elongation. *New Phytologist* 203: 437–448.

Ye, X., Al-Babili, S., Klöti, A. et al. (2000). Engineering the provitaminA (β-carotene) biosynthetic pathway into (carotenoid-free) rice endosperm. *Science* 287: 303–305.

Zhang, F., Liu, X., Zuo, K. et al. (2011). Molecular cloning and characterization of a novel Gossypium barbadense L. RAD-like gene. *Plant Molecular Biology Reporter* 29: 324–333.

Zhang, G., Cheng, Z., Zhang, X. et al. (2011). Double repression of soluble starch synthase genes SSIIa and SSIIIa in rice (Oryza sativa L.) uncovers interactive effects on the physicochemical properties of starch. *Genome* 54: 448–459.

Zhang, T., Hu, Y., Jiang, W. et al. (2015). Sequencing of allotetraploid cotton (*Gossypium hirsutum* L. acc. TM-1) provides a resource for fiber improvement. *Nature Biotechnology* 33: 531–537.

Zhang, Y., Liang, Z., Zong, Y. et al. (2016). Efficient and transgene-free genome editing in wheat through transient expression of CRISPR/Cas9 DNA or RNA. *Nature Communications* 7: 1–8.

Zhang, K., Raboanatahiry, N., Zhu, B., and Li, M. (2017). Progress in genome editing technology and its application in plants. *Frontiers in Plant Science* 8: 177.

Zhang, J., Zhang, H., Botella, J.R., and Zhu, J.K. (2018). Generation of new glutinous rice by CRISPR/Cas9-targeted mutagenesis of the waxy gene in elite rice varieties. *Journal of Integrative Plant Biology* 60: 369–375.

Zhou, H., Liu, B., Weeks, D.P. et al. (2014). Large chromosomal deletions and heritable small genetic changes induced by CRISPR/Cas9 in rice. *Nucleic Acids Research* 42: 10903–10914.

Zhu, J., Song, N., Sun, S. et al. (2016). Efficiency and inheritance of targeted mutagenesis in maize using CRISPR–Cas9. *Journal of Genetics and Genomics* 43: 25–36.

4

Genome Engineering for Enriching Fe and Zn in Rice Grain and Increasing Micronutrient Bioavailability

Conrado Dueñas, Jr., Gela Myan B. Mota, Norman Oliva, and Inez H. Slamet-Loedin

Trait and Genome Engineering Cluster, Strategic Innovation Platform, International Rice Research Institute, Manila, Philippines

CHAPTER MENU

4.1 Introduction

Alleviating micronutrient deficiency is an essential part of the second of the Sustainable Development Goals (SDG) (UN n.d.). Two of the most common micronutrient deficiencies of collectively known hidden hunger are the deficiencies of iron (Fe) and zinc (Zn). These deficiencies are prevalent among households with low socioeconomic status and vulnerable sections of the population in developing countries (Seshadri 2001). Iron is required for essential cellular functions and is critically important in muscles, and the brain, as well as red blood cells (NIH n.d.). Consequently, one of the most common causes of anemia is Fe deficiency anemia (IDA). Similar to Fe, Zn also plays a role in critical physiological processes. These micronutrients are crucial for growth and recovery. Therefore, Fe and Zn deficiencies can lead to stunted growth, increased susceptibility to disease and infection and delayed recovery time, reduced mental capacity, and increased prevalence of maternal, neonatal, and childhood complications (NIH n.d.; Schlemmer et al. 2009)

Numerous approaches have been employed in different national programs to prevent and overcome hidden hunger. These approaches or solutions have benefits and drawbacks. The first approach, and widely considered the best among these, is to increase the diversity

Genome Engineering for Crop Improvement, First Edition. Edited by Santosh Kumar Upadhyay.

of one's diet. However, some pockets of society have limited access to diverse diets due to low socioeconomic conditions. Supplementation and food fortification are chemical-based approaches to alleviate hidden hunger, and the impact depends on the coverage and frequency of the intervention. Because of the pitfalls of other methods, scientists use another food-based sustainable approach, which is increasing the nutrient value of staples through biofortification to complement other approaches.

Since breeding to increase Fe nutrient content is hindered due to the limited genetic variation of Fe concentration in polished rice, transgenic approaches were used, while Zn breeding is currently ongoing in many countries. This approach has been extensively reviewed by Bashir et al. (Bashir et al. 2013), Ludwig and Slamet-Loedin (2019), (Slamet-Loedin et al. (2015), and Tsakirpaloglou et al. (2019) for Fe, Zn, and their bioavailability, respectively. In summary, genes responsible for uptake, translocation, and storage of Fe and Zn were randomly inserted into the rice genome to improve Fe and Zn concentration in the grain by overexpression or ectopic expression. The best performing transgenic lines were selected among the random insertions. By contrast, this review will focus instead on the possibility to use site-directed nuclease (SDN)-1, -2, and -3 strategies, and how they could be applied to enhance Fe and Zn uptake and translocation in elite rice cultivars by editing the homeostasis genes related to these two metals. The SDN-1 approach is amenable to increasing the expression of metal chelator genes by modifying their regulatory element or by silencing a particular gene or element function as a repressor of the expression of the micronutrient in the grain. This type of editing can lead to an altered spatial or expression level, thereby resulting in a higher Fe and Zn concentration in the grain.

In comparison, SDN-2 offers potential for allele replacement from the wild-type allele to the novel allele of a relevant gene identified through genome-wide association study (GWAS) or other breeding methods (i.e. breeding to wild rice or landraces), whereas the SDN-3 approach requires an insertion of another copy of the endogenous gene of interest or transgenes in the targeted locus. CRISPR-Cas9 and TALEN genome-editing tools can be applied to edit elite rice cultivars for Fe and Zn biofortification traits using the three nuclease techniques.

4.2 Genes Related to Uptake of Fe and Zn from the Soil

Rice uses both System II and System I for the uptake of Fe in the soil. In gramineous plants under System II, mugineic acid (MA) family phytosiderophores are secreted and form complexes with Fe^{3+} for uptake in Fe-deficient plants (Takagi 1976; Römheld and Marschner 1986). 2′-deoxymugineic acid (DMA) is an example of MA, which is released by the rice roots in the soil to capture insoluble Fe^{3+}. DMA, like other MA, consists of carboxyl, amine, and a hydroxyl functional group, which functions as a ligand (Sugiura and Nomoto 1984). For System I, rice directly uptakes Fe^{2+} through OsIRT1 and OsIRT2 transporter genes since rice and its wild relatives are well adapted under submerged conditions, for which Fe^{2+} is more abundant than Fe^{3+}(Ishimaru et al. 2006). These highly homologous genes encode proteins that have similar features of transport regulated by Fe and also Zn (Bughio et al. 2002). On the other hand, nicotianamine (NA) chelates metal cations such as Fe^{3+}, including Fe^{2+} (vonWirén et al. 1999). This transporter is expressed in the plasma membrane of root epidermal cells and is, therefore, likely to be involved in Zn uptake, as shown in Figure 4.1 (Olsen and Palmgren 2014).

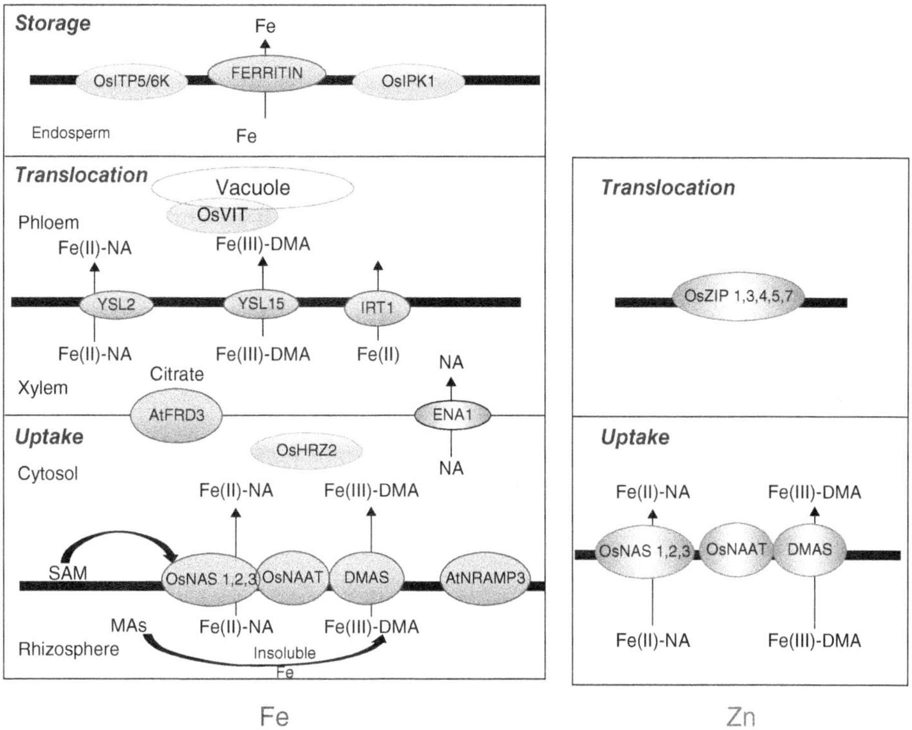

Figure 4.1 Current most promising candidate genes to increase grain Fe and Zn or their bioavailability by genome engineering.
Presented are candidate genes for SDN1 and 3. Pink colored genes are upregulated, blue, colored are for silencing, and gray zinc related genes for allele replacement.
Abbreviations: OsITP5/6K, inositol 1,3,4-triskisphosphate 5/6-kinase; OsIPK1, inositol 1,3,4,5,6-pentakisphosphate 2-kinase; OsHRZ2, hemerythrin motif-containing novel gene (RING)-and Zn-finger protein 2; YSL1, yellow stripe 1-like; IRT1, iron-regulated transporter; OsNRAMP, natural resistance-associated macrophage protein 1; OsFRD, ferric reductase defective 3; ENA, efflux transporter of nicotianamie; OsVIT, vacuolar iron transporter; OsFER, ferritin; MA, mugineic acid; DMA, 2′-deoxymugineic acid; NA, nicotianamine; SAM, S-adenosyl-L-methionine; NAS, nicotianamine synthase; NAAT, nicotianamine aminotransferase; DMAS, deoxymugineic acid synthase.
(For detailed information please see the colour figure in online version.)

Zinc, on the other hand, is immobilized under submerged soil in which rice is usually grown. This soil condition includes the soil pH, bicarbonate concentrations, soil moisture, and low Fe and aluminum oxides (Ptashnyk et al. 2011; Cakmak and Kutman 2018). Similar to Fe, Zn uptake by rice roots is facilitated by secretions of DMA (Dotaniya et al. 2013; Banakar et al. 2017). Though the order of affinity of DMA with Zn and Fe species is $Fe^{2+} < Zn^{2+} < Fe^{3+}$ (Murakami et al. 1989), Fe oxidation by roots releases oxygen, which causes a reduction in ph. This releases Zn from the highly insoluble fractions over a limited time (Kirk and Bajita 1995).

4.2.1 Translocation of Fe and Zn

After Fe and Zn are absorbed from the soil, they are transported to a continuous network of interconnected plant-cell protoplasts. From the rhizosphere, plants take up Fe and Zn through transporters. In rice, several Zn-regulated transporters and Fe-regulated transporter-like proteins have been reported, such as OsIRT1, OsIRT2, OsZIP1, OsZIP3, OsZIP4, and OsZIP5 (Ramesh et al. 2003; Ishimaru et al. 2007; Yang et al. 2009). The ZIP gene family was shown to be involved in metal uptake and transport in plants (Eide et al. 1996; Korshunova et al. 1999; Vert et al. 2001; Connolly et al. 2002; Vert et al. 2002). ZIP proteins are predicted to have eight transmembrane domains, with their amino- and carboxyl-terminal ends situated on the outer surface of the plasma membrane. These domains provide a potential metal-binding domain rich in histidine residues. The most-conserved region of the ZIP proteins contains a fully conserved histidine that may form part of an intramembranous metal-binding site involved in transport (Guerinot 2000). OsZIP4 has been identified to be responsible for Zn translocation to plant parts that require Zn. It is expressed in both roots and shoots, suggesting that it transports Zn intracellularly or between plant tissues (Figure 4.1)(Grotz et al. 1998; Guerinot 2000).

4.2.2 Storage of Fe and Zn in the Grain

The mechanisms involved in phloem movement of Fe and Zn to the developing seed are not well understood among cereals. However, a model has been proposed based on information about seed apoplastic barriers in barley and wheat. In barley, Fe and Zn are unloaded symplastically from the phloem and flow to highly specialized nucellar projection transfer cells instead of the seed coat (Olsen and Palmgren 2014). It was also reported that barley possesses a single apoplastic barrier between the maternal transfer cells and the filial endosperm (Borg et al. 2009), whereas an additional apoplastic barrier for Fe and Zn exists through the stem and rachis and the developing grain in wheat (Wang et al. 2011).

4.3 Fe and Zn Biofortification using the SDN-1 Approach

4.3.1 Increasing Grain Fe and Zn by Gene Silencing

SDN-1 type of editing creates a double-stranded break in the genome of a plant. DNA repair of this break can lead to a mutation or deletion, causing a change leading to silencing or reducing the expression of a gene. SDN-1 has been successfully used in increasing disease resistance, improving product quality and yield, as well as removing allergens (van de Wiel et al. 2017).

Rice vacuolar Fe transporter genes (*OsVIT1* and *OsVIT2*) play an important function in Fe^{2+} and Zn^{2+} transport into vacuoles across the tonoplast. The potential example of using a CRISPR-based approach is to knockdown *OsVIT2* to achieve an increase in grain Fe, similar to the published T-DNA insertion silencing of this gene (Bashir et al. 2013) in different rice cultivars. Silencing *OsVIT1* and *OsVIT2* significantly increased Fe and Zn accumulation in brown rice while decreasing Fe and Zn in flag leaves (Zhang et al. 2012). Similarly, an *OsVIT2* gene knockdown significantly enhanced Fe and Zn accumulation in polished rice (Bashir et al. 2013). These reports indicate a potential strategy for Fe and Zn biofortification via a silencing approach. However, Zhang et al. (Zhang et al. 2012) detected the accumulation of Cd^{2+} in brown rice in *OsVIT1* and *OsVIT2* gene knockdown when grown in Cd^{2+}-polluted soil. Thus, further understanding of the regulatory mechanism is required for safe biofortification using this strategy.

Another approach for gene silencing to increase micronutrient content was a knockdown of the *OsHRZ2* gene reported by Kobayashi et al. (Kobayashi et al. 2013). Rice hemerythrin motif-containing novel gene (RING) and Zn-finger protein (*OsHRZ1* and *OsHRZ2*) ubiquitin ligases are Fe-binding sensors that negatively regulate Fe acquisition under conditions of Fe sufficiency. *OsHRZ2* knockdown lines grown in pots with calcareous soil or normal soil and fields with normal soil showed increased Fe and Zn accumulation in both brown rice and polished rice without an apparent yield penalty (Kobayashi et al. 2013).

4.3.2 Improving Bioavailability by Gene Silencing

Myo-inositol 1,2,3,4,5,6-hexa*kis*phosphate (InsP6) or phytic acid (PA) is a potent chelator of metal cations and accumulates as mixed salts referred to as phytate. It acts as an antinutrient as it reduces the bioavailability of essential metals such as Fe^{2+} and Zn^{2+} in grains. Several reports have been made to reduce PA concentration in rice grains by down-regulating the genes involved in the PA biosynthetic pathway using RNAi technology. 1D-*myo*-inositol-3-phosphate synthase (MIPS) is the enzyme that catalyzes the first step in PA biosynthesis. Silencing rice *MIPS* (*RINO1*) has been reported to reduce the PA content of rice grains and simultaneously increase inorganic phosphate (Pi) (Feng and Yoshida 2004; Kuwano et al. 2006; Kuwano et al. 2009; Ali et al. 2013). Ali et al. (2013) reported an increase in Fe and Zn contents of milled rice due to a reduction in PA content. Apart from being an essential enzyme in PA biosynthesis, MIPS is a precursor for de novo synthesis of *myo*-inositol. Hence, suppression of *MIPS* also resulted in a decrease in *myo*-inositol in rice seeds (Ali et al. 2013). To avoid disrupting related essential biosynthetic pathways, the enzymes involved at a later stage in PA biosynthesis are a potential target for a silencing strategy leading to a reduction in PA in the grain.

Inositol 1,3,4,5,6-penta*kis*phosphate 2-kinase (IPK1) is the enzyme that catalyzes the last step of PA biosynthesis. Down-regulation of *IPK1* effectively reduced the PA content of rice grains. It increased Pi without significantly affecting other inositol phosphate kinase genes involved in PA biosynthesis as well as seed *myo*-inositol content (Ali et al. 2013). Additionally, the contents of Fe and Zn were increased in milled low-phytate rice. This report shows that a reduction in seed PA content with a simultaneous increase in Pi can be successfully achieved without disturbing the healthy growth and development of rice plants. Targeting *IPK1* for silencing can be an appropriate strategy for developing low-phytate rice and, consequently, increasing the bioavailability of Fe and Zn.

Another group of enzymes that can be targeted is inositol 1,3,4-tris*kis*phosphate 5/6-kinases (ITP5/6 Ks). These enzymes are involved in the intermediate steps of PA biosynthesis. Down-regulation of *ITP5/6K-1* significantly reduced seed PA content with a concomitant increase in Pi without interfering with seed germination and plant growth (Karmakar et al. 2019). *RINO1* expression and seed *myo*-inositol content were elevated while the expression of other *ITP5/6K* homologs was not altered. Moreover, Fe and Zn content in polished grains increased. Targeting *ITP5/6K-1* for silencing can be an alternative strategy for improving the bioavailability of Fe and Zn. Jiang et al. (Jiang et al. 2019) generated knockout mutants of an *ITP5/6K* homolog, *ITP5/6K-6*, using the CRISPR-Cas9 system. These mutants had significantly lower PA content in rice grains with a simultaneous increase in Pi. However, *ITP5/6K-6* gene knockout also significantly impaired plant growth and tolerance of abiotic stress. Thus, the use of the CRISPR-Cas9 system on the *ITP5/6K-6* gene may be more useful if knockdown lines are generated instead of a total knockout of *ITP5/6K-6* expression.

4.4 Fe and Zn Biofortification Using the SDN-2 Approach

Allele replacement using genome editing is one of the promising applications for crop improvement and is often mentioned as one of the major applications of SDN-2. This approach has been used successfully in rice using either homologous recombination (HR) (Sun et al. 2016) or non-homologous end joining (NHEJ) intron-mediated site-specific gene replacement via the CRISPR-Cas9 system (Li et al. 2016). Most successful reports of HR mutation in rice involve editing of the acetolactate synthase (ALS) gene, in which the expressed enzyme is a target of *ALS*-inhibiting herbicides. Precise replacement of the wild-type *ALS* gene with the intended mutant version conferred higher herbicide resistance and functioned as a selectable marker for the edited rice plants (Li and Xia 2020).

Breeding work on biofortification reported several potential quantitative trait loci for Fe and Zn in rice grains. Natural resistance-associated macrophage protein (*OsNRAMP*), nicotianamine synthase (*OsNAS*), *OsZIP*, yellow stripe-like (*OsYSL*), ferritin (*OsFER*), and Zn-induced facilitator-like (*OsZIFL*) genes were identified near QTLs for Fe and Zn (Descalsota et al. 2018; Swamy et al. 2018) and were identified as crucial for the accumulation of Fe and Zn in the polished grain (Swamy et al. 2018). Among these genes, the *OSNAS* gene family (Inoue et al. 2003) has been well characterized and extensively used to increase grain Zn through the transgenesis approach (Johnson et al. 2011). A novel functional allele from this gene family, together with the *ZIP* gene family (Guerinot 2000), is a potential target for allele replacement to increase grain Fe and Zn in rice (Table 4.1).

Another Zn transporter family that is the heavy metal ATPase from rice (OsHMA3) is known to play a role in maintaining Zn homeostasis (Cai et al. 2019). OsHMA3 is suggested to play a role in rice roots by Zn detoxification and storage by gathering the element into vacuoles depending on the environmental Zn concentration. Replacing a non-functional allele of *OsHMA3* with a functional allele by gene editing may be a strategy to increase Zn transport in rice (Table 4.1). Also, a study by Neeraja et al. (2018) has shown that phosphate transporter-exporter (*PHO*), proton-coupled peptide transporter (*POT*), and vacuolar iron transporter (*VIT*) showed enhanced and significant differential expression in landraces with high Zn in their panicles and grains (Table 4.1).

In *Arabidopsis*, a gene identified as *ZIF1* was proposed to have a role in a novel mechanism of Zn sequestration, possibly by transport of a Zn ligand or a Zn-ligand complex into vacuoles (Haydon and Cobbett 2007). The study identified three allelic Zn-sensitive mutants of *Arabidopsis*. Shoots of *ZIF1* mutants showed increased accumulation of Zn but not with other metal ions. A homologous *ZIF1* gene could also possibly exist in rice and be targeted for allelic replacement (Table 4.1).

Table 4.1 Potential genes and strategies that can be used for editing to increase Fe and Zn concentrations and their bioavailability in rice grain using SDN-1, SDN-2, and SDN-3.

Editing	Target gene	Function	Gene/approach references
SDN-1	*OsVIT1*	Fe and Zn translocation	Zhang et al. (2012)
	OsVIT2	Fe and Zn translocation	Zhang et al. (2012) and Bashir et al. (2013)
	OsHRZ2	Fe and Zn uptake and translocation regulators	Kobayashi et al. (2013)
	OsIPK1	PA biosynthesis	Ali et al. (2013)
	OsITP5/6K	PA biosynthesis	Karmakar et al. (2019) and Jiang et al. (2019)
SDN-2	*OsNAS*	Synthesis of nicotianamine for transport of Fe and Zn	Inoue et al. (2003) and Swamy et al. (2018)
	OsZIP	Translocation of Zn	Ishimaru et al. (2005) and Descalsota et al. (2018)
SDN-2, SDN-3 marker-free targeted multi-gene insertion	*OsNRAMP*	Transport of Fe and Zn	Nevo and Nelson (2006) Boonyaves et al. (2017) and Swamy et al. (2018)
	OsHMA	Tonoplast-localized transporter for Zn	Olsen and Palmgren (2014) and Cai et al. (2019)
	OsPHO	Micronutrient concentration in panicle	Saenchai et al. (2016) and Neeraja et al. (2018)
	OsPOT	Transport of Zn-small protein complexes	Zhao et al. (2010) and Neeraja et al. (2018)
	OsZIFL	Sequestration and transport of Fe and Zn	Haydon and Cobbett (2007) and Swamy et al. (2018)
	AtNAS1	NAS biosynthesis, phloem transport, pollen development, and pollen tube growth	Boonyaves et al. (2017)
	AtFRD3	Metal Fe transport, response to abscisic acid and Fe ion homeostasis	Roschzttardtz et al. (2011)
SDN-3 marker-free targeted multi-gene insertion	*AtNRAMP3*	Cellular metal ion homeostasis and response to Zn ion	Thomine et al. (2003)
	PVFerritin	Fe storage in a readily non-toxic and bioavailable form	Spence et al. (1991)
	SoyferH1 + *OsNAS2*	Fe storage in a readily non-toxic and bioavailable form	Trijatmiko et al. (2016)
SDN-3 *cis*genic insertion	*OsNAS*	Nicotianamine chelator biosynthesis	Johnson et al. (2011)

Another strategy in Zn biofortification is to increase the bioavailability of Zn in rice grain. PA acts as a potent chelator of metal cations to form phytate and is considered an anti-nutrient as it reduces the bioavailability of essential micronutrients. At least 15 putative genes are involved in PA biosynthesis and transport in rice (Perera et al. 2018). Replacement with a non-functional allele for a combination of these genes could reduce the amount of PA in the grain.

The challenge for allele replacement using genome editing is the low efficiency of HR (Eid and Mahfouz 2016). The main limitation is that HR is usually active only in the interphase of the cell cycle, in contrast to non-HR which is predominantly active in the cell. Post-mitotic cells are, therefore, resistant to targeted editing by HR. A recent development in genome editing called Prime Editing could be a better alternative. This new genome editing method directly creates new genetic information at a specified DNA site using a catalytically impaired Cas9 endonuclease fused to an engineered reverse transcriptase, coding with a prime editing guide RNA (pegRNA) that both specifies the target site and encodes the desired edit.

4.5 Fe and Zn Biofortification Using the SDN-3 Approach

Targeted delivery and integration of a functional gene unit or multiple genes by SDN-3 at a predetermined plant genomic safe harbor (GSH) locus provides a desirable alternative to insertions at random sites by conventional plant transformation methods (Dong et al. 2020). GSHs are sites in the genome that can support expression while minimizing any unwanted interaction that could lead to dysregulation of any endogenous gene in the host cell (Papapetrou and Schambach2016). This is a targeted chromosomal region for which disruption of transgene insertion in the genome causes no adverse effects on the host organism (Cantos et al. 2014; Dong et al. 2020).

Numerous studies regarding Fe and Zn biofortification through a transgenic approach have been conducted since the work of Goto et al. (1999) with the ferritin gene from soybean. In principle, all potential Fe and Zn homeostasis genes (Slamet-Loedin et al. 2015; Ludwig and Slamet-Loedin 2019) can be overexpressed in the GSH by SDN-3-type genome editing as shown in Table 4.1. To date, seven studies showed that reaching the attained breeding target of 13 ppm for Fe and 28 ppm for Zn can be achieved through simultaneous overexpression of genes related to uptake or translocation, and storage (multi-gene insertion). The studies above are highlighted in this review based on the two criteria. First, the values for Fe and Zn should be obtained from milled or polished grains because rice is mainly eaten as milled. Second, the values of Fe concentration in the wild-type control are, on average, around 2 ppm (Swamy et al. 2011). Controls with values higher than this known baseline could be due to low milling degree and/or possible contamination upon processing. For ease of regulatory processes in specific countries and potential wider public acceptance, only rice genes can be inserted. However, this may compromise the potential content of Zn and, especially, Fe.

Increasing nicotiana mine (NA) production by overexpression of NAS could increase Fe up to fourfold in glasshouse conditions (Johnson et al. 2011). This was higher than overexpression of NAAT (nicotiana mine aminotransferase) or DMA (2'deoxymugeneic acid) in the MA pathway that can increase grain Fe concentration by only one- to twofold of Fe (Aung et al. 2013; Bashir et al. 2017). To further increase the polished grain mineral content to more than fivefold of Fe and more than the double-fold increase in Zn in polished grain, multi-gene insertion approaches were applied in several studies (Trijatmiko et al. 2016; Boonyaves et al. 2017; Wu et al. 2018). Trijatmiko et al. (2016) combined OsNAS2 with SOYFER-H1 to reach 15 ppm for Fe and 47 ppm for Zn in polished rice grain. Other approaches combined the overexpression of the natural resistance-associated macrophage protein (AtNRAMP3) gene and ferric reductase defective 3 (FRD3) gene from *Arabidopsis* with OsNAS and also reached 11–13 ppm for Feand 33–48 ppm for Zn (Boonyaves et al. 2017; Wu et al. 2018; Wu et al. 2019). It can also be inferred that ectopic expression of non-rice ferritin especially from legumes (PvFERRITIN, SOYFERH)(Trijatmiko et al. 2016; Boonyaves et al. 2017; Wu et al. 2018; Wu et al. 2019) in rice endosperm is more effective than ectopic expression of OsFER (Oliva et al. 2014). Alternatively, a *cis*genic insertion of a rice gene is an option (Table 4.1). Insertion of OsNAS at the GSH region in the rice genome to increase the production of NA is a promising alternative to ease the regulatory process before product release; however, the grain micronutrient content is not likely to be as high as with the insertion of multiple transgenes.

4.6 Future Thrust and Implications of SDN-1, -2, and -3

SDN-1, -2, and -3 offer novel solutions that could help to overcome hidden hunger by improving efficiency and precision in plant breeding by providing variation to the existing gene pool, especially for nutrition. This new breeding approach could improve the nutritive content of rice amid the rapidly increasing population, decreasing agricultural area, and the diverse changing climate across the world on a global scale. Currently, some countries do not regulate plants derived from the new SDN-1 breeding technique. For example, in the US, the US Department of Agriculture (USDA) only regulates the newly generated plant as a genetically modified organism (GMO) when it contains a plant pest sequence. Therefore, the clean mutation products obtained by SDN-1 or even SDN-2 are not categorized as GMOs. On a case-by-case basis, countries such as Argentina, Australia, and Japan do not regulate SDN-1 as a GMO when no transgenes are present in the final product and only a few base-pair alterations took place. SDN-1, which uses NHEJ cell repair that induces insertions or deletions (INDELs) in the genomic sequence, is similar to the principles of mutation. However, the EU considers all types of genome editing as GMOs due to the process to develop the product, even though mutation by chemical agents or other techniques is exempted from any legislation concerning GMOs by Annex IB of Directive 2001/18/EC and the EFSA GMO unit opinion regarding SDN-1 as a form of mutagenesis (Schaart et al. 2016; Sprink et al. 2016; van de Wiel et al. 2017).

For SDN-2 plant products, some countries such as Brazil, Argentina, and Japan review the products on a case-by-case basis; they could be considered as non-GMO when there is no transgene presence and repaired through HR by the cell using a genomic template from the same species with minimal base-pair change. This product is considered similar to introgression and backcross breeding. This template could contain only one or several small sequence changes in the genomic code (NBTPlatform n.d.). This approach of precise insertion of fragments has become increasingly popular (Lackner et al. 2015; Kuscu et al. 2017). As mentioned above, the USDA does not regulate SDN-2 and SDN-3 if the final product does not contain a plant pest sequence. Regulatory consideration of SDN-3 varies from country to country. When the inserted genes come from non-sexually compatible species in most cases, they are considered as transgenic. If the inserted gene or its element, derived from a sexually compatible source, regardless of whether it is *cis*genic or intragenic, in a few countries, it could be considered as non-GMO on a case-by-case basis. These new breeding techniques open a new horizon to alleviate hidden hunger complementing transgenic approaches to reach those in need with a less cumbersome regulatory process.

References

Ali, N., Paul, S., Gayen, D. et al. (2013). Development of low phytate rice by RNAi mediated seed-specific silencing of inositol 1,3,4,5,6-pentakisphosphate 2-kinase gene (IPK1). *PLoS One* 8 (7): e68161.

Ali, N., Paul, S., Gayen, D. et al. (2013). RNAi mediated down regulation of myo-inositol-3-phosphate synthase to generate low phytate rice. *Rice* 6 (1): 1–12.

Aung, M.S., Masuda, H., Kobayashi, T. et al. (2013). Iron biofortification of Myanmar rice. *Frontiers in Plant Science* 4: 158.

Banakar, R., Alvarez Fernandez, A., Díaz-Benito, P. et al. (2017). Phytosiderophores determine thresholds for iron and zinc accumulation in biofortified rice endosperm while inhibiting the accumulation of cadmium. *Journal of Experimental Botany* 68 (17): 4983–4995.

Bashir, K., Takahashi, R., Nakanishi, H., and Nishizawa, N.K. (2013). The road to micronutrient biofortification of rice: progress and prospects. *Frontiers in Plant Science* 4: 15.

Bashir, K., Takahashi, R., Akhtar, S. et al. (2013). The knockdown of OsVIT2 and MIT affects iron localization in rice seed. *Rice* 6: 31.

Bashir, K., Nozoye, T., Nagasaka, S. et al. (2017). Paralogs and mutants show that one DMA synthase functions in iron homeostasis in rice. *Journal of Experimental Botany* 68 (7): 1785–1795.

Boonyaves, K., Wu, T.Y., Gruissem, W., and Bhullar, N.K. (2017). Enhanced grain iron levels in rice expressing an iron-regulated metal transporter, nicotianamine synthase, and ferritin gene cassette. *Frontiers in Plant Science* 8: 130.

Borg, S., Brinch-Pedersen, H., Tauris, B., and Holm, P.B. (2009). Iron transport, deposition and bioavailability in the wheat and barley grain. *Plant and Soil* 325 (1-2): 15–24.

Bughio, N., Yamaguchi, H., Nishizawa, N.K. et al. (2002). Cloning an iron regulated metal transporter from rice. *Journal of Experimental Botany* 53 (374): 1677–1682.

Cai, H., Huang, S., Che, J. et al. (2019). The tonoplast-localized transporter OsHMA3 plays an important role in maintaining Zn homeostasis in rice. *Journal of Experimental Botany* 70 (10): 2717–2725.

Cakmak, I. and Kutman, U.B. (2018). Agronomic biofortification of cereals with zinc: a review. *European Journal of Soil Science* 69 (1): 172–180.

Cantos, C., Francisco, P., Trijatmiko, K.R. et al. (2014). Identification of "safe harbor" loci in indica rice genome by harnessing the property of zinc-finger nucleases to induce DNA damage and repair. *Frontiers in Plant Science* 5: 302.

Connolly, E.L., Fett, J.P., and Guerinot, M.L. (2002). Expression of the IRT1 metal transporter is controlled by metals at the levels of transcript and protein accumulation. *The Plant Cell* 14 (6): 1347–1357.

Descalsota, G., Swamy, B., Zaw, H. et al. (2018). Genome-wide association mapping in a rice MAGIC plus population detects QTLs and genes useful for biofortification. *Frontiers in Plant Science* 9: 1347. https://doi.org/10.3389/fpls.2018.01347.

Dong, O.X., Yu, S., Jain, R. et al. (2020). Marker-free carotenoid-enriched rice generated through targeted gene insertion using CRISPR-Cas9. *Nature Communications* 11: 1178. https://doi.org/10.1038/s41467-020-14981-y.

Dotaniya, M.L., Meena, H.M., Lata, M., and Kumar, K. (2013). Role of phytosiderophores in iron uptake by plants. *Agricultural Science Digest-A Research Journal* 33 (1): 73–76.

Eid, A. and Mahfouz, M.M. (2016). Genome editing: the road of CRISPR/Cas9 from bench to clinic. *Experimental and Molecular Medicine* 48 (10): e265–e265.

Eide, D., Broderius, M., Fett, J., and Guerinot, M.L. (1996). A novel iron-regulated metal transporter from plants identified by functional expression in yeast. *Proceedings of the National Academy of Sciences of the United States of America* 93 (11): 5624–5628.

Feng, X. and Yoshida, K.T. (2004). Molecular approaches for producing low-phytic-acid grains in rice. *Plant Biotechnology* 21 (3): 183–189.

Goto, F., Yoshihara, T., Shigemoto, N. et al. (1999). Iron fortification of rice seed by the soybean ferritin gene. *Nature Biotechnology* 17 (3): 282–286.

Grotz, N., Fox, T., Connolly, E. et al. (1998). Identification of a family of zinc transporter genes from *Arabidopsis* that respond to zinc deficiency. *Proceedings of the National Academy of Sciences of the United States of America* 95 (12): 7220–7224.

Guerinot, M.L. (2000). The ZIP family of metal transporters. *Biochimica et Biophysica Acta (BBA) – Biomembranes* 1465 (1-2): 190–198.

Haydon, M.J. and Cobbett, C.S. (2007). A novel major facilitator superfamily protein at the tonoplast influences zinc tolerance and accumulation in *Arabidopsis*. *Plant Physiology* 143 (4): 1705–1719.

Inoue, H., Higuchi, K., Takahashi, M. et al. (2003). Three rice nicotianamine synthase genes, *OsNAS1*, *OsNAS2*, and *OsNAS3*, are expressed in cells involved in long distance transport of iron and differentially regulated by iron. *The Plant Journal* https://doi.org/10.1046/j.1365-313X.2003.01878.x.

Ishimaru, Y., Suzuki, M., Kobayashi, T. et al. (2005). OsZIP4, a novel zinc-regulated zinc transporter in rice. *Journal of Experimental Botany* 56 (422): 3207–3214.

Ishimaru, Y., Suzuki, M., Tsukamoto, T. et al. (2006). Rice plants take up iron as an Fe3+ phytosiderophore and as Fe2+. *The Plant Journal* 45 (3): 335–346.

Ishimaru, Y., Masuda, H., Suzuki, M. et al. (2007). Overexpression of the OsZIP4 zinc transporter confers disarrangement of zinc distribution in rice plants. *Journal of Experimental Botany* 58 (11): 2909–2915.

Jiang, M., Liu, Y., Liu, Y. et al. (2019). Mutation of inositol 1,3,4-trisphosphate 5/6-kinase6 impairs plant growth and phytic acid synthesis in rice. *Plants* 8 (5): 114.

Johnson, A.A., Kyriacou, B., Callahan, D.L. et al. (2011). Constitutive overexpression of the OsNAS gene family reveals single-gene strategies for effective iron- and zinc-biofortification of rice endosperm. *PLoS One* 6 (9): e24476.

Karmakar, A., Bhattacharya, S., Sengupta, S. et al. (2019). RNAi-mediated silencing of ITPK gene reduces phytic acid content, alters transcripts of phytic acid biosynthetic genes, and modulates mineral distribution in rice seeds. *Rice Science* 27 (2).

Kirk, G.J.D. and Bajita, J.B. (1995). Root induced iron oxidation, pH changes and zinc solubilization in the rhizosphere of lowland rice. *New Phytologist* 131 (1): 129–137.

Kobayashi, T., Nagasaka, S., Senoura, T. et al. (2013). Iron-binding haemerythrin RING ubiquitin ligases regulate plant iron responses and accumulation. *Nature Communications* 4 (1): 1–12.

Korshunova, Y.O., Eide, D., Clark, W.G. et al. (1999). The IRT1 protein from *Arabidopsis thaliana* is a metal transporter with a broad substrate range. *Plant Molecular Biology* 40 (1): 37–44.

Kuscu, C., Parlak, M., Tufan, T. et al. (2017). CRISPR-STOP: gene silencing through base-editing-induced nonsense mutations. *Nature Methods* 14 (7): 710–712.

Kuwano, M., Ohyama, A., Tanaka, Y. et al. (2006). Molecular breeding for transgenic rice with low-phytic-acid phenotype through manipulating myo-inositol 3-phosphate synthase gene. *Molecular Breeding* 18 (3): 263–272.

Kuwano, M., Mimura, T., Takaiwa, F., and Yoshida, K.T. (2009). Generation of stable 'low phytic acid' transgenic rice through antisense repression of the 1D-myo-inositol 3-phosphate synthase gene (RINO1) using the 18-kDa oleosin promoter. *Plant Biotechnology Journal* 7: 96–105.

Lackner, D.H., Carré, A., Guzzardo, P.M. et al. (2015). A generic strategy for CRISPR-Cas9-mediated gene tagging. *Nature Communications* 6 (1): 1–7.

Li, S. and Xia, L. (2020). Precise gene replacement in plants through CRISPR/Cas genome editing technology: current status and future perspectives. *aBIOTECH* 1: 58–73. https://doi.org/10.1007/s42994-019-00009-7.

Li, J., Meng, X., Zong, Y. et al. (2016). Gene replacements and insertions in rice by intron targeting using CRISPR–Cas9. *Nature Communications* 6 (1): 1–7.

Ludwig, Y. and Slamet-Loedin, I.H. (2019). Genetic biofortification to enrich rice and wheat grain iron: from genes to product. *Frontiers in Plant Science* 10: 833.

Murakami, T., Ise, K., Hayakawa, M. et al. (1989). Stabilities of metal complexes of mugineic acids and their specific affinities for iron (III). *Chemistry Letters* 18 (12): 2137–2140.

NBTPlatform (2014).Factsheet: Site-Directed Nuclease. https://www.nbtplatform.org/background-documents/factsheets/factsheet-site-directed-nucleases.pdf (accessed 17 July 2020.

Neeraja, C.N., Kulkarni, K.S., Babu, P.M. et al. (2018). Transporter genes identified in landraces associated with high zinc in polished rice through panicle transcriptome for biofortification. *PLoS One* 13 (2): e0192362.

Nevo, Y. and Nelson, N. (2006). The NRAMP family of metal-ion transporters. *Biochimica et Biophysica Acta* 1763: 609–620. https://doi.org/10.1016/j.bbamcr.2006.05.007.

NIH (2020).National Institute of Health. Iron. https://ods.od.nih.gov/factsheets/Iron-HealthProfessional (accessed 17 July 2020).

Oliva, N., Chadha-Mohanty, P., Poletti, S. et al. (2014). Large-scale production and evaluation of marker-free *indica* rice IR64 expressing phytoferritin genes. *Molecular Breeding* 33: 23–37.

Olsen, L.I. and Palmgren, M.G. (2014). Many rivers to cross: the journey of zinc from soil to seed. *Frontiers in Plant Science* 5: 30.

Papapetrou, E.P. and Schambach, A. (2016). Gene insertion into genomic safe harbors for human gene therapy. *Molecular Therapy* 24 (4): 678–684.

Perera, I., Seneweera, S., and Hirotsu, N. (2018). Manipulating the phytic acid content of rice grain toward improving micronutrient bioavailability. *Rice* 11 (1): 4.

Ptashnyk, M., Roose, T., Jones, D.L., and Kirk, G.J. (2011). Enhanced zinc uptake by rice through phytosiderophore secretion: a modelling study. *Plant, Cell and Environment* 34 (12): 2038–2046.

Ramesh, S.A., Shin, R., Eide, D.J., and Schachtman, D.P. (2003). Differential metal selectivity and gene expression of two zinc transporters from rice. *Plant Physiology* 133 (1): 126–134.

Römheld, V. and Marschner, H. (1986). Evidence for a specific uptake system for iron phytosiderophores in roots of grasses. *Plant Physiology* 80 (1): 175–180.

Roschzttardtz, H., Séguéla-Arnaud, M., Briat, J.F. et al. (2011). The FRD3 citrate effluxer promotes iron nutrition between symplastically disconnected tissues throughout *Arabidopsis* development. *The Plant Cell* 23 (7): 2725–2737.

Saenchai, C., Bouain, N., Kisko, M. et al. (2016). The involvement of OsPHO1;1 in the regulation of iron transport through integration of phosphate and zinc deficiency signaling. *Frontiers in Plant Science* 7: 396. https://doi.org/10.3389/fpls.2016.00396.

Schaart, J.G., van de Wiel, C.C., Lotz, L.A., and Smulders, M.J. (2016). Opportunities for products of new plant breeding techniques. *Trends in Plant Science* 21 (5): 438–449.

Schlemmer, U., Frølich, W., Prieto, R.M., and Grases, F. (2009). Phytate in foods and significance for humans: food sources, intake, processing, bioavailability, protective role and analysis. *Molecular Nutrition & Food Research* 53 (S2): S330–S375.

Seshadri, S. (2001). Prevalence of micronutrient deficiency particularly of iron, zinc and folic acid in pregnant women in South East Asia. *British Journal of Nutrition* 85 (S2): S87–S92. https://doi.org/10.1049/BJN2000299.

Slamet-Loedin, I.H., Johnson-Beebout, S.E., Impa, S., and Tsakirpaloglou, N. (2015). Enriching rice with Zn and Fe while minimizing Cd risk. *Frontiers in Plant Science* 6: 121.

Spence, M.J., Henzl, M.T., and Lammers, P.J. (1991). The structure of a *Phaseolus vulgaris* cDNA encoding the iron storage protein ferritin. *Plant Molecular Biology* 17 (3): 499–504.

Sprink, T., Eriksson, D., Schiemann, J., and Hartung, F. (2016). Regulatory hurdles for genome editing: process- vs. product-based approaches in different regulatory contexts. *Plant Cell Reports* 35 (7): 1493–1506.

Sugiura, Y. and Nomoto, K. (1984). Phytosiderophores structures and properties of mugineic acids and their metal complexes. In: *Siderophores from Microorganisms and Plants* (eds. A. Chimiak, R.C. Hider, et al.), 107–135. Berlin, Heidelberg: Springer.

Sun, Y., Zhang, X., Wu, C. et al. (2016). Engineering herbicide-resistant rice plants through CRISPR/Cas9-mediated homologous recombination of acetolactate synthase. *Molecular Plant* 9: 628–631.

Swamy, B.P.M., Kaladhar, K., Anuradha, K. et al. (2011).Enhancing iron and zinc concentration in rice grains using wild species. *CCMB ADNAT Convention and International Symposium on Genomics and Biodiversity*, Hyderabad, India (23–25 February 2011). India: ADNAT.

Swamy, B.P.M., Descalsota, G., Nha, C. et al. (2018). Identification of genomic regions associated with agronomic and biofortification traits in DH populations of rice. *PLoS One* 13 (8): e0201756.

Takagi, S.I. (1976). Naturally occurring iron-chelating compounds in oat- and rice-root washings: I. activity measurement and preliminary characterization. *Soil Science and Plant Nutrition* 22 (4): 423–433.

Thomine, S., Lelièvre, F., Debarbieux, E. et al. (2003). AtNRAMP3, a multispecific vacuolar metal transporter involved in plant responses to iron deficiency. *The Plant Journal* 34 (5): 685–695.

Trijatmiko, K.R., Dueñas, C., Tsakirpaloglou, N. et al. (2016). Biofortifiedindica rice attains iron and zinc nutrition dietary targets in the field. *Scientific Reports* 6: 19792.

Tsakirpaloglou, N., Mallikarjuna, S., Acuin, C., and Slamet-Loedin, I.H. (2019). Biofortified Zn and Fe Rice: potential contribution for dietary mineral and human health. In: *Nutritional Quality Improvement in Plants. Concepts and Strategies in Plant Sciences* (eds. P. Jaiwal, A. Chhillar, D. Chaudhary and R. Jaiwal), 1–24. Cham: Springer.

UN (n.d.).Goal 2: Zero Hunger. https://www.un.org/sustainabledevelopment/hunger (accessed 17 July 2020).

Vert, G., Briat, J.F., and Curie, C. (2001). *Arabidopsis* IRT2 gene encodes a root-periphery iron transporter. *The Plant Journal* 26 (2): 181–189.

Vert, G., Grotz, N., Dédaldéchamp, F. et al. (2002). IRT1, an *Arabidopsis* transporter essential for iron uptake from the soil and for plant growth. *The Plant Cell* 14 (6): 1223–1233.

vonWirén, N., Klair, S., Bansal, S. et al. (1999). Nicotianamine chelates both FeIII and FeII. Implications for metal transport in plants. *Plant Physiology* 119 (3): 1107–1114.

Wang, S., Yin, L., Tanaka, H. et al. (2011). Wheat-*Aegilops* chromosome addition lines showing high iron and zinc contents in grains. *Breeding Science* 61 (2): 189–195.

van de Wiel, C.C.M., Schaart, J.G., Lotz, L.A.P., and Smulders, M.J.M. (2017). New traits in crops produced by genome editing techniques based on deletions. *Plant Biotechnology Reports* 11 (1): 1–8.

Wu, T.Y., Gruissem, W., and Bhullar, N.K. (2018). Facilitated citrate-dependent iron translocation increases rice endosperm iron and zinc concentrations. *Plant Science* 270: 13–22.

Wu, T.Y., Gruissem, W., and Bhullar, N.K. (2019). Targeting intracellular transport combined with efficient uptake and storage significantly increases grain iron and zinc levels in rice. *Plant Biotechnology Journal* 17 (1): 9–20.

Yang, X., Huang, J., Jiang, Y., and Zhang, H.S. (2009). Cloning and functional identification of two members of the ZIP (Zrt, Irt-like protein) gene family in rice (*Oryza sativa* L.). *Molecular Biology Reports* 36 (2): 281–287.

Zhang, Y., Xu, Y.H., Yi, H.Y., and Gong, J.M. (2012). Vacuolar membrane transporters OsVIT1 and OsVIT2 modulate iron translocation between flag leaves and seeds in rice. *The Plant Journal* 72 (3): 400–410.

Zhao, X., Huang, J., Yu, H. et al. (2010). Genomic survey, characterization and expression profile analysis of the peptide transporter family in rice (*Oryza sativa* L.). *BMC Plant Biology* 10: 92.

5

Development of Carotenoids Rich Grains by Genome Engineering

Vikrant Gautam[1], Gurwinder Singh Rana[2], Pankaj Kumar[3], and Santosh Watpade[4]

[1] *ICAR-NBPGR, New Delhi, 110012, India*
[2] *AFC, New Delhi, India*
[3] *ICAR-CPRI, Shimla, 171001, India*
[4] *ICAR-IARI Regional Station (CHC), Shimla, 171004, India*

CHAPTER MENU

5.1 Introduction

All photosynthetic organisms such as plants, algae, cyanobacteria, some of non-photosynthetic bacteria and fungi synthesize carotenoid (Figure 5.1), which is a group of isoprenoid metabolites (Liang et al. 2018; Rodriguez-Concepcion et al. 2018). Carotenoids are an important natural product to animals and also very essential pigments to plants, it acts as both primary and specialized metabolites to fulfill multifaceted functions in plants.

Carotenoid performs a multitude of functions such as playing an important role in photosynthesis in plants; photoprotection acts as precursors to phytohormones such as ABA and strigolactones, providing color and aroma to flowers and fruits for attracting insects and other organisms for pollination and helps in seed dispersal, protects the seed from deterioration, responds to environmental stresses, such as photooxidative stress, improves the sexual behavior and reproduction of animals, and protects animals from predation as well as parasitism. For humans, carotenoids act as a dietary source of provitamin A, and is used commercially as feed additives to enhance pigmentation of fish and eggs and colorizing agents for human food, cosmetics, and pharmaceutical products.

Genome Engineering for Crop Improvement, First Edition. Edited by Santosh Kumar Upadhyay.

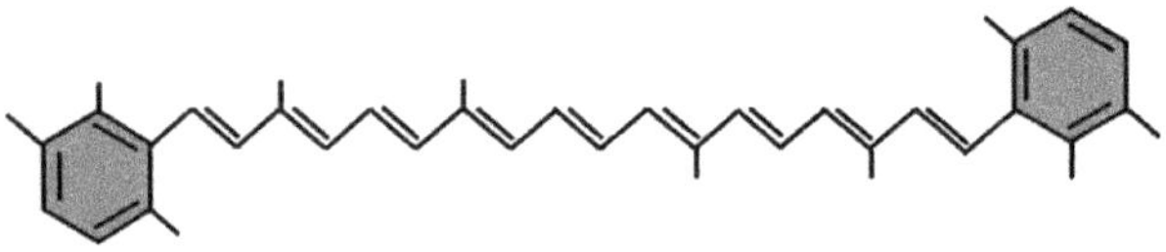

Figure 5.1 General structure of a carotenoid: polyene tail with double bonds, possible terminal rings.

Major staple cereals often contain very small amounts of carotenoids in their grains. Animals and humans rely upon diet such as fruits, vegetables, etc. as sources of these compounds as they are unable to synthesize carotenoids. Therefore, genetic manipulation to increase carotenoid content in cereal grain is very essential. In this chapter, particular attention is given on carotenoid metabolism and regulation in non-green plant tissues, as well as genetic manipulation in staple cereals such as rice, maize, and wheat (Zhai et al. 2016).

Decades ago, yield was the primary focus of the breeders but with surplus production of staple cereals biofortification of staple cereals becomes an important aspect in food production. To improve public health, breeding crops for enhanced level of vitamins, minerals, protein and healthier fats is the most practical means. Biofortification is the process by which the nutritional quality of food crops is improved through agronomic practices, conventional plant breeding or modern biotechnology. Biofortification aims to increase nutrient levels in crops during plant growth rather than through manual means during processing of the crops. Thus, need for understanding the regulatory mechanisms of carotenoids is a very important scientific pursuit for improving the quality of staple food (Zhai et al. 2016).

Cereals are considered as a staple food for the majority of population but most of them contain a far less quantity of carotenoids. Hence, enhancing carotenoid accumulations in cereals through genome engineering tools can overcome the problem of vitamin A deficiency for the greater population (Zhai et al. 2016). Important cereal grains like rice, wheat and maize fulfill the loric requirement of more than 50% of global nutritional requirement and are targeted to combat micronutrient deficiencies by major biofortification programs (Brinch-Pedersen et al. 2007). Carotenoid biosynthesis in wheat has received much less attention compared with maize and rice.

Partially iron-biofortified rice has improved the iron stores of reproductive-age women in the Philippines, iron-biofortified pearl millet has increased the iron stores and reversed iron deficiency in school children in India and iron-biofortified beans have improved the iron stores in women in Rwanda. Evidence for pro-vitamin A rich crops is more difficult to interpret as pro-vitamin A carotenoids are first absorbed in the body and then converted into the active form of vitamin A according to the body's need for the nutrient. There is broad evidence (both efficacy and effectiveness) that provitamin A biofortified orange sweet potato reduces vitamin A deficiency in children in Mozambique, Uganda and in South Africa, with an additional study in Bangladesh showing increased pro-vitamin A concentration but not vitamin A status. A study with pro-vitamin A biofortified yellow cassava in Kenya showed an increase in vitamin A status and a greater increment in pro-vitamin A concentrations in school children. For biofortified orange maize, results are ambiguous because the laboratory tests used different display degrees of sensitivity and accuracy. Therefore, the two studies conducted in independent-population groups in

Zambia showed different results: pro-vitamin A concentrations increased in both; the active form of vitamin A did not increase in either study partly because vitamin A deficiency was initially low; and only the study that used a more sensitive test for vitamin A status proved increased vitamin A stores in the school children who consumed the biofortified maize. Moreover, the study that did not improve the levels of vitamin A in the blood of the children found a positive effect in visual adaptation to twilight conditions (dusk/dawn) of ambient light illumination, which is a functional indicator of improved vitamin A status in the body. While more trials with biofortified crops with higher provitamin A densities in other populations are being conducted, evidence is accumulating in favor of biofortified crops being able to improve the micronutrient status of different populations.

In this chapter, we briefly try to describe the pathways, main flux-controlling steps for carotenoid accumulation in plants and the strategies by which we can produce the carotenoid rich cereal grains. Nutritional quality of cereal grains can be increased by applying various techniques among which most trending are as follows:

a) Nutritional quality improvement through Pathway engineering
b) Crop improvement through Genetic Engineering techniques
c) Increase in carotenoids through Clustered regularly interspaced short palindromic repeats (CRISPR)
d) Improvement of Carotenoid in Grain Crops through RNAi

5.2 Nutritional Quality Improvement Through Pathway Engineering

5.2.1 Strategies for Metabolic Engineering and Synthetic Biology in Plants

Synthetic biology and metabolic engineering in plants are regulated by the involvement and manipulation of rate-limiting enzyme or genes (one or more) in metabolic or synthetic pathways, and all the corresponding genes that are involved in the entire metabolic pathways. By understanding the behavior and role of the enzymes and gene in the metabolic pathway we can employ methods to deliver and express multiple genes in host plants that affect the development of plant synthetic metabolic engineering.

5.2.2 Strategies for Synthetic Metabolic Engineering

Metabolic pathway involved a series of chemical chain reaction triggered by different sets of enzymes and genes which catalyzes the substrate conversion to target products and by-product under regulatory conditions. Several strategies are applied to scale up the synthesis of target compounds that are as follow:

1) Upstream genes increased expression in target pathway for encoding rate limiting and multiple enzymes to provide a sufficient supply of precursors for increasing metabolic flux through manipulation in the targeted metabolic pathway.
2) Enzyme inhibition or gene knockout strategies for inhibiting the expression that are involved in catabolic/competitive pathway for synthesis of target product to eliminate

the diversification by synthesis of intermediates and to prevent the target metabolites decomposition.
3) Activation of multiple pathways by endogenous key genes with the help of transcription factors overexpression for metabolite synthesis.
4) Synthesis of the targeted products will be increased by combining all the above methods. In addition to the above-mentioned strategies CRISPR/dCas9-based activation/repression systems gives a new site in metabolic engineering manipulation (Zalatan et al. 2015; Li et al. 2017).

5.3 Crop Improvement through Genetic Engineering Techniques

Strategies for multigene transformation genetic engineering of complex metabolic pathways depend on expression of target genes, as well as restricted metabolic flux pathway, also ensures by entire genes in a pathway. Different strategies have been developed for this purpose including traditional plant-breeding methods (sexual gametic exchange and retransformation) and co-transformation (using physical, chemical and natural transformation techniques), and also multigene vector transformation (single T-DNA region linked with multiple gene expression cassettes), polycistronic transgenes and plastid transformation (Dafny-Yelin and Tzfira 2007; Bock 2013; Farre et al. 2014). Multigene vector transformation has the most promising advantages among all the other strategies (Dafny-Yelin et al. 2007), which makes the multiple transgenes become integrated and inherited as a single unit. Single T-DNA region of binary vectors having multiple target genes assembly of binary vectors is required for Agrobacterium-mediated transformation, or in single-plasmid vectors, for other transformation methods. Another effective method is plastid transformation which requires transferring multiple genes into the plastid genomes and used for certain purposes. Therefore, multigene vector transformation system has numerous advantages and this approach is more widely used in crop improvement related projects.

5.3.1 Plant Multigene Transformation Vector Systems

Natural vector system (Agrobacterium binary vectors) for transferring the T-DNA is an important tool for plant genetic transformation. Single-binary vector has a significant advantage for delivering the multiple genes rather than the application of the other multigene transformation strategies as mentioned above. All genes assembled into the multigene vector T-DNA region by single transformation can be integrated simultaneously into a single chromosomal site of the host plant genome which is further inherited together. Due to this advantage one has to select or identify only a small number of ideal transgenic events to produce transgenic with desirable trait or set of genomes. It's a challenging and tedious task to assemble and transform a construct carrying multiple genes for large transformation using multigene vector system by applying a number of approaches to deliver and stack a number of genes into individual plants. The first step is to learn the metabolic synthesis pathways and to understand the key nodes in the natural organisms, for the reconstruction of the metabolic pathways using molecular biology techniques to assemble, transfer and design the artificial pathways (Figure 5.2) into the model organisms, and to

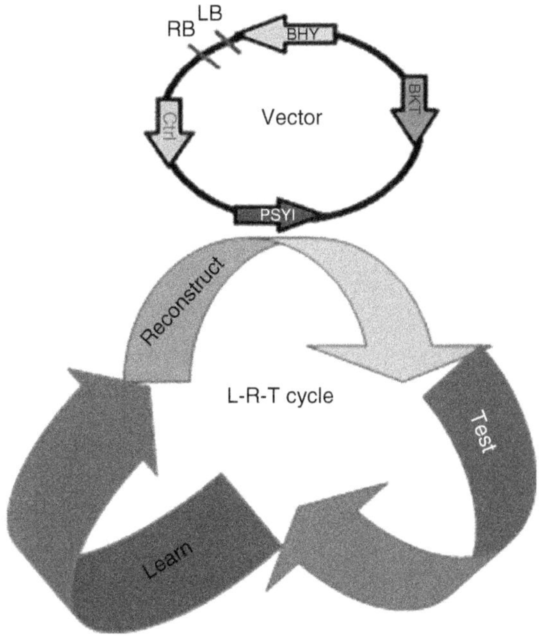

Figure 5.2 Diagrammatic representation about Learn/Reconstruct/Test cycle.

check the efficacy of the products formed in reconstructed pathways following stable transformation in target organisms.

a) Learn: metabolic pathway to understand role of genes and enzyme
b) Reconstruct: metabolic pathways by using molecular biology techniques
c) Test: for stable transformation of desirable trait/product.

First, check the original hypothesis is refined with all data processed analyzed and the next cycle can begin to test the new hypothesis. This is done by employing the molecular scissors; the rare cutting homing endonucleases and zinc-finger nucleases (ZFNs) (Zeevi et al. 2012), type IIS Restriction enzymes (REs) (Golden Gate cloning) (Engler et al. 2008), Gateway recombination (Chen et al. 2006; Wakasa et al. 2006), Cre/loxP recombination (Lin et al. 2003; Zhu et al. 2017), homologous recombination in yeast (Shih et al. 2016) and Gibson Assembly (Gibson et al. 2008). Large DNA-carrying capacity and easy manipulation employed by multigene vector systems (e.g. using artificial chromosomes as vector backbone) have more advantages, an example is the recently developed Trans- Gene Stacking II (TGS II) system (Zhu and Qian 2017; Zhu et al. 2017; Shen et al. 2019). Multigene can be cloned into binary plasmid by using restriction enzyme-based assembly of various gene expressions by using rare-cutting restriction enzymes. pAUX/pRCS and pSAT use rare-cutting homing endonucleases are examples of a commonly used vector system (Goderis et al. 2002; Tzfira et al. 2005; Dafny-Yelin and Tzfira 2007), and their upgrade pRCS11.1 vector system using homing endonucleases and ZFNs (Zeevi et al. 2012). These vectors also have limitations such as low overall efficiency and a small number of genes can be inserted or assembled. Due to the low cutting efficiency of the rare-cutting endonucleases and low specificity it is a time-consuming process. More commonly used methods in microbial metabolic engineering that can assemble small fragments of DNA or several genes are the Bio Brick method using isocaudamers (e.g. Xba I and Spe I) and Golden gate-related systems using type IIS restriction enzymes (e.g. BsaI, Esp3I and Bbs I that are hexa cutters and produce palindromic sticky ends) (Knight 2003; Engler et al. 2008; Emami et al. 2013; Sarrion-Perdigones et al. 2013). Plant genomic DNA sequence has frequent presence of these restriction enzyme sites which limits this method in plants (Liu et al. 2013; Ghareeb et al. 2016).

This system contains a transformation-competent artificial chromosome, first transgene-stacking system that uses a combination of the Cre/loxP-mediated recombination system, and two homing endonucleases (I-Sce I and PI-Sce I) were developed (Lin et al. 2003) that can overcome the limitation of the restriction ligation method. This is done by using site-specific recombinase for multigene stacking, mainly including Cre/loxP system and Gateway recombination. This vector system is based on acceptor vector and two donor delivery vectors in which target vector genes are used alternately to recombine with acceptor vector for sequential stacking of multiple genes in an acceptor vector. Utilization of an early vector system for assembling multigene still remains as challenge due to its operational difficulty in using homing endonuclease digestion and linker ligation for the plasmid circularization.

An easy and highly efficient multigene assembly and transformation vector system (TGS II) to combat such a problem was developed (Zhu et al. 2017). It is the only the reported multigene vector system that has a marker gene-free competent, uses the Cre-mediated recombination system with the wild type loxP sites and mutant loxP sites (for irreversible recombination) and can automatically remove backbone of donor vector during the multigene assembly cycles in a special *Escherichia coli* strain in which Cre enzyme is expressed.

Additionally, an acceptor vector that assembled target genes through Gateway reaction is able to select a marker gene-deletion construct to obtain marker-free transgenic plants. Cre/loxP mediated multigene systems also have other systems, i.e. conjugational transfer for multigene assembly using Cre/loxP recombination and Gateway recombination for deletion of the donor vector backbone *in vivo* (Chen et al. 2010) known as Multiple round *in vivo* site specific assembly MISSA system, an *In planta* transgene strategy for recombinase-mediated transformation which requires multirounds of plant transformation (Ow 2011). Experimentally, the TGS II system is simpler and highly efficient for multiple gene assembly, thus more suitable for engineering complex metabolic pathways (Zhu and Qian 2017; Fang et al. 2018). The system was successfully used previously to synthesize anthocyanins, canthaxanthin, and astaxanthin in rice endosperm by stacking and transferring 10 foreign genes (31 kb) involved in biosynthesis of anthocyanin, 4 and 6 ketocarotenoid biosynthetic foreign genes (15–19 kb), respectively, into rice (Zhu et al. 2017,2018). New rice germplasm with high light efficiency and productivity was developed by using the TGS II system to stack multiple genes of photorespiration (Shen et al. 2019).

Though metabolic engineering proves as useful tool for enhancement of carotenoid biosynthesis, the transgenic lines can induce unpredictable results with undiscovered feedback mechanisms. Further, legal requirements and public acceptability is one of the major issues for commercialization of genetically engineered crops in various countries. Though by consuming only 75 g of Golden Rice II grains one can fulfill the daily requirement of β -carotene, unfortunately, it has not yet been released in any country (Paine et al. 2005).

Lutein is the major carotenoid in staple cereal wheat; it accounts for 80–90% of total carotenoids. Whereas small amounts of zeaxanthin, a-carotene, β-cryptoxanthin, and β-carotene are also present (Abdel-Aal et al. 2007; Digesù et al. 2009). Carotenoids are variably present in seed. Endosperm contains higher leutin whereas the outer layer of kernel consists of zeaxanthin and β-carotene (Hentschel et al. 2002; Borrelli et al. 2008). Carotenoid presence is very low in cultivated wheat as compared to its primitive wild relatives, landraces and synthetic hexaploid such as einkorm ($2n = 14$); Khorasan and durum wheat ($2n = 28$) contain a higher amount of leutin (5.4–7.4 mg/g) as compared to cultivated wheat (1.9 mg/g) (Hidalgo et al. 2006).

Doubled-haploid population of bread wheat was used for carotenoid biosynthesis study during grain development (Howitt et al. 2009). At the early stages of grain development, higher levels zeaxanthin, antheraxanthin, and violaxanthin were present than lutein. Lutein and zeaxanthin amounts were detected most at 10 days post-anthesis (DPA), and remained significantly unchanged during endosperm development, carotenoids such as zeaxanthin, antheraxanthin, and violaxanthin declined gradually and were undetectable in mature grains.

Provitamin A biofortification in crops, give rise to "golden" crops (Giuliano 2017). Use of the classical transgenic approach based on biofortification technology helps to develop a higher level of provitamin in crops like "Golden rice" (Potrykus 2012) and sorghum (Lipkie et al. 2013). Carotenoid biosynthesis and metabolism in the plants provide the understanding that has been dissected and understood at the substantial level. CrtB is a bacterial phytoene synthase gene, whose overexpression and upregulation seems to be a beneficial strategy to increase high provitamin A content into important cereal crops (Giuliano et al. 2008) (Table 5.1). Total carotenoid content of 1.6 μg/g dry weight was found to be increased in

Table 5.1 Major cereal crops, gene used for increasing Vit A content and method used.

Crop	Gene used	Increase in Vit A content	Transformation method	References
Rice	*Phytoene synthase* (PSY) from daffodil, *Phytoene desaturase* (CrtI) gene from *Erwinia uredovora, Phytoene synthase* (PSY) from maize, CrtI from the *Erwinia uredovora* psy and lycopene β-cyclase (β-lcy) both from daffodil	1.6 μg/g DW 37 μg/g DW 1.6 μg/g DW	Agrobacterium tumefaciens	Ye et al. (2000) and Paine et al. (2005)
Wheat	maize psy1 gene encoding *Phytoene synthase*, Bacterial crtI, CrtB or CrtI	4.96 μg/g DW 3.21 μg g/1 of seed DW	Bombardment method	Cong et al. (2009)
Maize	Bacterial crtB and crtI psy1 (maize)	9.8 μg/g DW 59.32 μg/g DW	Bombardment method	Naqvi et al. (2009) and Aluru et al. (2011)

[a] DW- Dry weight.

rice endosperm by over-expression of the daffodil phytoene PSY gene coupled with bacterial crtI gene and/or the daffodil LCYB gene (Ye et al. 2000). Total 37 μg/g carotenoid is increased in Golden rice-2 by the expression of maize PSY1which is replaced by the daffodil gene (Paine et al. 2005). Similarly, β-carotene in the maize endosperm increased up to 34-fold by the overexpression of bacterial crtB and crtI genes under the control of a "super γ-zein promoter" (Aluru et al. 2008). In wheat endosperms, a 10.8-fold increase in the total carotenoids was obtained by transformation with maize y1 gene encoding phytoene synthase driven by an endosperm-specific 1Dx5 promoter together with the bacterial phytoene desaturase crtI gene under the control of constitutive promoter CaMV 35S (Cong et al. 2009). Co-transformation of the bacterial CrtB and CrtI genes enhanced the provitamin A level (Wang et al. 2014).

Potato, canola, tomato, carrot, and cauliflower were also engineered for higher β-carotene content (Shahriari et al. 2013). In cereal grains, carotenoid become degraded during storage that is mediated by oxygen and lipoxygenase enzyme (Carrera et al. 2007; Che et al. 2016). Upsurge carotenoid stability during storage was observed in genetic engineering to enhance accumulation of vitamin E through downregulation of enzyme lipoxygenease (Gayen et al. 2015). However, single-point interventions in genetic engineering to increase vitamin availability generally do not seem to improve the yields of the target vitamin (Zorrilla-Lopez et al. 2013). To overcome this challenge, where genes encoding pathway enzymes are modified and over expressed, multiple transgene expression and combinatorial engineering approaches can be used to improve complex traits in crops (Zhu et al. 2008; Farre et al. 2012). Earlier improvement of complex traits in rice, tomato and potato was done by inserting 2–3 transgenes cloned in a single Agrobacterium construct. Increased level of ascorbate, β-carotene, and folate in elite inbred transgenic corn plants was obtained by (Naqvi et al. 2009) specifically in the endosperm by modification of three separate metabolic pathways simultaneously. They recorded 169-fold increased amount of β-carotene, 6-fold in ascorbate, and double the normal amount of folate in transgenic kernels.

Recently, in other crops like canola and Lilium Agrobacterium, mediated transformation was done by assembling up to seven different bacterial carotenoid genes (from isopentenyl pyrophosphate isomerase (IPI) to CrtW (Giuliano 2014). Combined expression of ZmPSY1, PaCRTI with AtDXS or AtOR with large carotenoid accumulation in rice endosperm hints that the supply of isoprenoid precursors and metabolic sink are imperative rate-limiting steps in carotenoid biosynthesis (Bai et al. 2016). Nuclear transformation, along with minipathway engineering, is an alternative and novel method for the rapid production of multiplex-transgenic plants that were used to dissect and modify a complex metabolic pathway (Zhu et al. 2008). This allows metabolic library generation for investigation of metabolite of interest. Zhu et al. (2008) used this technology in maize to overexpressed Zmpsy1 and found an increase in carotenoid levels elevated 53-fold in total endosperm, whereas expression of two sequential enzymes (ZmPSY1 and PaCRTI in Ph-3) results in a three times increase in carotenoid levels.

Gene silencing can also be used to increase provitamin content in the food crop by redirecting metabolic flux toward carotenoids. Zhang et al. (2015) suggested that gene silencing (block strategy) in combination with overexpression (push strategy) can alter genome to address vitamin A deficiency. Transgenic wheat that accumulated β-carotene up to 5.06 μg/g was developed through simultaneous application of push (CrtB) and block strategy (carotenoid hydroxylase) to enhance and arrest the metabolic flux, respectively. A critical role is played by carotenoids in animal and human health. Higher living beings such as animals and humans are unable to synthesize carotenoids de novo, and get this from their diet which are sources of these compounds.

Major staple cereal grains often contain small amounts of carotenoids. Therefore, there is considerable interest for genetic manipulation of carotenoid content in cereal grain. Significant progress has been made in three aspects:

1) Carotenoid regulation in non-green plant tissues is controlled by seven carotenogenes, including phytoene synthase, β-cyclase, and ε-cyclase controlling biosynthesis, 1-deoxyxylulose-5-phosphate synthase influencing isoprenoid precursor supply, 1-hydroxy-2-methyl-2-(E)-butenyl 4-diphosphate reductase and carotenoid cleavage dioxygenases (CCD) responsible for degradation, and orange gene-conditioning sequestration sink.
2) Metabolic engineering or marker-assisted breeding were used to develop provitamin A-biofortified crops, such as rice and maize.
3) In wheat quantitative trait loci for carotenoid, content was identified on chromosomes 3B, 7A, and 7B along with 23 loci and eight carotenogenes, and 10 gene-specific markers for carotenoid accumulation that were used in wheat improvement.

Carotenoid metabolism comprehensive and deeper understanding helps in regulatory mechanisms in crops that will be beneficial for improving carotenoid content. New and emerging techniques for genomic selection and gene for transformation for provitamin A biofortification can play a critical role to combat nutritional deficiencies.

5.4 Improvement of Carotenoid in Grain Crops through CRISPR/Cas9

Genome editing is a technique in which DNA is inserted, deleted, modified or replaced in the genome of an organism. Genome editing is made possible due to engineered nucleases like meganucleases, ZFNs, transcription-activator-like effector nucleases (TALENs) and CRISPR associated protein-9 (Cas-9) (Stella and Montoya 2016).

The plant transformation tools used to incorporate gene of interest at random sites in the genome of a plant result in the yield penalties and other undesirable changes (Kamthan et al. 2016). Some regions of chromosome can integrate the gene of interest without any undesirable changes and penalties in the host; these regions are called as genomic safe harbors (GSH) (Papapetrou and Schambach 2016). By using CRISPR/Cas9, it is possible to integrate the gene of interest in GSH by using sequence-specific nucleases, e.g. Cas9 and a guide RNA (gRNA). Genome-editing technologies like CRISPR/Cas9 for reverse genetics has opened new possibilities of research in the life sciences, including grain-quality improvement. Currently, CRISPR/Cas9 technology is widely used by researchers for genome editing to attain the preferred objectives, because of its simple targeting. Many genes, related to various aspects of grain quality, have been successfully edited via CRISPR/Cas9 technology over the past few years (Svitashev et al. 2015; Dong et al. 2020). Dong et al. (2020) report the use of an optimized CRISPR-Cas9-based method to insert 5.2 kb carotenoid biosynthesis cassette in rice. Marker-free rice plant with increased carotenoid content in seed was obtained with no yield penalty. It demonstrates that marker-free targeted gene insertion is possible by using CRISPR-Cas9 genome editing. In this study, gene insertion targets in the host plant were selected by analyzing the morphological records and the whole-genome sequencing data of a host plant. CRISPR-PLANT gRNA design platform was used for selection of specific sites near GSH. Golden Rice 2 (GR2) cassette was selected for increasing carotenoid in rice. To reduce the size of the insert, the selectable marker gene and the T-DNA border sequences were not included in the modified cassette. Delivery of the carotenoid cassette into rice at genomic targets was done by particle bombardment. Donor plasmid pAcc-B containing 5.2 kb carotenoid cassette was assembled. Homology arms, which consist of 794 bp and 816 bp of Kitaake genomic sequence to the left and right of the Cas9 cleavage site at the gRNA B target (Target B), respectively. The homology arms were included to facilitate the possibility of homology-directed repair (HDR) CRISPR plasmid pCam1300-CRISPRB was constructed consisting of a Cas9p module. Equal mass of the donor plasmid pAcc-B and the CRISPR plasmid pCam1300-CRISPR-B were mixed and delivered by particle bombardment. Integration of carotenoid cassette was confirmed through PCR.

5.5 Improvement of Carotenoid in Grain Crops Through RNAi

RNAi is a biological process where sequence of RNA is used for inactivation of specific sequence of mRNA thus resulting in inhibition of expression of a particular gene. Accumulation of carotenoid might be affected by enzymes which degrade or convert it. Silencing of these genes could be used for blocking the degradation or conversion of carotenoid. Ko et al. (2018) suppressed 3 CCD genes viz., OsCCD1, OsCCD4a and OsCCD 4b through RNAi and found increased carotenoid content in rice. Maize is used as a staple food in different parts of world, but it contains a low level of pro-vitamin A carotenoid β-carotene. Blocking of hydroxylation reactions which result in production of zeaxanthin from β-carotene through RNAi improve the endosperm β-carotene content (Berman et al. 2017). Two genes viz., *ZmBCH1* and *ZmBCH2* encoding non-heme di-iron carotenoid β-hydroxylases were silenced through RNAi. In all hybrids in which *ZmBCH2 was silenced* β-carotene content of the endosperm increased significantly.

5.6 Future Perspectives and Conclusion

Modern biotechnological tools can be effectively used for improvement of nutritional quality in food crops. Usually, grains are used as staple food in most part of the world. Most of the grain crops are deficient in the β carotene which results vitamin A deficiency related diseases in the human population. Increase in β carotene, which is an important precursor for vitamin A, has been the main focus of the research over the past few years. The main advent with the modern technology is the need for less material for quality improvement in the case of technical application like CRISPR and RNAi. In future, more studies to improve the genetic structure for improvement of quality grains will help the world to lead a healthier life.

References

Abdel-Aal, E.M.S., Young, J.C., Rabalski, I. et al. (2007). Identification and quantification of seed carotenoids in selected wheat species. *Journal of Agricultural and Food Chemistry* 55: 787–794.

Aluru, M., Xu, Y., Guo, R. et al. (2008). Generation of transgenic maize with enhanced provitamin A content. *Journal of Experimental Botany* 59: 3551–3562.

Aluru, M.R., Rodermel, S.R., and Reddy, M.B. (2011). Genetic modification of low phytic acid 1-1 maize to enhance iron content and bioavailability. *Journal of Agricultural and Food Chemistry* 59: 12954–12962.

Bai, C., Capell, T., Berman, J. et al. (2016). Bottlenecks in carotenoid biosynthesis and accumulation in rice endosperm are influenced by the precursor-product balance. *Plant Biotechnology Journal* 14: 195–205.

Berman, J., Zorrilla-Lopez, U., Sandmann, G. et al. (2017). The silencing of carotenoid β-Hydroxylases by RNA interference in different maize genetic backgrounds increases the β-Carotene content of the endosperm. *International Journal of Molecular Sciences* 18 (12): 2515.

Bock, R. (2013). Strategies for metabolic pathway engineering with multiple transgenes. *Plant Molecular Biology* 83: 21–31.

Borrelli, G.M., DeLeonardis, A.M., Platani, C., and Troccoli, A. (2008). Distribution along durum wheat kernel of the components involved in semolina colour. *Journal of Cereal Science* 48: 494–502.

Brinch-Pedersen, H., Borg, S., Tauris, B., and Holm, P.B. (2007). Molecular genetic approaches to increasing mineral availability and vitamin content of cereals. *Journal of Cereal Science* 46: 308–326.

Carrera, A., Echenique, V., Zhang, W. et al. (2007). A deletion at the Lpx-B1 locus is associated with low lipoxygenase activity and improved pasta color in durum wheat (*Triticum turgidum* ssp. durum). *Journal of Cereal Science* 45: 67–77.

Che, P., Zhao, Z.Y., Glassman, K. et al. (2016). Elevated vitamin E content improves all-trans β-carotene accumulation and stability in biofortified sorghum. *Proceedings of the National Academy of Sciences of the United States of America* 113: 11040–11045.

Chen, Q.J., Zhou, H.M., Chen, J., and Wang, X.C. (2006). A Gateway based platform for multigene plant transformation. *Plant Molecular Biology* 62: 927–936.

Chen, Q.J., Xie, M., Ma, X.X. et al. (2010). MISSA is a highly efficient in vivo DNA assembly method for plant multiple-gene transformation. *Plant Physiology* 153: 41–51.

Cong, L., Wang, C., Chen, L. et al. (2009). Expression of phytoene synthase1 and carotene desaturase crtI genes result in an increase in the total carotenoids content in transgenic elite wheat (Triticum aestivum L.). *Journal of Agricultural and Food Chemistry* 57: 8652–8660.

Dafny-Yelin, M. and Tzfira, T. (2007). Delivery of multiple transgenes to plant cells. *Plant Physiology* 145: 1118–1128.

Dafny-Yelin, M., Chung, S.M., Frankman, E.L., and Tzfira, T. (2007). pSAT RNA interference vectors: a modular series for multiple gene down-regulation in plants. *Plant Physiology* 145: 1272–1281.

Digesù, A.M., Platani, C., Cattivelli, L. et al. (2009). Genetic variability in yellow pigment components in cultivated and wild tetraploid wheats. *Journal of Cereal Science* 50: 210–218.

Dong, O.X., Shu, Y., Jain, R. et al. (2020). Marker-free carotenoid-enriched rice generated through targeted gene insertion using CRISPR-Cas9. *Nature Communications* 11: 1178.

Emami, S., Yee, M.C., and Dinneny, J.R. (2013). A robust family of golden gate agrobacterium vectors for plant synthetic biology. *Frontiers in Plant Science* 4: 339.

Engler, C., Kandzia, R., and Marillonnet, S. (2008). A one pot, one step, precision cloning method with high throughput capability. *PLoS One* 3: e3647.

Engler, C., Kandzia, R., and Marillonnet, S. (2008). A one pot, one step, precision cloning method with high throughput capability. *PLoS One* 3: e3647.

Fang, X., Mao, Y., and Chen, X. (2018). Engineering purple rice for human health. *Science China. Life Sciences* 61: 365–367.

Farre, G., Naqvi, S., Sanahuja, G. et al. (2012). Combinatorial genetic transformation of cereals and the creation of metabolic libraries for the carotenoid pathway. In: *Transgenic Plants: Methods and Protocols. Methods in Molecular Biology* (eds. M.D. Jim and C.W. Andy), 419–435. Humana Press.

Farre, G., Blancquaert, D., Capell, T. et al. (2014). Engineering complex metabolic pathways in plants. *Annual Review of Plant Biology* 65: 187–223.

Gayen, D., Ali, N., Sarkar, S.N. et al. (2015). Down-regulation of lipoxygenase gene reduces degradation of carotenoids of golden rice during storage. *Planta* 242: 353–363.

Ghareeb, H., Laukamm, S., and Lipka, V. (2016). COLORFUL-circuit: a platform for rapid multigene assembly, delivery, and expression in plants. *Frontiers in Plant Science* 7: 246.

Gibson, D.G., Benders, G.A., Axelrod, K.C. et al. (2008). One-step assembly in yeast of 25 overlapping DNA fragments to form a complete synthetic *Mycoplasma genitalium* genome. *Proceedings of the National Academy of Sciences of the United States of America* 105: 20404–20409.

Giuliano, G. (2014). Plant carotenoids: genomics meets multi-gene engineering. *Current Opinion in Plant Biology* 19: 111–117.

Giuliano, G. (2017). Provitamin A biofortification of crop plants: a gold rush with many miners. *Current Opinion in Biotechnology* 44: 169–180.

Giuliano, G., Tavazza, R., Diretto, G. et al. (2008). Metabolic engineering of carotenoid biosynthesis in plants. *Trends in Biotechnology* 26: 139–145.

Goderis, I.J., de Bolle, M.F., Francois, I.E. et al. (2002). A set of modular plant transformation vectors allowing flexible insertion of up to six expression units. *Plant Molecular Biology* 50: 17–27.

Hentschel, V., Kranl, K., Hollmann, J. et al. (2002). Spectrophotometric determination of yellow pigment content and evaluation of carotenoids by high-performance liquid chromatography in durum wheat grain. *Journal of Agricultural and Food Chemistry* 50: 6663–6668.

Hidalgo, A., Brandolini, A., Pompei, C., and Piscozzi, R. (2006). Carotenoids and tocols of einkorn wheat (*Triticum monococcum* ssp. Monococcum L.). *Journal of Cereal Science* 44: 182–193.

Howitt, C.A., Cavanagh, C.R., Bowerman, A.F. et al. (2009). Alternative splicing, activation of cryptic exons and amino acid substitutions in carotenoid biosynthetic genes are associated with lutein accumulation in wheat endosperm. *Functional & Integrative Genomics* 9: 363–376.

Kamthan, A., Chaudhuri, A., Kamthan, M., and Datta, A. (2016). Genetically modified (GM) crops: milestones and new advances in crop improvement. *Theoretical and Applied Genetics* 129: 1639–1655.

Knight, T. (2003).Idempotent vector design for standard assembly of Biobricks, DSpace@ MIT. http://dspace.mit.edu/handle/1721.1/21168 1/21168.

Ko, M.R., Song, M.H., Kim, J.K. et al. (2018). RNAi-mediated suppression of three carotenoid-cleavage dioxygenase genes, OsCCD1, 4a, and 4b, increases carotenoid content in rice. *Journal of Experimental Botany* 69 (21): 5105–5116.

Li, Z., Zhang, D., Xiong, X. et al. (2017). A potent Cas9-derived gene activator for plant and mammalian cells. *Nature Plants* 3: 930–936.

Liang, M.H., Zhu, J., and Jiang, J.G. (2018). Carotenoids biosynthesis and cleavage related genes from bacteria to plants. *Critical Reviews in Food Science and Nutrition* 58: 2314–2333.

Lin, L., Liu, Y.-G., Xu, X., and Li, B. (2003). Efficient linking and transfer of multiple genes by a multigene assembly and transformation vector system. *Proceedings of the National Academy of Sciences of the United States of America* 100: 5962–5967.

Lipkie, T.E., De Moura, F.F., Zhao, Z.Y. et al. (2013). Bioaccessibility of carotenoids from transgenic provitamin A biofortified sorghum. *Journal of Agricultural and Food Chemistry* 61: 5764–5771.

Liu, W., Yuan, J.S., and Stewart, C.N. Jr. (2013). Advanced genetic tools for plant biotechnology. *Nature Reviews Genetics* 14: 781–793.

Naqvi, S., Zhu, C.F., Farre, G. et al. (2009). Transgenic multivitamin corn through biofortification of endosperm with three vitamins representing three distinct metabolic pathways. *Proceedings of the National Academy of Sciences of the United States of America* 106: 7762–7767.

Ow, D.W. (2011). Recombinase-mediated gene stacking as a transformation operating system. *Journal of Integrative Plant Biology* 53: 512–519.

Paine, J.A., Shipton, C.A., Chaggar, S. et al. (2005). Improving the nutritional value of golden rice through increased pro-vitamin A content. *Nature Biotechnology* 23: 482–487.

Paine, J., Shipton, C., Chaggar, S. et al. (2005). Improving the nutritional value of golden rice through increased pro-vitamin A content. *Nature Biotechnology* 23: 482–487.

Papapetrou, E.P. and Schambach, A. (2016). Gene insertion into genomic safe harbors for human gene therapy. *Molecular Therapy* 24: 678–684.

Potrykus, I. (2012). "Golden Rice", a GMO-product for public good, and the consequences of GE-regulation. *Journal of Plant Biochemistry and Biotechnology* 21: 68–75.

Rodriguez-Concepcion, M., Avalos, J., Bonet, M.L. et al. (2018). A global perspective on carotenoids: metabolism, biotechnology, and benefits for nutrition and health. *Progress in Lipid Research* 70: 62–93.

Sarrion-Perdigones, A., Vazquez-Vilar, M., Palacı´, J. et al. (2013). GoldenBraid 2.0: a comprehensive DNA assembly framework for plant synthetic biology. *Plant Physiology* 162: 1618–1631.

Shahriari, Z., Heidari, B., Cheraghi, M., and Shahriari, A.G. (2013). Biofortification of staple food crops, engineering the metabolic pathways. *International Research Journal of Applied and Basic Sciences* 5: 287–290.

Shen, B.R., Wang, L.M., Lin, X.L. et al. (2019). Engineering a new chloroplastic photorespiratory bypass to increase photosynthetic efficiency and productivity in rice. *Molecular Plant* 12: 199–214.

Shih, P.M., Vuu, K., Mansoori, N. et al. (2016). A robust gene-stacking method utilizing yeast assembly for plant synthetic biology. *Nature Communications* 7: 13215.

Stella, S. and Montoya, G. (2016). The genome editing revolution: a CRISPR-Cas TALE off target story. *BioEssays* 38 (Suppl. 1): S4–S13.

Svitashev, S., Young, J.K., Schwartz, C. et al. (2015). Targeted mutagenesis, precise gene editing, and site-specific gene insertion in maize using Cas9 and guide RNA. *Plant Physiology* 169: 931–945.

Tzfira, T., Tian, G.W., Lacroix, B. et al. (2005). pSAT vectors: a modular series of plasmids for autofluorescent protein tagging and expression of multiple genes in plants. *Plant Molecular Biology* 57: 503–516.

Wakasa, Y., Yasuda, H., and Takaiwa, F. (2006). High accumulation of bioactive peptide in transgenic rice seeds by expression of introduced multiple genes. *Plant Biotechnology Journal* 4: 499–510.

Wang, C., Zeng, J., Li, Y. et al. (2014). Enrichment of provitamin A content in wheat (Triticum aestivum L.) by introduction of the bacterial carotenoid biosynthetic genes CrtB and CrtI. *Journal of Experimental Botany* 65: 545–556.

Ye, X., Al-Babili, S., Kloti, A. et al. (2000). Engineering the provitamin A (beta-carotene) biosynthetic pathway into (carotenoid-free) rice endosperm. *Science* 287: 303–305.

Ye, X., Al-Babili, S., Kloti, A. et al. (2000). Engineering the provitamin A (b-carotene) biosynthetic pathway into (carotenoid free) rice endosperm. *Science* 287: 303–305.

Zalatan, J.G., Lee, M.E., Almeida, R. et al. (2015). Engineering complex synthetic transcriptional programs with CRISPR RNA scaffolds. *Cell* 160: 339–350.

Zeevi, V., Liang, Z., Arieli, U., and Tzfira, T. (2012). Zinc finger nuclease and homing endonuclease-mediated assembly of multigene plant transformation vectors. *Plant Physiology* 158: 132–144.

Zhai, S.N., Xia, X.C., and He, Z.H. (2016). Carotenoids in staple cereals: metabolism, regulation, and genetic manipulation. *Frontier in Plant Sciences.* 7: 1197. https://doi.org/10.3389/fpls.2016.01197.

Zhang, Y., Butelli, E., Alseekh, S. et al. (2015). Multilevel engineering facilitates the production of phenylpropanoid compounds in tomato. *Nature Communications* 6: 8635.

Zhu, L. and Qian, Q. (2017). Development of "Purple Endosperm Rice" by engineering anthocyanin biosynthesis in endosperm: significant breakthrough in Transgene Stacking System, new progress in rice biofortification (in Chinese). *Chin Bull Bot* 52: 539–542.

Zhu, C., Naqvi, S., Breitenbach, J. et al. (2008). Combinatorial genetic transformation generates a library of metabolic phenotypes for the carotenoid pathway in maize. *Proceedings of the National Academy of Sciences of the United States of America* 105: 18232–18237.

Zhu, Q., Yu, S., Zeng, D. et al. (2017). Development of "purple endosperm rice" by engineering anthocyanin biosynthesis in the endosperm with a high-efficiency transgene stacking system. *Molecular Plant* 10: 918–929.

Zhu, Q., Zeng, D., Yu, S. et al. (2018). From golden rice to aSTARice: bioengineering astaxanthin biosynthesis in rice endosperm. *Molecular Plant* 11: 1440–1448.

Zorrilla-Lopez, U., Masip, G., Arj, G. et al. (2013). Engineering metabolic pathways in plants by multigene transformation. *The International Journal of Developmental Biology* 57: 565–576.

6

CRISPR-Cas9 System for Agriculture Crop Improvement

Ashish Sharma[1,2], *Poorwa Kamal Badola*[1,2], *and Prabodh Kumar Trivedi*[1,2,3]

[1] *CSIR-National Botanical Research Institute, Council of Scientific and Industrial Research (CSIR-NBRI), Lucknow, Uttar Pradesh, India*
[2] *Academy of Scientific and Innovative Research (AcSIR), Ghaziabad, Uttar Pradesh, India*
[3] *CSIR-Central Institute of Medicinal and Aromatic Plants (CSIR-CIMAP), Lucknow, Uttar Pradesh, India*

CHAPTER MENU

6.1 Introduction

Efficient genome-editing technologies are crucial for improving and enhancing the crop production to meet the never-ending challenges of producing sufficient and nutritious food for an exponentially increasing world population (Belhar et al. 2013). Extreme weather and climate change also affect agricultural productivity and yield worldwide. Therefore, it is crucial to ensure future food security and to improve crop productivity. Genetic diversity is of significant importance for trait improvement in plants to produce new and improved, desired varieties (Osakabe et al. 2016). Modern genetic tools and methods allow researchers to be more efficient in manipulating the genetic diversity that will undoubtedly enhance the efficiency and effectiveness of plant breeding. Gene-editing technologies are developing rapidly (Cook et al. 2014). In the recent past, a new breakthrough technology for

Genome Engineering for Crop Improvement, First Edition. Edited by Santosh Kumar Upadhyay.

genome editing, Clustered Regularly Interspaced Short Palindromic Repeats (CRISPR)/Cas9, i.e. CRISPR and CRISPR associated (Cas) endonuclease, have been discovered which, has created milestones in plant molecular biology research. Its ease and simplicity, along with its unique DNA cleaving mechanism, to precisely and efficiently edit and modify the target DNA sequences in the vast array of genome make it most suitable for plant research (Charpentier 2014; Jinek et al. 2012; Soda et al. 2017). This book chapter provides an insight into the recent advances of diverse applications of the CRISPR/Cas9 in plant research and also highlights its potential in crop improvement.

6.2 Genome Engineering

Genome engineering is the ability to create specific changes at the desired genomic sites in complex organism that are of fundamental importance in multiple domains of biological sciences.

6.3 Tools for Genome Engineering

Recent advances in genome-editing technologies are based on programmable nucleases. The emergence of programmable sequence-specific nucleases (SSNs) created a platform in genome manipulation. Though they were demonstrated to be successful gene-editing techniques, many associated drawbacks significantly restrict their application.

SSNs can generate double-stranded breaks (DSBs) at specific chromosomal locations. The resulting DSBs can be repaired by the error-prone non-homologous end joining (NHEJ) pathway, producing insertions, deletions, and point mutations. DSBs can also result in homologous recombination (HR), another independent pathway between chromosomal DNA and foreign donor DNA through the HR pathway (Symington and Gautier 2011).

6.3.1 Meganucleases

Meganucleases are characterized by their ability to recognize and cleave large DNA sequences (12–40 bp) (Souza 2012; Stoddard 2006). However, there are very few chances of finding the exact meganuclease required to act on a specific DNA sequence (Smith et al. 2006).

6.3.2 Zinc-Finger Nuclease (ZFN)

Zinc-Finger Nuclease (ZFNs) comprises of zinc-finger mediated DNA binding domain for DNA recognition and a nuclease-activity domain of FokI for DNA cleavage. ZFNs act through DNA/protein recognition (Kim et al. 1996). The ZFNs exploit FokI endonuclease which has separable recognition and non-specific cleavage domains. To enhance the efficiency, FokI nucleases have been engineered that can only function as heterodimers and have increased catalytic activity. Initially, ZFNs have been successfully employed to modify various plant genomes (Lloyd et al. 2005; Petolino 2015; Townsend et al. 2009; Zhang et al. 2010). Though ZFNs were used to improve agronomic traits in several crop plants, its wider use is restricted as the constructs are costly and not easy to manipulate, owing to its complex technical procedure, which restricts their application in plants (Ramirez et al. 2008).

6.3.3 Transcription Activator-like Effector Nucleases (TALENs)

These are artificial restriction enzymes generated by fusing a specific DNA-binding domain to a non-specific DNA cleaving domain. The DNA binding domains, which can be designed to bind any desired DNA sequence, comes from TAL effectors, DNA-binding proteins excreted by plant pathogenic bacteria Xanthomanos. Plant pathogen Xanthomonas can secrete TALEs upon infection of various host species having DNA-binding domains content a series of 33–35-amino-acid. Similar to ZFNs, the sequence-independent FokI nuclease, found in *Flavobacterium okeanokoites*, functions as the site-specific nuclease for TALEN. (Boch et al. 2009; Boch and Bonas 2010; Christian et al. 2010; Li et al. 2011). The high similarity of TALE recognition sequences, limits their wide use in genome engineering as tedious procedures are required to generate programmable proteins that target the genomic DNA at specific sites.

6.3.4 CRISPR

The CRISPR system is derived from *Streptococcus pyogenes*. It employs a Cas9 nuclease protein to generate the double-strand DNA breaks. The expression of Cas9, along with sgRNA, can introduce site-specific double-strand breaks (DSBs) into the DNA of a living cell from various organisms. CRISPR/Cas9 can induce targeted mutations very efficiently based on base-pairing of the engineered single-guide RNAs (sgRNAs) to the target DNA sites and thus it is much easier to manipulate than ZFNs and TALENs (Jinek et al. 2012).

Compared to ZFNs and TALENs, the CRISPR/Cas system exhibits significantly higher genome-editing efficiencies in a variety of organisms. It has been successfully applied to genome editing in many organisms including plants (Charpentier 2014). CRISPR/Cas system employs use of RNA–DNA base pairing to define the target specificity. It demonstrates key advantages over ZFNs and TALENs such as simplicity, targeting specificity and multiplex ability.

6.3.4.1 The History of CRISPR/Cas System

In 1987, short repeats were first discovered in the genome of *E. coli* K12, which is now recognized as a spacer (Ishino et al. 1987). Later, researchers found that the similar interspaced repeats also existed widely in the genomes of bacteria and the archaea (Mojica et al. 2005). Later, Jansen et al. performed excessive studies on the short, repetitive DNA sequences in prokaryotes, characterized the CRISPR locus and coined the term as the CRISPR. They also identified four CRISPR-associated (Cas) genes, suggesting that CRISPR/Cas system might play a significant role in the biological processes (Jansen et al. 2002; Mojica et al. 2005). In 2005, the association of these elements to the immune system against the invading DNA was revealed (Bolotin et al. 2005; Jansen et al. 2002; Mojica et al. 2005; Pourcel et al. 2005).

In 2007, Barrangou et al. determined that CRISPR-induced immunity was used to safeguard the bacteria from invading phage (Barrangou et al. 2007). In 2008, it found that the CRISPR system could hinder the transfer of exogenous plasmid (Marraffini and Sontheimer 2008). Overall, CRISPR loci have been reported to be found in about 40% of bacterial and in most archaeal species with genomes sequenced (Bhaya et al. 2011). These important findings opened a new avenue to understand the functional mechanism of the CRISPR/Cas system for the advancement of life sciences. In 2010, researchers suggested that spacer sequences guide the Cas9 to cleave target DNA (Garneau et al. 2010). In 2011, it found that a duplex structure was generated by tracrRNA and crRNA in association with Cas9 (Deltcheva et al. 2011). In 2012, it was clearly demonstrated that Cas9 was an RNA-guided endonuclease (Mojica et al. 2005). In 2013, efficient and specific genome editing was successfully achieved for the first time using CRISPR/Cas9 in eukaryotic cells (Cong et al. 2013). Subsequently in 2014, the crystal structure of Cas9 was addressed and the interactions between Cas9 and gRNA and target DNA were characterized (Nishimasu et al. 2014). Recently, a new gene editing system, CRISPR/Cpf1was found, as a single RNA-guided endonuclease lacking tracrRNA. It utilizes a T-rich protospacer-adjacent motif and cleaves DNA via a staggered DNA double-strand break (DSB) (Zetsche et al. 2015).

6.4 CRISPR/Cas Beyond Genome Editing

Target sequence -specific sgRNA fused with Cas9 induces DSBs at a precise position in the genome and facilitates random insertions or deletions by NHEJ (Voytas and Gao 2014). In plants, targeting genes to make it nonfunctional or down-regulated is well established. CRISPR/Cas9 is most frequently utilized for efficient RNA-guided genome editing in various organisms including several plant species, such as Arabidopsis, tobacco, wheat, maize, petunia, soybean, tomato and populus (Cai et al. 2015; Cermak et al. 2015; Fan et al. 2015; Li et al. 2013a,b; Liang et al. 2014; Nekrasov et al. 2017; Upadhyay et al. 2013; Zhang et al. 2016). These reports demonstrated that the CRISPR/Cas9 targeted mutagenesis is efficient and highly specific to the plants. Phytoene desaturase, a gene for carotenoid biosynthesis pathway is mutated by using sgRNA in *Nicotiana benthamiana* (Nekrasov et al. 2013), MILDEW RESISTANT LOCUS O gene (SlMlo1) with double gRNAs in tomato (Nekrasov et al. 2017). Expression of Mlo genes in plants make them susceptible to fungal pathogens causing powdery mildew disease. CRISPR-Cas9 technology prevailed rapid generation of transgene free genetically altered tomato varieties with targeted deletion in SlMlo1 locus (Nekrasov et al. 2017). CRISPR-Cas9 induced targeted mutation in rice transcription factor encoding gene, OsERF922, and enhanced its resistance to rice blast fungal pathogen (Wang et al. 2016). Thermo-sensitive genic male sterile rice has also been raised for hybrid rice production by knocking out OsTMS5 gene using CRISPRCas9 system (Zhou et al. 2016).

6.4.1 CRISPR/Cas9 System for Multiple Gene Editing

In plants, genome multiple gene families are present and editing of multiple genes of a family simultaneously is required for characterization of the gene. Multiple gene editing is achieved by using Cas9 with multiple sgRNAs specific for different target genes (Xie et al. 2015). CRISPR/Cas9-mediated multiple gene editing in plants having their own promoter is reported (Ma et al. 2015; Schiml et al. 2014). This strategy has been successfully demonstrated to produce up to eight sgRNA that cleave eight target sites in DNA (Xie et al. 2015). CRISPR-Cas9 system has been efficiently utilized in conferring powdery mildew resistance to hexaploid bread wheat by simultaneous knock-out of three homoeoalleles encoding

MILDEW-RESISTANCE LOCUS (MLO) proteins (Wang et al. 2014). In a recent report, a chimeric guide RNA (cgRNA) was used for targeting two different genes inositol oxygenase (INOX) and phytoene desaturase (PDS) in wheat (Upadhyay et al. 2013).

6.4.2 CRISPR/Cas9 as Transcriptional Modulation

Recently, it has been shown that the type II-A CRISPR system of *S. pyogenes*, which is known to direct DNA cleavage, could be engineered to repress gene expression (Bikard et al. 2013; Qi et al. 2013). Cas9, have two catalytic sites, an RuvC-like domain and an HNH nuclease domain, each cleaving one strand of the target DNA (Gasiunas et al. 2012; Mojica et al. 2005). Therefore, CRISPR system with modified Cas9, i.e. dCas9 allowed researchers to manipulate not only gene structure but also its regulation. Piatek et al. successfully used modified CRISPR-Cas9 system for modulating gene expression in plants. For down regulation of the target gene, dCas9 was fused with transcriptional repressor domain (Piatek et al. 2015).

The binding of dCas9 to the promoter region of a target gene blocks transcription initiation (Figure 6.1), while binding to the coding strand of the open reading frame was shown to inhibit transcription elongation (Bikard et al. 2013; Qi et al. 2013). Therefore, co-expression of dCas9 and its guide RNA provides a simple way to down expression of any gene. Likewise, for up-regulation of the target gene, modified cas9, i.e. dCas9 can also be used. This was achieved simply by fusing different transcription activator domains (Zhang et al. 2012).

6.4.3 CRISPR/Cas9 as a Visualization Tool

Traditional DNA labeling techniques are a laborious procedure of sample fixation and are not appropriate for live imaging. GFP-tagged CRISPR-dCas9 is one of the most useful tools for fluorescent labeling of native repetitive DNA sequences and visualization of various gene interactions in different chromatin states (Chen et al. 2016; Hu et al. 2014). Fluorescently tagged Cas9 is a potent tool for live-cell-imaging of DNA (Chen et al. 2013). Dreissig et al. reported an imaging technique based on the use of CRISPR-Cas9 system in

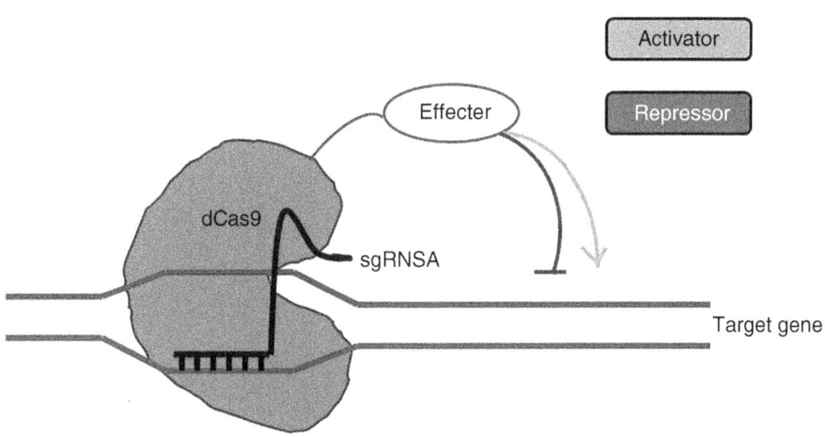

Figure 6.1 dCas9 as a transcriptional regulator: Transcriptional activator or repressor effector can be fuse with modulated Cas9 (dCas9) for target gene regulation.

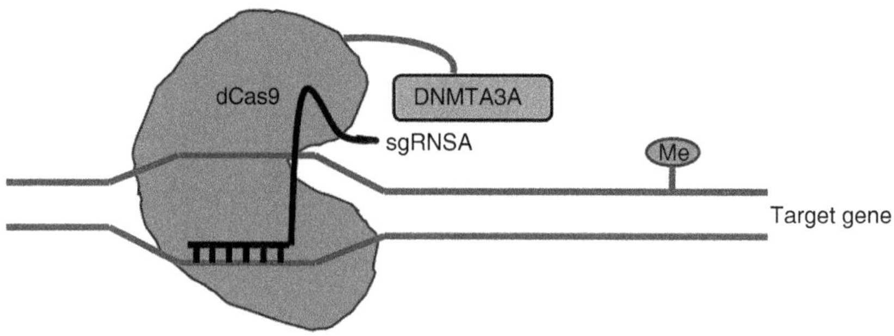

Figure 6.2 dCas9 DNMT3A system for targeted DNA methylation.

Table 6.1 Epigenetic regulation by dCas9.

Modification type	Function	Effect on Gene Expression
Histone Modifications	Acetylation	Activation
	Demethylation	Repression
DNA Methylation/Demthylation	Demethylation	Activation
	Methylation	Repression

studying telomere dynamics during the interphase stage of the cell cycle and also demonstrated the use of this system in visualizing DNA-protein interaction *in vivo* in plants. Nevertheless, potential of CRISPR-Cas9-based imaging technique in plant molecular cytogenetics still needs to be studied further in greater depth (Dreissig et al. 2017).

6.4.4 CRISPR/Cas9 as an Epigenetic Regulator

With the development of genome-editing technologies, we are able to understand how gene is regulated in better way, with the development for epigenome editing now enabling direct study of functional relevance of precise epigenetic modifications and gene regulation (Thakore et al. 2015). It has been reported that a CRISPR/Cas9-based tool was developed for specific DNA methylation using modified Cas9 (dCas9) nuclease and catalytic domain of the DNA methyltransferase DNMT3A targeted by co–expression of a sgRNA (Figure 6.2). These modifications of Cas9 expanded the applications of CRISPR-Cas9 from a gene targeting tool to a more versatile transcription regulatory tool as shown in the Table 6.1, (Vojta et al. 2016; Xu et al. 2016).

6.5 CRISPR/Cas and Crop Improvement

CRISPR is used to produce modified plants, which will hopefully be used to greatly improve agronomical traits. Various research groups independently showed that CRISPR-Cas9 technology can efficiently edit genomes of various plants, such as Arabidopsis, tomatoes, rice, and wheat (Li et al. 2013a,b; Shan et al. 2013).

Rice (*Oryza sativa* L.) is the most important food crop in the world, feeding the majority of the global population. Improvement of grain quality is one of the key targets for rice breeders along with high yield. Many genes responsible for rice grain quality and appearance have been identified and have the potential to be edited with CRISPR/Cas9 technology. Editing of the four yield-related genes Gn1a, DEP1, GS3, and IPA1 genes in the rice offered new avenues for plant-breeding strategies to improve different agronomic traits such as grain number and size, panicle structure indicating the application of this technique for improvement of agronomic traits in rice (Li et al. 2016). In another study, three major genes GW2, GW5, and TGW6 that negatively regulate rice grain weight were simultaneously mutated using the CRISPR/Cas9-mediated multiplex genome editing system to increase seed size by up to 30% in the triple mutant lines (Xu et al. 2016).

Wheat is one of the most widely cultivated crops and reflects a major component of the human diet. Transgene-free low-gluten wheat has also been developed via CRISPR/Cas9-mediated genetic engineering This indicates that CRISPR/Cas9 can be successfully applied to engineer cereal crops for higher yields (Sánchez-León et al. 2018).

To augment grain size and yield, TaGASR7, TaNAC2, TaGW2, and TaDEP1 genes of wheat were targeted and knocked-out using the CRISPR/Cas9 system. Editing of these genes resulted in increased grain weight and grain area along with more grain length grain width as compared to wild type (WT) (Wang et al. 2018). ZmIPK gene of maize that regulates in phytic acid synthesis was edited using CRISPR (Liang et al. 2014). In barley, the endo-N-acetyl-b-D-glucosaminidase (ENGase) gene was knocked out using CRISPR (Kapusi et al. 2017).

In tomato, the number of locules derived from the flower carpels render a major effect on tomato fruit size. Locule number is controlled by numerous quantitative trait loci (QTL), several of which are identified. CRISPR-Cas9 technology has been used to rapidly develop larger tomato by destructing the classical CLAVATA-WUSCHEL (CLV-WUS) stem-cell circuit (Rodríguez-Leal et al. 2017). Developing fruits that exhibit a shelf life without affecting quality and other biological aspects is crucial and one of the major challenges for tomato breeders globally. Studies have shown that the inactivation of ripening inhibitor (RIN) by CRISPR can lead to incomplete ripening with a longer shelf life (Ito et al. 2015). Silencing pectate lyase (PL) and alcobaca (ALC) confers genetic improvement of the tomato by controlling the rate of softening to extend its shelf life without affecting its flavor, suggesting that the CRISPR system has great potential for fruit crop improvement. (Uluisik et al. 2016; Yu et al. 2017). CRISPR/Cas9 was also harnessed to regulate multiple genes associated with the carotenoid metabolic pathway of tomato, to increase the lycopene content in fruits by understanding the roles of specific key genes in carotenoid metabolism (D'Ambrosio et al. 2018). Color reflects another significant aspect that influences consumer perception and preferences for fresh tomatoes. In this context, researchers have successfully cultivated yellow, pink and purple tomatoes by targeting phytoene synthase 1 (PSY1), MYB transcription factor 12 (MYB12), and Anthocyanin 2 (ANT2), respectively (Cermak et al. 2015; Deng et al. 2018). In addition, CRISPR-Cas technology has been utilized in plants for several years and it is continuously employed in new species and targets. For example, powdery mildew locus O (MLO) genes were targeted in tomatoes using CRISPR-Cas9 (Nekrasov et al. 2017), producing acrylamide free potatoes (Halterman et al. 2016), non-browning apples, mushrooms and potatoes by mutating Polyphenol oxidase (PPO) genes (Halterman et al. 2016; Nekrasov et al. 2017; Nishitani et al. 2016) and low phytic acid in maize pioneered the work of targeted genome editing in sweet orange using Cas9/sgRNA (Liang et al. 2014; Wang et al. 2014).

Overall, the CRISPR/Cas9-based genome editing system offers huge possibilities to improve the cereal crops to overcome the adverse effects of climate change and help to strengthen the food security in developing nations. Research done on various crops since the discovery of CRISPR/Cas9 technology in the plant world is highlighted in Table 6.2.

Table 6.2 Plant gene edited using CRISPR/Cas9 for improved agronomically important traits.

Plant	Target gene and function	References
Rice *(Oryza sativa)*	Rice blast disease OsERF922 (ethylene responsive factor transcription factor)	Wang et al. (2016)
	OsWaxy; amylase Synthase	
	Gn1a, DEP1, GS3, and IPA1; regulators of grain	Ma et al. (2015)
	number, panicle architecture	Li et al. (2016)
	OsSWEET; disease susceptibility gene	Zhou et al. (2014)
	Bacterial blight of rice OsSWEET11, OsSWEET14 (rice bacterial blight susceptibility genes)	Jiang et al. (2013)
Maize *(Zea mays)*	Liguleless1 (LIG1) gene, male fertility genes (Ms26 and Ms45), and acetolactate synthase (ALS) genes (ALS1 and ALS2)	Svitashev et al. (2016)
	Argonaute 18, dihydroflavonol 4-reductase	Char et al. (2017)
Wheat *(Triticum aestivum)*	Inositol oxygenase (inox) gene	Upadhyay et al. (2013)
	TaMLO homologs; repress resistance pathway to powdery mildew	Wang et al. (2014)
	Powdery mildew disease TaMLO-A1 (wheat mildew resistance locus1)	Wang et al. (2014)
Potato *(Solanum tuberosum)*	GBSS; starch synthase gene	Andersson et al. (2017)
Tomato *(Solanum lycopersicum)*	ANT1; Anthocyanin Biosynthesis	Cermak et al. (2015)
	SIAGO7; biogenesis of trans-acting short interfering RNAs	Brooks et al. (2014)
Cotton *(Gossypium hirsutum)*	Cloroplastos alterados 1 (GhCLA1) and vacuolar	Chen et al. (2017)
	HC-pyrophosphatase (GhVP) genes, Cotton leaf curl disease CLCuD IR and Rep regions	Iqbal et al. (2016)
Poplar *(Populus tomentosa)*	MYB115; proanthocyanidin biosynthesis	Wang et al. (2017)
Soybean *(Glycine max)*	ALS1; encode acetolactate synthase involved in amino acid Biosynthesis	Li et al. (2015)
Lotus *(Lotus japonicas)*	SNF (symbiotic nitrogen fixation) related genes	Wang et al. (2016)
Camelina *(Camelina sativa)*	FAD2 gene; key enzyme for synthesis of polyunsaturated fatty acids	Jiang et al. (2016)

6.6 Application of Genome Engineering Tools in Metabolic Engineering

Numerous metabolic pathways in plants are known that are responsible for the biosynthesis of complex metabolites. Metabolic engineering is commonly defined as the diversion of one or more enzymatic reactions to generate and alter compounds in plants, and generally improve the production of original metabolite. There are reports of genome engineering for the enhancement in the production of valuable plant-derived metabolites. A number of plant pathways were transformed in heterologous hosts to synthesize valuable metabolites. For example, *Nicotiana tabacum* (tobacco) plants were used to overproduce high value triterpenoids and other molecules using *Saccharomyces cerevisiae* to produce the plant derived highly valuable opiate morphine (Crocoll et al. 2016; Geisler et al. 2013; Polturak et al. 2016). Various efforts on metabolic engineering are on-going to reduce lignin content in plants, as it is the important limiting factor of conversion of biomass to fermentable sugars. A key enzyme in lignin biosynthesis, CCR (cinammoyl-CoA reductase), which catalyzes the change of hydroxycinna-moyl-CoA esters to the respective aldehydes, and studies have suggested that down regulation of CCR causes easy production of biofuel (bioethanol) (Van Acker et al. 2014). In tomato, the level of phenylpropanoid was upregulated by transforming an *Arabidopsis thaliana* transcription factor (AtMYB12), a major regulatory transcription factor, of this pathway (Zhang et al. 2015). In transgenic lines, rutin production enhanced significantly and led to the biopesticidal activity by over expression of AtMYB12 in tobacco (Pandey et al. 2012). Studies have also suggested that co-expression of genes have potential for a role in metabolic engineering. Co-expression of two genes, Arabidopsis MYB12 (AtMYB12) and soybean isoflavone synthase (GmIFS1), in tobacco led to the enhanced biosynthesis of isoflavones and flavonols which have medicinal value (Pandey et al. 2014). Transformation of tomato genes (*SmHMGR)* and/or *SmGGDS genes*, as well as *SmDXS* in *Salvia miltiorrhiza* showed enhancement in the production of tanshinone (Kai et al. 2011). New genome engineering tools like CRISPR/Cas9 genome-editing tools are emerging and, by using this technique, it has already been proven that this technique has a potential role in metabolic engineering which has been shown earlier in many medicinal as well as agronomically important crops. Recently our group has demonstrated that mutation in miR858 or small peptide encoded by the pri-miR858a through CRISPR/Cas9 approach leads to enhanced flavonoid biosynthesis in Arabidopsis (Sharma et al. 2020).

6.7 Future Prospective

CRISPR-Cas9 system has emerged as a breakthrough genome engineering tool in plant-based agriculture research. Continuous manipulation and improvements in CRISPR/Cas9 system has shown its utmost potential beyond the DNA cleavage. Modifications in CRISPR-Cas9 are being explored continuously for novel applications such as RNA cleavage, gene regulation, epigenetic modification and silencing of multiple genes. CRISPR-Cas9 has become a very efficient, simple, and quick tool for developing agronomically improved crops and removal of any undesirable traits. CRISPR/Cas9 system or modified CRISPR/Cas9 therefore provide a convenient and simpler way to generate new varieties of crops with multiple pathogen resistance and other agronomically important traits.

With the pace at which this technology is developing, it is sure that, with its ethical and efficient use, the world can witness a new era of improved crop varieties in all aspects such as like quality, nutrition, pest resistance, etc. One of the best applications of CRISPR-Cas9 system would be realized in terms of marker-free plants with enhanced agronomic traits. Besides the significant role of CRISPR-Cas9 system, there are many questions that need to be answered. An important question is whether the "marker-free plants" are considered as GMOs or normal plants.

References

Doudna, J.A. and Charpentier, E. (2014). Genome editing, the new frontier of genome engineering with CRISPR-Cas9. *Science* 346: 1258096.

Andersson, M., Turesson, H., Nicolia, A. et al. (2017). Efficient targeted multiallelic mutagenesis in tetraploid potato (Solanum tuberosum) by transient CRISPR-Cas9 expression in protoplasts. *Plant Cell Reports* 36 (1): 117–128.

Barrangou, R., Fremaux, C., Deveau, H. et al. (2007). CRISPR provides acquired resistanceagainst viruses in prokaryotes. *Science* 315 (5819): 1709–1712.

Belhaj, K., Chaparro-Garcia, A., Kamoun, S., and Nekrasov, V. (2013). Plant genome editing made easy: targeted mutagenesis in model and crop plants using the CRISPR/Cas system. *Plant Methods* 9: 39.

Bhaya, D., Davison, M., and Barrangou, R. (2011). CRISPR-Cas systems in bacteria and archaea: versatile small RNAs for adaptive defense and regulation. *Annual Review of Genetics* 45: 273–297.

Bikard, D., Jiang, W., Samai, P. et al. (2013). Programmable repression and activation of bacterial gene expression using an engineered CRISPR-Cas system. *Nucleic Acids Research* 41: 7429–7437.

Boch, J. and Bonas, U. (2010). Xanthomonas AvrBs3 family-type III effectors: discovery and function. *Annual Review of Phytopathology* 48: 419–436.

Boch, J., Scholze, H., Schornack, S. et al. (2009). Breaking the code of DNA binding specificity of TAL-type III effectors. *Science* 326: 1509–1512.

Bolotin, A., Quinquis, B., Sorokin, A., and Ehrlich, S.D. (2005). Clustered regularly interspaced short palindrome repeats (CRISPRs) have spacers of extrachromosomal origin. *Microbiology* 151 (8): 2551–2561.

Brooks, C., Nekrasov, V., Lippman, Z.B., and Van Eck, J. (2014). Efficient gene editing in tomato in the first generation using the clustered regularly interspaced short palindromic repeats/CRISPR-associated9 system. *Plant Physiology* 166: 1292–1297.

Cai, Y., Chen, L., Liu, X. et al. (2015). CRISPR/Cas9- mediated genome editing in soybean hairy roots. *PLoS One* 10 (8): e0136064.

Cermak, T., Baltes, N.J., Cegan, R. et al. (2015). High frequency, precise modification of the tomato genome. *Genome Biology* 16: 232–246.

Char, S.N., Neelakandan, A.K., Nahampun, H. et al. (2017). An agrobacterium-delivered CRISPR/Cas9 system for high-frequency targeted mutagenesis in maize. *Plant Biotechnology Journal* 15 (2): 257–268.

Chen, B., Gilbert, L.A., Cimini, B.A. et al. (2013). Dynamic imaging of genomic loci in living human cells by an optimized CRISPR/Cas system. *Cell* 155: 1479–1491.

Chen, B., Hu, J., Almeida, R. et al. (2016). Expanding the CRISPR imaging toolset with Staphylococcus aureus Cas9 for simultaneous imaging of multiple genomic loci. *Nucleic Acids Research* 44: 75.

Chen, X., Lu, X., Shu, N. et al. (2017). Targeted mutagenesis in cotton (Gossypium hirsutum L.) using the CRISPR/Cas9 system. *Scientific Reports* 7: 44304–44311.

Christian, M., Cermak, T., Doyle, E.L. et al. (2010). Targeting DNA double-strand breaks with TAL effector nucleases. *Genetics* 186: 757–761.

Cong, L., Ran, F.A., Cox, D. et al. (2013). Multiplex genome engineering using CRISPR/Cas systems. *Science* 339 (6121): 819–823.

Cook, C., Martin, L., and Bastow, R. (2014). Opportunities in plant synthetic biology. *Journal of Experimental Botany* 65: 1921–1926.

Crocoll, C., Mirza, N., Reichelt, M. et al. (2016). Optimization of engineered production of the glucoraphanin precursor dihomomethionine in Nicotiana benthamiana. *Frontiers in Bioengineering and Biotechnology* 16: 4–14.

D'Ambrosio, C., Stigliani, A.L., and Giorio, G. (2018). CRISPR/Cas9 editing of carotenoid genes in tomato. *Transgenic Research* 27 (4): 367–378.

Deltcheva, E., Chylinski, K., Sharma, C.M. et al. (2011). CRISPR RNA maturation by trans-encoded small RNA and host factor RNase III. *Nature* 471 (7340): 602–607.

Deng, L., Wang, H., Sun, C. et al. (2018). Efficient generation of pink-fruited tomatoes using CRISPR/Cas9 system. *Journal of Genetics and Genomics* 45: 51.

Dreissig, S., Schiml, S., Schindele, P. et al. (2017). Live cell CRISPR imaging in plants reveals dynamic telomere movements. *Plant Journal* 91: 565–573.

Fan, D., Liu, T., Li, C. et al. (2015). Efficient CRISPR/Cas9-mediated targeted mutagenesis in populus in the first generation. *Scientific Reports* 5: 12217.

Garneau, J.E., Dupuis, M.È., Villion, M. et al. (2010). The CRISPR/Cas bacterial immune system cleaves bacteriophage and plasmid DNA. *Nature* 468 (7320): 67–71.

Gasiunas, G., Barrangou, R., Horvath, P., and Siksnys, V. (2012). Cas9-crRNA ribonucleoprotein complex mediates specific DNA cleavage for adaptive immunity in bacteria. *Proceedings of the National Academy of Sciences* 109: 2579–2586.

Geisler, K., Hughes, R.K., Sainsbury, F. et al. (2013). Biochemical analysis of a multifunctional cytochrome P450 (CYP51) enzyme required for synthesis of antimicrobial triterpenes in plants. *Proceedings of the National Academy of Sciences* 110 (35): E3360–E3367.

Halterman, D., Guenthner, J., Collinge, S. et al. (2016). Biotech potatoes in the 21st century: 20 years since the first biotech potato. *American Journal of Potato Research* 93 (1): 1–20.

Hu, J., Lei, Y., Wong, W.K. et al. (2014). Direct activation of human and mouse Oct4 genes using engineered TALE and Cas9 transcription factors. *NucleicAcids Research* 42: 4375–4390.

Iqbal, Z., Sattar, M.N., and Shafiq, M. (2016). CRISPR/Cas9: a tool to circumscribe cotton leaf curl disease. *Frontiers in Plant Science* 7: 475.

Ishino, Y., Shinagawa, H., Makino, K. et al. (1987). Nucleotide sequence of the iap gene, responsible for alkaline phosphatase isozyme conversion in Escherichia coli, and identification of the gene product. *Journal of Bacteriology* 169 (12): 5429–5433.

Ito, Y., Nishizawa-Yokoi, A., Endo, M. et al. (2015). CRISPR/Cas9-mediated mutagenesis of the RIN locus that regulates tomato fruit ripening. *Biochemical and Biophysical Research Communications* 467: 76–82.

Jansen, R., van Embden, J.D.A., Gaastra, W., and Schouls, L.M. (2002). Identification of genes that are associated with DNA repeats in prokaryotes. *Molecular Microbiology* 43 (6): 1565–1575.

Jiang, W., Zhou, H., Bi, H. et al. (2013). Demonstration of CRISPR/Cas9/sgRNA-mediated targeted gene modification in Arabidopsis, tobacco, sorghum and rice. *Nucleic Acids Research* 41 (20): e188–e188.

Jiang, W.Z., Henry, I.M., Lynagh, P.G. et al. (2016). Significant enhancement of fatty acid composition in seeds of the allohexaploid, Camelina sativa, using CRISPR/ Cas9 gene editing. *Plant Biotechnology Journal* 15: 648–657.

Jinek, M., Chylinski, K., Fonfara, I. et al. (2012). A programmable dual-RNA-guided DNA endonuclease in adaptive bacterial immunity. *Science* 337 (6096): 816–821.

Kai, G., Xu, H., Zhou, C. et al. (2011). Metabolic engineering tanshinone biosynthetic pathway in salvia miltiorrhiza hairy root cultures. *Metabolic Engineering* 13 (3): 319–327.

Kapusi, E., Corcuera-Gómez, M., Melnik, S., and Stoger, E. (2017). Heritable genomic fragment deletions and small indels in the putative ENGase gene induced by CRISPR/Cas9 in barley. *Frontiers in Plant Science* 8: 540.

Kim, Y.G., Cha, J., and Chandrasegaran, S. (1996). Hybrid restriction enzymes: zinc finger fusions to Fok I cleavage domain. *Proceedings of the National Academy of Sciences* 93: 1156–1160.

Li, T., Huang, S., Jiang, W.Z. et al. (2011). TAL nucleases (TALNs): hybrid proteins composed of TAL effectors and FokI DNA-cleavage domain. *Nucleic Acids Research* 39: 359–372.

Li, J.F., Norville, J.E., Aach, J. et al. (2013a). Multiplex and homologous recombination-mediated genome editing in Arabidopsis and Nicotiana benthamiana using guide RNA and Cas9. *Nature Biotechnology* 31 (8): 688–691.

Li, D., Qiu, Z., Shao, Y. et al. (2013b). Heritable gene targeting in the mouse and rat using a CRISPR-Cas system. *Nature Biotechnology* 31 (8): 681–683.

Li, Z., Liu, Z., Xing, A. et al. (2015). Cas9-guide RNA directed genome editing in soybean. *Plant Physiology* 169: 960–970.

Li, M., Xiaoxia, L., Zhou, Z. et al. (2016). Reassessment of the four yield-related genes Gn1a, DEP1, GS3, and IPA1 in rice using a CRISPR/Cas9 system. *Frontiers in Plant Science* 7: 377.

Liang, Z., Zhang, K., Chen, K., and Gao, C. (2014). Targeted mutagenesis in Zea mays using TALENs and the CRISPR/Cas system. *Journal of Genetics and Genomics* 41: 63–68.

Liang, Z., Zhang, K., Chen, K., and Gao, C. (2014). Targeted mutagenesis in Zea mays using TALENs and the CRISPR/Cas system. *Journal of Genetics and Genomics* 41: 63–68.

Lloyd, A., Plaisier, C.L., Carroll, D., and Drews, G.N. (2005). Targeted mutagenesis using zinc-finger nucleases in Arabidopsis. *Proceedings of the National Academy of Sciences* 102: 2232–2237.

Ma, X., Zhang, Q., Zhu, Q. et al. (2015). A robust CRISPR/Cas9 system for convenient, high-efficiency multiplex genome editing in monocot and dicot plants. *Molecular Plant* 8: 1274–1284.

Marraffini, L.A. and Sontheimer, E.J. (2008). CRISPR interference limits horizontal gene transfer in staphylococci by targeting DNA. *Science* 322 (5909): 1843–1845.

Mojica, F.J.M., Díez-Villaseñor, C., García-Martínez, J., and Soria, E. (2005). Intervening sequences of regularly spaced prokaryotic repeats derive from foreign genetic elements. *Journal of Molecular Evolution* 60 (2): 174–182.

Nekrasov, V., Staskawicz, B., Weigel, D. et al. (2013). Targeted mutagenesis in the model plant Nicotiana benthamiana using Cas9 RNA-guided endonuclease. *Nature Biotechnology* 31 (8): 691–693.

Nekrasov, V., Wang, C., Win, J. et al. (2017). Rapid generation of a transgene free powdery mildew resistant tomato by genome deletion. *Scientific Reports* 7 (1): 482.

Nekrasov, V., Wang, C., Win, J. et al. (2017). Rapid generation of a transgene free powdery mildew resistant tomato by genome deletion. *Scientific Reports* 28 7 (1): 482.

Nishimasu, H., Ran, F.A., Hsu, P.D. et al. (2014). Crystal structure of Cas9 in complex with guide RNA and target DNA. *Cell* 156 (5): 935–949.

Nishitani, C., Hirai, N., Komori, S. et al. (2016). Efficient genome editing in apple using a CRISPR/Cas9 system. *Scientific Reports* 6 (1): 1–8.

Osakabe, Y., Watanabe, T., Sugano, S.S. et al. (2016). Optimization of CRISPR/Cas9 genome editing to modify abiotic stress responses in plants. *Scientific Reports* 6: 26685.

Pandey, A., Misra, P., Chandrashekar, K., and Trivedi, P.K. (2012). Development of AtMYB12-expressing transgenic tobacco callus culture for production of rutin with biopesticidal potential. *Plant Cell Reports* 31 (10): 1867–1876.

Pandey, A., Misra, P., Khan, M.P. et al. (2014). Co-expression of Arabidopsis transcription factor, at MYB 12, and soybean isoflavone synthase, gm IFS 1, genes in tobacco leads to enhanced biosynthesis of isoflavones and flavonols resulting in osteoprotective activity. *Plant Biotechnology Journal* 12 (1): 69–80.

Petolino, J.F. (2015). Genome editing in plants via designed zinc finger nucleases. *In Vitro Cellular & Developmental Biology. Plant* 51 (1): 1–8.

Piatek, A., Ali, Z., Baazim, H. et al. (2015). RNA-guided transcriptional regulation in planta via synthetic dCas9-based transcription factors. *Plant Biotechnology Journal* 13: 578–589.

Polturak, G., Breitel, D., Grossman, N. et al. (2016). Elucidation of the first committed step in betalain biosynthesis enables the heterologous engineering of betalain pigments in plants. *New Phytologist* 210: 269–283.

Pourcel, C., Salvignol, G., and Vergnaud, G. (2005). CRISPR elements in *Yersiniapestis* acquire new repeats by preferential uptake of bacteriophage DNA, and provide additional tools for evolutionary studies. *Microbiology* 151: 653–663.

Qi, L.S., Larson, M.H., Gilbert, L.A. et al. (2013). Repurposing CRISPR as an RNA-guided platform for sequence-specific control of gene expression. *Cell* 152: 1173–1183.

Ramirez, C.L., Foley, J.E., Wright, D.A. et al. (2008). Unexpected failure rates for modular assembly of engineered zinc fingers. *Nature Methods* 5: 74–375.

Rodríguez-Leal, D., Lemmon, Z.H., Man, J. et al. (2017). Engineering quantitative trait variation for crop improvement by genome editing. *Cell* 171: 470–480.

Sánchez-León, S., Gil-Humanes, J., Ozuna, C.V. et al. (2018). Low-gluten, nontransgenic wheat engineered with CRISPR/Cas9. *Plant Biotechnology Journal* 16: 902–910.

Schiml, S., Fauser, F., and Puchta, H. (2014). The CRISPR/Cas system can be used as nuclease for in planta gene targeting and as paired nickases for directed mutagenesis in Arabidopsis resulting in heritable progeny. *The Plant Journal* 80: 1139–1150.

Shan, Q., Wang, Y., Li, J. et al. (2013). Targeted genome modification of crop plants using a CRISPR-Cas system. *Nature Biotechnology* 31 (8): 686–688. https://doi.org/10.1038/nbt.2650.

Sharma, A., Badola, P., Bhatia, C. et al. (2020). Primary transcript of miR858 encodes regulatory peptide and controls flavonoid biosynthesis and development in Arabidopsis. *Nature Plants* 6: 1262–1274.

Smith, J., Grizot, S., Arnould, S. et al. (2006). A combinatorial approach to create artificial homing endonucleases cleaving chosen sequences. *Nucleic Acids Research* 34 (22): 149.

Soda, N., Verma, L., and andGiri, J. (2017). CRISPR-Cas9 based plant genome editing: significance, opportunities and recent advances. *Plant Physiology and Biochemistry* 131: 2–11.

Souza, N. (2012). Primer: Genome editing with engineered nucleases. *Nature Methods* 9 (1): 27.

Stoddard, B.L. (2005). Homing endonuclease structure and function. *Quarterly Reviews in Biophysics* 38 (1): 49–95.

Svitashev, S., Young, J.K., Schwartz, C. et al. (2016). Targeted mutagenesis, precise gene editing, and site-specific gene insertion in maize using Cas9 and guide RNA. *Plant Physiology* 169: 931–945.

Symington, L.S. and Gautier, J. (2011). Double-strand break end resection and repair pathway choice. *Annual Review of Genetics* 45: 247–271.

Thakore, P.I., D'ippolito, A.M., Song, L. et al. (2015). Highly specific epigenome editing by CRISPR-Cas9 repressors for silencing of distal regulatory elements. *Nature Methods* 12: 1143–1149.

Townsend, J.A., Wright, D.A., Winfrey, R.J. et al. (2009). High frequency modification of plant genes using engineered zinc-finger nucleases. *Nature* 459: 442–445.

Uluisik, S., Chapman, N.H., Smith, R. et al. (2016). Erratum: corrigendum: genetic improvement of tomato by targeted control of fruit softening. *Nature Biotechnology* 34: 1072.

Upadhyay, S.K., Kumar, J., Alok, A., and Tuli, R. (2013). RNA-guided genome editing for target gene mutations in wheat. *G3: Genes, Genomes, Genetics* 3 (12): 2233–2238.

Van Acker, R., Leple, J.C., Aerts, D. et al. (2014). Improved saccharification and ethanol yield from field-grown transgenic poplar deficient in cinnamoyl-CoA reductase. *Proceedings of the National Academy of Sciences* 111: 845–850.

Vojta, A., Dobrini, P., Tadi, V. et al. (2016). Repurposing the CRISPR-Cas9 system for targeted DNA methylation. *Nucleic Acids Research* 44 (12): 5615–5628.

Voytas, D.F. and Gao, C. (2014). Precision genome engineering and agriculture: opportunities and regulatory challenges. *PLoS Biology* 12 (6).

Wang, Y., Cheng, X., Shan, Q. et al. (2014). Simultaneous editing of three homoeoalleles in hexaploid bread wheat confers heritable resistance to powdery mildew. *Nature Biotechnology* 32: 947–951.

Wang, F., Wang, C., Liu, P. et al. (2016). Enhanced rice blast resistance by CRISPR/Cas9-targeted mutagenesis of the ERF transcription factor gene OsERF922. *PLoS One* 11: e0154027.

Wang, L., Ran, L., Hou, Y. et al. (2017). The transcription factor MYB115 contributes to the regulation of proanthocyanidin biosynthesis and enhances fungal resistance in poplar. *New Phytologist* 215 (1): 351–367.

Wang, W., Pan, Q., He, F. et al. (2018). Transgenerational CRISPR-Cas9 activity facilitates multiplex gene editing in allopolyploid wheat. *The CRISPR Journal* 1 (1): 165–174.

Xie, K., Minkenberg, B., and Yang, Y. (2015). Boosting CRISPR/Cas9 multiplex editing capability with the endogenous tRNA-processing system. *Proceedings of the National Academy of Sciences* 112: 3570–3575.

Xu, X., Tao, Y., Gao, X. et al. (2016). A CRISPR-based approach for targeted DNA demethylation. *Cell Discovery* 2: 16009.

Xu, R., Yang, Y., Qin, R. et al. (2016). Rapid improvement of grain weight via highly efficient CRISPR/Cas9-mediated multiplex genome editing in rice. *Journal of Genetics and Genomics* 43: 529.

Yu, Q.H., Wang, B., Li, N. et al. (2017). CRISPR/Cas9-induced targeted mutagenesis and gene replacement to generate long-shelf life tomato lines. *Scientific Reports* 7: 1–9.

Zetsche, B., Gootenberg, J.S., Abudayyeh, O.O. et al. (2015). Cpf1 is a single RNA-guided endonuclease of a class 2CRISPR-Cas system. *Cell* 163 (3): 759–771.

Zhang, F., Maeder, M.L., Unger-Wallace, E. et al. (2010). High frequency targeted mutagenesis in Arabidopsis thaliana using zinc finger nucleases. *Proceedings of the National Academy of Sciences* 107 (26): 12028–12033.

Zhang, J., Rouillon, C., Kerou, M. et al. (2012). Structure and mechanism of the CMR complex for CRISPR mediated antiviral immunity. *Molecular Cell* 45: 303–313.

Zhang, Y., Butelli, E., Alseekh, S. et al. (2015). Multi-level engineering facilitates the production of phenylpropanoid compounds in tomato. *Nature Communications* 6: 8635.

Zhang, Y., Liang, Z., Zong, Y. et al. (2016). Efficient and transgene-free genome editing in wheat through transient expression of CRISPR/Cas9 DNA or RNA. *Nature Communications* 7: 12617.

Zhou, H., Liu, B., Weeks, D.P. et al. (2014). Large chromosomal deletions and heritable small genetic changes induced by CRISPR/Cas9 in rice. *Nucleic Acids Research* 42: 10903–10914.

Zhou, H., He, M., Li, J. et al. (2016). Development of commercial thermo-sensitive genic male sterile rice accelerates hybrid rice breeding using the CRISPR/Cas9-mediated TMS5 editing system. *Scientific Reports* 6: 37395.

7

Contribution of Crop Biofortification in Mitigating Vitamin Deficiency Globally

Siddhant Chaturvedi[1,2], *Roni Chaudhary*[1], *and Siddharth Tiwari*[1]

[1] *National Agri-Food Biotechnology Institute (NABI), Department of Biotechnology, Ministry of Science and Technology (Government of India), Mohali, Punjab, India*
[2] *Department of Biotechnology, Panjab University, Chandigarh, India*

CHAPTER MENU

7.1 Introduction

Vitamins are the group of highly complex organic compounds those which are usually synthesized by plants and microorganisms (Uribe et al. 2017). Vitamins are required in trace amounts for normal metabolism, growth, and development, and their inadequate uptake causes various disorders (Albahrani and Greaves 2016). Vitamin was discovered by the Funk while working on polished rice (Uribe et al. 2017). Based on solubility, vitamins are divided into the water- and fat-soluble groups (Albahrani and Greaves 2016). These groups show a significant difference in physical, chemical, and biological properties. Vitamin B consists of multiple compounds, so it has given to numerical nomenclature.

Genome Engineering for Crop Improvement, First Edition. Edited by Santosh Kumar Upadhyay.

Typically, vitamins are not synthesized by the animal body. Though, some of the vitamins, such as vitamin D (calciferol) and vitamin B3 (niacin) synthesize in the human body from sun exposure and tryptophan protein, respectively (Nair and Maseeh 2012). Vitamins B1 (thiamin), B2 (riboflavin), K2 (menaquinones), and B7 (biotin) are synthesized by the endogenous bacteria, found in the small intestine, but their quantity is not sufficient to fulfill the daily recommended requirements. Therefore, additional dietary intake of vitamins is dependent on other sources (Semba 2012). Vitamin deficiency is associated with groups of diseases and known as "deficiency disease." Malnourishment is a severe health problem in both developing and developed countries. However, it is prevalent in developing countries (FAO and UNICEF 2017). Nearly 794.60 million people in the world are facing the condition of malnutrition, and most of them belong to developing countries (FAO and UNICEF 2017). Besides malnourishment, people are affected by the invisible form of malnutrition, commonly known as micronutrient malnutrition or hidden hunger (Gillespie et al. 2016). Vitamin A deficiency (VAD) alone is reported to affect 250 million people of the developing countries that results in night blindness and a weak immune system (World Health Organization 2019). The large population of the economically weak section in developing countries is dependent on a monotonous diet. They prefer eating staple cereals such as rice, maize, and wheat. The grains of these crop plants contain a meager quantity of vitamins and minerals that can be enhanced by biofortification (White and Broadley 2005).

The term biofortification refers to the process of increasing the nutritional quality of crops. Various techniques such as plant breeding (marker-assisted selection [MAS], quantitative trait loci [QTL]), genetic engineering, and agro-economic intervention are being used for biofortification (Lockyer et al. 2018; Zhu et al. 2007). Biofortification of staple crops is an ideal approach that could provide long-term sustainable and cost-effective solutions to combat malnourishment and micronutrient deficiencies. Different strategies have been adopted for micronutrient biofortification in food crops (White and Broadley 2005). One of the simplest ways to achieve biofortification is by adding inorganic compounds in the form of fertilizers (Lyons et al. 2004). However, this method cannot be extensively used in all crops. In addition, excessive use of fertilizers also deteriorates the soil quality, eventually making the soil sterile (Lyons et al. 2004; Zhu et al. 2007). Conventional breeding is another approach to improve the nutrient content of plants (White and Broadley 2005). The main drawback of this technology is that it cannot be applied to all crops, and the identification of useful traits and breeding of desirable varieties is tedious (Zhu et al. 2007). The genetic-engineering approach, offers a fast way to introduce elite characters in any crop plant. This method gives a chance to introduce a new character into a plant genome. Various crops have been biofortified for vitamins by different approaches; these are highlighted in this chapter.

7.2 Effect of Vitamins on Human Health and Their Sources

Vitamins are considered as micronutrients because they are required in low or trace amounts for normal human metabolism. Vitamins, essential amino acids, minerals and fatty acids are unavoidable for human health as they play important roles in the neurotransmitter production as well as in several metabolic biosynthesis pathways (Castro et al. 2016; El-Ansary et al. 2018; MacFabe et al. 2007; Obrenovich et al. 2015). Insufficient intake of vitamins and minerals has been considered the major contributing factor to various health problems such as anemia, scurvy, rickets and hypothyroidism. These health issues are linked with the deficiency of iron, vitamin C, iodine, and vitamin D (Cannell 2017; Kočovská et al. 2012). It is suggested that global social progress should be linked to provide food security to all populations. It has reduced from 20% in the 1990s to 10% in 2016, but it still persists in a population of more than 800 million worldwide (Food and Agriculture Organization of the United Nations 2017). Indeed, over the last two decades, the production of major crops like wheat, rice, and potato has been significantly increased, but they are insufficient to provide an adequate amount of micronutrients (Food and Agriculture Organization of the United Nations 2017). Anemia has demonstrated the physiological impact of micronutrient malnutrition. It is associated with the deficiencies of micronutrients, such vitamin B1, B9, B12, and iron (Imdad and Bhutta 2012; Stabler 2013). The human body requires several micronutrients that act either as co-enzymes or as an active electron/proton carrier in the breakdown process of the different biomolecules. Vitamins play a major role in energy metabolism and also regulate several enzyme-substrate reactions inside the body (Uribe et al. 2017). The following are the major vitamins that are essential for human health and can be obtained from different dietary sources.

7.2.1 Vitamin A (Retinol)

Vitamin A is a fat-soluble vitamin which includes the nutritionally active form (retinoic acid and retinol) and plant-derived carotenoids. Each form of vitamin A is recognized by a common skeleton, which is known as the retinyl group, comprised of the β-ionone ring and an isoprenoid side chain. Dairy and meat sources primarily contain retinyl esters, which are converted into retinol and retinoic acid. On the other hand, in plants, provitamin A (precursor of vitamin A) is derived from carotenoids, particularly β-carotene. These carotenoids are converted into retinal by enzyme β-β-carotene-15,15′-monooxygenase (BCO) (Chichili et al. 2005). Vitamin A plays an important role in the reproductive system, eye vision, epithelial tissues, and bone formation (Clagett-Dame and Knutson 2011). VAD results in night-blindness, hyperkeratosis, and keratomalacia (Masri et al. 2015). It also inhibits cell growth and differentiation (Kim et al. 2006; Manna and Aggarwal 2000). VAD is the major cause of preventable blindness in developing and low-income countries. The world health organization (WHO) reported that around 250 million preschool children were VAD, and nearly 250 thousand to 500 thousand VAD children become blind (World Health Organization 2019). VAD in different age group populations is also responsible for impaired growth and weakened immunity, which causes susceptibility to many diseases (Harding et al. 2017). Approximately 77.9% of children in developing countries are reported to suffer from autism spectrum disorder (ASD) due to VAD (Yamamoto-Sasaki et al. 2019). The supplementation of vitamin A increases the oxytocin level via the CD38 process pathway, which results in a significant increase in the brain and social activities in ASD patients (Liu et al. 2017). Vitamin A and retinoic acid are required for the proper development of immunity against pathogens by promoting immunoglobulin response and phagocytic function (Masri et al. 2015). The recommended dietary allowance (RDA) of vitamin A varies for different age groups, person, and gender (Table 7.1).

Table 7.1 Recommended dietary allowance (RDA) of vitamin A (retinol equivalents [RE], µg/day).

Group	Subgroup	RDA (µg/day)
Adult man	—	600
Adult women	—	600
Pregnant women	—	800
Lactating women	—	850
Adolescents	10–17 year	600
Children	7–9 year	500
	1–6 year	400
Infants	6–12 month	400
	0–6 month	375

The information is adapted from the Indian Council of Medical Research (ICMR 2009) and modified for presentation. *Source:* Modified from Rao (2010).

7.2.2 Vitamin D (Calciferol)

Vitamin D is a group of sterols, which are usually produced by the body under sunlight. Ergocalciferol (vitamin D2) and cholecalciferol (vitamin D3) are the two most important but inactive forms of vitamin D. The active form of vitamin D is 1,25-dihydroxyvitamin D3 (1,25-$[OH]_2$ D3) (Hilger et al. 2014; Mithal et al. 2009). Vitamin D3 is synthesized from 7-dehydrocholesterol under the influence of ultraviolet light on the skin. In the liver, they both (vitamin D2 and D3) are converted to 25-hydroxycholecalciferol. It is the most abundant and the primary storage form of vitamin D in the liver (Bischoff et al. 1999; Hilger et al. 2014). Finally, the 25-hydroxycholecalciferol is converted to 1,25-dihydroxycholecalciferol (1,25-dihydroxyvitamin D3, or 1,25-$[OH]_2$ D3) in the kidney which is the biologically active form of vitamin D (Mao et al. 2016). Deficiency of vitamin D can cause osteoporosis, osteomalacia, and rickets (Mithal et al. 2009). RDA of vitamin D is given in Table 7.2. Vitamin D can reduce reactive oxygen species (ROS) production by reducing angiotensin II (Dong et al. 2012). Vitamin D is responsible for influencing the synthesis of pro-inflammatory cytokines such as IL-6 and TNF-a adhesion, C-reactive protein), COX-1, and phagocytic receptors such as CD36 (Brewer et al. 2011). Vitamin D increases IkappaBalpha (IκBα) mRNA stability by reducing IκBα phosphorylation (Calton et al. 2015; Janjetovic et al. 2009). Thereby, it may be useful as an anti-inflammatory therapeutic agent.

7.2.3 Vitamin E (Tocopherol)

Vitamin E is a group of compounds that are generated by plants from homogentisic acid. The plant produces eight forms (alpha-, beta-, gamma-, and delta-tocopherol and alpha-, beta-, gamma-, and delta-tocotrienols) of vitamin E. The alpha-tocopherol is the most crucial form of vitamin E to meet human nutritional requirements (Whatham et al. 2008). RDA of vitamin E is differentiated according to age group (Table 7.3). Significant sources of

Table 7.2 Recommended dietary allowance (RDA) for vitamin D.

Group	RDA (IU/day)	25OHD (nmol/l)
1–70 year (male + female)	600	20
Pregnant & lactating women	800	20
71+ year (male + female)	800	20

Source: Adapted and modified from (Ross et al. 2011).

Table 7.3 Recommended dietary allowance (RDA) for vitamin E.

Group	Subgroup	RDA (mg/day)
Adult 14 + (male + female)		15
	Lactating women	19
Children	9–13 year	11
	4–8 year	7
	1–3 year	6
Infants	7–12 months	5
	0–6 months	4

Source: Adapted and modified from the Office of Dietary Supplementary (ODS), National Institute of Health (NIH 2019).

vitamin E are nuts, seeds and vegetable oils. Vitamin E acts as an antioxidant and plays a role in biomembrane formation. Vitamin E also acts as a major lipid-soluble antioxidant and protects plasma and red blood cells against peroxidative damage (Traber and Atkinson 2007). Vitamin E deficiency is responsible for progressive neurological disorders, cardiomyopathy in children and spinocerebellar ataxia (Euch-Fayache et al. 2014; Grimm et al. 2016). Vitamin E deficiency is directly linked with malnutrition. A study in India revealed that a malnourished child greatly suffers from vitamin E deficiency (Traber 2013). Vitamin E deficiency is also associated with obesity (Traber 2013).

7.2.4 Vitamin K (Phylloquinone and Menaquinone)

Vitamin K is a fat-soluble vitamin which carries a chemical structure of 2-methyl 1–4 naphthoquinone. Vitamin K is derived from the German word "Koagulation," which means able to clot blood or prevent hemorrhage (Schwalfenberg 2017). Natural vitamin K found in two forms that are phylloquinone (vitamin K1) and menaquinone (vitamin K2) (Halder et al. 2019; McCabe et al. 2013). The major source of vitamin K1 is derived from plants, such as green leafy vegetables, kiwi, spinach, asparagus, grapes, and avocado. Vitamin K2 is mainly extracted from bacterial and animal sources like dairy products, meat, and fermented foods. Natto is a Japanese dish famous for containing the highest amount of vitamin K2 and is obtained from the fermentation of soybean by *Bacillus natto* (McCabe et al. 2013). Recommended dietary intake for vitamin K is mentioned in Table 7.4.

Table 7.4 Recommended adequate intake (AI) for vitamin K.

Group	Subgroup	AI[a] (mcg/day)[b]
Adult 19+ (male + female)	Male	120
	Female	90
Adolescent	14–18 year (M + F)	75
	10–13 year (M + F)	60
Children	4–9 year	55
	1–3 year	30
Infants	7–12 months	2.5
	0–6 months	2.0

[a] AI is used when insufficient data are available to prove the RDA.
[b] mcg stands for micrograms. *Source:* Adapted and modified from the Office of Dietary Supplementary (ODS), National Institute of Health (NIH 2019).

Prothrombin is a widely known vitamin K-dependent protein that helps in blood clotting (Ufer 2005). Matrix gal-protein is another vitamin K-dependent protein present in bone, cartilage, smooth muscles and involve reducing the abnormal calcification (Schurgers 2013). Osteocalcin protein provides support in mineralization of bones (Coates et al. 2010). In the human body, vitamin K is circulated in the form of lipoprotein in the blood (Erdman Jr et al. 2012). Vitamin K is metabolized very quickly and excreted from the body through urine and feces (Erdman Jr et al. 2012; Shearer and Newman 2008). Vitamin K deficiency leads to osteoporosis and coronary heart disease (Geleijnse et al. 2004). Osteoporosis is a condition with porous and fragile bones and is a major concern in the world, and generally, females are more affected than males (National Institutes of Health n.d.).

7.2.5 Vitamin B Complex

Vitamin B complex is composed of eight (B1, B2, B3, B5, B6, B7, B9, and B12) water-soluble vitamins (Yahia et al. 2019). All forms of vitamin B are synthesized by plants, except vitamin B12 which is synthesized by bacteria and yeast (Kennedy 2014). These vitamins are classified on the basis of their water solubility and cellular co-enzyme function rather chemical and structural similarity (Kennedy 2016). All forms of vitamin B act as a co-enzyme in many enzymatic processes and play a role in many physiological functions (Kennedy 2016). Vitamin Bs are directly involved in different catabolic and anabolic processes (Mock 2007). Vitamin B complex deficiency leads to various disorders, like neurodegenerative disease, Wernicke Korsakoff syndrome, anxiety, psychotic movement, encephalopathy and cognitive decline (Maguire et al. 2014; Tanaka et al. 2004). Source, role in human body and deficiency disorders of different vitamin B are summarized in Table 7.5.

7.2.6 Vitamin C (Ascorbic Acid)

Vitamin C is water-soluble and commonly known as L-ascorbic acid or ascorbate. It is commonly present in several juicy fruit crops. It is considered to be the essential antioxidant in

Table 7.5 Vitamin B complex, dietary source, role, and deficiency disorders.

Vitamin B complex	Other names of vitamin B complex	Preferred source	Role in the human body	Deficiency disorder/disease
Vitamin B1	Thiamine	Nuts, whole grain cereals, organ meats, pork, and liver	Act as co-enzyme precursor for thiamine pyrophosphate, help in nucleic acid and fatty acid synthesis	Wet and dry beriberi, Wernicke Korsakoff syndrome, tingling and numbness
Vitamin B2	Riboflavin	Flesh foods, poultry, green vegetables, nuts, legumes, and dairy products	The precursor of co-enzyme flavin adenine mononucleotide and flavin adenine dinucleotide, necessary for mitochondrial function.	Mucocutaneous lesions of the mouth, angular stomatitis, glossitis, abnormal function of psychomotor, anemia, reduced phagocytosis
Vitamin B3	Niacin or nicotinic acid	Almonds, cape gooseberry, avocado, pulses, milk, and egg	Essential in bioenergetics and redox reactions of metabolism.	Pellagra, vomiting, diarrhea, headache and memory loss
Vitamin B5	Pantothenic acid	Nuts, broccoli, Peas, mushroom, beans, potatoes	The precursor of co-enzyme A, important in lipid and carbohydrate metabolism	Restlessness, fatigue, apathy, numbness, paresthesias, muscle cramps
Vitamin B6	Pyridoxine	fish, beef liver, potatoes, and other starchy vegetables, and fruit (except citrus fruits)	The precursor of co-enzyme pyridoxal phosphate, required in transamination, decarboxylation and deamination reaction	Peripheral neuritis, epileptiform convulsions, anemia, glossitis, and seborrheic dermatitis and neurological degeneration in the child.
Vitamin B7	Biotin	eggs, fish, meat, seeds, nuts, sweet potatoes, sunflower seeds, almonds	Required in the metabolic reaction of deamination of amino acids and decarboxylation reaction	Biotinidase, alopecia, scaly erythematous dermatitis, aciduria, hearing and vision problems, paresthesias, myalgias, and mild depression
Vitamin B9	Folic acid/folate	Green fruits and vegetables, dairy products, meat	The precursor of constituents of RNA and DNA synthesis required in the metabolism of amino acids like tyrosine, glutamic acids, histidine, and choline.	megaloblastic anemia, fatigue, gray hair, mouth sores, tongue swelling growth and development related problems
Vitamin B12	Cobalamin	Meat, fish, poultry, egg, milk and milk product	Act as a cofactor for methionine synthase and L-methylmalonyl-CoA mutase, help in DNA synthesis.	abnormal hemopoiesis, megaloblastic anemia, neurological manifestations and degeneration of the spinal cord

Table 7.6 Recommended dietary allowance (RDA) of Vitamin C (mg/day).

Group	Subgroup	RDA (mg/day)
Adult 19+ (male + female)	Male	40
	Female	40
Adolescent	14–18 year (M + F)	40
	10–13 year (M + F)	40
Children	4–9 year	40
	1–3 year	40
Infants	7–12 months	25
	0–6 months	25

Source: Adapted and modified from the Indian Council of Medical Research (ICMR 2009) (Rao 2010).

extracellular fluids (Padayatty and Levine 2016). Vitamin C has shown the antioxidant property and scavenges various oxidants and free radicals like hydrogen peroxide, hypochlorite, peroxy radicals, hydroxy radical, and $^1O_2$ (Smirnoff 2018). It is synthesized by the plant using L-galactose or galacturonic acid as a precursor, and due to the presence of 2,3-enediol moiety, it shows the antioxidant and acidic properties (Macknight et al. 2017). Vitamin C can be quickly obtained from fruits (citrus, kiwifruits, strawberry) and vegetables (spinach and broccoli) and so forth (Yahia et al. 2019). The RDA of vitamin C varies from nation to nation. In the USA, it is 90 mg/day for males and 75 mg/day for the female, while in India, it is 40 mg/day for both males and females (Macknight et al. 2017). Recommended dietary intake for vitamin C in India is mentioned in Table 7.6. Vitamin C deficiency causes scurvy disease along with anemia. Severe deficiency of vitamin C leads to diseases like anasarca, hemolysis, jaundice, and convulsions (Mayland et al. 2005). Ascorbic acid protects biomembranes from oxidation by enhancing the activities of tocopherol (Macknight et al. 2017).

The various sources are available for different vitamins, and those are summarized below in Table 7.7.

7.3 Plan Biofortification to Overcome Vitamin Deficiency

Biofortification or "biological fortification" is a process by which crops accumulate more nutrition than the normal crops. Biofortification can be achieved by the three methods. (i) Conventional biofortification, (ii) agronomic biofortification, and (iii) transgenic/genetic engineering biofortification (Lockyer et al. 2018). These biofortification methods have specific strengths and limitations (Table 7.8).

Biofortified crops can be grown in the field if they fulfill definite criteria as assigned by the regulatory authority in a specific country. The bioavailability of a specific nutrient should be considered as an important parameter in biofortified crops (Zhu et al. 2007). Micronutrients play various important roles in the human body, and their deficiency can create stunted growth, cognitive impairments and weak growth, which can also lead to morbidity and mortality. Biofortification of staple crops can significantly overcome vitamin deficiency. The plant breeding and genetic engineering approaches are widely used for crop improvement programs and their specific applications in vitamin biofortification are summarized below:

Table 7.7 Most common sources and solubility of different vitamins.

Vitamin	Solubility type	Common sources	References
A	Fat-soluble	*Moringa oleifera*, *Codium fragile*, *Gracilaria chilensis*, many dark green vegetables (broccoli, spinach, turnip greens, Swiss chard), orange vegetables (carrots, pumpkins, sweet potato), orange flesh fruits (apricot, cantaloupe, mango, nectarine, orange, papaya, peach, persimmon, pineapple), red pepper, tomato	(Masri et al. 2015; Yahia et al. 2019)
B complex	Water-soluble	Fruits (green and, citrus fruits), vegetables (spinach, cauliflower, avocados, peas, and beans), algae (*Dunaliella tertiolecta*, *Tetraselmis suecia*, and *Chlorella* spp.)	(Kennedy 2016; Maguire et al. 2014; Tanaka et al. 2004)
C	Water-soluble	Fruits (citrus fruits, guava, pineapples, strawberry, watermelon, cantaloupe and kiwifruit), vegetables (broccoli, cabbage, potato, and tomato), moringa, algae (*Porphyra umbilicalis*), and *Eisenia arborea*	(Macknight et al. 2017; Masri et al. 2015; Yahia et al. 2019)
D	Fat-soluble	Sunlight, the flesh of fatty fish, liver oils, beef liver, cheese and egg yolks	(Mattila et al. 1994; Ovesen et al. 2003)
E	Fat-soluble	Avocado, nuts (almonds, cashew nuts, filberts, macadamia nut, peanuts, pistachio, walnuts), lentils, chickpeas, and green leafy vegetables	(Masri et al. 2015; Ortiz-Monasterio et al. 2007; Yahia et al. 2019)
K	Fat-soluble	Green leafy vegetables and some plant oils, animal-derived foods (butter, eggs yolks, and lard), and by bacteria in the intestinal tract	(Halder et al. 2019; McCabe et al. 2013)

7.3.1 Biofortification Through Breeding

Plant breeding is a traditional and robust method for the improvement of several important traits in different plant species. However, the conventional breeding approach is a tedious process to identify useful traits and breed them into sexually compatible elite cultivars (Kumar et al. 2019). Another pitfall of conventional breeding technique includes the complexity of trait selection (Collard and Mackill 2007; Winter and Kahl 1995). Therefore, DNA-based markers are recognized as an important tool to overcome the challenges of conventional breeding as they provide simpler screening and selection of events even at the seedling stage. Thus, MAS reduced the size of the population for screening, perform in lesser time and thus, hasten the breeding process (Collard and Mackill 2007; Fernie and

Table 7.8 Advantages and disadvantages of major biofortification approaches.

Strength and limitation	Mineral Fertilization	Conventional breeding/ mutagenesis	Transgenic/genetic engineering
Strength: ➔	It is a simple, cost-effective and rapid method.	This method uses the native property of crop.	It is rapid, no constrained of the gene pool. Target/ random integration and expression is possible in any part of the plants, direct application to elite cultivar.
limitation: ➔	It works for minerals but not applicable for organic materials. Application is highly specific to crop and cultivars and unable to target edible parts of the plant.	Dependent on the available gene pool, long-time process, needed introgression of traits only from wild and close related varieties. Intellectual property constraints.	Political and socio-economic issues with transgenic crops need to be addressed, strict regulatory guidelines.

Schauer 2009). Different molecular breeding strategies are being used to enhance vitamin content in staple crops. The maize, carrot and tomato lines were used to identify QTL of carotene and tocopherol content(Santos and Simon 2002; et al. 2004). Mutagenesis has been used to produce lysine-enriched maize lines (Gibbon and Larkins 2005). One of the most promising breeding programs known as HarvestPlus was launched by Consultative Group on International Agricultural Research (CGIAR) along with the International Center for Tropical Agriculture (CIAT) and the International Food Policy Research Institute for biofortification of staple food crops.

7.3.1.1 Food Crop Biofortification by Breeding

The major emphasis of breeding has given in provitamin A biofortification in several staple crop plants. A few studies have also shown the increase of vitamin E and C content in crop plants. In 2005, the Indian Agriculture Research Institute (IARI) released the durum wheat (HI8627) variety with ahigh provitamin A content.

Maize is a staple crop and widely used for feeding animals, industrial processing and as a human food (Sun et al. 2019). A natural variation found in provitamin A content in maize was exploited to develop a number of provitamin A biofortified varieties. Provitamin, A enriched maize varieties such as, GV662A, GV664A, and GV665A (Zambia), Sammaz 38, Sammaz 39 (Nigeria), ZS242 (Zimbabwe), CSIR-CRI Honampa (OPV) (Ghana), have been developed with the help of HarvestPlus biofortification program (Matsungo et al. 2018; Pixley et al. 2013). The provitamin A enriched (21.50 μg/g) maize variety (Vivek QPM9) has also been released by IARI in India (Garg et al. 2018).

Sweet potato is largely produced in developing countries and improved for provitamin A using a breeding approach. Six (Kabode, Naspot 12O, Naspot 13O, Kakamega, Vita, Ejumula,) and three (Twatasha, Kokota, and Chiwoko) varieties of sweet potato with high provitamin A content have been released in Uganda and Zambia, respectively (Garg et al. 2018). In India, provitamin A enriched cauliflower (Pusa Betakesari) and mango (Amarpali, Pusa Pratibha, Pusa, Pusa Arunima, Pusa Surya) have been released for commercial cultivation (Garg et al. 2018).

7.3.2 Biofortification Through Genetic Engineering

The Biofortification of micronutrients through genetic engineering is a convenient approach that can easily be accompanied by the native agronomic traits of plants (Zhu et al. 2007). Genetic engineering through the transgenic method allows crop improvement by metabolic pathway activation or pyramiding of genes (Bock 2013). In contrast to plant breeding, genetic engineering is independent of sexual compatibility issue, and a diverse gene pool can be used as the source for a crop improvement program (Kumar et al. 2019).

The development of "Golden Rice" with high provitamin A content in rice grain is considered as the landmark discovery in the genetic engineering field. The transgenic method is then extended for various micronutrients biofortification in various crops such as maize, wheat, potato, and soybean. For instance, folate (vitamin B9) biofortification is achieved by *GTP-cyclohydroxylase 1* (At *GTPCH 1*) and *aminodeoxy chorismate synthase* (At *ADCS*) modulation in rice and tomato (Diaz de la Garza et al. 2007). Vitamin E biofortification is reported in canola, maize, and soybean (Beyer 2010; Naqvi et al. 2009). The study in Arabidopsis is suggested that vitamin B6 (pyridoxine) level could be modulated by the expression of *pyridoxine biosynthesis* (*PDX1* and *PDX2*) gene (Tambasco-Studart et al. 2007). The multi-vitamins enriched maize was developed by genetic engineering of three metabolic pathways in order to increase the β-carotene, ascorbic acid, and folate content. The vitamin biofortification studies carried out by genetic engineering in selected food crops are described below.

7.3.2.1 Transgenic Rice (*Oryza sativa*)

Golden Rice is a well-established example of biofortification for provitamin A in rice (Bhullar and Gruissem 2013; Ye et al. 2000). The Golden Rice has shown that genetically engineering of carotenoid biosynthesis pathway could be allowed to produce β-carotene in the endosperm of rice (Ye et al. 2000). Naturally, β-carotene synthesis does not take place in rice grain because it lacks some of the pathway enzymes in the endosperm (Bhullar and Gruissem 2013). It was found that the bacterial (*Erwinia uredovora*) derived *carotene desaturase (crtI)* along with *phytoene synthase* (*PSY*) is sufficient to convert geranylgeranyl diphosphate (GGPP) into β-carotene (Burkhardt et al. 1997; Ye et al. 2000). Subsequently, the *PSY* gene from maize, together with bacterial *crtI*, is demonstrated as an efficient combination for up to 31 μg/g β-carotene accumulation in rice endosperm (Tang et al. 2009).

Folate (vitamin B9) biosynthesis requires the parabenzoic acid (PABA) and pterin as precursors. Parabenzoic acid is intermediate of the plastid chorismate pathway, and pterin is derived from guanosine triphosphate (GTP), which is present in cytosol (Storozhenko et al. 2005). Arabidopsis derived *GTP-cyclohydrolaseI* (At *GTPCH I*) and aminodeoxy chorismate synthase (At *ADCS*) were expressed in rice to increase the folate biosynthesis (Diaz de la Garza et al. 2007; Storozhenko et al. 2005). The co-expression of both of these genes has shown 15–100-fold high folate level in different transgenic rice lines as compared to non-transgenic plants. It was also reported that when *ADCS* expressed alone, PABA level increased to 49-fold, but folate level was decreased to six-fold than in the wild type (Storozhenko et al. 2005). It suggested that the feedback mechanism might control the folate synthesis in rice. The other study showed that *ADCS* overexpression in tomato enhanced the folate accumulation (Diaz de la Garza et al. 2007). It was suggested that the 100 g of transgenic rice overexpressing both *ADCS* and *GTPCHI* genes were able to meet the daily requirements for folate (Storozhenko et al. 2005).

7.3.2.2 Transgenic Maize (*Zea mays*)

Maize is one of the most promising staple food crops in developing and developed countries. Different studies have shown an increase in the vitamin and mineral content in maize by using of transgenic approach., Provitamin A level has been increased in maize endosperm by the overexpression of carotenoid pathway genes and bacterial *crtB* genes (Aluru et al. 2008). Vitamin E biofortification in maize endosperm is achieved by overexpression of the *homogentisic acid geranylgeranyl transferase (*HGGT*)* gene, which has increased the tocotrienol and tocopherol levels (Jiang et al. 2017). L- ascorbic acid (vitamin C) has increased in maize by overexpression of *dehydroascorbate reductase (DHAR)* gene, which convert oxidized form into a reduced form of ascorbic acid (Chen et al. 2003). Three distinct metabolic pathways were genetically engineered in maize to enhance multivitamins levels in endosperm (Naqvi et al. 2009). The transgenic maize has enhanced 169-foldβ-carotene, twofold folate and nearly six-fold ascorbic acid than the non-transformed control plant (Naqvi et al. 2009).

7.3.2.3 Transgenic Wheat (*Triticum aesitivum*)

Wheat is an important staple food crop and widely grown in the world. Pro-vitamin A content has been enhanced in wheat by the expression of bacterially derived *crtB* and *crtI* genes (Cong et al. 2009; Wang et al. 2014). Wheat accumulates a very small amount of folate, and as a major crop, it should be enriched with folate (Garg et al. 2018). The co-expression of soybean derived *8gGCHI*, and *ADCS* genes in wheat have shown 2.3-fold enhanced folate in wheat (Garg et al. 2018).

7.3.2.4 Transgenic Soybean (*Glycine max*)

Soybean is a major source of high-quality protein and vegetable oil. Soybean is targeted to increase the β-carotene level by expressing bacteria-derived *crtBI* gene (Schmidt et al. 2015). Other bacterial derived orthologs (*crtB*, *crtw*, *bktI*) of *PSY* gene have been expressed in soybean to enhance the provitamin A content (Pierce et al. 2015). Vitamin E content in soybean is enhanced by the co-expression of *2-methyl-6 phytyl benzoquinol methyltransferase* gene (Van Eenennaam et al. 2003).

7.3.2.5 Transgenic Potato (*Solanum tuberosum*)

Potato is the world's fourth most calorie providing source so that any nutritional enhancement can be of benefit to millions of people globally. Various methods have been used so far to increase the provitamin A content in potato. The β-carotene is increased by the expression of three genes *PSY*, *PDS*, and *lycopene β-cyclase* (Ducreux et al. 2005). The RNAi approach was used to silence the *β-carotene hydroxylase* (BCH) gene that showed an enhanced level of β-carotene in potato (Van Eck et al. 2007). The overexpression of *Or* gene from the orange cauliflower mutant has enhanced the carotenoid content in potato tuber (Garg et al. 2018). The vitamin C enhancement was also reported in potato by overexpression of strawberry derived *D-galacturonic acid reductase (GalUR)* gene (Upadhyaya et al. 2009).

7.4 Conclusion

It is well established that biofortification is a holistic and efficient strategy to enhance the nutritional value of different crop plants. It can significantly contribute to fighting the battle with malnourishment. Different biofortification strategies like agronomic practice, plant breeding and biotechnological approaches have demonstrated their capabilities to enhance the different minerals and vitamins in several crop plants. Biotechnological approaches such as transgenic, cisgenic, and the most recent advancement in genome editing are some of the most feasible tools of crop biofortification that can overcome the limitations of agronomic and plant-breeding strategies. Very recently, we have successfully used CRISPR/Cas9-based genome editing technique to edit *lycopene epsilon-cyclase (LCYε)* gene in the cultivated banana variety and showed enhanced accumulation of β-carotene content that is up to six fold in fruit-pulp as compared to unedited control plants (Kaur et al. 2020). Therefore, the generation of vitamin-enriched food-fruit crops has the great potential to tackle malnutrition-related disorders in millions of people who are mostly living in low-income and developing countries.

Acknowledgments

The authors express gratitude to the Biotechnology Industry Research Assistance Council (BIRAC) for a banana biofortification project grant and to the Department of Biotechnology (DBT), Government of India for the grant (BT/PR25789/GET/119/97/2017) under the scheme of Genome Engineering Technology and Their Applications. The authors would like to acknowledge National Agri-Food Biotechnology Institute (NABI) for research facility, and to the DBT-eLibrary Consortium (Del-CON) for providing access to online journals. SC thankful to Panjab University Chandigarh and RC thankful to Regional Center for Biotechnology (RCB) for PhD registration.

References

Albahrani, A.A. and Greaves, R.F. (2016). Fat-soluble vitamins: clinical indications and current challenges for chromatographic measurement. *Clinical Biochemist Reviews* 37: 27–47.

Aluru, M., Xu, Y., Guo, R. et al. (2008). Generation of transgenic maize with enhanced provitamin A content. *Journal of Experimental Botany* 59 (13): 3551–3562.

Beyer, P. (2010). Golden rice and 'Golden 'crops for human nutrition. *New Biotechnology* 27 (5): 478–481.

Bhullar and Gruissem (2013). Nutritional enhancement of rice for human health: the contribution of biotechnology. *Biotechnology Advances* 31: 50–57.
Bischoff, H.A., Stahelin, H.B., Urscheler, N. et al. (1999). Muscle strength in the elderly: its relation to vitamin D metabolites. *Archives of Physical Medicine and Rehabilitation* 80: 54–58.
Bock, R. (2013). Strategies for metabolic pathway engineering with multiple transgenes. *Plant Molecular Biology* 83 (1–2): 21–31.
Brewer, C.L., D Michos, E., and P Reis, J. (2011). Vitamin D in atherosclerosis, vascular disease, and endothelial function. *Current DrugTargets* 12 (1): 54–60.
Burkhardt, P.K., Beyer, P., Wünn, J. et al. (1997). Transgenic rice (Oryza sativa) endosperm expressing daffodil (Narcissus pseudonarcissus) phytoene synthase accumulates phytoene, a key intermediate Provitamin a biosynthesis. *The Plant Journal* 11: 1071–1078.
Calton, E.K., Keane, K.N., Newsholme, P., and Soares, M.J. (2015). The impact of vitamin D levels on inflammatory status: a systematic review of immune cell studies. *PLoS One* 10 (11): e0141770.
Cannell, J.J. (2017). Vitamin D and autism, what's new? *Reviews in Endocrine & Metabolic Disorders* 18 (2): 183–193.
Castro, E.M., Van Regenmortel, T., Vanhaecht, K. et al. (2016). Patient empowerment, patient participation and patient-centeredness in hospital care: a concept analysis based on a literature review. *Patient Education and Counseling* 99 (12): 1923–1939.
Chen, Z., Young, T.E., Ling, J. et al. (2003). Increasing vitamin C content of plants through enhanced ascorbate recycling. *Proceedings of the National Academy of Sciences of the United States of America* 100: 3525–3530. https://doi.org/10.1073/pnas.0635176100.
Chichili, G.R., Nohr, D., Schäffer, M. et al. (2005). β-Carotene conversion into vitamin a in human retinal pigment epithelial cells. *Investigative Ophthalmology & Visual Science* 46 (10): 3562–3569.
Matsungo, T.M., Musamadya, G., Tagwireyi, J. et al. (2018). A review of the landscape and approach for biofortification initiatives in Zimbabwe. *Central African Journal of Medicine* 64 (7/9): 77–81.
Clagett-Dame, M. and Knutson, D. (2011). Vitamin a in reproduction and development. *Nutrients* 3 (4): 385–428.
Coates, P.M., Betz, J.M., Blackman, M.R. et al. (eds.) (2010). *Encyclopedia of Dietary Supplements*. CRC Press.
Collard, B.C. and Mackill, D.J. (2007). Marker-assisted selection: an approach for precision plant breeding in the twenty-first century. *Philosophical Transactions of the Royal Society, B: Biological Sciences* 363 (1491): 557–572.
Cong, L., Wang, C., Chen, L. et al. (2009). Expression of phytoene synthase1 and carotene desaturase crtI genes result in an increase in the total carotenoids content in transgenic elite wheat (Triticum aestivum L.). *Journal of Agricultural and Food Chemistry* 57 (18): 8652–8660. https://doi.org/10.1021/jf9012218.
Diaz de la Garza, R., Gregory, J.F., and Hanson, A.D. (2007). Folate biofortification of tomato fruit. *Proceedings of the National Academy of Sciences of the United States of America* 104: 4218–4222.
Dong, J., Wong, S.L., Lau, C.W. et al. (2012). Calcitriol protects renovascular function in hypertension by down-regulating angiotensin II type 1 receptors and reducing oxidative stress. *European Heart Journal* 33 (23): 2980–2990.

Ducreux, L.J., Morris, W.L., Hedley, P.E. et al. (2005). Metabolic engineering of high carotenoid potato tubers containing enhanced levels of β-carotene and lutein. *Journal of Experimental Botany* 56 (409): 81–89.

El-Ansary, A., Cannell, J.J., Bjørklund, G. et al. (2018). In the search for reliable biomarkers for the early diagnosis of autism spectrum disorder: the role of vitamin D. *Metabolic Brain Disease* 33 (3): 917–931.

Erdman, J.W. Jr., MacDonald, I.A., and Zeisel, S.H.e. (2012). *Present Knowledge in Nutrition*. Wiley.

Euch-Fayache, G.E., Bouhlal, Y., Amouri, R. et al. (2014). Molecular, clinical and peripheral neuropathy study of Tunisian patients with ataxia with vitamin E deficiency. *Brain* 137 (2): 402–410.

FAO, I. and UNICEF, W. (2017). *The State of Food Security and Nutrition in the World*. Rome, Italy: Food and Agriculture Organization of the United Nations.

Fernie, A.R. and Schauer, N. (2009). Metabolomics-assisted breeding: a viable option for crop improvement? *Trends in Genetics* 25 (1): 39–48.

Food and Agriculture Organization of the United Nations (2017). *The State of Food Security and Nutrition in the World 2017: Building Resilience for Peace and Food Security*. FAO.

Garg, M., Sharma, N., Sharma, S. et al. (2018). Biofortified crops generated by breeding, agronomy, and transgenic approaches are improving lives of millions of people around the world. *Frontiers in Nutrition* 5: 12.

Geleijnse, J.M., Vermeer, C., Grobbee, D.E. et al. (2004). Dietary intake of menaquinone is associated with a reduced risk of coronary heart disease: the Rotterdam study. *The Journal of Nutrition* 134 (11): 3100–3105.

Gibbon, B.C. and Larkins, B.A. (2005). Molecular genetic approaches to developing quality protein maize. *Trends in Genetics* 21 (4): 227–233.

Gillespie, S., Hodge, J., Yosef, S., and Pandya-Lorch, R. (eds.) (2016). *Nourishing Millions: Stories of Change in Nutrition*. Washington, D.C.: International Food Policy Research Institute.

Grimm, M., Mett, J., and Hartmann, T. (2016). The impact of vitamin E and other fat-soluble vitamins on Alzheimer s disease. *International Journal of Molecular Sciences* 17 (11): 1785.

Halder, M., Petsophonsakul, P., Akbulut, A.C. et al. (2019). Vitamin K: double bonds beyond coagulation insights into differences between vitamin K1 and K2 in health and disease. *International Journal of Molecular Sciences* 20 (4): 896.

Harding, K.L., Aguayo, V.M., and Webb, P. (2017). Hidden hunger in South Asia: a review of recent trends and persistent challenges. *Public Health Nutrition* 21: 1–11.

Hilger, J., Friedel, A., Herr, R. et al. (2014). A systematic review of vitamin D status in populations worldwide. *British Journal of Nutrition* 111 (1): 23–45.

ICMR, I. (2009). *Nutrient requirements and recommended dietary allowances for Indians. In A report of the expert group of the Indian Council of Medical Research*. India: National Institute of Nutrition Hyderabad.

Imdad, A. and Bhutta, Z.A. (2012). Maternal nutrition and birth outcomes: effect of balanced protein-energy supplementation. *Paediatric and Perinatal Epidemiology* 26: 178–190.

Janjetovic, Z., Zmijewski, M.A., Tuckey, R.C. et al. (2009). 20-hydroxycholecalciferol, product of vitamin D3 hydroxylation by P450scc, decreases NF-κB activity by increasing IκBα levels in human keratinocytes. *PLoS One* 4 (6): e5988.

Jiang, L., Wang, W., Lian, T., and Zhang, C. (2017). Manipulation of metabolic pathways to develop vitamin-enriched crops for human health. *Frontiers in Plant Science* 8: 937.

Kaur, N., Alok, A., Shivani et al. (2020). CRISPR/Cas9 directed editing of lycopene epsilon-cyclase modulates metabolic flux for β-carotene biosynthesis in banana fruit. *Metabolic Engineering* 59: 76–86.

Kennedy, D.O. (2014). *Plants and the Human Brain*. Oxford University Press.

Kennedy, D.O. (2016). B vitamins and the brain: mechanisms, dose and efficacy – a review. *Nutrients* 8 (2): 68.

Kim, K.N., Pie, J.E., Park, J.H. et al. (2006). Retinoic acid and ascorbic acid act synergistically in inhibiting human breast cancer cell proliferation. *The Journal of Nutritional Biochemistry* 17: 454–62.11.

Kočovská, E., Fernell, E., Billstedt, E. et al. (2012). Vitamin D and autism: clinical review. *Research in Developmental Disabilities* 33 (5): 1541–1550.

Kumar, S., Palve, A., Joshi, C., and Srivastava, R.K. (2019). Crop biofortification for iron (Fe), zinc (Zn) and vitamin a with transgenic approaches. *Heliyon* 5 (6): e01914.

Liu, J., Liu, X., Xiong, X.Q. et al. (2017). Effect of vitamin a supplementation on gut microbiota in children with autism spectrum disorders-a pilot study. *BMC Microbiology* 17 (1): 204.

Lockyer, S., White, A., and Buttriss, J.L. (2018). Biofortified crops for tackling micronutrient deficiencies–what impact are these having in developing countries and could they be of relevance within Europe? *Nutrition Bulletin* 43 (4): 319–357.

Lyons, G.H., Stangoulis, J.C., and Graham, R.D. (2004). Exploiting micronutrient interaction to optimize biofortification programs: the case for inclusion of selenium and iodine in the HarvestPlus program. *Nutrition Reviews* 62 (6): 247–252.

MacFabe, D.F., Cain, D.P., Rodriguez-Capote, K. et al. (2007). Neurobiological effects of intraventricular propionic acid in rats: possible role of short chain fatty acids on the pathogenesis and characteristics of autism spectrum disorders. *Behavioural Brain Research* 176 (1): 149–169.

Macknight, R.C., Laing, W.A., Bulley, S.M. et al. (2017). Increasing ascorbate levels in crops to enhance human nutrition and plant abiotic stress tolerance. *Current Opinion in Biotechnology* 44: 153–160.

Maguire, F., Henriquez, F.L., Leonard, G. et al. (2014). Complex patterns of gene fission in the eukaryotic folate biosynthesis pathway. *Genome Biology and Evolution* 6 (10): 2709–2720.

Manna, S.K. and Aggarwal, B.B. (2000). All-trans-retinoic acid upregulates TNF receptors and potentiates TNF-induced activation of nuclear factors-kappaB, activated protein-1 and apoptosis in human lung cancer cells. *Oncogene* 19: 2110–2119.

Mao, X., Xing, X., Xu, R. et al. (2016). Folic Acid and vitamins D and B12 correlate with homocysteine in Chinese patients with type-2 diabetes mellitus, hypertension, or cardiovascular disease. *Medicine* 95 (6): e2652.

Masri, O.A., Chalhoub, J.M., and Sharara, A.I. (2015). Role of vitamins in gastrointestinal diseases. *World Journal of Gastroenterology: WJG* 21 (17): 5191.

Mattila, P.H., Piironen, V.I., Uusi-Rauva, E.J., and Koivistoinen, P.E. (1994). Vitamin D contents in edible mushrooms. *Journal of Agricultural and Food Chemistry* 42: 2449–2453.

Mayland, C.R., Bennett, M.I., and Allan, K. (2005). Vitamin C deficiency in cancer patients. *Palliative Medicine* 19 (1): 17–20.

McCabe, K., Adams, M., and Holden, R. (2013). Vitamin K status in chronic kidney disease. *Nutrients* 5 (11): 4390–4398.

Mithal, A., Wahl, D.A., Bonjour, J.P. et al. (2009). Global vitamin D status and determinants of hypovitaminosis D. *Osteoporosis International* 20 (11): 1807–1820.

Mock, D.M. (2007). Biotin. In: *Handbook of Vitamins*, 4e (eds. J. Zempleni, R.B. Rucker, D.B. McCormick and J.W. Suttie). Boca Raton, FL, USA: CRC Press.

Nair, R. and Maseeh, A. (2012). Vitamin D: the "sunshine" vitamin. *Journal of Pharmacology and Pharmacotherapeutics* 3 (2): 118–126.

Naqvi, S., Zhu, C., Farre, G. et al. (2009). Transgenic multivitamin corn through biofortification of endosperm with three vitamins representing three distinct metabolic pathways. *Proceedings of the National Academy of Sciences of the United States of America* 106 (19): 7762–7767. https://doi.org/10.1073/pnas.0901412106.

National Institutes of Health (2019).Office of Dietary Supplements. https://ods.od.nih.gov/factsheets (accessed 17 July 2020).

Obrenovich, M.E., Shola, D., Schroedel, K. et al. (2015). The role of trace elements, thiamin (e) in autism and autistic spectrum disorder. *Frontiers in Bioscience* 7: 263–277.

Ortiz-Monasterio, J.I., Palacios-Rojas, N., Meng, E. et al. (2007). Enhancing the mineral and vitamin content of wheat and maize through plant breeding. *Journal of Cereal Science* 46 (3): 293–307.

Ovesen, L., Brot, C., and Jakobsen, J. (2003). Food contents and biological activity of 25-hydroxyvitamin D: a vitamin D metabolite to be reckoned with? *Annals of Nutrition & Metabolism* 47: 107–113.

Padayatty, S.J. and Levine, M. (2016). Vitamin C: the known and the unknown and goldilocks. *Oral Diseases* 22 (6): 463–493.

Pierce, E.C., LaFayette, P.R., Ortega, M.A. et al. (2015). Ketocarotenoid production in soybean seeds through metabolic engineering. *PLoS One* 10 (9): e0138196.

Pixley, K., Palacios-Rojas, N., Babu, R. et al. (2013). Biofortification of maize with provitamin a carotenoids. In: *Editor. Carotenoids and Human Health* (ed. S.A. Tanumihardjo), 271–292. New York: Springer Science.

Rao, B.N. (2010). Nutrient requirement and safe dietary intake for Indians. *Bulletin of the Nutrition Foundation of India* 31 (1).

Ross, A.C., Manson, J.E., Abrams, S.A. et al. (2011). The 2011 report on dietary reference intakes for calcium and vitamin D from the Institute of Medicine: what clinicians need to know. *The Journal of Clinical Endocrinology & Metabolism* 96 (1): 53–58.

Santos, C. and Simon, P. (2002). QTL analyses reveal clustered loci for accumulation of major provitamin a carotenes and lycopene in carrot roots. *Molecular Genetics and Genomics* 268 (1): 122–129.

Schmidt, M.A., Parrott, W.A., Hildebrand, D.F. et al. (2015). Transgenic soya bean seeds accumulating β-carotene exhibit the collateral enhancements of oleate and protein content traits. *Plant Biotechnology Journal* 13 (4): 590–600.

Schurgers, L.J. (2013). Vitamin K: key vitamin in controlling vascular calcification in chronic kidney disease. *Kidney International* 83 (5): 782–784.

Schwalfenberg, G.K. (2017). Vitamins K1 and K2: the emerging group of vitamins required for human health. *Journal of Nutrition and Metabolism* 2017: 6254836.

Semba, R.D. (2012). The discovery of the vitamins. *International Journal for Vitamin and Nutrition Research* 82 (5): 310–315.

Shearer, M.J. and Newman, P. (2008). Metabolism and cell biology of vitamin K. *Thrombosis and Haemostasis* 100 (10): 530–547.

Smirnoff, N. (2018). Ascorbic acid metabolism and functions: a comparison of plants and mammals. *Free Radical Biology and Medicine* 122: 116–129.

Stabler, S.P. (2013). Vitamin B12 deficiency. *New England Journal of Medicine* 368 (2): 149–160.

Storozhenko, S., Ravanel, S., Zhang, G.F. et al. (2005). Folate enhancement in staple crops by metabolic engineering. *Trends in Food Science & Technology* 16 (6–7): 271–281.

Sun, G., Feng, X., Yin, R. et al. (2019). Corn (Zea mays L.): a low methylmercury staple cereal source and an important biospheric sink of atmospheric mercury, and health risk assessment. *Environment International* 131: 104971.

Tambasco-Studart, M., Tews, I., Amrhein, N., and Fitzpatrick, T.B. (2007). Functional analysis of PDX2 from Arabidopsis, a glutaminase involved in vitamin B6 biosynthesis. *Plant Physiology* 144 (2): 915–925.

Tanaka, T., Tateno, Y., and Gojobori, T. (2004). Evolution of vitamin B6 (pyridoxine) metabolism by gain and loss of genes. *Molecular Biology and Evolution* 22 (2): 243–250.

Tang, G., Qin, J., Dolnikowski, G.G. et al. (2009). Golden Rice is an effective source of vitamin a. *The American Journal of Clinical Nutrition* 89: 1776–1783.

Traber, M.G. (2013). Mechanisms for the prevention of vitamin E excess. *Journal of Lipid Research* 54 (9): 2295–2306.

Traber, M.G. and Atkinson, J. (2007). Vitamin E, antioxidant and nothing more. *Free Radical Biology and Medicine* 43 (1): 4–15.

Ufer, M. (2005). Comparative pharmacokinetics of vitamin K antagonists. *Clinical Pharmacokinetics* 44 (12): 1227–1246.

Upadhyaya, C.P., Young, K.E., Akula, N. et al. (2009). Over-expression of strawberry D-galacturonic acid reductase in potato leads to accumulation of vitamin C with enhanced abiotic stress tolerance. *Plant Science* 177 (6): 659–667.

Uribe, N.G., García-Galbis, M.R., and Espinosa, R.M.M. (2017). New advances about the effect of vitamins on human health: vitamins supplements and nutritional aspects. In: *Functional Food: Improve Health through Adequate Food* (ed. M.C. Hueda), 57. InTech.

Van Eck, J., Conlin, B., Garvin, D.F. et al. (2007). Enhancing beta-carotene content in potato by RNAi-mediated silencing of the beta-carotene hydroxylase gene. *American Journal of Potato Research* 84 (4): 331.

Van Eenennaam, A.L., Lincoln, K., Durrett, T.P. et al. (2003). Engineering vitamin E content: from Arabidopsis mutant to soy oil. *The Plant Cell* 15 (12): 3007–3019.

Wang, C., Zeng, J., Li, Y. et al. (2014). Enrichment of Provitamin a content in wheat (Triticum aestivum L.) by introduction of the bacterial carotenoid biosynthetic genes CrtB and CrtI. *Journal of Experimental Botany* 65 (9): 2545–2556. https://doi.org/10.1093/jxb/eru138.

Whatham, A., Bartlett, H., Eperjesi, F. et al. (2008). Vitamin and mineral deficiencies in the developed world and their effect on the eye and vision. *Ophthalmic and Physiological Optics* 28 (1): 1–12.

White, P.J. and Broadley, M.R. (2005). Biofortifying crops with essential mineral elements. *Trends in Plant Science* 10 (12): 586–593.

Whitfield, K.C., Bourassa, M.W., Adamolekun, B. et al. (2018). Thiamine deficiency disorders: diagnosis, prevalence, and a roadmap for global control programs. *Annals of the New York Academy of Sciences* 1430 (1): 3.

Winter, P. and Kahl, G. (1995). Molecular marker technologies for plant improvement. *World Journal of Microbiology and Biotechnology* 11 (4): 438–448.

Wong, J.C., Lambert, R.J., Wurtzel, E.T., and Rocheford, T.R. (2004). QTL and candidate genes phytoene synthase and ζ-carotene desaturase associated with the accumulation of carotenoids in maize. *Theoretical and Applied Genetics* 108 (2): 349–359.

World Health Organization(2019). Quality and regulatory considerations for the use of vitamin A supplements in public health programmes for infants and children aged 6–59 months. Technical Note.

Yahia, E.M., García-Solís, P., and Celis, M.E.M. (2019). Contribution of fruits and vegetables to human nutrition and health. In: *Postharvest Physiology and Biochemistry of Fruits and Vegetables* (ed. E.M. Yahia), 19–45. Woodhead Publishing.

Yamamoto-Sasaki, M., Yoshida, S., Takeuchi, M. et al. (2019). Association between antidepressant use during pregnancy and autism spectrum disorder in children: a retrospective cohort study based on Japanese claims data. *Maternal Health, Neonatology and Perinatology* 5 (1): 1.

Ye, X., Al-Babili, S., Kloti, A. et al. (2000). Engineering the Provitamin a (β-carotene) biosynthetic pathway into (carotenoid-free) rice endosperm. *Science* 287: 303–305.

Zhu, C., Naqvi, S., Gomez-Galera, S. et al. (2007). Transgenic strategies for the nutritional enhancement of plants. *Trends in Plant Science* 12: 548–555. https://doi.org/10.1016/j.tplants.2007.09.007.

8

Genome Editing Approaches for Trait Improvement in the Hairy Root Cultures of the Economically Important Plants

Pravin Prakash[1], *Rakesh Srivastava*[1], *Ajay Kumar*[1], *Gurminder Kaur*[2], *and Praveen Chandra Verma*[1]

[1] *Molecular Biology and Biotechnology Division, Council of Scientific and Industrial Research, National Botanical Research Institute (CSIR-NBRI), Lucknow, Uttar Pradesh, India*
[2] *Institute of Bioscience and Technology, Shri Ramswaroop Memorial University, Barabanki, Uttar Pradesh, India*

CHAPTER MENU

8.1 Introduction

The *in vitro* culture of hairy roots is a highly effective approach for the enhanced production of plant secondary metabolites. The secondary metabolites play a variety of functions during the plant growth and developmental stages. Several plant secondary metabolites are utilized by the pharmaceutical as well as other industries for the production of drugs, cosmetics, dyes, fragrances, flavors, nutrition, and dietary supplements. Therefore, there are enormous scientific curiosities as well as industrial implications of the plant secondary metabolites. However, these secondary metabolites are produced in significantly less amount by the plants, and their biosynthesis is critically influenced by the plant's physiological, growth, and developmental conditions (Isah 2019; Yang et al. 2018). The hairy root cultures provide significant advantages over the traditional plant propagation methods to achieve the enhanced quantity of the desired metabolite in the required time frame. However, there are several limitations, which cause a decline in the production of the metabolite during the *in vitro* hairy root culture process (Eapen and Mitra 2001; Hu and

Genome Engineering for Crop Improvement, First Edition. Edited by Santosh Kumar Upadhyay.

Du 2006; Srivastava et al. 2016). Several genetic manipulation approaches have been used to improve the yield of the desired metabolites through the plant hairy root cultures. The emergence of the genome-editing tools offers significant advantages to genomically edit the hairy root lines to generate the economically desired important metabolites.

The specific targeting of a gene within the plant genome is crucial to dissect its functional aspects and the generation of genetically engineered cells that can serve as biofactories for the production of commercially important metabolites. The precise double-strand breaks (DSBs) at the particular genomic targets and their repair mechanisms form the basis of genome editing. The mutation generally occurs by the DSBs at the DNA-cleavage site, which is mediated by the error-prone non-homologous end-joining pathway. Another effect by the DSBs could be the integration between the external DNA and the chromosomal DNA, which is mediated by the homologous recombination pathway. The precise creation of the DSBs at the target loci by the sequence-specific nucleases, such as the meganucleases (MGNs), the transcription activator-like effector (TALE) nucleases (TALENs), and the zinc-finger nucleases (ZFNs) have proven them as important tools for the targeted genome editing in animals and plants (Gaj et al. 2013). Besides the sequence-specific nucleases, the CRISPR (Clustered Regularly Interspaced Short Palindromic Repeats)/Cas (CRISPR-associated), the CRISPR/Cas system, has recently emerged as another potent genome editing method in different organisms including the plants (Feng et al. 2013; Pennisi 2013; Rai et al. 2019; Segal 2013). Altogether, based on the experimental requirements, any tool among the MGNs, ZFNs, TALENs, and CRISPR/Cas9 could be selected and successfully applied for the targeted genome editing. This book chapter aims to provide insights into the recent advancements and the application of genome engineering strategies in the plant hairy root cultures for the enhanced production of the desired secondary metabolites.

8.2 Secondary Metabolites and Hairy Root Culture: An Insight

On the basis of their functions in the fundamental metabolic processes, the plant metabolites are classified into two groups, viz. the primary and secondary metabolites. Primary metabolites participate in the necessary cellular processes; therefore, these metabolites are comparatively similar in all living cells. The plant secondary metabolites are the output of the secondary pathways and perform diverse functions in plants (Bennett and Wallsgrove 1994; Hussein and El-Anssary 2018; Srivastava et al. 2018). The plant secondary metabolites usually fall into three key categories, viz. alkaloids (contains nitrogen, structurally diverse), phenolics (contains aromatic benzene ring along with one or more hydroxyl groups), and terpenoids (derived from the isopentenyl pyrophosphate units, biosynthesized through the mevalonate and/or non-mevalonate pathways (Guerriero et al. 2018; Hussein and El-Anssary 2018; Srivastava et al. 2018). The alkaloids are generally toxic in nature and exhibit pharmacological activities (Debnath et al. 2018). Terpenes are the biggest group of secondary metabolites, biosynthesized from the acetyl-CoA through the mevalonic acid (MVA), and the methylerythritol 4-phosphate (MEP) biosynthesis pathways. Besides several other roles, the terpenes have been shown to exhibit potent antioxidant properties, which indicates their potential pharmaceutical significance (González-Burgos and Gómez-Serranillos 2012; Tholl 2015). Another group of secondary metabolites, the plant phenolics, besides playing important roles in the plants, also exhibit pharmacological activities (Cheynier et al. 2013).

Secondary metabolites are hydrophilic or lipophilic in nature and, therefore, require different cellular and signaling mechanisms for their transport and storage. Hydrophilic secondary metabolites are generally deposited in the vacuoles, whereas, the lipophilic substances are localized in the resin ducts, trichomes, or the glandular hairs (Pagare et al. 2015). Several pharmaceutically and commercially important secondary metabolites are accumulated in the plant roots (Guillon et al. 2006). However, harvesting of the roots is harmful to the plants; therefore, the hairy root culture is preferred for the production of the secondary metabolites. Hairy roots production is induced through the interaction between a gram-negative soil bacterium *Agrobacterium rhizogenes* (also known as *Rhizobium rhizogenes*), and the host plant. Hairy roots production is the consequence of genetic transformation by *A. rhizogenes*, and the hairy root cultures show various appealing characteristics like high genetic stability, rapid growth rate, and development in the hormone-free culture media. Hairy roots exhibit high growth and branching rate without the use of the exogenous treatment of hormones, and the lack of geotropism (Eapen and Mitra 2001; Guillon et al. 2006).

Further, the hairy root cultures generate the secondary metabolites for prolonged durations, contrasting the natural roots. The genetic stability and the potential of hairy roots cultures mimic the parent plants and make them unique for the biosynthesis of secondary metabolites over the plant suspension and cell/callus cultures. The hairy root cultures are also being developed as a cost-effective adoption for the mass production and expression of the heterologous proteins. Hence, the natural plant organ culturing is switching toward the hairy root culture, and it is being considered as an attractive substitute for the production of various valuable secondary metabolites. To enhance the secondary metabolites accumulation in hairy root culture, media compositions (macro and microelements, growth regulators, and vitamins) and culture conditions should be optimized. Along with these, pH of the medium, temperature, and the light intensity, also plays a crucial role in secondary metabolite production and biomass accumulation (Charlwood and Charlwood 1991; Guillon et al. 2006; Nartop 2018). The medium composition is one of the most important factors for increasing the biomass and the yield of secondary metabolites. The Murashige and Skoog (MS), Nitsch and Nitsch (NN), Linsmaier and Skoog (LS), and Gamborg B5 are the commonly used medium for plant tissue culture. Their effectiveness relies on the composition of major and minor elements, the sucrose content (carbon source), the nitrate and phosphate content, and the plant growth regulators (Nartop 2018). Among the media components, the abiotic or biotic elicitation also affects the secondary metabolites production in the hairy root cultures as the secondary metabolites are mostly the defensive compounds induced after an attack on the plant. The abiotic elicitors are non-biological factors and can be classified into three categories, viz. the chemical, physical, and hormonal elicitors (Nartop 2018; Pandey et al. 2016). Recently, the copper, gold, and zinc nanoparticles have also been used as elicitors.

8.3 Genome Editing Process in Plants

Editing the plant genome exactly at the defined target loci implicates the modification/alteration of the DNA sequence by explicitly creating the DSBs at the site of interest. The DSBs are usually deleterious; however, the organisms have different DNA damage repair mechanisms to maintain the gene functionality. The two fundamental DNA repair mechanisms, i.e. the homologous recombination and the non-homologous end-joining, are employed by the plant cells (Puchta 2005). Accumulating studies have suggested that during the past few years, significant methodological advancements have been made in creating the sequence-specific DSBs in the DNA. The four sequence-specific nuclease-based approaches utilizing the meganucleases, ZFNs, TALENs, and CRISPR/Cas9 have been successfully used for targeting the genomic DNA at specific loci. All four methods are indistinguishable, as each one facilitates the experimental design to accurately recognize and precisely cleave the desired DNA sequence (Figure 8.1).

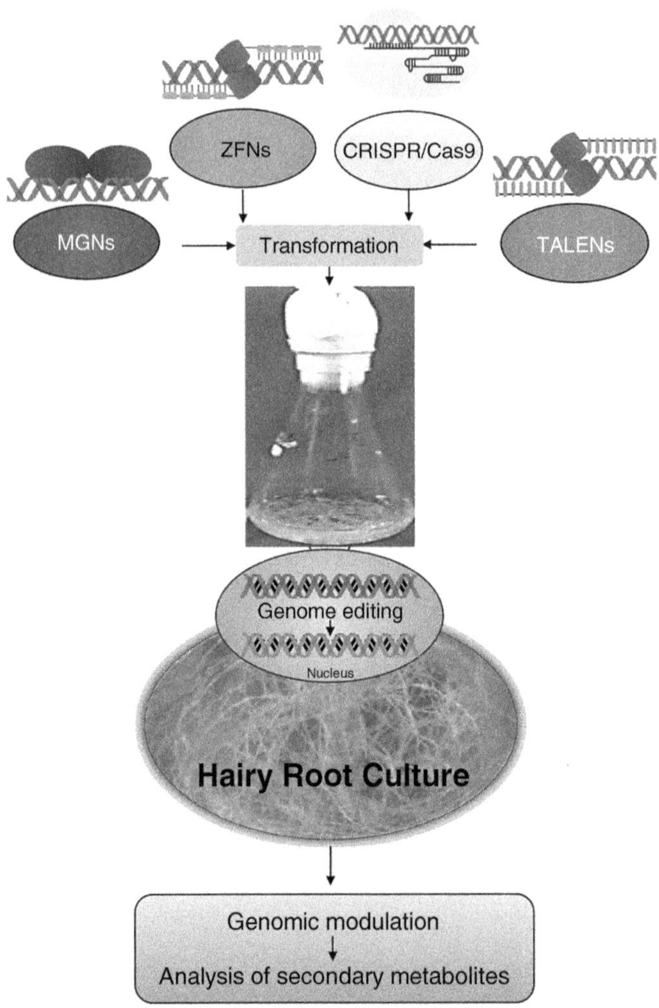

Figure 8.1 Outline model of the genome engineering approaches to modify the hairy root cultures for the enhanced production of secondary metabolites. *Source:* Praveen Chandra Verma. MGNs- Meganucleases; ZFNs- Zinc-finger nucleases; TALENs- Transcription activator-like effector nucleases; CRISPR/Cas9- Clustered regularly interspaced short palindromic repeats/CRISPR-associated protein 9.

8.3.1 Meganucleases (MGNs)

The MGNs, also known as the LAGLIDADG homing endonucleases (LHEs), are unique enzymes encoded by the mobile introns or inteins and are naturally present in various organisms (Arnould et al. 2011; Epinat et al. 2003). The MGNs may occur as either a monomer (a single peptide) or a homodimer (two identical protein subunits). In single-peptide MGNs, a linker sequence facilitates the joining of two tandem repeat monomers, whereas the homodimers consist of approximately 160–200 amino acid long two identical protein subunits (Stoddard 2011). MGNs can recognize and effectively cleave ~20–30 bp target DNA sequences (Curtin et al. 2012) and have been successfully used for the plant-genome engineering. In plants, the meganuclease I-SceI was successfully used to create targeted DSBs in tobacco (Puchta 1999). The use of a re-engineered nuclease has been shown to effectively cleave an endogenous target sequence in cotton (D'Halluin et al. 2013). The *Chlamydomonas reinhardtii* I-CreI endonuclease derived engineered homing endonuclease, PB1, has also been used successfully for the targeted DNA modification in Arabidopsis (Antunes et al. 2012). Considering the high DNA cleavage specificity of MGNs at targeted genomic loci, they represent themselves as potential candidates for plant hairy root cultures. However, the major drawback of MGNs is their challenging re-design.

8.3.2 Zinc-Finger Nucleases (ZFNs)

The Cys_2His_2zinc-finger proteins, which bind to the major groove of DNA double helix, have been demonstrated to act as site-specific nucleases and are termed as ZFNs (Kim et al. 1996). The ZFNs are engineered artificial endonucleases consisting of a pre-designed specific DNA-binding zinc-finger protein (ZFP) fused to the DNA cleavage domain of the *Fok*I restriction enzyme (Carroll 2011; Dreyer and Cathomen 2012; Kim and Kim 2011; Urnov et al. 2010). The ZFNs can induce the DNA DSBs at specific sites in the genome, which then undergo the cellular DNA repair process and lead to highly efficient mutagenesis at the targeted locus. The ZFNs have been shown to effectively manipulate the genome of several plants, including tobacco (Cai et al. 2009), *Zea mays* (Shukla et al. 2009), Arabidopsis (Osakabe et al. 2010; Zhang et al. 2010), and soybean (Curtin et al. 2011). The ZFNs allow the precise targeting of the targeted genomic loci in plants and, therefore, are better candidates for generating edited hairy root lines. However, making the highly specific ZFNs against the target loci is challenging.

8.3.3 Transcription Activator-like Effector Nucleases (TALENs)

The transcription activator-like effectors (TALEs), in association with a nuclease, known as TALENs, have been shown to cut the DNA precisely (Boch 2011). Like ZFNs, TALENs can also introduce DSBs at the specific sites within the genome. The composition of TALENs includes the TALEs and the *Fok*I endonuclease (Joung and Sander 2013). The TALEs are naturally occurring highly conserved repeats found in the *Xanthomonas* genus of the plant pathogenic bacteria (Mew et al. 1993). The TALEs, which act as DNA-binding domains, could be customized for the targeted mutagenesis, and allow the targeted modification and regulation of the desired genes (Joung and Sander 2013; Sun and Zhao 2013). TALENs have been used successfully to modify endogenous genes in Arabidopsis protoplasts (Cermak et al. 2011) and rice (Li et al. 2012). Designing of target-specific TALENs is not as challenging as MGNs and ZFNs. The cloning of TALE motifs is also easy. Therefore, the TALENs also represent themselves as potential candidates for genomic modification of the hairy root lines.

8.3.4 Clustered Regularly Interspaced Short Palindromic Repeats (CRISPR)/CRISPR-Associated Protein 9 (CRISPR/Cas9)

Though the MGNs, ZFNs, and TALENs have been successfully used for genome engineering in several plants, to date, there is no report of their use in the genome editing of the plant hairy root lines. The use of any of the genome-editing tool depends upon the specific experimental requirements and the availability of technical expertise. The MGNs, ZFNs, and TALENs demand more technical knowledge for target design than CRISPR/Cas9-based approaches. Another major limitation observed with the use of MGNs, ZFNs, and TALENs is the lack of multiplexing and less efficiency than the CRISPR/Cas9-based approaches. However, depending on the specific experimental requirements and the availability of the technical expertise, the MGNs, ZFNs, and TALENs represent themselves as potential candidates for the genome engineering of the plant hairy root cultures.

In the year 2002, Ruud Jansen and colleagues coined the term Clustered Regularly Interspaced Short Palindromic Repeats (CRISPR) (Jansen et al. 2002). The CRISPR found in bacterial and archeal cells protect them from the invading harmful foreign viral and plasmid DNA. CRISPR-associated (Cas) proteins assist in this process by exhibiting the endonuclease activity. The Cas proteins are selectively guided to the specific cleavage sites by the CRISPR RNAs (crRNAs). Thus, the CRISPR/Cas system provides a natural defense against the foreign genetic elements to the bacterial and archaeal cells and considered to be the immune system providing the acquired immunity to the prokaryotes (Barrangou et al. 2007; Horvath and Barrangou 2010). The *Streptococcus pyogenes* type II CRISPR system mainly performs its function by utilizing three genes, one encoding the Cas9 endonuclease and two non-coding RNAs forming the precursor crRNA (pre-crRNA) and the *trans*-activating crRNA (tracrRNA). The pre-crRNA and the small tracrRNA hybridization lead to the formation of a duplex mature crRNA. A ribonucleoprotein complex is formed with the crRNA by the activity of the Cas9 endonuclease, which specifically recognizes a seed sequence present upstream to the protospacer adjacent motif (PAM) having the canonical sequence 5'-NGG-3′ (Barrangou et al. 2007; Deltcheva et al. 2011; Marraffini and Sontheimer 2008; Jinek et al. 2012; Jinek et al. 2014; Nishimasu et al. 2014; Sternberg et al. 2014). With the advancements in CRISPR research, a single or multiple guide RNAs can be used along with an engineered hairpin structure that resembles and functions as a crRNA-tracrRNA complex. The Cas9 endonuclease consists of HNH and RuvC-like domains, which create DSB by cleaving one of the DNA strands at the same position. The PAM sequence (5′-NGG-3′) is crucial for determining the binding specificity of the Cas9 enzyme. The Cas9 introduces the DSBs three nucleotides upstream to the PAM sequence (Hsu et al. 2013; Jinek et al. 2012; Jinek et al. 2014; Nishimasu et al. 2014; Sternberg et al. 2014). The CRISPR/Cas9 system is more specific and efficient compared to the engineered nucleases and has superseded the MGNs, ZFNs, and TALENs as genome-editing tools. The CRISPR/cas9 system has proven to be simple, cost-effective, and highly specific for genome editing. It has been successfully applied in yeast (Dicarlo et al. 2013), zebrafish (Hwang et al. 2013), *Caenorhabditis elegans* (Friedland et al. 2013), humans (Cho et al. 2013; Mali et al. 2013) and plants (Belhaj et al. 2013; Fauser et al. 2014; Feng et al. 2013; Jiang et al. 2013; Jinek et al. 2014; Li et al. 2013; Mao et al. 2013; Miao et al. 2013; Nekrasov et al. 2013; Shan et al. 2013; Xie and Yang 2013).

8.4 Plant Hairy Root Culture as a Model for Genome Engineering

Agrobacterium rhizogenes or *Rhizobium rhizogenes* is a soil-borne gram-negative bacterium that induces the hairy root production in monocot and eudicot plants. Its ability to transform the plant genomes makes it a preferred tool for genome engineering through the hairy root culture in plants. The hairy root culture has demonstrated its effectiveness and utility for the fast screening of gene function and production of desired metabolites through the targeted gene editing-mediated pathway engineering in several plant species (Table 8.1). The efficient genome-editing system, especially CRISPR/Cas9, has quickly been developed as a robust technology for selected gene editing and mutagenesis in the hairy root cultures (Figure 8.1 **and** Table 8.1). In pomegranate (*Punica granatum* L.), two UDP-dependent glycosyltransferases (UGTs), *PgUGT84A23*, and *PgUGT84A24*, involved in the formation of β-glucogallin from gallic acid and UDP-glucose, were successfully targeted and edited by using the CRISPR/Cas9 approach. Metabolite analysis showed the accumulation of 3-O- and 4-O-glucosides of gallic acid in the edited hairy root lines (Chang et al. 2019). Two soybean *phytoene desaturase* (PDS) genes, *GmPDS11* and *GmPDS18*, were edited by using the TALEN and CRISPR/Cas9 tools. The hairy root transformation of soybean showed the efficiency of CRISPR/Cas9 over TALENs. The application of *the GmU6-16g-1* promoter displayed higher mutagenesis frequencies than the promoter of the *AtU6-26*. The editing of *PDS* showed albino and dwarf buds phenotype in the soybean cotyledon nodes (Du et al. 2016). By using the CRISPR/Cas9 approach, the cowpea symbiosis receptor-like kinase (*VuSYMRK*) gene was successfully edited with ~67% mutagenic efficiency in the hairy root-transformed plants. In *VuSYMRK*-edited hairy root lines, the nodule formation was observed to be completely inhibited (Ji et al. 2019). In *Glycine max* (L.) Merrill, single guide RNAs (sgRNAs) that target homozygous transgene (*bar*), encoding phosphinothricin N-acetyltransferase, and two endogenous soybean genes (*GmFEI2* and *GmSHR*) were designed and successfully used for *A. rhizogenes*-mediated soybean hairy root transformation. The CRISPR/Cas9 system was found to be highly efficient for targeting both the endogenous (*GmFEI2* and *GmSHR*), and the exogenous gene (*bar*) in the soybean hairy roots (Cai et al. 2015). *Brassica carinata FASCICLIN-LIKE ARABINOGALACTAN PROTEIN 1* (*BcFLA1*) gene regulates the root-hair elongation. CRISPR/Cas9-mediated editing of *the BcFLA1* gene resulted in shorter root hairs in *B. carinata* (Kirchner et al. 2017). The CRISPR/Cas9-mediated multiplex mutations in soybean have been observed to have a profound effect on the secondary metabolite isoflavone content. The sgRNAs were designed to target the *G. max flavanone-3-hydroxylase* (F3H) and *flavone synthase* II (FNS II) genes involved in the isoflavonoid

Table 8.1 List of edited genes in different plant species by using the hairy root cultures as a model system.

Sr. No.	Gene	Genome engineering approach	Plant Species	References
1.	*UDP-dependent glycosyltransferases (UGTs) PgUGT84A23* and *PgUGT84A24*	CRISPR/Cas9	*Punica granatum*	Chang et al. (2019)
2.	*Phytoene desaturase (PDS) GmPDS11* and *GmPDS18*	TALEN and CRISPR/Cas9	*Glycine max*	Du et al. (2016)
3.	Cowpea symbiosis receptor-like kinase (*VuSYMRK*)	CRISPR/Cas9	*Vigna unguiculata*	Ji et al. (2019)
4.	*Bar, GmFEI2* and *GmSHR*	CRISPR/Cas9	*Glycine max*	Cai et al. (2015)
5.	*FASCICLIN-LIKE ARABINOGALACTAN PROTEIN 1 (BcFLA1)*	CRISPR/Cas9	*Brassica carinata*	Kirchner et al. (2017)
6.	*Flavanone-3-hydroxylase (F3H)* and *flavone synthase II (FNS II)*	CRISPR/Cas9	*Glycine max*	Zhang et al. (2020)
7.	*Rosmarinic acid synthase (SmRAS)*	CRISPR/Cas9	*Salvia miltiorrhiza*	Zhou et al. (2018)
8.	*UDP-glucose dependent glycosyltransferases (UGT71L1)*	CRISPR/Cas9	*Populus tremula x Populus alba*	Fellenberg et al. (2020)
9.	*LATE ELONGATED HYPOCOTYL (LHY) GmLHY1a, GmLHY1b, GmLHY2a*, and *GmLHY2b*	CRISPR/Cas9	*Glycine max*	Cheng et al. (2019)
10.	*Fatty Acid Desaturase 2GmFAD2–1A* and *GmFAD2–1B*	CRISPR/Cas9	*Glycine max*	Do et al. (2019)
11.	*GmMYB118*	CRISPR/Cas9	*Glycine max*	Du et al. (2018)
12.	*Green fluorescent protein (GFP), Glyma07g14530, Glyma01g38150* and *Glyma11g07220*	CRISPR/Cas9	*Glycine max*	Jacobs et al. (2015)
13.	*Diterpene synthase (SmCPS1)*	CRISPR/Cas9	*S. miltiorrhiza*	Li et al. (2017)
14.	*Glyma.03g163500, Glyma.10g246300, Glyma.10g037100, Glyma.13g123500, Glyma.19g164900, Glyma.19g164800, Glyma.20g148400, Glyma.20g146200*, and *Glyma.20g148200*	CRISPR/Cas9	*Glycine max*	Li et al. (2019)
15.	*Glyma06g14180, Glyma08g02290* and *Glyma12g37050*	CRISPR/Cas9	*Glycine max*	Sun et al. (2015)
16.	*CYP716A51*	CRISPR/Cas9	*Lotus japonicus*	Suzuki et al. (2019)
17.	*GmLCLa1, GmLCLa2, GmLCLb1*, and *GmLCLb2*	CRISPR/Cas9	*Glycine max*	Wang et al. (2020)
18.	*ahFAD2A* and *ahFAD2B*	CRISPR/Cas9	*Arachis hypogaea*	Yuan et al. (2019)

synthesis. Editing of *GmF3H1*, *GmF3H2*, and *GmFNSII-1* in soybean hairy roots not only increased the isoflavone content but also reduced the soya bean mosaic viral load (Zhang et al. 2020). *Salvia miltiorrhiza* is a Chinese herb with important medicinal properties. The *rosmarinic acid synthase* (*SmRAS*) gene, which participates in the phenolic acid biosynthetic pathway, was edited in *S. miltiorrhiza*. The CRISPR/Cas9 approach was found to successfully edit the genomes in 50% of the *S. miltiorrhiza* hairy root lines. The *SmRAS* gene expression was reduced, and the rosmarinic acid and lithospermic acid contents were decreased. The content of rosmarinic acid precursor, 3,4-dihydroxyphenyllactic acid, was found to be increased (Zhou et al. 2018). A UDP-glucose dependent glycosyltransferases (UGT71L1) was successfully knocked out by CRISPR/Cas9 in poplar hairy root cultures. The salicortin, tremulacin, and tremuloidin content were completely lost, and the salicin content was partially reduced. Thus, by using genome editing, UGT71L1 was observed to be essential for the majority of the salicinoids biosynthesis (Fellenberg et al. 2020). *G. max LATE ELONGATED HYPOCOTYL* (LHY) genes *GmLHY1a*, *GmLHY1b*, *GmLHY2a*, and *GmLHY2b* were edited by using the CRISPR/Cas9 approach. The soybean hairy roots were transformed with the prepared constructs by using the *A. rhizogenes* K599 strain, which resulted in high-efficiency mutation frequencies observed in the transgenic hairy roots. These soybeans hairy roots-tested *GmLHY* sgRNAs were then used to make stably transformed plants. Reduced plant height and shortened internodes were observed in the *GmLHY* quadruple mutants. The gibberellic acid (GA3) content and the expression levels of the GA3 metabolic pathway genes were significantly reduced in *GmLHY* mutants in comparison to the wild-type controls (Cheng et al. 2019). The CRISPR/Cas9 gene-editing tool was successfully applied to regulate the monounsaturated fat content in soybean seeds. The *Fatty Acid Desaturase 2* (*GmFAD2*) gene was targeted by designing two sgRNAs targeting two homeologous soybean genes *GmFAD2–1A* and *GmFAD2–1B*. These constructs were shown to edit the transgenic soybean hairy roots efficiently. These tested constructs were then used to generate stable soybean transformants. The CRISPR-edited plants showed over 80% increases in the oleic acid content, while the linoleic acid content was decreased to 1.3–1.7% (Do et al. 2019). In *G. max*, the role of MYB family transcription factor *GmMYB118* in drought and salt stress was elucidated by CRISPR/Cas9-assisted genome editing. The CRISPR-edited plants were observed to be more susceptible to drought and salt stress in comparison to the control plants. Compared to the control plants, the proline and chlorophyll content was also found to be reduced in the CRISPR-edited plants (Du et al. 2018). In another study in soybean, a single copy, as well as its homologs gene pair, was targeted by using the CRISPR system. The single-copy gene *Glyma07g14530*, encoding a putative glucosyl-transferase, and the homeologous genes, *Glyma01g38150* and *Glyma11g07220*, were successfully edited by using the CRISPR system in soybean (Jacobs et al. 2015). Hairy root cultures were used for the functional validation of a diterpene synthase gene (*SmCPS1*), associated with tanshinone biosynthesis in *S. miltiorrhiza*, by using the CRISPR/Cas system. *A. rhizogenes*-mediated transformation and the metabolite analysis of transgenic hairy root lines showed the complete absence of tanshinones, particularly cryptotanshinone, tanshinone I, and tanshinone IIA in the homozygous mutants (Li et al. 2017). In soybean, nine different major seed storage protein genes were targeted by using the CRISPR/Cas approaches. The *A. rhizogenes*-mediated hairy root hairy root transformation system was utilized to check the efficiency of sgRNAs designed to edit these soybean storage protein genes. Mutations, in the ratio of 3.87–43.7%, were successfully detected in three storage protein genes (Li et al. 2019). The CRISPR/Cas9 system was used to target and edit three soybean genes *Glyma06g14180*, *Glyma08g02290*, and *Glyma12g37050*. The sgRNAs were designed to precisely target these genes, and the soybean hairy roots were transformed to analyze the mutation efficiencies in the transgenic lines. *Glyma06g14180* and *Glyma08g02290* exhibited the biallelic mutations in the transgenic hairy roots, whereas the off-target activities were observed for *Glyma06g14180* and *Glyma12g37050* (Sun et al. 2015). In *Lotus japonicus*, the role of *CYP716A51* in triterpenoid biosynthesis was studied by using the CRISPR/Cas9 system. Complete loss of C-28 oxidized triterpenoids was observed in the CRISPR-edited *cyp716a51* transgenic hairy root lines (Suzuki et al. 2019). CRISPR-mediated editing of *G. max GmLCLa1*, *GmLCLa2*, *GmLCLb1*, and *GmLCLb2* genes revealed the molecular basis of the circadian clock in soybean. The knock-out mutants of these genes were obtained by *A. rhizogenes*-mediated transformation of the sgRNAs-harboring vector constructs in the soybean hairy roots. In transgenic soybean, extreme short-period circadian rhythm and late-flowering was observed (Wang et al. 2020). Peanut (*Arachis hypogaea* L.), an allotetraploid crop, contains oleic acid in seeds. By using the hairy root cultures as a model, the CRISPR/Cas9-mediated editing of two homoeologous genes, *ahFAD2A* and *ahFAD2B*, which encodes fatty acid desaturases, was successfully achieved. Mutations in the coding region of *ahFAD2B* could potentially be utilized to increase the oleic acid content in peanut (Yuan et al. 2019).

8.5 Conclusions

Recent advances in plant research have made it easy to precisely target specific genes in the genomes by the application of genome engineering. The CRISPR/Cas9, TALENs, and ZFNs are the preferred approaches to achieve the targeted mutagenesis of selected genes through genome engineering. The CRISPR/Cas9 approach has been successfully applied to target single copy as well as homoeologous genes (homoeoalleles). The effective targeting of homologous genes is important for species that harbor more than one genome. The generation of stable transgenic plants through genome engineering takes significant time. It is generally considered to be logical and beneficial to elucidate the function of the gene of interest before using it for stable plant transformation. Hairy root cultures offer a significant advantage to determine the function of a gene easily and quickly as the hairy root cultures require less time to grow and produce the metabolites in comparison to the stable plant transformation. Moreover, the growth conditions and media composition can be easily regulated in hairy root cultures, which is often challenging in field conditions, while generating stable transgenic plants. Genome engineering can also be utilized to overcome the most common problems associated with hairy root cultures such as browning, contamination, etc. The *A. rhizogenes*-mediated transformation of hairy root cultures has been optimized and successfully utilized to determine the gene function through the genome engineering approaches and to manipulate several important metabolic pathways for the enhanced production of commercially important secondary metabolites.

Acknowledgements

Authors are also thankful to CSIR, New Delhi for financial support in the form of "FBR Genome Editing Network Project" (MLP-007).

Institute's Manuscript Number is CSIR-NBRI_MS/2020/06/23.

References

Antunes, M.S., Smith, J.J., Jantz, D., and Medford, J.I. (2012). Targeted DNA excision in Arabidopsis by a re-engineered homing endonuclease. *BMC Biotechnol.* 12: 86.

Arnould, S., Delenda, C., Grizot, S. et al. (2011). The I-CreI meganuclease and its engineered derivatives: applications from cell modification to gene therapy. *Protein Eng. Des. Sel.* 24: 27–31.

Barrangou,R.and Van Der Oost, J.(eds.). (2013). *CRISPR-Cas systems. RNA-Mediated Adaptative Immunity in Bacteria and Archaea*, 6e. Berlin; New York: Springer.

Barrangou, R., Fremaux, C., Deveau, H. et al. (2007). CRISPR provides acquired resistance against viruses in prokaryotes. *Science* 315: 1709–1712.

Belhaj, K., Chaparro-Garcia, A., Kamoun, S., and Nekrasov, V. (2013). Plant genome editing made easy: targeted mutagenesis in model and crop plants using the CRISPR/Cas system. *Plant Methods* 9: 39.

Bennett, R.N. and Wallsgrove, R.M. (1994). Secondary metabolites in plant defence mechanisms. *New Phytol* 127: 617–633.

Boch, J. (2011). TALEs of genome targeting. *Nat. Biotechnol.* 29: 135–136.

Cai, C.Q., Doyon, Y., Ainley, W.M. et al. (2009). Targeted transgene integration in plant cells using designed zinc finger nucleases. *Plant Mol. Biol.* 69: 699–709.

Cai, Y., Chen, L., Liu, X. et al. (2015). CRISPR/Cas9-mediated genome editing in soybean hairy roots. *PLoS One* 10: e0136064.

Carroll, D. (2011). Genome engineering with zinc-finger nucleases. *Genetics* 188: 773–782.

Cermak, T., Doyle, E.L., Christian, M. et al. (2011). Efficient design and assembly of custom TALEN and other TAL effector-based constructs for DNA targeting. *Nucleic Acids Res.* 39: e82.

Charlwood, B.V. and Charlwood, K.A. (1991). Terpenoid production in plant cell culture. In: *Ecological Chemistry and Biochemistry of Plant Terpenoids* (eds. J.B. Harborne and F.A. Tomás-Barberán), 95–132. Oxford: Clarendon Press.

Chang, L., Wu, S., and Tian, L. (2019). Effective genome editing and identification of a regiospecific gallic acid 4-O-glycosyltransferase in pomegranate (*Punica granatum* L.). *Hortic Res* 6: 1–15.

Cheng, Q., Dong, L., Su, T. et al. (2019). CRISPR/Cas9-mediated targeted mutagenesis of *GmLHY* genes alters plant height and internode length in soybean. *BMC Plant Biol.* 19: 562.

Cheynier, V., Comte, G., Davies, K.M. et al. (2013). Plant phenolics: recent advances on their biosynthesis, genetics, and ecophysiology. *Plant Physiol. Biochem.* 72: 1–20.

Cho, S.W., Kim, S., Kim, J.M., and Kim, J.S. (2013). Targeted genome engineering in human cells with the Cas9 RNA-guided endonuclease. *Nat. Biotechnol.* 31: 230–232.

Curtin, S.J., Zhang, F., Sander, J.D. et al. (2011). Targeted mutagenesis of duplicated genes in soybean with zinc-finger nucleases. *Plant Physiol.* 156: 466–473.

Curtin, S.J., Voytas, D.F., and Stupar, R.M. (2012). Genome engineering of crops with designer nucleases. *The Plant Genome* 5: 42–50.
Debnath, B., Singh, W.S., Das, M. et al. (2018). Role of plant alkaloids on human health: a review of biological activities. *Biology Materials Today Chemistry* 9: 56–72.
Deltcheva, E., Chylinski, K., Sharma, C.M. et al. (2011). CRISPR RNA maturation by trans-encoded small RNA and host factor RNase III. *Nature* 471: 602–607.
D'Halluin, K., Vanderstraeten, C., Van Hulle, J. et al. (2013). Targeted molecular trait stacking in cotton through targeted double-strand break induction. *Plant Biotechnol. J.* 11: 933–941.
Dicarlo, J.E., Norville, J.E., Mali, P. et al. (2013). Genome engineering in Saccharomyces cerevisiae using CRISPR-Cas systems. *Nucleic Acids Res.* 41: 4336–4343.
Do, P.T., Nguyen, C.X., Bui, H.T. et al. (2019). Demonstration of highly efficient dual gRNA CRISPR/Cas9 editing of the homeologous GmFAD2-1A and GmFAD2-1B genes to yield a high oleic, low linoleic and alpha-linolenic acid phenotype in soybean. *BMC Plant Biol.* 19: 311.
Dreyer, A.K. and Cathomen, T. (2012). Zinc-finger nucleases-based genome engineering to generate isogenic human cell lines. *Methods Mol. Biol.* 813: 145–156.
Du, H., Zeng, X., Zhao, M. et al. (2016). Efficient targeted mutagenesis in soybean by TALENs and CRISPR/Cas9. *J. Biotechnol.* 217: 90–97.
Du, Y.T., Zhao, M.J., Wang, C.T. et al. (2018). Identification and characterization of GmMYB118 responses to drought and salt stress. *BMC Plant Biol.* 18: 320.
Eapen, S. and Mitra, R.J.P.-I.N.S.P.B. (2001). Plant hairy root cultures: prospects and limitations. *Proceedings of the Indian National Science Academy* 67: 107–120.
Epinat, J.C., Arnould, S., Chames, P. et al. (2003). A novel engineered meganuclease induces homologous recombination in yeast and mammalian cells. *Nucleic Acids Res.* 31: 2952–2962.
Fauser, F., Schiml, S., and Puchta, H. (2014). Both CRISPR/C as-based nucleases and nickases can be used efficiently for genome engineering in a rabidopsis thaliana. *Plant J.* 79: 348–359.
Fellenberg, C., Corea, O., Yan, L.H. et al. (2020). Discovery of salicyl benzoate UDP-glycosyltransferase, a central enzyme in poplar salicinoid phenolic glycoside biosynthesis. *Plant J.* 102: 99–115.
Feng, Z., Zhang, B., Ding, W. et al. (2013). Efficient genome editing in plants using a CRISPR/Cas system. *Cell Res.* 23: 1229–1232.
Friedland, A.E., Tzur, Y.B., Esvelt, K.M. et al. (2013). Heritable genome editing in C. elegans via a CRISPR-Cas9 system. *Nat. Methods* 10: 741–743.
Gaj, T., Gersbach, C.A., and Barbas, C.F. 3rd (2013). ZFN, TALEN, and CRISPR/Cas-based methods for genome engineering. *Trends Biotechnol.* 31: 397–405.
González-Burgos, E. and Gómez-Serranillos, M.P. (2012). Terpene compounds in nature: a review of their potential antioxidant activity. *Curr Med Chem* 19: 5319–5341.
Guerriero, G., Berni, R., Munoz-Sanchez, J.A. et al. (2018). Production of plant secondary metabolites: examples, tips and suggestions for biotechnologists. *Gene* 9: 309.
Guillon, S., Tremouillaux-Guiller, J., Pati, P.K. et al. (2006). Harnessing the potential of hairy roots: dawn of a new era. *Trends Biotechnol.* 24: 403–409.
Horvath, P. and Barrangou, R. (2010). CRISPR/Cas, the immune system of bacteria and archaea. *Science* 327: 167–170.
Hsu, P.D., Scott, D.A., Weinstein, J.A. et al. (2013). DNA targeting specificity of RNA-guided Cas9 nucleases. *Nat. Biotechnol.* 31: 827–832.

Hu, Z.B. and Du, M. (2006). Hairy root and its application in plant genetic engineering. *J. Integr. Plant Biol.* 48: 121–127.

Hussein, R.A. and El-Anssary, A.J.H.M. (2018). Plants secondary metabolites: the key drivers of the pharmacological actions of medicinal plants. In: *Herbal Medicine* (ed. P.F. Builders). London, UK: IntechOpen.

Hwang, W.Y., Fu, Y., Reyon, D. et al. (2013). Efficient genome editing in zebrafish using a CRISPR-Cas system. *Nat. Biotechnol.* 31: 227–229.

Isah, T. (2019). Stress and defense responses in plant secondary metabolites production. *Biol. Res.* 52: 39.

Jacobs, T.B., LaFayette, P.R., Schmitz, R.J., and Parrott, W.A. (2015). Targeted genome modifications in soybean with CRISPR/Cas9. *BMC Biotechnol* 15: 16.

Jansen, R., Embden, J.D., Gaastra, W., and Schouls, L.M. (2002). Identification of genes that are associated with DNA repeats in prokaryotes. *Mol Microbiol* 43: 1565–1575.

Ji, J., Zhang, C., Sun, Z. et al. (2019). Genome editing in cowpea Vigna unguiculata using CRISPR-Cas9. *Int. J. Mol. Sci.* 20: 2471.

Jiang, W.Z., Zhou, H.B., Bi, H.H. et al. (2013). Demonstration of CRISPR/Cas9/sgRNA-mediated targeted gene modification in Arabidopsis, tobacco, sorghum and rice. *Nucleic Acids Res.* 41: e188.

Jinek, M., Chylinski, K., Fonfara, I. et al. (2012). A programmable dual-RNA-guided DNA endonuclease in adaptive bacterial immunity. *Science* 337: 816–821.

Jinek, M., Jiang, F., Taylor, D.W. et al. (2014). Structures of Cas9 endonucleases reveal RNA-mediated conformational activation. *Science* 343: 1247997.

Joung, J.K. and Sander, J.D. (2013). TALENs: a widely applicable technology for targeted genome editing. *Nat. Rev. Mol. Cell Biol.* 14: 49–55.

Kim, S. and Kim, J.S. (2011). Targeted genome engineering via zinc finger nucleases. *Plant Biotechnol. Rep.* 5: 9–17.

Kim, Y.G., Cha, J., and Chandrasegaran, S. (1996). Hybrid restriction enzymes: zinc finger fusions to Fok I cleavage domain. *Proc. Natl. Acad. Sci. U. S. A.* 93: 1156–1160.

Kirchner, T.W., Niehaus, M., Debener, T. et al. (2017). Efficient generation of mutations mediated by CRISPR/Cas9 in the hairy root transformation system of Brassica carinata. *PLoS One* 12: e0185429.

Li, T., Liu, B., Spalding, M.H. et al. (2012). High-efficiency TALEN-based gene editing produces disease-resistant rice. *Nat. Biotechnol.* 30: 390–392.

Li, J.F., Norville, J.E., Aach, J. et al. (2013). Multiplex and homologous recombination-mediated genome editing in Arabidopsis and Nicotiana benthamiana using guide RNA and Cas9. *Nat. Biotechnol.* 31: 688–691.

Li, B., Cui, G., Shen, G. et al. (2017). Targeted mutagenesis in the medicinal plant salvia miltiorrhiza. *Sci. Rep.* 7: 43320.

Li, C., Nguyen, V., Liu, J. et al. (2019). Mutagenesis of seed storage protein genes in soybean using CRISPR/Cas9. *BMC Res. Notes* 12: 176.

Mali, P., Yang, L.H., Esvelt, K.M. et al. (2013). RNA-guided human genome engineering via Cas9. *Science* 339: 823–826.

Mao, Y., Zhang, H., Xu, N. et al. (2013). Application of the CRISPR-Cas system for efficient genome engineering in plants. *Mol. Plant* 6: 2008–2011.

Marraffini, L.A. and Sontheimer, E.J. (2008). CRISPR interference limits horizontal gene transfer in staphylococci by targeting DNA. *Science* 322: 1843–1845.

Mew, T.W., Alvarez, A.M., Leach, J., and Swings, J. (1993). Focus on bacterial blight of rice. *Plant Dis.* 77: 5–12.

Miao, J., Guo, D.S., Zhang, J.Z. et al. (2013). Targeted mutagenesis in rice using CRISPR-Cas system. *Cell Res.* 23: 1233–1236.

Mojica, F.J. and Rodriguez-Valera, F. (2016). The discovery of CRISPR in archaea and bacteria. *FEBS J.* 283: 3162–3169.

Nartop, P. (2018). Engineering of biomass accumulation and secondary metabolite production in plant cell and tissue cultures. In: *Plant Metabolites and Regulation under Environmental Stress* (eds. P. Ahmad, M.A. Ahanger, V.P. Sing, et al.), 169–194. London: Elsevier.

Nekrasov, V., Staskawicz, B., Weigel, D. et al. (2013). Targeted mutagenesis in the model plant Nicotiana benthamiana using Cas9 RNA-guided endonuclease. *Nat. Biotechnol.* 31: 691–693.

Nishimasu, H., Ran, F.A., Hsu, P.D. et al. (2014). Crystal structure of Cas9 in complex with guide RNA and target DNA. *Cell* 156: 935–949.

Osakabe, K., Osakabe, Y., and Toki, S. (2010). Site-directed mutagenesis in Arabidopsis using custom-designed zinc finger nucleases. *Proc. Natl. Acad. Sci. U. S. A.* 107: 12034–12039.

Pagare, S., Bhatia, M., Tripathi, N. et al. (2015). Secondary metabolites of plants and their role: overview. *Current Trends in Biotechnology and Pharmacy* 9: 293–304.

Pandey, V., Srivastava, R., Akhtar, N. et al. (2016). Expression of Withaniasomnifera steroidal glucosyltransferase gene enhances Withanolide content in hairy roots. *Plant Mol. Biol. Report.* 34: 681–689.

Pennisi, E. (2013). The CRISPR craze. *Science* 341: 833–836.

Pichersky, E., Noel, J.P., and Dudareva, N. (2006). Biosynthesis of plant volatiles: Nature's diversity and ingenuity. *Science* 311: 808–811.

Puchta, H. (1999). Use of I-Sce I to induce DNA double-strand breaks in Nicotiana. *Methods Mol. Biol.* 113: 447–451.

Puchta, H. (2005). The repair of double-strand breaks in plants: mechanisms and consequences for genome evolution. *J. Exp. Bot.* 56: 1–14.

Rai, K.M., Ghose, K., Rai, A. et al. (2019). Genome engineering tools in plant synthetic biology. In: *Current Developments in Biotechnology and Bioengineering* (eds. S.P. Singh, A. Pandey, G. Du and S. Kumar), 47–73. Elsevier.

Segal, D.J. (2013). Bacteria herald a new era of gene editing. *elife* 2: e00563.

Shan, Q., Wang, Y., Li, J. et al. (2013). Targeted genome modification of crop plants using a CRISPR-Cas system. *Nat. Biotechnol.* 31: 686–688.

Shukla, V.K., Doyon, Y., Miller, J.C. et al. (2009). Precise genome modification in the crop species Zea mays using zinc-finger nucleases. *Nature* 459: 437–441.

Srivastava, V., Mehrotra, S., and Mishra, S. (2016). Biotransformation through hairy roots: perspectives, outcomes, and major challenges. In: *Transgenesis and Secondary Metabolism*, Reference Series in Phytochemistry (ed. S. Jha), 1–24. Cham: Springer International Publishing.

Srivastava, R., Rai, K.M., and Srivastava, R. (2018). Plant biosynthetic engineering through transcription regulation: an insight into molecular mechanisms during environmental stress. In: *Biosynthetic Technology and Environmental Challenges Energy, Environment and Sustainability* (eds. S. Varjani, B. Parameswaran, S. Kumar and S. Khare), 51–72. Singapore: Springer.

Sternberg, S.H., Redding, S., Jinek, M. et al. (2014). DNA interrogation by the CRISPR RNA-guided endonuclease Cas9. *Nature* 507: 62–67.

Stoddard, B.L. (2011). Homing endonucleases: from microbial genetic invaders to reagents for targeted DNA modification. *Structure* 19: 7–15.

Sun, N. and Zhao, H. (2013). Transcription activator-like effector nucleases (TALENs): a highly efficient and versatile tool for genome editing. *Biotechnol. Bioeng.* 110: 1811–1821.

Sun, X., Hu, Z., Chen, R. et al. (2015). Targeted mutagenesis in soybean using the CRISPR-Cas9 system. *Sci. Rep.* 5: 10342.

Suzuki, H., Fukushima, E.O., Shimizu, Y. et al. (2019). Lotus japonicus triterpenoid profile and characterization of the CYP716A51 and LjCYP93E1 genes involved in their biosynthesis in planta. *Plant Cell Physiol.* 60: 2496–2509.

Tholl, D. (2015). Biosynthesis and biological functions of terpenoids in plants. *Adv Biochem Eng Biotechnol* 148: 63–106.

Urnov, F.D., Rebar, E.J., Holmes, M.C. et al. (2010). Genome editing with engineered zinc finger nucleases. *Nat. Rev. Genet.* 11: 636–646.

Wang, Y., Yuan, L., Su, T. et al. (2020). Light- and temperature-entrainable circadian clock in soybean development. *Plant Cell Environ.* 43 (3): 637–648.

Xie, K. and Yang, Y. (2013). RNA-guided genome editing in plants using a CRISPR-Cas system. *Mol. Plant* 6: 1975–1983.

Yang, L., Wen, K.S., Ruan, X. et al. (2018). Response of plant secondary metabolites to environmental factors. *Molecules* 23: 762.

Yuan, M., Zhu, J., Gong, L. et al. (2019). Mutagenesis of FAD2 genes in peanut with CRISPR/Cas9 based gene editing. *BMC Biotechnol.* 19: 24.

Zhang, F., Maeder, M.L., Unger-Wallace, E. et al. (2010). High frequency targeted mutagenesis in Arabidopsis thaliana using zinc finger nucleases. *Proc. Natl. Acad. Sci. U. S. A.* 107: 12028–12033.

Zhang, P., Du, H., Wang, J. et al. (2020). Multiplex CRISPR/Cas9-mediated metabolic engineering increases soya bean isoflavone content and resistance to soya bean mosaic virus. *Plant Biotechnol. J.* 18 (6): 1384–1395.

Zhou, Z., Tan, H., Li, Q. et al. (2018). CRISPR/Cas9-mediated efficient targeted mutagenesis of RAS in salvia miltiorrhiza. *Phytochemistry* 148: 63–70.

9

Phytic Acid Reduction in Cereal Grains by Genome Engineering

Potential Targets to Achieve Low Phytate Wheat

Ajay K. Pandey[1], *Sipla Aggarwal*[2], *Varsha Meena*[1], *and Anil Kumar*[1]

[1] *Department of Biotechnology, National Agri-Food Biotechnology Institute, Mohali, Punjab, India*
[2] *Department of Biotechnology, National Institute of Plant Genome Research, Jawaharlal Nehru University Campus, New Delhi, India*

CHAPTER MENU

9.1 Introduction

Among all the cereal crops, hexaploid wheat (*Triticum aestivum* L.) is an important crop that contributes 20% of all the food calories consumed globally (Breiman and Graur 1995). Though, wheat is a good source of mineral ions (Fe^{2+} and Zn^{2+}) but the presence of certain antinutritional components limits their bioavailability (Coulibaly et al. 2011). It has been suggested that the micronutrient bioavailability in the cereal crops can also be done by targeting genes that could enhance the total bioavailability by decreasing the concentrations of few antinutrients such as phytate, polyphenols, and tannins. Of these antinutrients, phytic acid (PA) is one of the major components that limit the bioavailable form of micronutrients including Fe and Zn, from cereal grains.

PA, also known as *myo*-inositol hexa*kis* phosphate (IP_6; PA), is the major storage form of phosphorus (P) in cereal grains. PA is an abundant natural product in all eukaryotes (Sasakawa et al. 1995). It comprises approximately 1–2% of the seed dry weight and accounts for more than 75% of total seed P (Raboy et al. 2000). In most cereals, approximately 70–80% of the total PA accumulates in the aleurone layer of the grains except maize (*Zea mays*). Though in less amount, PA also accumulates in other plant tissues such as

Genome Engineering for Crop Improvement, First Edition. Edited by Santosh Kumar Upadhyay.

roots, tubers, leaves, and pollen (Cosgrove and Irving 1980; Raboy 1997). PA is a naturally occurring compound formed during maturation of plant seeds. The amount of phytate in different cereals is highly variable. PA is higher in foods grown using modern high-phosphate fertilizers than those grown in natural compost (Srivastava 1955).

Chemically, phytate has six negatively charged ions, making it a potent chelator of mineral cations such as iron, zinc, calcium and potassium. Due to this inherent ability of PA, total bioavailability of these nutritionally important mineral ions is reduced (García-Estepa et al. 1999). Therefore, developing grains of cereal crops and in wheat, PA accumulates in protein storage bodies as mixed salts called phytate that form insoluble complexes with mineral cations (Fe^{2+} and Zn^{2+}). In contrast, during germination, grain phytase and phosphatases are activated and release the stored P from the *myo*-inositol ring in membranous compartments referred as vacuoles (Raboy et al. 2000). This release of phosphate and other minerals is beneficial to the developing seedlings because they are unable to absorb the nutrients from its environment at this developmental stage (Loewus and Murthy 2000).

Though PA is detected in almost all the eukaryotes, in plants it is known to be involved in developmental and signaling processes including cell wall biosynthesis, phosphatidyl inositol signaling, auxin storage and transport, and production of stress-related molecules (Loewus and Murthy 2000; Hegeman et al. 2001; Doria et al. 2009). In developing grains, the biosynthesis and accumulation of PA can be detected from a few days after anthesis (DAA) and increases during grain maturation (Bhati et al. 2014). Thus, PA acts as a reservoir of *myo*-inositol, phosphorus, and mineral cations, thereby aiding in plant growth at the seedling stage (Loewus and Murthy 2000). The plant seed accumulates PA in protein storage bodies as mixed salts of cations (Fe^{2+} and Zn^{2+}) called phytate therefore, drastically reducing the total bioavailability of these nutritionally important mineral ions (García-Estepa et al. 1999). Additionally, monogastric animals do not utilize PA efficiently due to lack of phytase enzyme. Thus, excretion of undigested PA in animal waste contributes to environmental phosphate pollution, where it accelerates eutrophication (Stevenson-Paulik et al. 2005). Therefore, reducing the PA content in cereal grains is a desired goal for the genetic improvement of several crops (Figure 9.1).

9.2 Genes Involved in Phytic Acid Biosynthesis

Classically, PA biosynthesis have been studied in a variety of organisms including slime mold (Stephens and Irvine 1990), duckweed (Brearley and Hanke 1996), yeast (Odom et al. 2000), maize (Shi et al. 2003), drosophila (Seeds et al. 2004), *Arabidopsis* (Stevenson-Paulik et al. 2005), and humans (Verbsky et al. 2005). Specifically in plants, PA biosynthesis utilizes two pathways which are either lipid-dependent or lipid-independent (Raboy 2003; Stevenson-Paulik et al. 2005). The first committed step involves the formation of inositol 3-phosphate (Ins3P) from glucose-6-phosphate by *myo*-inositol-3-phosphate synthase (MIPS). Free inositol can be produced through the dephosphorylation of Ins3P by inositol monophosphatase (IMP) (Loewus and Murthy 2000). Further, in lipid-independent pathway, Ins3P is converted to IP_6 via sequential and ordered phosphorylation of the remaining five positions of the inositol ring by a number of kinases namely inositol tris/

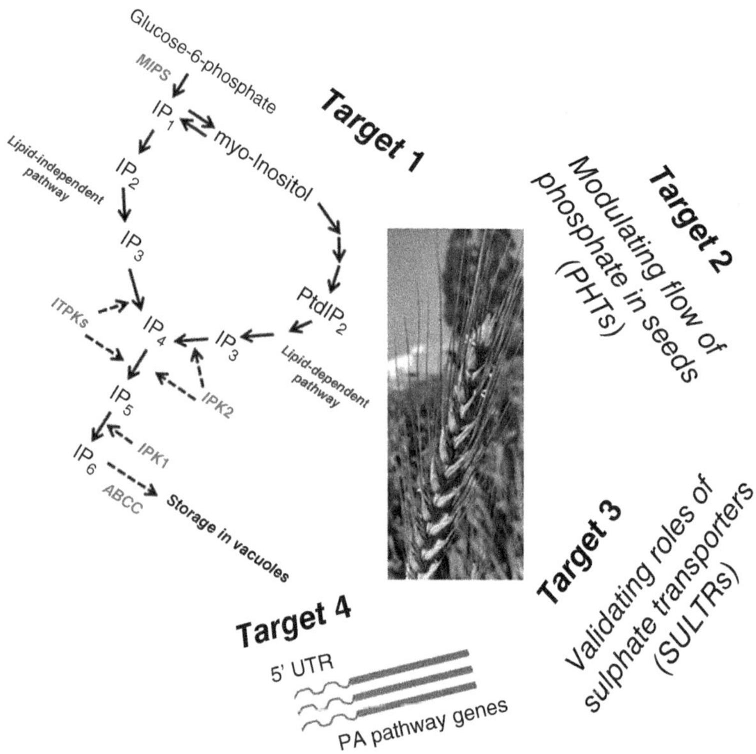

Figure 9.1 Schematic representation of the target areas that are important to achieve the lowering of PA content in cereal grains especially in wheat. As shown, in total, four important target areas have been identified Target 1: metabolic engineering of PA pathway genes or its transport in the storage tissue. Target 2: Deciphering the role of phosphate transporters (PHTs) to modulate the flow of Pi in cereal grains. Target 3: Validating the upcoming role of sulfate transporter in wheat and Target 4: Studying the role of untranslated region (UTR) of PA pathway genes to address their regulation. *Source:* Ajay K. Pandey.

tetraphosphate kinase (ITPK) and inositol-penta*kis* phosphate 2-kinase (IPK1) (Stiles et al. 2008; Rasmussen et al. 2010). The lipid-independent pathway primarily involves ITPKs that belong to a large family of ATP-grasp proteins. This enzyme family is best known for its Ins (1,3,4) P_3 5/6-kinase and Ins (3,4,5,6) P_4 1-kinase activities. Lipid-independent pathway involving sequential phosphorylation has been described previously in both duckweed (Brearley and Hanke 1996) and slime mold (Stephens and Irvine 1990) and their pathways are relatively similar except some intermediates.

Lipid-dependent pathway has been described in the budding yeast *Saccharomyces cerevisiae* (York et al. 1999; Odom et al. 2000) as well as in the plants, *Arabidopsis* Stevenson-Paulik et al. 2002) and maize (Shi et al. 2003). Phospholipase C (PLC) is a lipid-dependent pathway enzyme involved in the formation of Inositol 1,4,5-triphosphate (IP_3) utilizing phosphatidylinositol-4,5-bisphosphate (PIP2) as a substrate (Suzuki et al. 2007; Rasmussen et al. 2010). Further, IP_3 is sequentially phosphorylated to IP_6 by the action of two kinases, inositol polyphosphate kinase (IPK2) and inositol-penta*kis*phosphate 2-kinase (IPK1). IPK2 is a multifunctional kinase that phosphorylates IP_3 at the 6th and the 3rd position. IPK1 is involved in the final step in PA biosynthesis. This step is common to both lipid-dependent and independent pathways. The genes encoding for the PA biosynthesis pathway have now been reported in *Glycine max*, *Zea mays*, *Arabidopsis thaliana*, *Oryza sativa* and *T. aestivum*.

9.3 Potential Targets and Strategies to Achieve Low Phytate Wheat

9.3.1 Targeting Phytic Acid Pathway Genes

Though researchers have targeted multiple stages of the PA biosynthesis pathway, the prime success has been obtained when targeted genes are either involved in the last step of PA biosynthesis or the transport of PA to the site of storage, i.e. vacuoles (Shi et al. 2003; Aggarwal et al. 2018). In this regard, two important genes *IPK1* and an encoded ABC-C type transporter have been shortlisted. In soybeans, targeting an ABC transporter (*MRP5*) showed very promising results that resulted in dual effect. First, lowering of PA was achieved, second, enhanced Fe and Zn accumulations were also observed, suggesting implementation of this strategy. Similar strategy was consistently proven to be successful in rice, beans and maize wherein lowering of PA was observed at drastic levels (Shi et al. 2003; Panzeri et al. 2011). This candidate gene *IPK1* also shows promising results for wheat. In wheat, RNAi-mediated targeting of *TaIPK1* resulted in PA lowering of up to ~56% with concomitant increase in Fe:PA and Zn:PA molar ratios. Hence, targeting *IPK1* for genome editing will be a suitable choice in wheat (Aggarwal et al. 2018). Since all the homoeologs are differentially expressed, it is therefore important to have consideration of the conserved regions of transcript deriving from all the three A, B, and D genome.

9.3.2 Manipulating Flux of Phosphate Uptake and Homeostasis

Phosphate (Pi) transport from the roots to the different plant tissues, primarily involves multiple phosphate transporters (PHTs) belonging to PHT1, PHT2, PHT3, or PHT4 sub-families (Nussaume et al. 2011). Studies in rice and *Arabidopsis* suggested the role of *PHT1* family genes in remobilization of Pi to the actively developing tissues such as in embryo development (Nagarajan et al. 2011; Zhang et al. 2015). Pi transport within the plant tissues is suspected to be contributed by PHT3 and PHT4 family members by regulating "intracellular Pi starvation" signaling (Hassler et al. 2012). Earlier, it has been proposed that the Pi translocation and subsequent loading in grains is a demand-driven process (Wang et al. 2015). During the early stages of maturation, Pi is transported in the seeds and rapidly converted to the bound form, commonly referred to as PA, a well-known antinutrient (Raboy 2003). In order to develop Pi-efficient crops with lowered seed PA, it is important to understand the allocation of Pi in developing grains. Pi is a major reservoir in seeds in the form of PA. The seed PA accounts for 70–80% of the total phosphorus in the grains. Recently, the loading of Pi has been addressed as a means to modulate the Pi content in the filial tissue and also their roles were speculated to modulate the content of PA. On this line, rice gene referred as *OsPht1; 8* was shown to play an important role in phosphorus redistribution from source to sink organs and allocation between embryo and endosperm of seeds (Li et al. 2015). Targeting this PHT also resulted in drastic reduction of PA in the developing seeds.

This suggests that the flow and accumulation of PA could be controlled by addressing the function of tissue specific PHT genes. This study indicates that it is imperative to study the role of transporters for Pi when addressed at a tissue-specific level. Though multiple studies have focused on seed Pi content but, not many attempts have been undertaken in wheat. (Wang et al. 2015). Recently, new *TaPHTs* and their probable site of function in wheat tissue were reported (Shukla et al. 2016; Teng et al. 2017). These PHTs could be studied in detail for their predominant activity in the seeds and could be then evaluated for their role to control the Pi flux movement in the developing tissue to control PA accumulation.

9.3.3 Emerging Role of Sulphate Transporters

Previously, it was suggested that a nonsense mutation in the functional domain of the gene encoding for the sulphate transporter resulted in generating low PA in barley species (Ye et al. 2011). This preliminary clue implies that P allocation in the plant tissue could be a complex trait and may be a functional result of other multiple transporters. This also suggests a homeostatic interaction of the two macronutrients P and S. Once such exemplary evidence has been pointed out recently. Using a map-based cloning an employment of complementation studies was done to identify the candidate gene in rice for low PA trait and was reported as a transcript encoding for a sulphate transporter *OsSULTR3;3*. This demonstrates that disruption in the function of *SULTR3; 3* gene results in grains with low PA and perturbed levels of P (Liu et al. 2007). Moreover, the disruption of *OsSULTR3;3* affects the transcription level of gene which are involved in homeostasis of S and P nutrients that reinforce the probable role in interaction between these macronutrients (Zhao et al. 2016). The role of sulfate transporter in wheat has not been explored. Wheat genome multiple number of transporter (Buchner et al. 2010). It will be worth performing a comparison of the rice SULTR3;3 and assess its role in monocots such as wheat for its conserved functionality. With the advancement of recent genome-editing tools it would be interesting to generate low phytate wheat by targeting the sulfate transporter. This could be an exciting and innovative strategy to achieve low PA besides developing a good germplasm resource to address the macro-interactions of P and S.

9.3.4 Targeting the Regulators of PA Pathway

One of the areas that is still untapped is the identification of the regulators for PA biosynthesis. Earlier, an attempt to study the regulation of the PA pathway genes was largely focused on the effect of exogenous treatments of the hormones to the seedlings (Aggarwal et al. 2015). This study provided a preliminary clue for the regulation of PA pathway genes. Whether this could be a strategy to fluctuate the levels of PA in grains remains to be tested. Regulatory regions including the promoters and the 5'UTR are important. The functional evidence regulating the gene expression was provided by using GUS reporter-based assays (Cominelli et al. 2018). Elegant experiments using a GUS-reporter assay in *Arabidopsis* was done for the promoters of *PvMRP1* and *PvMRP5*. Their work also demonstrated that this regulation could be species-specific and these genes performing their function in a tissue-specific manner. Further detailed elucidation in the characterization of the cis-elements for the genes involved in PA biosynthesis is required.

9.4 Evolution of Genome Engineering for Trait Development in Wheat

Generating variation in the crop germplasm has been a constant endeavor to breed new genetic crops for developing important traits. Earlier, ethyl methane sulfonate and radiation-based approaches were exploited to create the genetic variation in the existing wheat germplasm. As a proof of concept, clustered regularly interspaced short palindromic repeats (CRISPR)-Cas9-mediated genome editing of wheat genes including *phytoene desaturase* were demonstrated by using suspension cells (Upadhyay et al. 2013). Subsequently, genome-editing tools, such as transcription activator-like effector nucleases (TALENs), and CRISPR-Cas9 technologies were implemented in hexaploid wheat (Wang et al. 2014).

With special reference to the traits related to lowering of PA and micronutrient mobilization in cereal crops, the first usage of the site-specific nucleases by using zinc-finger nucleases (ZFNs) (Shukla et al. 2009). This very first approach that was used for editing of crop gene to achieve lowering of PA in grains was done in maize. Using this approach, maize plant was produced by editing *IPK1*; which encodes for inositol-1,3,4,5,6-penta*kis*phosphate 2-kinase catalyzing the last step of PA biosynthesis pathway. This editing resulted in the favorable outcome in maize (Shukla et al. 2009). This also set the precedence for the utilization of the editing tools in cereal crops for trait development with special focus on the PA biosynthesis. Subsequent to ZFNs, TALENs have been introduced that provided more specificity for the targeted editing. Of special mention is the usage of TALENs in wheat for mutating three homeologous genes for *TaMLO* to provide tolerance for powdery mildew infections (Wang et al. 2014).

Despite this, there are many drawbacks of TALENs and ZFN, as their limitation in the range of target sites, designing construct is time consuming and a tedious process as well as not being cost effective. CRISPR/Cas systems, mainly based on exploitation of bacterial immune system and controlled by RNA–DNA interactions. CRISPR-mediated editing has many advantages over ZFNs and TALENs, including cost effectiveness, design flexibility, range of target sites, easier to deliver in cells due to short sgRNA sequence, less time consuming, and ability to target multiple sites by using multiple sgRNAs (multiplexing). The first attempt to introduce CRISPR/Cas9 technology in bread wheat was done by targeting the A genome homoeoallele of *TaMLO (*Wang et al. 2014*)*. Since the target site was chosen on the conserved regions of the transcript arising from A, B, and D genome; a catalog of mutations was expected. As expected, in this study, they have also reported editing in all three homoeologs. Since, the CRISPR/Cas9 construct are integrated in the genome, one could anticipate that the leaky expression might have induced undesirable changes in the genome. This limitation was overcome by the researchers by using transient expression of CRISPR/Cas9 DNA, or through *in-vitro* transcripts of the Cas9-coding sequence and guide RNA, followed by selection of the putative transformed calli in absence of herbicide selection (Zhang et al. 2016). Subsequently, a DNA-free CRISPR/Cas9 system using CRISPR/Cas9 ribonucleoproteins (RNPs) has also been developed (Liang et al. 2017). Since RNPs could be rapidly degraded by endogenous proteases in plant cells, CRISPR/Cas9 is only expressed for a short time, which should reduce the frequency of mosaicism and off-target effects in the regenerated plants. Completely transgene-free mutants were indeed obtained by using the CRISPR/Cas9 RNP-mediated genome-editing system in bread wheat, and the frequency of off-target mutations was significantly reduced. By utilizing the above methods for targeting the genes that are mentioned in the Table 9.1, and thereby manipulating them,

Table 9.1 Enlisting the genes that are good potential target for achieving low PA levels in crop plants and model species *Arabidopsis*.

Candidate gene/s	Plant species	Targeted gene	Approaches used	Percentage lowering of PA	References
Inositol pentakisphosphate kinase-1 (IPK1)	*Arabidopsis thaliana*	*AtIpk2β-1*	T-DNA insertion	35%	Stevenson-Paulik et al. (2005)
		AtIpk1–1	T-DNA insertion	83%	
	Zea mays	*ZmIPK1*	ZFN	Not mentioned	Shukla et al. (2009)
	Oryza sativa	*OsIPK1*	RNAi	50–69%	Ali et al. (2013)
	Glycine max	*GmIPK2*	RNAi	45%	Punjabi et al. (2018)
	Triticum aestivum	*TaIPK1*	RNAi	28–56%	Aggarwal et al. (2018)
Inositol tetrakisphosphate kinase (ITPK)	*Zea mays*	*lpa2–1*	EMS Mu insertion	50%	Raboy et al. (2000)
MIPS	*Glycine max*	*GmMIPS1*	RNAi	94.5%	Nunes et al. (2006)
	Oryza sativa	*RINO1*	antisense	68%	Kuwano et al. (2009)
	Arabidopsis thaliana	*atips1*	T-DNA insertion	Not reported	Kim and Tai (2011)
		atips2	T-DNA insertion		
MRP	*Zea mays*	*lpa1–1*	Gene silencing	85%	Shi et al. (2007)
	Oryza sativa	OsMRP5	T-DNA insertion	Os-lpa-XS110–2 ~20% 4A-02500 (~90%)	Xu et al. (2009)
	Glycine max	*lpa1*	EMS	------	Gillman et al. (2009)
	Triticum aestivum	*TaABCC13*	RNAi	28–22%	Bhati et al. (2016)

could affect the lowering of PA in grains or in a tissue-specific manner. Though, some of the genes have been reported to achieve low PA grains, yet, it will be interesting to assess the alternative strategies that could indirectly perturb the PA levels.

9.5 Future Implications

Rapid progress has been made in developing the tools for reverse genetics in hexaploid wheat such as genome-editing mediated CRISPR/Cas9 and base editing (Zong et al. 2017). It is assessed that editing tools, coupled with the applications of nano-biotechnology, could escape the tedious and time-consuming wheat transformation procedures. The enlisted candidate genes for achieving the lowering of PA are an excellent resource for generating new germplasm that could subsequently be introgressed into the elite wheat cultivars with high productivity and rust resistance. This will help in addressing the larger goals of micronutrient biofortification by enhancing its bioavailability in humans.

Acknowledgements

Authors thank Executive Director, NABI for facilities and support. The work done by the authors at NABI was supported through NABI-CORE grant or through extramural grant by DBT to AKP. IWGSC is acknowledged for the resource development for wheat genomics. AKP also acknowledges his previous lab members for their research contribution in the area of inositol phosphate biology in wheat.

References

Aggarwal, S., Shukla, V., Bhati, K.K. et al. (2015). Hormonal regulation and expression profiles of wheat genes involved during Phytic acid biosynthesis pathway. *Plants (Basel, Switzerland)* 4 (2): 298–319.

Aggarwal, S., Kumar, A., Bhati, K.K. et al. (2018). RNAi-mediated downregulation of inositol Pentakisphosphate Kinase (IPK1) in wheat grains decreases Phytic acid levels and increases Fe and Zn accumulation. *Front. Plant Sci.* 9: 259.

Ali, N., Paul, S., Gayen, D. et al. (2013). Development of low phytate rice by RNAi mediated seed-specific silencing of inositol 1,3,4,5,6-pentakisphosphate 2-kinase gene (IPK1). *PLoS One* 8 (7): e68161.

Bhati, K.K., Aggarwal, S., Sharma, S. et al. (2014). Differential expression of structural genes for the late phase of phytic acid biosynthesis in developing seeds of wheat (Triticum aestivum L.). *Plant Sci.* 224: 74–85.

Bhati, K.K., Alok, A., Kumar, A. et al. (2016). Silencing of ABCC13 transporter in wheat reveals its involvement in grain development, phytic acid accumulation and lateral root formation. *J. Exp. Bot.* 67 (14): 4379–4389.

Brearley, C.A. and Hanke, D.E. (1996). Metabolic evidence for the order of addition of individual phosphate esters in the myo-inositol moiety of inositol hexakisphosphate in the duckweed Spirodela polyrhiza L. *Biochem. J.* 314 (Pt 1), no. Pt 1: 227–233.

Breiman, A. and Graur, D. (1995). Wheat Evolution. *Isr. J. Plant Sci.* 43 (2): 85–98.

Buchner, P., Parmar, S., Kriegel, A. et al. (2010). The sulfate transporter family in wheat: tissue-specific gene expression in relation to nutrition. *Mol. Plant* 3 (2): 374–389.

Cominelli, E., Confalonieri, M., Carlessi, M. et al. (2018). Phytic acid transport in Phaseolus vulgaris: a new low phytic acid mutant in the PvMRP1 gene and study of the PvMRPs promoters in two different plant systems. *Plant Sci.* 270: 1–12.

Cosgrove, D.J. and Irving, G.C.J. (1980). *Inositol Phosphates: Their Chemistry, Biochemistry, and Physiology*. Elsevier Scientific Publication.

Coulibaly, A., Kouakou, B., and Chen, J. (2011). Phytic acid in cereal grains: structure, healthy or harmful ways to reduce Phytic acid in cereal grains and their effects on nutritional quality. *Am. J. Plant Nutr. Fertil. Technol.* 1 (1): 1–22.

Doria, E., Galleschi, L., Calucci, L. et al. (2009). Phytic acid prevents oxidative stress in seeds: evidence from a maize (Zea mays L.) low phytic acid mutant. *J. Exp. Bot.* 60 (3): 967–978.

García-Estepa, RosaMa, Guerra-Hernández, E., and García-Villanova, B. (1999). Phytic acid content in milled cereal products and breads. *Food Res. Int.* 32 (3): 217–221.

Gillman, J.D., Pantalone, V.R., and Bilyeu, K. (2009). The low Phytic acid phenotype in soybean line CX1834 is due to mutations in two homologs of the maize low Phytic acid gene. *Plant Genome* 2: 179–190.

Hassler, S., Lemke, L., Jung, B. et al. (2012). Lack of the Golgi phosphate transporter PHT4;6 causes strong developmental defects, constitutively activated disease resistance mechanisms and altered intracellular phosphate compartmentation in Arabidopsis. *Plant J.* 72 (5): 732–744.

Hegeman, C.E., Good, L.L., and Grabau, E.A. (2001). Expression of D-myo-inositol-3-phosphate synthase in soybean. Implications for phytic acid biosynthesis. *Plant Physiol.* 125 (4): 1941–1948.

Kim, S.-I. and Tai, T.H. (2011). Identification of genes necessary for wild-type levels of seed phytic acid in Arabidopsis thaliana using a reverse genetics approach. *Mol. Gen. Genomics.* 286 (2): 119–133.

Kuwano, M., Mimura, T., Takaiwa, F., and Yoshida, K.T. (2009). Generation of stable 'low phytic acid' transgenic rice through antisense repression of the 1D-myo-inositol 3-phosphate synthase gene (RINO1) using the 18-kDa oleosin promoter. *Plant Biotechnol. J.* 7 (1): 96–105.

Li, Y. et al. (2015). Phosphate transporter OsPht1;8 in rice plays an important role in phosphorus redistribution from source to sink organs and allocation between embryo and endosperm of seeds. *Plant Sci.* 230: 23–32.

Liang, Z. et al. (2017). Efficient DNA-free genome editing of bread wheat using CRISPR/Cas9 ribonucleoprotein complexes. *Nat. Commun.* 8: 14261.

Liu, Q.-L., Xu, X.-H., Ren, X.-L. et al. (2007). Generation and characterization of low phytic acid germplasm in rice (Oryza sativa L.). *Theor. Appl. Genet.* 114 (5): 803–814.

Loewus, F.A. and Murthy, P.P.N. (2000). Myo -inositol metabolism in plants. *Plant Sci.* 150: 1–19.

Nagarajan, V.K., Jain, A., Poling, M.D. et al. (2011). Arabidopsis Pht1; 5 mobilizes phosphate between source and sink organs and influences the interaction between. *Plant Physiol.* 156 (July): 1149–1163.

Nunes, A.C.S., Vianna, G.R. et al. (2006). RNAi-mediated silencing of the myo-inositol-1-phosphate synthase gene (GmMIPS1) in transgenic soybean inhibited seed development and reduced phytate content. *Planta* 224 (1): 125–132.

Nussaume, L., Kanno, S., Javot, H. et al. (2011). Phosphate import in plants: focus on the PHT1 transporters. *Front. Plant Sci.* 2: 83.

Odom, A.R., Stahlberg, A., Wente, S.R., and York, J.D. (2000). A role for nuclear inositol 1,4,5-trisphosphate kinase in transcriptional control. *Science* 287 (5460): 2026–2029.

Odom, A.R., Stahlberg, A., Wente, S.R., and York, J.D. (2000). A role for nuclear inositol 1,4,5-trisphosphate kinase in transcriptional control. *Science* 287 (5460): 2026–2029.

Panzeri, D., Cassani, E., Doria, E. et al. (2011). A defective ABC transporter of the MRP family, responsible for the bean lpa1 mutation, affects the regulation of the phytic acid pathway, reduces seed myo-inositol and alters ABA sensitivity. *New Phytol.* 191 (1): 70–83.

Punjabi, M., Bharadvaja, N., Jolly, M. et al. (2018). Development and evaluation of low Phytic acid soybean by siRNA triggered seed specific silencing of inositol polyphosphate 6−/3−/5-kinase gene. *Front. Plant Sci.* 9: 804.

Raboy, V. (1997). Accumulation and storage of phosphate and minerals. Cellular and molecular biology of plant seed development. In: *Cellular and Molecular Biology of Plant Seed Development* (eds. B.A. Larkins and I.K. Vasil), 441–477. Dordrecht: Springer Netherlands.

Raboy, V., Gerbasi, P.F., Young, K.A. et al. (2000). Origin and seed phenotype of maize low phytic acid 1-1 and low phytic acid 2-1. *Plant Physiol.* 124 (1): 355–368.

Raboy, V. (2003). Myo-Inositol-1,2,3,4,5,6-hexakisphosphate. *Phytochemistry* 64 (6): 1033–1043.

Rasmussen, S.K., Ingvardsen, C.R., and Torp, A.M. (2010). Mutations in genes controlling the biosynthesis and accumulation of inositol phosphates in seeds. *Biochem. Soc. Trans.* 38 (2): 689–694.

Rasmussen, S.K., Ingvardsen, C.R., and Torp, A.M. (2010). Mutations in genes controlling the biosynthesis and accumulation of inositol phosphates in seeds. *Biochem. Soc. Trans.* 38 (2): 689 LP–694.

Sasakawa, N., Sharif, M., and Hanley, M.R. (1995). Metabolism and biological activities of inositol pentakisphosphate and inositol hexakisphosphate. *Biochem. Pharmacol.* 50 (2): 137–146.

Seeds, A.M., Sandquist, J.C., Spana, E.P., and York, J.D. (2004). A molecular basis for inositol polyphosphate synthesis in drosophila melanogaster. *J. Biol. Chem.* 279 (45): 47222–47232.

Shi, J., Wang, H., Wu, Y. et al. (2003). The maize low-phytic acid mutant lpa2 is caused by mutation in an inositol phosphate kinase gene. *Plant Physiol.* 131 (2): 507–515.

Shi, J., Wang, H., Wu, Y. et al. (2003). The maize low-Phytic acid mutant lpa2 is caused by mutation in an inositol phosphate kinase gene. *Plant Physiol.* 131 (2): 507–515.

Shi, J., Wang, H., Schellin, K. et al. (2007). Embryo-specific silencing of a transporter reduces phytic acid content of maize and soybean seeds. *Nat. Biotechnol.* 25 (8): 930–937.

Shukla, V.K., Doyon, Y., Miller, J.C. et al. (2009). Precise genome modification in the crop species Zea mays using zinc-finger nucleases. *Nature* 459 (7245): 437–441.

Shukla, V., Kaur, M., Sipla, A. et al. (2016). Tissue specific transcript profiling of wheat phosphate transporter genes and its association with phosphate allocation in grains. *Scientific Reports* 6: 392–393.

Srivastava, B.N., Biswas, T.D., and Das, N.B. (1955). Influence of fertilizer and manures on the content of phytin and other forms of phosphorus in wheat and their relation to soil phosphorus. *J. Indian Soc. Soil Sci.* 3 (1): 33–40.

Stephens, L.R. and Irvine, R.F. (1990). Stepwise phosphorylation of myo-inositol leading to myo-inositol hexakisphosphate in Dictyostelium. *Nature* 346 (6284): 580–583.

Stevenson-Paulik, J., Odom, A.R., and York, J.D. (2002). Molecular and biochemical characterization of two plant inositol polyphosphate 6−/3−/5-kinases. *J. Biol. Chem.* 277 (45): 42711–42718.

Stevenson-Paulik, J., Bastidas, R.J., Chiou, S.-T. et al. (2005). Generation of phytate-free seeds in Arabidopsis through disruption of inositol polyphosphate kinases. *Proc. Natl. Acad. Sci. U. S. A.* 102 (35): 12612–12617.

Stiles, A.R., Qian, X., Shears, S.B., and Grabau, E.A. (2008). Metabolic and signaling properties of an Itpk gene family in Glycine max. *FEBS Lett.* 582 (13): 1853–1858.

Suzuki, M., Tanaka, K., Kuwano, M., and Yoshida, K.T. (2007). Expression pattern of inositol phosphate-related enzymes in rice (Oryza sativa L.): implications for the phytic acid biosynthetic pathway. *Gene* 405 (1–2): 55–64.

Teng, W., Zhao, Y.Y., Zhao, X.Q. et al. (2017). Genome-wide identification, characterization, and expression analysis of PHT1 phosphate transporters in wheat. *Front. Plant Sci.* 8: 543.

Upadhyay, S.K., Kumar, J., Alok, A., and Tuli, R. (2013). RNA-guided genome editing for target gene mutations in wheat. *G3 (Bethesda).* 3 (12): 2233–2238.

Verbsky, J.W., Chang, S.-C., Wilson, M.P. et al. (2005). The pathway for the production of inositol hexakisphosphate in human cells. *J. Biol. Chem.* 280 (3): 1911–1920.

Wang, Y., Cheng, X., Shan, Q. et al. (2014). Simultaneous editing of three homoeoalleles in hexaploid bread wheat confers heritable resistance to powdery mildew. *Nat. Biotechnol.* 32 (9): 947–951.

Wang, F., Rose, T., Jeong, K. et al. (2015). The knowns and unknowns of phosphorus loading into grains, and implications for phosphorus efficiency in cropping systems. *J. Exp. Bot.* 67 (5): 1221–1229.

Xu, X.-H., Zhao, H.J., Liu, Q.L. et al. (2009). Mutations of the multi-drug resistance-associated protein ABC transporter gene 5 result in reduction of phytic acid in rice seeds. *Theor. Appl. Genet.* 119 (1): 75–83.

Ye, H., Zhang, Q.X., Broughton, S. et al. (2011). A nonsense mutation in a putative sulphate transporter gene results in low phytic acid in barley. *Funct. Integr. Genomics* 11 (1): 103–110.

York, J.D., Odom, A.R., Murphy, R. et al. (1999). A phospholipase C-dependent inositol polyphosphate kinase pathway required for efficient messenger RNA export. *Science* 285 (5424): 96–100.

Zhang, F., Sun, Y., Pei, W. et al. (2015). Involvement of OsPht1;4 in phosphate acquisition and mobilization facilitates embryo development in rice. *Plant J.* 82 (4): 556–569.

Zhang, Y., Liang, Z., Zong, Y. et al. (2016). Efficient and transgene-free genome editing in wheat through transient expression of CRISPR/Cas9 DNA or RNA. *Nat. Commun.* 7: 12617.

Zhao, H., Frank, T., Tan, Y. et al. (2016). Disruption of OsSULTR3;3 reduces phytate and phosphorus concentrations and alters the metabolite profile in rice grains. *New Phytol.* 211 (3): 926–939.

Zong, Y., Wang, Y., Li, C. et al. (2017). Precise base editing in rice, wheat and maize with a Cas9-cytidine deaminase fusion. *Nat. Biotechnol.* 35 (5): 438–440.

10

Genome Engineering for Nutritional Improvement in Pulses

Chirag Uppal, Ajinder Kaur, and Chhaya Sharma

School of Agricultural Biotechnology, Punjab Agricultural University, Ludhiana, Punjab, India

CHAPTER MENU

10.1 Introduction

Several biomolecules, such as amino acids, fatty acids, vitamins and minerals are essential to maintain the normal physiology of the body. Humans and animals are unable to synthesize all of them and, therefore, these biomolecules are strictly needed to be supplied with the diet. Most of the essential amino acids, fatty acids and vitamins are either lacking or completely missing in crops used for human food and animal feed, e.g. soybean is deficient in Vitamin D and methionine; chickpea is deficient in cysteine and lysine. For years, a number of people who suffered from generalized malnutrition were being tracked throughout the world, but only over the past decade, nutritionists have gathered evidence about the significance of malnutrition. More than half of the world's population is suffering from diseases caused by inadequate intake of essential amino acids, vitamins and minerals (Zhu et al. 2010). These problems are pronounced in developing countries where most of the food intake relies on a single crop, which often provides an inadequate nutrition (Graham and Gregorio 2001). The poor people suffer the most from this hidden hunger, particularly

Genome Engineering for Crop Improvement, First Edition. Edited by Santosh Kumar Upadhyay.

young children and women. These people spend 60–80% of their earnings on food and depend heavily on economically cheap and highly nutritious staple crops such as pulses. Pulses, the world's most important staple crop, are a good source of protein, but poor in essential nutrients (Toenniessen 2002). Therefore, over the past two decades, several biotechnological strategies aimed at enhancing the nutritional quality of pulses have been developed and evaluated. This chapter covers examples of pulse crops that have been improved nutritionally through biotechnological strategies. These strategies range from modifying amino acid composition, modifying seed oil composition, over-expression of heterologous or homologous proteins, enhancing minerals and vitamin content, eliminating allergenic compounds and development of edible vaccines. The chapter reviews the progress and potential of different biotechnological advances viz., molecular breeding, genetic transformation and genome editing that are employed worldwide for improving nutritional level in pulses.

10.2 Need for Nutritional Improvement in Pulses

Pulses, also known as grain legumes (Figure 10.1), are the major staple crop all over the world providing animals and humans with a nutrient rich diet. It is a food source that does not even need post-harvest processing or industrial processing. Pulses constitute one of the most important crops of the agriculture system and are second only to cereals in providing

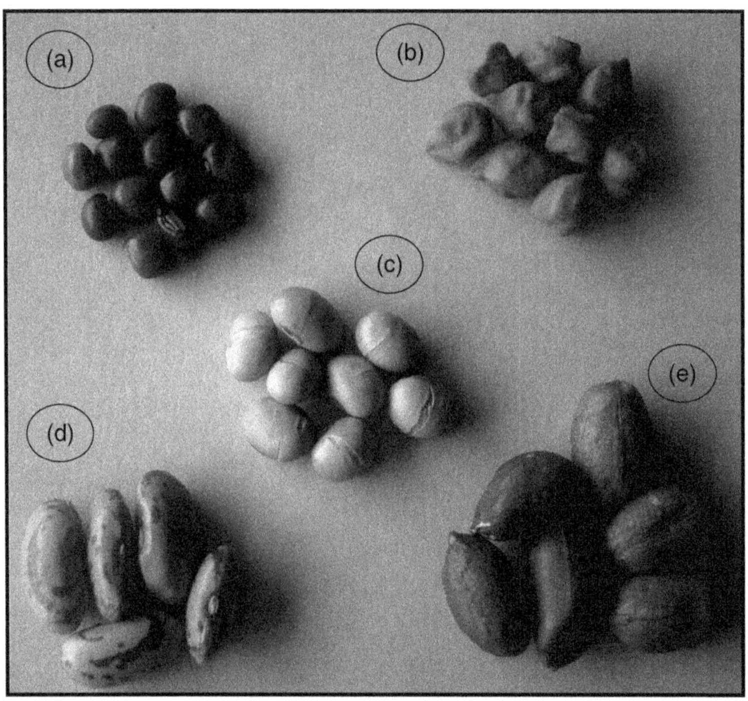

Figure 10.1 Pulses (a) Pigeon pea (b) Chickpea. (d) Rajmash and legumes (c) Soybean (e) Peanut used for genetic transformation to improve nutritional quality. *Source:* Ajinder Kaur.

food for the world. Cereals, being starchy crops, provide only calories to fulfill energy requirements, but lack nutritive diversity to ensure a balanced diet (Kearney 2010). Pulses, such as beans, peas, lentils and chickpeas constitute a major staple food of human beings as these are cheap, easily available, rich in protein, complex carbohydrates, fiber, high-grade fat, vitamins and minerals, etc. Many pulses are utilized as fodders, green manures and forages, for example *Lupinus* and *Trifolium* (International Legume Database and Information Service (ILDIS) 2004). Pulses are a poor man's only source of protein and contain 18–40% protein per kg which is more than an animal-based diet like milk, eggs and meat. Also, pulses' proteins are easy to digest compared to animal-based proteins. Carbohydrates in pulses promote overall body growth. Increased intake of legume protein has weight-loss benefits, for example, grain legume α-amylase inhibitors have been reported to prevent and treat obesity and diabetes. Pulses contain different anti-carcinogenic and anti-inflammatory compounds, like protease inhibitors, saponins, phytosterols, lectins and phytates (Mathers 2002). Serine protease inhibitors of soybean have been shown to arrest mammalian tumors (Kennedy 1998). Pulses fix atmospheric nitrogen, are easy to cultivate, highly non-perishable, provide food security and generate income. The significance of pulses is rapidly growing due to nutrition-rich diet demand of the increasing population and the need to reduce risks related to intake of an animal-based food diet.

However, pulses are deficient in most of the essential amino acids, oils, minerals and vitamins, etc. resulting in people suffering from micronutrient deficiencies (MiND), especially those relying primarily on pulses as their staple diet (Ritchie et al. 2018). More than two billion people worldwide, with nearly 200 million living in India, are suffering from hidden hunger due to micronutrient deficiencies in their daily diet (Food and Agricultural Organization 2013, 2017). In this regard, there is a need to identify and promote biotechnological interventions in pulses as these can be grown using less water, are more adaptable to the changing climate, and can thus help to curb climate change and feed the growing population (Chibarabada et al. 2017). The present chapter discusses the role played of biotechnology in improving nutritional composition of pulses through genome engineering.

10.3 Nutritional Defects in Pulses Targeted for Genetic Engineering

Fruits, vegetables and meat are the major source of micronutrients and proteins, respectively. Animal-based diet remains unaffordable, while fruits and vegetables are comparatively cheaper when in season, but expensive when out of season. Therefore, most of the low- or middle-income countries depend solely on pulses for meeting their nutritional requirements (Iqbal et al. 2006). However, pulses are deficient in sulfur-containing amino acids, lysine, minerals (especially iron, zinc and calcium) and ultimately lead to malnutrition.

Further, pulses also contain compounds detrimental to our health referred to as antinutrient factors (ANFs). ANFs are the compounds that are synthesized by the plants as a part of their defense mechanism. Synthesis of ANF compounds lead to inactivation of some useful nutrients and ultimately decrease the nutritive value of the crop. Several common ANFs are phytates, oxalates, saponins, lectins, Arah2 proteins, raffinose family of oligosaccharides (RFO) and trypsin inhibitors (TI), etc. ANFs have several harmful effects on humans, for example, phytates impair iron, calcium and zinc absorption, saponins increase intestinal permeability and cause leaky gut, RFO are responsible for flatulence, Arah2 proteins cause allergy (Agarwal 2016). These reasons make pulses an excellent target for genetic engineering to fortify pulses with sulfur-containing and other essential amino acids, essential fatty oils and even vaccines. Similarly, genetic approaches, such as RNA interference (RNAi) are employed to make pulses free of antinutritional or anti-allergic compounds. RNAi is a naturally occurring mechanism which is used to down-regulate or silent the target genes (Fire et al. 1998). In RNAi technology, the gene product is not a protein but an RNAi molecule. RNAi, also known as post-transcriptional gene silencing (PTGS), is a conserved biological response which uses non-coding RNAs like short interfering RNA (siRNA), micro RNA (miRNA) to suppress the activity of other intrinsic genes. In natural phenomenon, these non-coding RNAs complex with special endonuclease proteins to form complexes, called as RISC (RNA induced silencing complex). These complexes degrade the endogenous and exogenous pathogenic RNA molecules and finally inhibit the expression of the specific gene.

10.4 Genome Engineering as an Alternate to Conventional Breeding

Conventional breeding involves crop improvement through crossing two distinct genotypes for increasing yield and physiochemical properties of the recipient genotype. Conventional breeding has several limitations, for example, transfer compatibility, absence of desired traits within the gene pool and masking effects of the environment during screening experiments (Prohens 2011). This makes genetic transformation the best alternate for the introduction or removal of several traits from the genome of pulses because of its ability to bypass several inter-kingdom, inter-generic or inter-specific gene transfer-related problems (Tiwari and Tuli 2012). Different genes for fortifying pulses with missing amino acids, minerals and vitamins that are present in alien germplasm can be introduced in pulses using several methods like *Agrobacterium*-mediated, biolistic and electroporation (Eapen 2008). *Agrobacterium tumefaciens*, also known as a natural genetic engineer, is a bacterial pathogen occurring in soil. *Agrobacterium* infects the plants at wound sites and cause crown-gall disease. *Agrobacterium* transfers its transfer (T)-DNA into host plant cells using a bacterial type IV secretion system (Horsch et al. 1985; Hwang et al. 2017). Biolistic, also known as particle bombardment, is commonly employed to transfer the gene of interest in nucleus, chloroplast or mitochondria. In this method, the gene of interest is first coated over the tungsten particles which are then accelerated to a high speed to bombard the recipient cells or tissues (Klein et al. 1987; Sanford et al. 1987). Electroporation is a direct gene transfer method in which the foreign gene is transferred to protoplast with the use of high voltage electric pulse (Fromm et al. 1986). The following section provides detail of commonly used transformation methods along with examples of several genetically engineered pulses with improved nutritional value. Table 10.1 summarizes the major genetic interventions in different pulse and legume crops.

Table 10.1 Major genome engineering events for nutritional improvement in different pulse and legume crops.

Plant material/ Cultivar	Transformation method	Explant	1. Promoter 2. Transgene	Selectable marker	Remarks	Reference(s)
a. PIGEONPEA						
Pigeon pea var. "Hyderabad"	*Agrobacterium tumefaciens* strain EHA105	Embryonic axis, Cotyledonary node	1) CaMV 35S 2) Two Kb fragment containing the coding region of Haemagglutinin protein antigen of Rinderpest virus	*npt II*	Cotyledonary nodes responded with higher transformation efficiency (67%) as compared to embryonic axes (51%). The highest level of expression of H protein in leaves of pigeon pea was 0.49% of the total soluble protein. The transgene was expressed in the progeny.	Satyavathi et al. 2003)
Pigeon pea var. 'Hyderabad'	*Agrobacterium tumefaciens* strain GV3 101	Cotyledonary node	1) CaMV 35S 2) Hemagglutinin-neuraminidase gene (HN) from an Indian isolate of *Peste des Petits ruminant* (PPR) virus	*npt II*	PCR-based transformation efficiency was 60–65%. HN protein expression in transgenic lines was confirmed by Western blot analysis. Plant-derived HN protein was biologically active as demonstrated by neuraminidase activity. Transgenic plants were fertile and PCR of T_1 plants confirmed the inheritance of the transgene.	Prasad et al. (2004)
Pigeonpea var. "ICPL 87"	*Agrobacterium*-mediated transformation and Biolistic method	Embryonic axis	1) phaseolin or *Arabidopsis* 2S2 promoter 2) *dhdps-r1* gene (mutant gene encoding DHDPS) from *N. sylvestris*	*npt II*	PCR analysis indicated that transgene was stably transmitted to the next generation and most of the transgenic lines contained a single transgene locus. Enhanced DHDPS activity (two to six-fold) was found in immature seeds at a late stage of maturation. Over expression of *dhdps-r1* led to an enhanced content (1.6–8.5 times) of free lysine in transgenic pigeon pea seeds as compared to wild type seeds.	Thu et al. (2007)

(*Continued*)

Table 10.1 (Continued)

Plant material/ Cultivar	Transformation method	Explant	1. Promoter 2. Transgene	Selectable marker	Remarks	Reference(s)
b. CHICKPEA						
Chickpea Two *desi* type cultivars, Semsen and Amethyst	*Agrobacterium tumefaciens* strain AGL1	One cotyledon attached to half of the embryonic axis	1) Seed specific promoter of a pea vicilin gene 2) A chimeric gene encoding sunflower seed albumin (SSA)	*npt II*	The methionine and cysteine content of chickpea seeds was increased by 16 and 8%, respectively that remained consistent in the next generations.	Chiaiese et al. (2004)
c. SOYBEAN						
Soybean	Electroporation	Protoplast	1) Nos 2) Chimeric genes coding for resistance to kanamycin and G428	*npt II*	Stably transformed calli were obtained from protoplasts, however no plants could be regenerated.	Christou et al. (1987)
Soybean *cv.* "Merrill"	Biolistic	Meristem	1) β-conglycinin promoter 2) *fad* 2-1 anitsense gene	*npt II*	Transgenic seeds had elevated oleic acid content (70–80% as compared to 13% in control). Southern analysis confirmed over expression of oleic acid (up to 80%), the expression was maintained in T_1 and T_2 generations. Northern analysis revealed consistent expression of mRNA level.	Christou et al. (1990)
A method has been developed for producing a plant exhibiting at least one modified phenotypic trait by suppressing expression of an endogenous gene in the plant. This phenomenon has been patented with the name "TransSwitch," as a method for deliberately turning-off endogenous genes.						Jorgensen and Napoli (1994)
Soybean *cv.* "Williams"	Biolistic	Meristem of immature seed	1) Phaseolin promoter 2) Two transgenes: a) *Corynebacterium dapA* gene [Lys insensitive dihydrodipicolinic acid synthase (DHDPS)] b) *E. coli lys C* gene 3) [Lys insensitive *E. coli* aspartokinase (AK)]	*npt II*	Expression of lysine insensitive forms of two key enzymes DHDPS and AK in soybean resulted in 100-fold increase in free lysine and fivefold increase in total lysine that was heritable.	Falco et al. (1995)

Soybean *cv.* "Merrill"	Biolistic	Meristem	1) β-conglycinin promoter 2) *fad 2-1* gene. They have compared antisense and TransSwitch in somatic embryos using *fad 2-1* gene	*npt II*	Produced soybean oil with high oleic acid content by silencing the expression of *fad 2-1* gene in developing soybean seeds. TransSwitch suppressed embryos had an oleic acid content tightly clustered around 70%, whereas the antisense embryos had oleic acid content ranging from 15 to 20%.	Kinney (1995)
Soybean *cv.* "Jack" and line "F173"	*Agrobacterium*-mediated transformation and Biolistic method	Embryogenic culture	1) β-phaseolin promoter 2) maize 15 kDa zein protein gene	*hpt*	On an average, there was 12 ± 20% and 15 ± 35% increase in methionine and cysteine, respectively in the transgenic lines, and the increase was heritable.	Dinkins et al. (2001)
Soybean *cv.* 'Jack'	Particle bombardment	Somatic embryo	1) β-conglycinin 2) cDNA coding a major soybean allergen, i.e. Gly m Bd 30 K, also referred to as P34 protein	*hpt*	Transgene-induced gene silencing was used to prevent the accumulation of P34 protein. Transgenic plants were free of allergen and were maintained in the next three generations in Mendelian fashion. Immunoblots of seed proteins confirmed the presence of P34 in the control and silencing of P34 expression in the transgenic lines.	Herman et al. (2003)
Soybean	*Agrobacterium tumefaciens* strain ABI	Meristem	1. An enhanced 35S promoter: PE35S 2. a. *At-VTE3* gene (2-methyl-6-phytylbenzoquinol methyltransferase) b. At-VTE4 (tocopherol methyltransferase	*npt II*	Transfection of soybean with *At-VTE3* gene resulted in 10% increase inα-tocopherol. However, co-expression of *At-VTE3* gene with *At-VTE4* gene resulted in accumulation of 95% α-tocopherol in transgenic soybean, thereby improving vitamin E content in soybean oil.	Van et al. (2003)

(*Continued*)

Table 10.1 (Continued)

Plant material/ Cultivar	Transformation method	Explant	1. Promoter 2. Transgene	Selectable marker	Remarks	Reference(s)
Soybean genotype "A3237"	*Agrobacterium tumefaciens* strain EHA101	Cotyledon	1) β-conglycinin promoter 2) cDNA clone of Δ^6-desaturase from *Borage officinalis*	*bar*	Transformation using Δ^6-desaturase gene was stable and heritable as confirmed by Northern analysis. Transgenic soybean showed significantly high level of γ-linolenic acid (GLA) and STA (ω-3 fatty acid) leading to improved nutritional quality.	Sato et al. (2004)
Soybean genotype "Thorne"	*Agrobacterium tumefaciens* strain EHA101	Cotyledon	1) CaMV 35S 2) The major K99 fimbrial subunit, FanC, a plant-optimized synthetic version of the gene (*synfanC*)	*bar*	The first step toward the development of a K99 edible vaccine to protect against enterotoxigenic *E. coli* (ETEC) that causes acute diarrhea in humans and farm animals. Mice immunized with protein extract derived from transgenic leaves developed significant level of antibody titers and produced antigen-specific CD4+ T lymphocytes which gave protection against diarrhea.	Piller et al. (2005)
Soybean *cv.* "Jack"	Biolistic method	Somatic embryo	1) Soybean seed-specific glycinin promoter 2) β-subunit of heat labile toxin of enterotoxigenic *E. coli* (LTB)	*hyg*	PCR and Western blot analyses confirmed the presence of LTB in T_0 plants that was stable in next generations. LTB accumulation was 1.8–2.4% of the total protein and was biochemically stable, functionally active and highly immunogenic. Significant anti-LTB titers were detected in pooled sera and fecal samples following transgenic administration.	Moravec et al. (2007)

Soybean *cv.* 'Jack'	Particle bombardment	Embryogenic suspension culture	1) Specific promoter from a soybean glycinin (*A1aB1b*) gene 2) A modified Gy1 (*A1aB1b*) proglycinin gene with a synthetic DNA encoding for four continuous methionines (V3-1)	*hpt*	SDS PAGE confirmed higher accumulation of glycinin in the seeds obtained from transgenic soybean as compared to control seeds.	Shemy et al. (2007)
Soybean *cv.* "Jack"	*Agrobacterium tumefaciens* strain EHA101	Cotyledon	1) CaMV 35S 2) ω-3 fatty acid desaturase (*fad3*) gene	*bar*	RNAi lines exhibited extremely low level of α-linolenic acid.	Flores et al. (2008)
Demonstrated the stability of transgenic soybean (36) seed powder and soymilk formulations as the absence of cold chain had no adverse effect on the formulations or the FanC antigen. FanC in transgenic soybean could be stored for more than four years at ambient temperature. FanC was stable during overseas shipment of transgenic seed powder and soymilk in the absence of refrigeration. The heat stability of a seed-derived antigen was up to 15 minutes at 85 °C.						Oakes et al. (2009)
Soybean *cv.* "A3525"	*Agrobacterium tumefaciens* strain ABI	Meristem	1) Two different seed-specific promoters 2) AK (Aspartate Kinase) from the enteric bacterium *Xenorhabdusbovienii* (Xb)	*gly*	Transgenic seed showed substantial increase in free threonine level (up to 100-fold) and other major free amino acids (FAAs), resulting in up to 3.5-fold increase in total FAA content.	Qi et al. (2011)
Soybean *cv.* 'Jack'	*Agrobacterium tumefaciens* strain KYRT1	Cotyledon	1) CaMV 35S 2) Synthesized gene (sORF7) of a modified-live Porcine Reproductive and Respiratory Syndrome Vaccine (PRRSV)	*hyg*	PCR and Southern blot confirmed the integration of the transgene into soybean genome. Western blot analysis confirmed the transgene expression. Intra gastric administration of female mice with transgenic soybean seeds elicited both humoral and mucosal immune responses. These were rather similar to antibody levels detected in mice administered with three doses of 10 μg pure PRRSV-ORF7.	Vimolmangkang et al. (2012)

(*Continued*)

Table 10.1 (Continued)

Plant material/ Cultivar	Transformation method	Explant	1. Promoter 2. Transgene	Selectable marker	Remarks	Reference(s)
Soybean *cv.* "Kwangan"	*Agrobacterium tumefaciens* strain EHA101	Cotyledon	1) CaMV 35S and β-conglycinin promoter 2) Recombinant gene PAC (Phytoene synthase-2A Carotene desaturase) gene from "New Golden Rice"	*ppt*	RT-PCR and qRT-PCR analyses confirmed the presence of PAC gene in the leaves of transgenic soybean when β-conglycinin promoter was used. Transgenic seeds accumulated 62-fold more carotenoid content as compared to non-transgenic seeds.	Kim et al. (2012)
Soybean genotype "Thorne"	*Agrobacterium tumefaciens* strain EHA101	Cotyledon	1) CaMV 35S 2) Non-toxic form of SEB (Staphylococcal enterotoxin B) carrying a triple mutation (L45R/Y89A/Y94A)	*bar*	Transfected a triple mutant of Staphylococcal enterotoxin B carrying L45R, Y89A, and Y94A modifications in soybean seeds (soy-mSEB). Vaccination of pigs with purified soy-mSEB resulted in high antibody titers against the native toxin in immunized animals. Antisera from immunized group were able to block native SEB super antigen activity in an *in vitro* neutralization assay. Similar results were obtained when immunized animals were challenged with a sub lethal dose of native toxin.	Hudson et al. (2013)
Soybean zcultivar "Zigongdongdou" (ZD)	*Agrobacterium* strain EHA105	Cotyledonary-node	1) Seed specific promoter "Legumin B4" 2) Met-insensitive form of Arabidopsis cystathionineγ-synthase (*AtD-CGS*) gene	glufosinate resistance	The expression of AtD-CGS was not observed in mature seeds, but in developing seeds as legumin B4 promoter is expressed during seed development. Western blot analyses revealed that soluble Met increased significantly in developing green seeds (3.8 to 7-fold). The total Met was increased by 1.8 to 2.3-fold. Total protein also increased with increased Met content showing positive relation between high Met content and synthesis of other amino acids.	Song et al. (2013)

Soybean *cv.* "Merrill"	Biolistic	Embryogenic culture	1) Promoters: a) CaMV 35S and b) β-conglycinin c) Cruciferin 2) Two transgenes: d) *MB-16 bc* (bacterial codon) e) *MB-16 sc* (soybean codon)	*hyg*	*MB-16 bc* gene under β-conglycinin promoter was stable up to T_3 generation. *MB-16 sc* gene under β-conglycinin promoter was stable up to T_5 generation. Methionine and cysteine increased significantly by 16.2% and 65.9%, respectively. The levels of threonine, lysine and leucine also increased.	Zhang et al. (2013)
Soybean genotype "Thorne"	*Agrobacterium tumefaciens* strain EHA101	Cotyledon	1) Either glycinin or β-conglycinin promoter 2) Methionine-rich 11 kDa δ-zein	*bar*	RNAi suppressed the expression of β-conglycinin, the abundant 7S seed storage protein of soybean. Transfection of 11 kDa δ-zein into β-conglycinin knockdown line did not enhance the accumulation of 11 kDa δ-zein. However, 3 to 16-fold increased accumulation of 11 kDa δ-zein was observed when the transgenic plants were grown in sulfur-rich medium.	Kim et al. (2014)
d. COMMON BEAN						
Common bean *cv.* "Olathe"	Biolistic	Embryonic axis	1) CaMV 35S 2) *2S albumin* gene from Brazilian nut, Antisense sequence of*AC1*, *AC2*, *AC3*, and *BC1* genes from Bean Golden Mosaic Gemini virus	*neo*	The average transformation frequency was 0.9%. PCR and Southern blot analyses confirmed the co-maintenance of foreign genes in T_2 and T_3 generations in Mendelian fashion.	Aragao et al. (1996)
Common bean *cv.* 'Olathe'	Biolistic	Embryonic axis	1) CaMV 35S 2) *2S albumin* gene from Brazilian nut (*be2s1* gene)	*neo*	ELISA and Western blotting confirmed the stable expression of the transgene. Transgenic bean accumulated 14–23% more methionine as compared to non-transformed bean.	Aragao et al. (1999)

(*Continued*)

Table 10.1 (Continued)

Plant material/ Cultivar	Transformation method	Explant	1. Promoter 2. Transgene	Selectable marker	Remarks	Reference(s)
e. PEANUT						
Peanut *cv*. "Tatu" (Valencia group)	Biolistic	Cotyledon	1) CaMV 35 S 2) *2S albumin* gene from Brazilian nut	*gus*	ELISA confirmed detection of both 2S storage protein and β-glucuronidase. Both peptides were present at similar levels in the seed extracts.	Lacorte et al. (1997)
Peanut	*Agrobacterium tumefaciens* strain EHA105	Shoot apex	1) CaMV 35 S 2) Haemagglutinin protein of Rinderpest viral gene	*npt II*	Transformation was stable. ELISA confirmed the expression level of H protein in the range of 0.2–1.3% of total soluble protein. Transgenic peanut induced both humoral and cell-mediated immunity without the need of any adjuvant.	Khandelwal et al. 2003a; Khandelwal et al. (2003b)
Peanut *cv*. "JL24"	Biolistic	Pre-cultured embryos	1) CaMV 35 S 2) *VP2* gene of Bluetongue virus	*npt II*	The presence of transgene was confirmed by Southern analysis. RT-PCR confirmed the expression of the transgene.	Athmaram et al. (2006)
Peanut genotype "Georgia Green"	Biolistic	Somatic embryo	1) CaMV 35S 2) *arah-2* RNAi plasmid constructed from pFGC1008 (AY310333)	*hpt*	*arah2* RNAi was stably integrated in the transgenic plants and the following generations. Transgenic plants showed significant reduction of Arah-2 and Arah-6 allergens. On feeding the patients with transgenic peanuts, there was significant reduction in the level of allergenicity due to reduced level of Ig-E in patients.	Chu et al. (2008)
Peanut genotype "Georgia Green"	*Agrobacterium tumefaciens* strain EHA105	Hypocotyl	1) CaMV 35S 2) PCR constructed inverted repeat of *arah-2* transgene used as RNAi	*npt II*	Transformation frequency was 44%. PCR and Southern analyses confirmed the elimination of Arah-2 allergen in transgenic plants. The allergenicity was also reduced significantly.	Dodo et al. (2008)

Peanut	*Agrobacterium tumefaciens* strain GV3 101	Shoot apex	1) CaMV 35 S 2) HN protein of PPR virus	*npt II*	Neuraminidase assay confirmed the presence of transgene in T_0 and T_1 generations suggesting stable transformation. Oral feeding of transgenic peanut induced cell-mediated secondary immune response in sheep in the absence of any adjuvant.	Khandelwal et al. (2011)
f. *Vicia* spp.						
Vicia narbonensis	*Agrobacterium tumefaciens* strain EHA 101	Shoot tip, epicotyl	1) CaMV 35S 2) *2S albumin* gene of Brazilian nut	*npt II*	The presence of foreign gene in transgenic plants was confirmed by Southern analysis. The expression level of transgene was high in leaves and roots of *Vicianarbonensis*. Foreign gene was inherited in Mendelian fashion in T_1 and T_2 progenies.	Saalbach et al. (1994)
Vicia faba L. and pea	*Agrobacterium rhizogenes*strain *RifR 15 834*	Epicotyl segment of 5–6 day-old seedling	1) CaMV 35S 2) *2S albumin* gene of Brazil nut	*npt II*	Southern and Western blotting confirmed the presence of 2S albumin in hairy roots of *Viciafaba*and pea. However, no fertile plants could be obtained.	
g. LUPINS						
Lupins *cv.* "Warrah"	*Agrobacterium tumefaciens*	Embryonic axis of immature seed	1) Seed specific promoter of a pea *vicilin* gene 2) Sunflower seed albumin gene (*ssa*) gene	*ppt*	In transgenic lupins, SSA protein was processed in the same way as in sunflower. Immunoblotting and Southern blotting confirmed stable transformation in T_0, T_1 and T_2 generations. Transgenic plants contained 94% more methionine as compared to non-transgenic plants and significantly increased the weight of rats.	Molvig et al. (1997)

(*Continued*)

Table 10.1 (Continued)

Plant material/ Cultivar	Transformation method	Explant	1. Promoter 2. Transgene	Selectable marker	Remarks	Reference(s)
Transgenic (Tg) lupin obtained by Molvig et al. Molvig et al. (1997)	Evaluated the nutritive value of Tg lupin in broiler chicken. On feeding the birds with 250 g/kg Tg lupin, there was a significant decrease in the requirement of supplemented methionine by 0.6 g/kg diet. On feeding Tg lupins, the apparent metabolic energy value increased from 9.42 to 10.18 mj/kg dry matter.					Ravindran et al. (2002)
Lupins *cv.* "Kalya"	*Agrobacterium tumefaciens* strain AGL1	Narrow leaf	1) CaMV 35S 2) Gene encoding for serine acetyltransferase (SAT1) from *Arabidopsis thaliana*	*ppt*	This work combined two approaches. First, SAT was transferred to lupin, producing transgenic events with fivefold increase in O-acetylserine (OAS) content and up to 26-fold increase in the concentration of free cysteine. Secondly, these transgenic lines were crossed with SSA transgenic line (58) which led to dramatic increase in free cysteine and OAS in F_1 generation. There was no significant increase in free methionine.	Tabe et al. (2010)
h. AZUKI BEAN						
Azuki bean *cv.* "Benidainagon" and *L.japonicus*	*Agrobacterium tumefaciens* EHA105	Hypocotyl	1) CaMV 35S 2) D6-fatty-acid desaturase gene form *Mortierelle alpine*	*npt II* and *hpt* for Azuki bean and *L. japonicas*, respectively	GLA (γ-linolenic acid) accumulated at levels of up to 0.44% and 1.26% of the total fatty acids in the leaves and seeds, respectively of a transgenic *V. angularis* line, and at 0.69% in the leaves of its progeny.	Chen et al. (2005)
Azuki bean *cv.* "Beni-dainagon"	*Agrobacterium tumefaciens* strain EHA105	Epicotyl of etiolated seedling	1) CaMV 35S 2) A rice gene (OASA1D) coding for modified α-subunit of anthranilate synthase	hygromycin B	The expression of OASA1D resulted in 6.5 to 16.5-fold free tryptophan content and 1.3 to 2.1-fold increase in total tryptophan content in transgenic seeds as compared to wild-type seeds.	Hanafy et al. (2006)

10.4.1 Biofortified Pigeonpea

Pigeonpea (*Cajanus cajan*) is the most popular grain legume in the world, especially in the Indian subcontinent. Pigeonpea is extensively used in intercropping systems in semi-arid regions. By providing protein to the poorest populations, pigeon pea is playing an important role in curbing malnutrition throughout the world. Despite containing a high protein content of about 21%, pigeon pea is considered as nutritionally poor as compared to the other legumes due to the low content of lysine. DHDPS (dihydrodipicolinate synthase) is an important regulatory enzyme for lysine biosynthesis. Even a small accumulation of lysine can inactivate DHDPS gene by feedback mechanism and stop the lysine biosynthesis which lead to low levels of lysine in pigeon pea. Therefore, mutant DHDPS gene (*dhdps-r1* from *Nicotiana sylvestris*) which is insensitive to lysine feedback inhibition has been transformed in pigeon pea. Transgenic pigeon pea showed two- to six-fold increase in DHDPS activity, which in turn led to 8.5-fold increase in free lysine content in the seeds of transgenic pigeon pea (Thu et al. 2007). Moreover, pigeon pea has also been used to produce edible vaccines. Haemagglutinin protein antigen of Rinderpest virus was stably transformed and expressed in pigeon pea with transformation efficiency of about 67% (Satyavathi et al. 2003). Similarly, Hemagglutinin-neuraminidase gene (HN) from an Indian isolate of *Peste des Petits ruminant*(PPR) virus has been stably transformed and expressed in transgenic pigeon pea. HN protein was biologically active in transgenic pigeon pea as demonstrated by neuraminidase activity (Prasad et al. 2004). A scheme for genetic transformation of pigeon pea developed at School of Agricultural Biotechnology, Punjab Agricultural University (PAU), Ludhiana, is given in Figure 10.2.

10.4.2 Biofortified Chickpea

Chickpea (*Cicero arietinum*), like other grain legumes, is deficient in sulfur amino acids. Transformation using gene encoding for sunflower seed albumin (SSA) increased the methionine and cysteine content of chickpea seeds by 16 and 8%, respectively (Chiaiese et al. 2004).

10.4.3 Biofortified Soybean

Soybean (*Glycine max*) is the most important legume crop widely used for food and feed purposes because of its high protein content, competitive cost, palatability and digestive traits. Despite being rich in high-quality protein (up to 50%) and vegetable oil (up to 20%), soybean lacks in several sulfur containing amino acids, lysine amino acids and essential fatty acids.

However, soybean oil contains 7–10% unstable fatty acids like α-linolenic acids (18:3) that undergoes hydrogenation and leads to the synthesis of undesirable *trans*-fatty acids contributing to poor nutrition quality. Therefore, genome engineering is applied to silence the ω-3 *fad3* gene to reduce the level of linolenic acid (Flores et al. 2008;Kinney 1995). In another experiment, soybean is transfected with the gene coding for Δ6-desaturase which is responsible for the conversion of α-linolenic acid and linoleic acid to γ-linolenic acid (GLA) and STA (ω-3 fatty acids) that led to improvement in the soybean seed oil quality

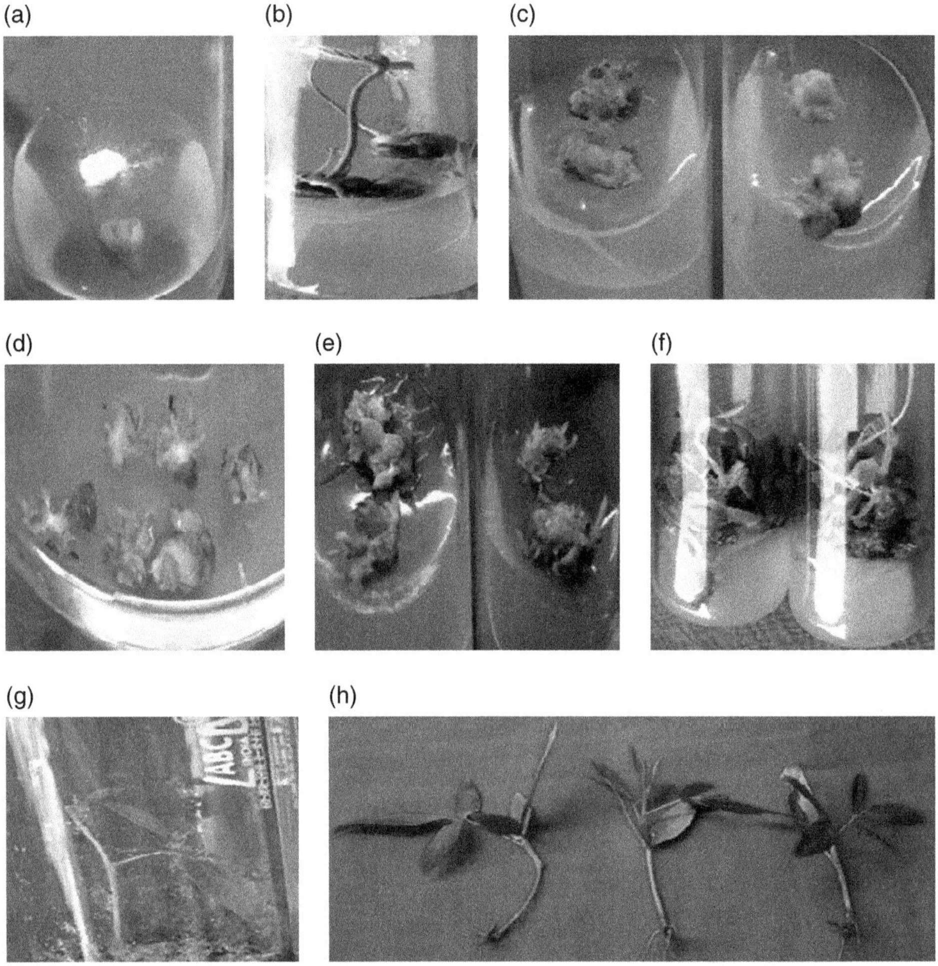

Figure 10.2 Scheme for *Agrobacterium*-mediated genetic transformation in pigeon pea (a) Seed culturing on 5 mg/l BAP. (b) Eight-day old seedling. (c) Multiple shoot induction on cotyledonary node explants (excised from de-capitated seedlings) after 10 days of culturing on MS + 5 mg/l BAP. (d) Agro infection for 10 minutes. (e) Co-cultivation. (f) Regeneration of putative transgenic shoots on MS + 5 mg/l BAP + 250 mg/l Cefotaxime + Selective agent. (g) Root induction in sand after treating with rooting powder. (h) Putative transgenic plants. *Source:* Ajinder Kaur.

(Sato et al. 2004). Similarly, soybean has been targeted to increase the oleic acid contents by suppressing *fad2-1* gene (fatty acid desaturase). *fad2-1* gene suppression led to increase in the oleic acid content of up to 80% (Christou et al. 1987, 1990). Apart from antisense approach of gene silencing, coordinated gene silencing of a sense transgene and a homologous endogenous gene (TransSwitch method) have also been reported to silence *fad2-1* gene. TransSwitch suppressed embryos appeared to have oleic acid content tightly clustered around 70%, whereas the antisense embryos had oleic acid content ranging from 15 to 20% (Jorgensen and Napoli 1994; Kinney 1995).

The importance of ω-3 fatty acid in soybean is clear from the fact that cultivars with improved linoleic, oleic, STA and GLA have been released by several private companies. For example, high oleic acid transgenic soybean cultivars viz., G94-19, G94-1, G168, Treus™ (DP305423), Plenish™ have been released by Dupont in Mexico, Philippines, Singapore Japan, New Zealand, USA, China, European Union, Australia, Canada, South Africa, South Korea, Taiwan and USA. Monsanto also released oleic acid transgenic varieties of soybean under the name of Vistive Gold™ (MON87705) in Canada, European Union, Australia, Columbia, Indonesia, Japan, Mexico, Singapore, South Korea, Taiwan, USA, Vietnam, New Zealand and Philippines. Similarly, STA-rich transgenic varieties (MON89788) have also been released by Monsanto.

Another important nutrient, i.e. lysine amino acid has been enhanced in soybean by expressing lysine insensitive forms of two key enzymes, DHDPS and AK that resulted in 100-fold increase in free lysine and fivefold increase in total lysine content (Falco et al. 1995) Similarly, methionine and cystiene contents have been increased up to 15% by transferring maize 15 kDa zein protein gene in soybean (Dinkins et al. 2001; Kim et al. 2014). Transformation of soybean with Met-insensitive form of Arabidopsis cystathionine γ-synthase (AtD-CGS) gene also resulted in a three-fold increase in methioine content (Song et al. 2013). In a different approach, methionine and cysteine were enhanced in soybean by introduction of a synthetic gene that coded for MB-16, an 11 kDa protein enriched in essential amino acids (Zhang et al. 2013). Similarly, transfection of soybean with AK (Aspartate Kinase) from the enteric bacterium *Xenorhabdusbovienii* resulted in 100-fold increase in threonine level (Qi et al. 2011). In another experiment, glycinin, a seed-storage protein content was increased to improve the nutritional quality of soybean (Shemy et al. 2007). Another important approach to improve the nutrition quality of soybean is by eliminating its allergenic compounds. Complementary DNA (cDNA) coding a major soybean allergen, i.e. Gly m Bd 30 K, also referred to as P34 protein has been used to silence the P34 allergen (Herman et al. 2003). Kim et al. (Kim et al. 2012) reported the production of provitamin-A (beta-carotene) rich soybean by over-expressing carotene desaturase enzyme. Soybean was fortified with vitamin E by increasing the content of δ-tocopherol and decreasing γ-tocopherol through co-expression of 2-methyl-6-phytyl benzoquinol methyltransferase genes from *Arabidopsis* mutant (Van et al. 2003).

Genome engineering in soybean has also led to the production of edible vaccine. Transgenic soybean plants are used as edible vaccine against enterotoxigenic *E. coli* (ETEC) bacterium that causes acute diarrhea in humans and farm animals. A plant-optimized synthetic version of the gene (*synfanC*) coding for K99 fimbrial subunit was used as transgene. Both B-cell and T-cell immunity were generated on feeding mice with *synfanC*-transgenic soybean (Piller et al. 2005). Further, these *synfanC*-transgenic soybean vaccines were reported to be stable for four years at ambient temperature and these did not lose their efficiency even at higher temperatures (Oakes et al. 2009). Similarly, soybean-derived vaccine against ETEC has been produced by expressing β-subunit of heat labile toxin of enterotoxigenic *E. coli* (LTB) in soybean (Moravec et al. 2007). Transgenic soybean expressing synthesized gene (sORF7) provided protection against porcine reproductive and respiratory syndrome (Vimolmangkang et al. 2012). Similarly, transfecting soybean with triple mutant of staphylococcal enterotoxin B, led to production of the soy-mSEB vaccine. Vaccination of pigs with purified soy-mSEB resulted in high antibody titers against the native toxin in immunized animals (Hudson et al. 2013).

10.4.4 Biofortified Common Bean

Common bean (*Phaseolus vulgaris*) is the third most important legume crop followed by soybean and peanut. Common bean shows great variation in growth habit that enables its production in a wide range of cropping systems and environments as in Africa, America, the Middle East, Europe and Asia (Blair et al. 2010). Common bean consumption has been reported to reduce risk of diseases, such as diabetes, obesity, cardiovascular problems, and colon, prostate and breast cancer (Thompson et al. 2009). Despite having an average protein content of 22%, beans have poor nutritive value because these lack sulfur-containing amino acids (Escribano et al. 1990). Transformation of common bean with Brazilian nut *2S Albumin* gene has been reported to increase the methionine content by 14–23% (Aragao et al. 1996, 1999).

10.4.5 Biofortified Peanut

Peanut (*Arachis hypogaea*), also known as ground nut, is the fifth most important legume crop in the world after soybean, cotton, sunflower and rapeseed. Peanut is an inexpensive and nutritious source of protein, carbohydrates and minerals. However, the major seed proteins of peanut, just like other leguminous crops, are deficient in essential sulfur-containing amino acids (SAA). Therefore, genes that code for essential SAA can be introduced in peanut through genome engineering. 2S albumin is a seed-storage protein with more than 18% methionine and 8% cysteine and is found extensively in Brazil nut (*Bertholletia excelsa*) (Gander et al. 1991). Lacorte et al. (Lacorte et al. 1997) used a biolistic approach to transform Brazil nut *2S albumin* gene placed under the control of CaMV 35S promoter in peanut. The 2S albumin protein was correctly processed and stored in peanut seeds. Further, peanut consumption causes Ig-E-mediated allergic reactions in 0.6% of the total world population. Peanut allergy is reported as the most common cause of severe and fatal food-induced anaphylactic reactions. It was estimated that about 2% of the Americans ≤18 years of age are affected by peanut allergy (Sicherer et al. 2003). Yet, to date, there is no cure for peanut allergy and the only approved management for peanut allergy is avoidance. A total of 11 peanut proteins (Ara h1–11) have been identified as allergens. Among these proteins, Ara h2 and Ara h6 are reported to be most potent immune allergens, and therefore a high percentage (80–90%) of peanut allergic patients' sera has Ig-E antibodies specific to Ara h2 and Ara h6 only (Flinterman et al. 2007). Also, allergy in about 90% of patients is caused only by Ara h2 protein (Nicolaou and Custovic 2011). Throughout the world, researchers have been employing an RNAi (RNA interference)-mediated gene silencing approach to eliminate the immunodominant Ara h1 and Ara h2 proteins from peanuts. Chu et al. (Chu et al. 2008) uses the biolistic method that led to stable transformation of the *Ara h2* RNAi plasmid in peanut. As *Ara h2* gene has sequence homologous to *Ara h6* gene, hence *Ara h2* RNAi plasmid was reported to silence both the Ara h2 and Ara h6 allergens. On feeding the patients with transgenic peanuts, reduced level of Ig-E in patients was shown which depicted significant reduction in the level of allergenicity. Similarly, *Agrobacterium*-mediated transformation method was used to silence Ara h2 allergens (Dodo et al. 2008). Genome engineering has also been employed to produce transgenic peanuts that are used as edible vaccines. The vaccines produced from the transgenic peanut are thermostable and are suitable to provide active immunity against different viral pathogens. Both humoral and cell-mediated responses were reported in cattle on feeding of the transgenic peanut generated by transfecting haemagglutinin gene of rinderpest viral (RPVH) gene via *Agrobacterium tumefacians* method (Khandelwal et al. 2003a,b). Further, resistance to bluetongue disease caused by Bluetongue Virus (BTV) has been developed in sheep on feeding transgenic peanut transformed with BTV-VP2 gene (Athmaram et al. 2006). Similarly, peanut transformation with *HN* gene expressing hemagglutinin–neuraminidase protein of Peste des petits ruminant virus (PPRV) provided active immunity to sheep on oral administration of transgenic peanut plants (Khandelwal et al. 2011).

10.4.6 Biofortified Faba Bean

Faba bean (*V. faba*) commonly known as broad bean, horse bean and field bean, is known for its positive nutritional properties. It is the most important staple and traditional diet of Mediterranean regions, Asia, Middle East, America and Africa. Faba bean is rich in polyphenols and other essential component like L-3,4-dihydroxyphenylalanine (L-DOPA, a precursor of dopamine). *V. Faba* intake has also been reported in the management of Parkinson disease, cardiovascular disease, diabetes and cancer (Turco et al. 2016). However, *V. faba* is deficient in sulfur-containing amino acids. Therefore, *2S albumin* gene has been transformed in *V. faba* and *Vicia narbonensis* via *A. tumefacians*-mediated transformation method. *V. narbonensis* showed stable transformation and expression of the 2S albumin gene. Southern and Western blotting confirmed the presence of 2S albumin in hairy roots of *V. faba*, however no fertile plants could be obtained (Saalbach et al. 1994)

10.4.7 Biofortified Lupins

Lupin (*Lupinus angustifolius*), particularly the narrow leaf lupin, is a major grain legume grown in Australia. Lupin is an important protein diet for cattle, swine, sheep and poultry. Like other grain legumes, lupins are also deficient in sulfur-containing amino acids contributing to poor nutritional quality. Therefore, the nutritional quality of lupins has been improved by transferring the sunflower seed albumin gene (*ssa*) gene via *A. tumefacians* method (Molvig et al. 1997). Transgenic lupins have increased the methionine content by 94%. Feeding of transgenic lupins to rats and birds significantly increased their weight, metabolic energy value and also lowered the requirement of the supplemented methionine in their diet (Ravindran et al. 2002). Similarly, a gene for serine acetyltransferase (SAT1) from *Arabidopsis thaliana* was transformed into lupins that led to fivefold and 26-fold increase in O-acetyl serine and free cysteine, respectively. Further, crossing these transgenic lupins with SSA transgenic line (Molvig et al. 1997) led to a dramatic increase in free cysteine and OAS in F_1 generation (Tabe et al. 2010).

10.4.8 Biofortified Azuki Bean

Azuki bean (*Vigna angularis*) is one of the major grain legumes. Its GLA (γ-linolenic acid) content has been increased by the expression of D6-fatty-acid desaturase gene from

Mortierelle alpine (Chen et al. 2005). Further, the expression of a rice gene (OASA1D) increased tryptophan content of azuki bean by 16.5-fold (Hanafy et al. 2006).

10.5 Conclusive Discussion

Biofortification of eco-friendly and economic crops like pulses enables us to feed the growing population and curb the global challenge of climate change. Global scarcity of water and arable land has shifted the focus to increasing the nutritional quality of food rather than increasing the quantity of food. This chapter has well established the role of biofortification in increasing not only the food quantity but also the nutritional quality of pulses. This chapter has provided detailed methodology and outcome of important genome engineering events pertaining to fortification of different varieties of pulses with various nutrients and micronutrients that are usually lacking in the diet of people from developed and developing world. Though a lot has been done in several pulse varieties, a lot is still needed to be done in several other varieties. For example, very little has been done to improve the nutritional content of the common bean, Azuki bean and chickpea, etc. This chapter will help the researchers to choose the appropriate methodology for their specific pulse/legume variety and help them to get the best outcome. New pulse crops developed must be examined for their nutritional benefits. All the non-profit agencies, commercial sectors and healthcare officials must act together so that genome engineering could be used in a proper and efficient manner with the aim of eliminating global hunger. In view of several constraints faced by GMOs, biofortified pulses hold a very bright future as these have potential to curb worldwide challenges of climate change, global hunger, water and arable land scarcity.

References

Agarwal, A. (2016). Duality of anti-nutritional factors in pulses. *Journal of Nutritional Disorders & Therapy* 6: 1.

Aragao, F.J.L., Barros, L.M.G., Brasileiro, A.C.M. et al. (1996). Inheritance of foreign genes in transgenic bean *(Phaseolus vulgaris* L.) co-transformed via particle bombardment. *Theoretical and Applied Genetics* 93: 142–150.

Aragao, F.J.L., Barros, L.M.G., De-Sousa, M.V. et al. (1999). Expression of a methionine-rich storage albumin from the Brazil nut (*Bertholletia excelsa* H.B.K., *Lecythidaceae*) in transgenic bean plants (*Phaseolus vulgaris* L., *Fabaceae*). *Genetics and Molecular Biology* 22: 445–449.

Athmaram, T.N., Bali, G., and Devaiah, K.M. (2006). Integration and expression of bluetongue VP2 gene in somatic embryos of peanut through particle bombardment method. *Vaccine* 24: 2994–3000.

Blair, M.W., González, L.F., Kimani, P.M., and Butare, L. (2010). Genetic diversity, inter-gene pool introgression and nutritional quality of common beans (*Phaseolus vulgaris* L.) from Central Africa. *Theoretical and Applied Genetics* 121: 237–248.

Chen, R., Tsuda, S., Matsui, K. et al. (2005). Production of gamma-linolenic acid in *Lotus japonicus* and *Vigna angularis* by expression of the D6-fatty-acid desaturase gene isolated from *Mortierelle alpine. Plant Science* 169: 599–605.

Chiaiese, P., Ohtsu, N.O., Molvig, L. et al. (2004). Sulphur and nitrogen nutrition influence the response of chickpea seeds to an added, transgenic sink for organic sulphur. *Journal of Experimental Botany* 55: 404.

Chibarabada, T.P., Modi, A., and Mabhaudhi, T. (2017). Expounding the value of grain legumes in the semi and arid tropics. *Sustainability* 9: 60.

Christou, P., Murphy, J.E., and Swain, W.F. (1987). *Proceedings of National Academy of Sciences United States of America* 84: 3962–3966.

Christou, P., McCabe, D.E., Martinell, B.J., and Swain, W.F. (1990). Soybean genetic engineering - commercial production of transgenic plants. *Trends in Biotechnology* 8: 145–151.

Chu, Y., Faustinelli, P., Ramos, M.L. et al. (2008). Reduction of IgE binding and non-promotion of *Aspergillus flavus* fungal growth by simultaneously silencing Arah2 and Arah6 in peanut. *Journal of Agriculture and Food Chemistry* 56: 11225–11233.

Dinkins, R.D., Reddy, M.S.S., Meurer, C.A. et al. (2001). Increased sulfur amino acids in soybean plants overexpressing the maize 15 kda zein protein. *In Vitro Cellular & Developmental Biology* 35: 742–747.

Dodo, H.W., Konan, K.N., Chen, F.C. et al. (2008). Alleviating peanut allergy using genetic engineering: the silencing of the immunodominant allergen Ara h 2 leads to its significant reduction and a decrease in peanut allergenicity. *Plant Biotechnology Journal* 6: 135–145.

Eapen, E. (2008). Advances in development of transgenic pulse crops. *Biotechnology Advances* 26: 162–168.

Escribano, M.R., Santalla, M., and De-Ron, A.M. (1990). Preliminary study of quality characters in populations of common bean from the northwestern Iberian Peninsula. *Estación Experimental de Aula Dei* 20: 189–198.

Falco, S.C., Guida, T., Locke, M. et al. (1995). Transgenic canola and soybean with increased lysine. *Nature Biotechnology* 13: 577–582.

Fire, A., Xu, S., Montgomery, M.K. et al. (1998). Potent and specific genetic interference by double-stranded RNA in *Caenorhabditis elegans*. *Nature* 391: 806–811.

Flinterman, A.E., van Hoffen, E., den Hartog Jager, C.F. et al. (2007). Children with peanut allergy recognize predominantly Ara h2 and Ara h6, which remains stable over time. *Clinical and Experimental Allergy* 37: 1221–1228.

Flores, T., Karpova, O., Su, X. et al. (2008). Silencing of GmFAD3 gene by siRNA leads to low alpha-linolenic acids (18:3) of fad3-mutant phenotype in soybean [*Glycine max* (Merr.)]. *Transgenic Research* 17: 839–850.

Food and Agricultural Organization (2013).*The State of Food and Agriculture*. Rome: FAO.

Food and Agricultural Organization (2017). Building Resilience for Peace and Food Security. Rome: FAO. The State of Food Security and Nutrition in the World. IFAD, UNICEF, WFP and WHO. Available from: http://www.fao.org/3/a-I7695e.pdf .

Fromm, M.E., Taylor, L.P., and Walbot, V. (1986). Expression of genes transferred into monocot and dicot plant cells by electroporation. *Proceedings of the National Academy of Sciences of the United States of America* 82: 5824–5828.

Gander, E.S., Homstrocm, K., De-Paiva, G.R. et al. (1991). Isolation, characterization and expression of a gene coding for a 2S albumin gene from *Bertholetia excelsa* (Brazil nut). *Plant Molecular Biology* 16: 437–448.

Graham, R.D. and Gregorio, G. (2001). Breeding for nutritional characteristics in cereals. *Novartis Foundation Symposia* 236: 205–214.

Hanafy, M.S., Rahman, S.M., Khalafalla, M.M. et al. (2006). Accumulation of free tryptophan in azuki bean (*Vigna angularis*) induced by expression of a gene (OAISA1D) for a modified α-subunit of rice anthranalite synthase. *Plant Science* 171: 670–676.

Herman, E.M., Helm, R.M., Jung, R., and Kinney, A.J. (2003). Genetic modification removes an immunodominant allergen from soybean. *Plant Physiology* 132: 36–43.

Horsch, R.B., Fry, J.E., Hoffman, N.L. et al. (1985). A simple and general method for transferring genes into plants. *Science* 227: 1229–1231.

Hudson, L.C., Seabolt, B.S., Odle, J. et al. (2013). Sublethal Staphylococcal enterotoxin B challenge model in pigs to evaluate protection following immunization with a soybean-derived vaccine. *Clinical and Vaccine Immunology* 20: 24.

Hwang, H., Yu, M., and Lai, E. (2017). Agrobacterium-mediated plant transformation: biology and applications. *The Arabidopsis Book* 15.

International Legume Database & Information Service (ILDIS) (2004).World Database of Legumes (version 9.00). Canberra, Australia.

Iqbal, A., Khalil, I.A., Nadia, A., and Khan, M.S. (2006). Nutritional quality of important food legumes. *Food Chemistry* 97: 331–335.

Jorgensen, R.A. and Napoli, C.A. (1994).Genetic engineering of novel plant phenotypes. US Patent 5,283,184, filed 17 April 1991 and issued 1 February 1994.

Kearney, J. (2010). Food consumption trends and drivers. *Philosophical Transactions of the Royal Society Biological Sciences* 365: 2793–2807.

Kennedy, A.R. (1998). Chemopreventive agents: protease inhibitors. *Pharmacology and Therapeutics* 78: 167.

Khandelwal, A., Sita, G.L., and Shaila, M.S. (2003a). Oral immunization of cattle with hemagglutinin protein of rinderpest virus expressed in transgenic peanut induces specific immune responses. *Vaccine* 21: 3282–3289.

Khandelwal, A., Sita, G.L., and Shaila, M.S. (2003b). Expression of hemagglutinin protein of rinderpest virus in transgenic tobacco and immunogenicity of plant-derived protein in a mouse model. *Virology* 308 (2): 207–215.

Khandelwal, A., Renukaradhyaa, G.J., Rajasekharb, M. et al. (2011). Immune responses to hemagglutinin-neuraminidase protein of peste des petits ruminants virus expressed in transgenic peanut plants in sheep. *Veterinary Immunology and Immunopathology* 140: 291–296.

Kim, W.S., Chronis, D., Juergens, M. et al. (2012). Transgenic soybean plants overexpressing O-acetylserine sulfhydrylase accumulate enhanced levels of cysteine and Bowman-Birk protease inhibitor in seeds. *Planta* 235: 13–23.

Kim, W., Jez, J.M., and Krishnan, H.B. (2014). Effects of proteome rebalancing and sulfur nutrition on the accumulation of methionine rich δ-zein in transgenic soybeans. *Fronteirs in Plant Science.* 5: 633.

Kinney, A.J. (1995).Improving soybean seed quality. Symposium on 'Induced Mutations and Molecular Techniques for Crop Improvement', Vienna.

Klein, T.M., Wolf, E.D., Wu, R., and Sanford, J.C. (1987). High-velocity microprojectiles for delivering nucleic acids into living cells. *Nature* 327: 70–73.

Lacorte, C., Aragao, F., Almeida, E. et al. (1997). Transient expression of GUS and the 2S albumin gene from Brazil nut in peanut (*Arachis hypogaea* L.) seed explants using particle bombardment. *Plant Cell Reproduction* 16: 619–623.

Mathers, J.C. (2002). Pulses and carcinogenesis: potential for the prevention of colon, breast and other cancers. *British Journal of Nutrition* 88: 273.

Molvig, L., Tabe, L.M., Eggum, B.O. et al. (1997). Enhanced methionine levels and increased nutritive value of seeds of transgenic lupins (*Lupinusangustifolius* L.) expressing a sunflower seed albumin gene. *Proceedings of National Academy of Sciences United States of America* 94: 8393–8398.

Moravec, T., Schmidt, M.A., Herman, E.M., and Woodford-Thomas, T. (2007). Production of *Escherichia coli* heat labile toxin (LT) B subunit in soybean seed and analysis of its immunogenicity as an oral vaccine. *Vaccine* 25: 1647–1657.

Nicolaou, N. and Custovic, A. (2011). Molecular diagnosis of peanut and legume allergy. *Current Opinion in Allergy and Clinical Immunology* 11: 222–228.

Oakes, J.L., Bost, K.L., and Piller, K.J. (2009). Stability of a soybean seed-derived vaccine antigen following long-term storage, processing and transport in the absence of a cold chain. *Journal of Science and Food Agriculture* 89: 2191–2199.

Piller, K.J., Clemente, T.E., Jun, S.M. et al. (2005). Expression and immunogenicity of an *Escherichia coli* K99 fimbrial subunit antigen in soybean. *Planta* 222: 6–18.

Prasad, V., Satyavathi, V.V., Sanjaya, K.M. et al. (2004). Expression of biologically active Hemagglutinin-neuraminidase protein of Peste des petits ruminants virus in transgenic pigeon pea [*Cajanus cajan* (L.) Millsp.]. *Plant Science* 166: 199–205.

Prohens, J. (2011). Plant breeding: a success story to be continued, thanks to the advances in genomics. *Frontiers in Plant Science* 2: 51.

Qi, Q., Huang, J., Crowley, J. et al. (2011). Engineered soybean seed with enhanced threonine levels: biochemical characterization and seed-specific expression of lysine-insensitive variants of aspartate kinases from the enteric bacterium *Xenorhabdusbovienii*. *Plant Biotechnology Journal* 9: 193–204.

Ravindran, V., Tabe, L.M., Molvig, L. et al. (2002). Nutritional evaluation of transgenic high-methionine lupins with broiler chickens. *Journal of the Science of Food and Agricluture* 82: 280–285.

Ritchie, H., Reay, D.S., and Higgins, P. (2018). Quantifying, projecting and addressing India's hidden hunger. *Frontiers in Sustainable Food Systems* 30: 1.

Saalbach, I., Pickardt, T., Machemehl, F. et al. (1994). Chimeric gene encoding the methionine-rich 2S albumin of the Brazil nut (*Bertholletiaexcelsa* H.B.K.) is stably expressed and inherited in transgenic grain legumes. *Molecular Genomics and Genetics* 242: 226–236.

Sanford, J.C., Klein, T.M., Wolf, E.D., and Allen, N. (1987). Delivery of substances into cells and tissues using a particle bombardment process. *Particulate Science and Technology* 5 (1): 27–37.

Sato, S., Xing, A., Ye, X. et al. (2004). Production of γ-linolenic acid and stearidonic acid in seeds of marker-free transgenic soybean. *Crop Science* 44: 646–652.

Satyavathi, V.V., Prasad, V., Khandelwal, A. et al. (2003). Expression of hemagglutin protein of rinderpest virus in transgenic pigeon pea [*Cajanus cajan* (L.) Millsp.] plants. *Plant Cell Reproduction* 21: 651–658.

Shemy, H.A., Khalafalla, M.M., Fujita, K., and Ishimoto, M. (2007). Improvement of protein quality in transgenic soybean plants. *Biology Plant* 51: 277–284.

Sicherer, S.H., Munoz-Furlong, A., and Sampson, H.A. (2003). Prevalence of peanut and tree nut allergy in the United States determined by means of a random digit dial telephone survey: a 5-year follow-up study. *Journal of Allergy and Clinical Immunology* 112: 1203–1207.

Song, S., Hou, W., Godo, I. et al. (2013). Soybean seeds expressing feedback-insensitive cystathionine-γ-synthase exhibit a higher content of methionine. *Journal of Experimental Botany* 64: 1917–1926.

Tabe, L., Wirtz, M., Molvig, L. et al. (2010). Overexpression of serine acetyl transferase produced large increases in O-acetylserine and free cysteine in developing seeds of a grain legume. *Journal of Experimental Botany* 61: 721–733.

Thompson, M.D., Brick, M.A., McGinley, J.N. et al. (2009). Chemical composition and mammary cancer inhibitory activity of dry bean. *Crop Science* 49: 179–186.

Thu, T.T., Dewaele, E., Trung, L.Q. et al. (2007). Increasing lysine levels in pigeon pea [*Cajanus cajan* (L.) Millsp.] seeds through genetic engineering. *Plant Cell, Tissue and Organ Culture* 91: 135–143.

Tiwari, S. and Tuli, R. (2012). Optimization of factors for efficient recovery of transgenic peanut (*Arachis hypogaea* L.). *Plant Cell, Tissue and Organ Culture* 109: 111–121.

Toenniessen, G.H. (2002).Crop genetic improvement for enhanced human nutrition. International Vitamin A Consultative Group Meeting, New York.

Turco, I., Ferretti, G., and Bachhetti, T. (2016). Review of the health benefits of faba bean (*Viciafaba* L.) polyphenols. *Journal of Food and Nutrition Research* 1: 11.

Van, E.A.L., Lincoln, K., Durrett, T.P. et al. (2003). Engineering vitamin E content: from *Arabidopsis* mutant to soy oil. *Plant Cell* 15: 3007–3019.

Vimolmangkang, S., Gasic, K., Soria-Guerra, R. et al. (2012). Expression of the nucleocapsid protein of porcine reproductive and respiratory syndrome virus in soybean seed yields an immunogenic antigenic protein. *Planta* 235: 513–522.

Zhang, L., Yang, X.D., Zhang, Y.Y. et al. (2013). Changes in oleic acid content of transgenic soybeans by antisense RNA-mediated post transcriptional gene silencing. *International Journal of Genomics* 2014: 921950.

Zhu, C.F., Naqvi, S., Gomez, S.G. et al. (2010). Vitamin, protein and essential mineral enhancement of cereal crops for food security. *GMO Biosafety Research* 1: 315–318.

11

The Survey of Genetic Engineering Approaches for Oil/Fatty Acid Content Improvement in Oilseed Crops

Kaushal Kumar Bhati[1], Riyazuddin Riyazuddin[2], Ashish Kumar Pathak[3], and Anuradha Singh[4]

[1] *Louvain Institute of Biomolecular Science, Catholic University of Louvain, Louvain-la-Neuve, Belgium*
[2] *Department of Plant Biology and Doctoral School in Biology, Faculty of Science and Informatics, University of Szeged, Szeged, Hungary*
[3] *Department of Botany, Charles University, Praha 2, Czech Republic*
[4] *French Associates Institute for Agriculture and Biotechnology of Drylands (FAAB), Ben-Gurion University of the Negev, Sede Boqer Campus, Sede Boqer, Israel*

CHAPTER MENU

11.1 Background

Oils are the most energy-dense plant reserves, providing both calories and essential fatty acids to consumers. In most oil-seed crops, oil production and accumulation take place during seed maturation (maturation is a phase of seed development) which progresses through the slowing down of cellular morphogenesis, enlargement, reserve accumulation and desiccation (Hajduch et al. 2010). Lipid bodies (oil bodies) are a major reserve of oil in seeds. They are highly dynamic subcellular structures in the cytoplasm and coordinate with a wide range of physiological processes, including membrane biogenesis, hormone signaling, and plant growth and development (Shimada et al. 2018). They are generally circular to ovoid, with varying diameter ranging from 0.5–2.5 μm and can exceed two to

Genome Engineering for Crop Improvement, First Edition. Edited by Santosh Kumar Upadhyay.

three times greater than 2.5 μm as can be seen for soybean (Song et al. 2017). Oil bodies are mostly observed in seed (dry seed, seed embryo, pollen, and seedling), various non-seed organs, tissues or cells (root, mesophyll, and bundle sheath cells of leaf, and culture cell) (Pyc et al. 2017) as well as vegetative organs of non-flowering plants, and algae (Huang 2018). Oil bodies have a specialized structure comprising triacylglycerol (TAG) enclosed by phospholipid monolayer and various integral membrane oleosin proteins, thus maintained as a small discrete structure (Tzen 2012).

11.1.1 Oil Biosynthesis Pathway

In seed, oil biosynthesis is localized in two subcellular compartments (plastids and endoplasmic reticulum) and composed of four steps: *de novo* fatty acid synthesis, acyl editing, TAG assembly, and oil drop formation. The *de novo* biosynthesis of the first three fatty acids (FA) (e.g. palmitic acid-16 : 0; stearic acid –18 : 0; and oleic acid-18 : 1) begin with sugar imported from the photosynthesis pathway. This is where sucrose first cleaved into hexose and/or triose via glycolysis and condensed to acetyl coenzyme A (acetyl-CoA) to deliver FA in plastid (Philip et al. 2013). During the next step, FA conjugated to acyl carrier protein (ACP) referred to as acyl-ACP that control the fatty acid chain length (up to 18 carbons) and the level of saturated FAs. The enzyme acetyl-CoA carboxylase (ACCase) catalyzed the formation of malonyl-CoA from acetyl-CoA, while fatty acid synthase (FAS) produce 16:0-ACP, which further elongates to 18:0-ACP by ketoacyl-ACP synthase (KASII). Efficient conversion to 18:0-ACP to 18:1-ACP catalyzed by stearoyl-ACP desaturase (SAD) enzyme. Interestingly, 16:0-ACP, 18:0-ACP and 18:1-ACP are committed substrates of two acyl-ACP thioesterases (FATA/B) for production of free FAs (Chapman and Ohlrogge 2012). The free FAs activated to long-chain acyl-CoA (LACS) on the chloroplast outer envelope while 16:0-CoA, 18:0-CoA and 18:1-CoA are transferred to the endoplasmic reticulum (ER).

The biosynthesis of polyunsaturated FAs and further elongation to very long-chain FAs (20:5 and 22:6) performed by membrane-bound enzymes in the endoplasmic reticulum (Bellou et al. 2014). Inter-membrane transport of fatty acids (18:1) can occur much more rapidly than diffusion through the cytosol via soluble carriers, and thus may provide an efficient plastid-ER acyl-transfer mechanism. In this connection, acyl chain at the Sn-2 carbon of phosphatidylcholine (PC) may act as a carrier of FAs through the action of lysophosphatidyl choline acyltransferase (LPCAT) and FA desaturase 2 (FAD2) and FAD3 sequentially convert 18:1 to 18:2 and 18:3 unsaturated FAs. Most commonly, grown oil crops majorly accumulate C16/C18 saturated and unsaturated fatty acids (C18:1, C18:2, and C18:3) as a component of their storage lipids (Chen and Smith 2012).

The final storage form of lipids in the seed is TAG. The fatty acid conversion to triacylglycerol catalyzed by acyl-transferase enzyme through Kennedy or acyl-CoA dependent pathway (Cagliari et al. 2011). This pathway was first identified in animals and later in the plants, which include some additional steps involving four acylation steps using acyl-CoA and substrate. The sequential acylation of glycerol-3-phosphate at Sn-1 and Sn-2 position are catalyzed by glycerol-3-phosphate acyltransferase (GPAT) and lysophosphatidic acid acyltransferase (LPAAT), yielding one of the central metabolite phosphatidic acid (PA) (Xu and Shanklin 2016). Further, dephosphorylated PA to diacylglycerol (DAG) considered as the final committed step in TGA synthesis, where acylation at the Sn-3 position mediated by diacylglycerol acyltransferase (DGAT). It is worth mentioning that there are two distinctive DGATs (DGAT1 and DGAT2) located in the ER membrane (Philip et al. 2013). However, their relative activity differed among tissue types. For example, *DGAT1* is expressed at similar levels in leaves, flowers, and developing seeds, whereas *DGAT2* is strongly induced in seeds during oil synthesis in *Brassica napus*, Castor and many other plant species (He et al. 2004; Lacey and Hills 1996; Shockey et al. 2006).

To this end, free TAGs undergo final packaging into oil bodies, enclosed by a monolayer of phospholipid (PL) embedded with integral membrane oleosin proteins (15–30 kDa) and the minor caleosin and steroleosin proteins (Song et al. 2017). Briefly, oleosins are co-translationally inserted into the ER and associate with the TAGs which accumulate between the two phospholipid leaflets (Schmidt and Herman 2008). The TAGs surrounded by the phospholipid monolayer, with the embedded proteins, eventually bud off to form the oil body (Figure 11.1). Thus, TAG works as a primary lipid source of energy reserve that supports the germination of seeds and starvation consequences.

Our current understanding of lipid metabolism and assimilation largely comes from the model plant Arabidopsis; however, efforts have also been made in commercial crops to unravel the key enzyme and their roles in lipid metabolism. Interestingly, the economic significance of oil has imparted a long-lasting impact on biotechnological interventions to improve the oil crops. In the following sections, we discuss the different genome engineering approaches used to improve the quality/quantitative trait associated with lipid/oil in two oil seed crops. Most of these efforts include targeting the rate limiting steps in lipid

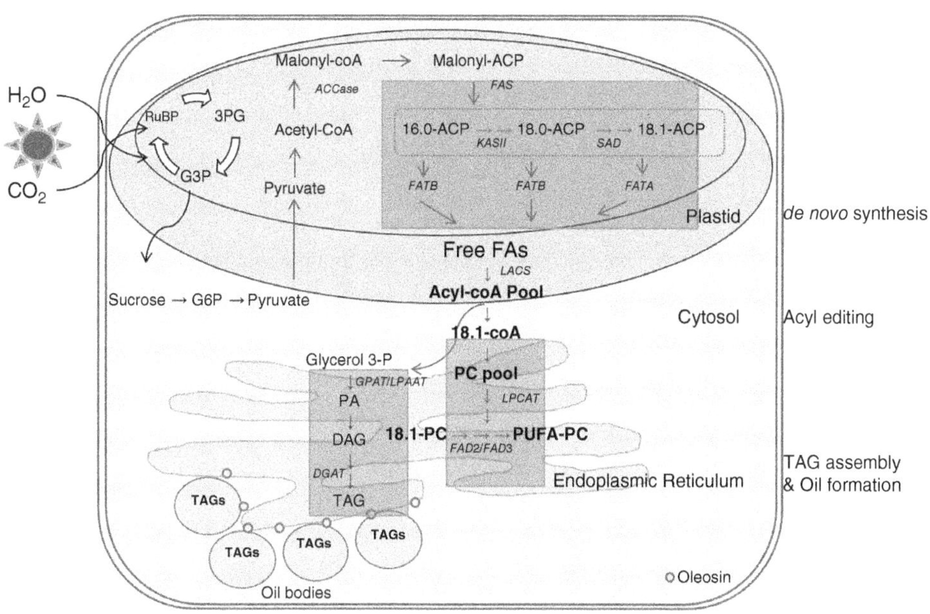

Figure 11.1 Schematic illustration of generalized lipid biosynthesis/storage pathway in plants. The information majorly driven from published reports cited in reference section. Major target areas exploited for metabolic engineering are highlighted by yellow boxes.

biosynthesis by overexpression of enzymes or reducing the activity of negative regulators. The introduction we put forward will help readers to relocate the enzymes/steps that are of prime interest to oil content improvement in crops (Figure 11.1). We included two crops representatives of animal feed (Soybean) and biofuel (Camelina). We believe that this tailormade information will provide a comprehensive outlook on what genome-engineered crops bring to socio-economic scenario.

11.2 Soybean: Triumph Oil Crop

Soybean (*Glycine max* L. Merril) is feedstock legume with global economic and societally important crops. The world soybean production in 2018 was over 360 million metric tons, and about 80% of the production was shared by the US, Brazil and Argentina, while India contributes for ~6.0% (11.0 million metric tons) of the world's soybean supply. Soybean is a significant source of high-quality plant protein (40%), carbohydrates (30%) and essential nutrients including vitamins, nutraceutical, and antioxidants both for humans and animals (Natarajan et al. 2013). Additionally, it is also a renewable and sustainable resource for high-energy biofuel requirements. In context of this chapter, we should know that soybean contributes to 61% the global seed oil production and is considered as the world's second most highly produced oil after palm oil. Soybean oil consumption accounts for 28% of world total oil consumption (www.soystats.com) (American Soybean Association 2019). Oil products are used in the formulation of cooking oils, ready meals, salad dressings, shortenings, spreads, and bakery products. The population explosion at the current time has resulting in a growing demand for seed oil in the food industry. Therefore, enhancing the functional and nutritional quality of seed oil content in soybean has become an important area of scientific research and a key target trait for soybean breeders.

11.2.1 Need for Genetic Modification

Soybean oil composed approximately 14% saturated fatty acids (10% palmitic acid-C16:0 and 4% stearic acid-C18:0), 20% monounsaturated fatty acids such as oleic acid (C18:1), and over 65% of polyunsaturated fatty acids (PUFAs) which consist of 55% linoleic acid (C18:2) and 15% linolenic acid (C18:3) (Clemente and Cahoon 2009). Additionally, there are various minor FAs such as myristic acid (C14:0), arachidic acid (C20:0), behenic acid (C22:0), erucic acid (C22:1). More importantly a non-fish source of omega-3 fatty acid in seed oil improves the marketable value of soybean for food and industrial applications (Clemente and Cahoon 2009).

It is a well-known fact how amounts and composition of fatty acids determine the functional and nutritional values of seed oil. The relatively high percentages of PUFAs (linoleic acid and linolenic acid) is undesirable for industrial application as these are susceptible to lipid oxidation. The oxidative instability becomes greater under high temperature conditions of soybean oil frying, that leads to rancidity and off flavors in food products (Pullen and Saeed 2012). Therefore, to enhance the oxidative stability and shelf life, soybean oil is often partially hydrogenated or saturated fatty acids content should be increased (Clemente and Cahoon 2009). Higher saturated fatty acid increases the fat melting point, which then stabilizes longer as a solid and provides the desired attribute for bakery products, shortenings, spreads, and margarines (Devi and Khachkar 2016). Some major sources of solid fats are palmitic acid and partially hydrogenated soybean oils (with poor cold-flow properties, reduced lubricity) (Kloof 2007). These solid fats as a dietary source elevates trans-fatty acids and low-density lipoprotein (LDL) cholesterol and increases the risk of coronary heart diseases in humans (Briggs et al. 2017). Studies showed that stearic acid (18:0) has a neutral effect on LDL. Thus, it is requisite to increase the stearic acid content over palmitic acid, increase the monounsaturated fatty acids level (high oleic acid oil; >80%), and reduce polyunsaturated fatty acid synthesis (low linolenic acid oil; <3%). Collectively, these metabolic readjustments will not only reduce the formation of *trans* fatty acid but also improve the nutritional value and practical utilization of soybean seed oil (Clemente and Cahoon 2009).

11.2.2 Elevated Stearic Acid Soybean Oil

Saturated fatty acid (including stearic acid) are synthesized by 3-keto-acyl-ACP II (KAS-II) and steroyl-ACP-thioesterase/Δ9-steroyl-ACP-desaturase enzymes. The KAS-II catalyzes the elongation of 16:0-ACP to 18:0-ACP, while steroyl-ACP-thioesterase breakdown 18:0-ACP to 18:0 and ACP. Further, Δ9-steroyl-ACP-desaturase (SACPD) converts 18:0-ACP to 18:1-ACP, which is a key step (Pantalone et al. 2002). Three SACPD genes, designated SACPD-A, SACPD-B, and SACPD-C have been reported in Soybean (Byfield et al. 2006; Zhang et al. 2008), and among them SACPD-C is exclusively expressed in developing seed. Several efforts have been made to alter the stearic acid composition by targeting the SACPD, for example, non-functional SACPD-C locus increases stearic acid while changing fatty acid composition in root nodules (Gillman et al. 2014). Further, more such examples are summarized in Table 11.1.

11.2.3 High Oleic and Mid-Oleic Acid Soybean Oil

In the lipid biosynthetic pathway, fatty acid desaturase-2 (FAD2) is responsible for the conversion of oleic acid (18:1) precursors to linoleic acid (18:2). FAD2 is considered as the best target increasing the oleic acid content in soybean seed oil. In soybean, genome duplication resulted into two microsomal FAD2-1 desaturases (FAD2-1A and FAD2-1B), those mainly expressed in developing seeds (Schlueter et al. 2017). Additionally, other *FAD2* genes were also reported such as *FAD2-2A*, *FAD2-2B*, and *FAD2-2C*, while their expressions were found to be widely in vegetative tissues of the soybean plant (Tang et al. 2005). There have been successful efforts to create high oleic-acid soybean varieties by disrupting *FAD2-1A* and *FAD2-1B* genes; this includes commercially available Plenish® and Vistive Gold® (Haun et al. 2014; Sweeney et al. 2017). There are diverse genetic quality improvement approaches currently in practice. More than 80% increase in oleic acid has been achieved using conventional breeding techniques like combinations of natural (Pham et al. 2010, 2011) and induced mutations (Dierking and Bilyeu 2009; Lakhssassi et al. 2017), genetic engineering via Ribozyme-terminated antisense approach (Buhr et al. 2002), and genome-editing technologies including TALLEN (Transcription-Activator like Effector Nucleases) (Haun et al. 2014) and clustered regularly interspaced short palindromic repeats/ CRISPR-associated protein 9 (CRISPR/Cas9) (al Amin et al. 2019). More

Table 11.1 Summary of various genetic improvement approaches employed to oil content improvement in Soybean.

Process/trait	Gene product	Modification technique	Mutant Source	Altered fatty acid composition (%)	References
High steric acid	SACPD-C	Sodium azide	A6	~28%	Zhang et al. (2008)
	SACPD-C	X-ray Fast neutron	MM106, M25, FN8, W82, MM106, KK24	10–15%	Gillman et al. (2014)
	SACPD-C	Breeding population	LLL-05-1 × TCJWB03–806–7-19 TCHM08–1087/755	~7%	Ruddle et al. (2013, 2014)
	SACPD-B/ SACPD-C	Gamma ray	Hfa180	~20%	Kim et al. (2019)
	SACPD-C	X-ray	KK24, MM106 and M25	No significant different in stearic acid content	Rahman et al. (1995, 1997)
	SACPD-C	EMS	RG7, 194D, PE980	~12–16%	Boersma et al. (2012),Gillman et al. (2014), Jeong et al. (2018)
High oleic acid	FAD2–1A FAD2–1B	Breeding population	Jake × PI 283327 M23 × PI 283327 M23 × PI 567189A PI 603452 × PI 283327	Fourfold (~85%) accumulation	Pham et al. (2010, 2011)
	FAD2–1A 17 D mutant allele	TILLING	EMS mutagenized populations	~ 50% significantly higher Oleic acid levels	Dierking and Bilyeu (2009), Lakhssassi et al. (2017)
	FAD2 FAD3 FAD2–1B	Gene silencing	Ribozyme termination	85% enhanced oleic acid and saturated fatty acids levels, less than 6%.	Buhr et al. (2002)
			RNAi	80% oleic acid and less than 10% polyunsaturated	Wagner et al. (2011)
			Antisense RNA	51.71% increase oleic acid and up to 3% reduction in palmitic acid	Zhang et al. (2014)
			RNAi	>80% enhanced oleic acid	Yang et al. (2018)

	Palmitoyl-ACP thioesterase and FAD2		Dual silencing	70% oleate and 11–19% stearate	Park et al. (2014)
	FAD2/FAD3	TALEN CRISPR/CAS9		>80% oleic acid and1.3–1.7% linoleic acid	Haun et al. (2014) al Amin et al. (2019) Do et al. (2019)
Mid-oleic acid	FAD2–1A and FAD2–1B *FAD2–1A*	X-ray mutagenesis	M23, KK21	50% enhanced oleic acid	Alt et al. (2005), Anai et al. (2008), Sandhu et al. (2007)
		Breeding population	N98–4445A	55% enhanced oleic acid and 3.5% linolenic acid	Burton et al. (2006)
Low and Ultra-low linolenic acid	FAD-3-2a FAD2–1A, 1B, and FAD3	Breeding population	C1640	3.4%linolenic acid	Chappell and Bilyeu (2006)
			A23	2.2%linolenic acid	Fehr et al. (1992)
			A26	1.0%linolenic acid	Ross et al. (2000)
			RG10	<3% linolenic acid	Reinprecht and Pauls (2016)
		Mutation breeding	*fan1, fan2, and fan3*	<1% linolenic acid	Bilyeu et al. (2003, 2005, 2006, 2011)
		TILLING	NMU-mutagenized populations	<2–3% α-linolenic acid	Held et al. (2019)
		TILLING	X-ray and EMS mutagenized populations	High oleate and < 2% α-linolenic acid	Hoshino et al. (2010, 2014)
		TALENs	*fad2–1a fad2–1b fad3a*	High oleic and low linolenic	Demorest et al. (2016)

importantly, none of these cases reported compromising the agronomic performance of the new genetic amid an increase in oleic acid (Table 11.1). In contrast, mid-oleic acid soybean genotypes with a partially blocked desaturation pathway from oleate to linoleate (lesion in one of the *FAD2-1* genes) are usually associated with yield drag and altered fatty acid profile in vegetative tissue (Alt et al. 2005).

11.2.4 Low and Ultra-Low Linolenic Acid Soybean Oil

A lower concentration of linolenic acid (low, <3% and ultra-low, 1%) is a desirable trait, as it improves the stability against oxidation and extends product shelf and frying life. In soybean, fatty acid desaturase (FAD3) enzymes (Alleles: *FAD3A*, *FAD3B*, and *FAD3C*) are responsible for the conversion of linoleic acids (18:2) to α-linolenic acids (18:3). Their expression is higher in developing seeds and FAD3A is primarily responsible for linolenic acid levels in oil (Bilyeu et al. 2003). Traditional breeding methods and targeted genetic engineering have created improved soybean with <3% and ultra-low linolenic acid and thus provided opportunities for consumers to reduce their intake of trans-fatty acids (Bilyeu et al. 2005, 2006; Demorest et al. 2016).

11.3 *Camelina sativa:* Biofuel and Future Ready Crop

Camelina is an economically important oil crop, not only in terms of nutrition but also as a potential biofuel supplement. Being an ancient member of the Brassicaceae genus, *Camelina sativa* L. Crantz – also addressed with distinctive names such as gold-of-pleasure, wild flax, false flax, Siberian oilseed, linseed dodder and German sesame. Camelina seeds contains approximately 35–45% of oil content, which is rich in polyunsaturated omega-3 and omega-6 fatty acids, as well as fat-soluble antioxidants such as the vitamin E-active tocopherols (Kurasiak-Popowska et al. 2018; Moser 2010). A higher omega-3 level and lower content of undesirable erucic acid (<4%; omega-9) Camelina crop received recent worldwide commercial interest (Nguyen et al. 2013; Zanetti et al. 2017). Camelina is equally attracting industries, farmers and the scientific community due to its agronomic superiorities such as a short growing season, cultivation in a variety of climatic and soil conditions, satisfactory seed yields, environmental adaptability to adverse conditions, resistant to most pests and combines with a unique oil suitable for a multitude of bio-based applications (i.e. biofuels, jet fuel, oleochemical compounds, feed, and food) (Berhow et al. 2014; Moser 2010). It is also worth mentioning the genetic transformation of Camelina is almost effortless and possible by a flower-dipping method similar to Arabidopsis. Thus, genetic improvement is also perusable without major optimization.

The lipid stored in Camelina seeds is Triacylglycerol. Tissue-specific transcriptomic and lipid composition analysis identified the typical lipid biosynthesis pathway, key enzymes, and gene networks associated with TAGs metabolism in Camelina (Abdullah et al. 2016). Camelina seeds oil contained a higher amount of polyunsaturated fatty acids (90%) such as α-linolenic acid (18:3, 30–43%), linoleic acid (18:2, 20–25%), oleic acid (18:1, 15–20%) and low levels of gondoic (20:1), and erucic acids (22:1) (less than 10%) (Zanetti et al. 2017). Camelina seeds contains high iodine content in addition to a significantly higher amount of oil. Iodine protects against rapid deterioration of the lubricating oil more than in other members of Brassicaceae family. After extraction of oil, left over part of seeds are used as Camelina meal which contains a respectable amount of protein (45–50%), residual oil (10–15%), fiber (13%), minerals (5%), and vitamins. Due to its nutritional value, FDA has approved Camelina meal as a cheap source of protein and lipids for livestock feed up to 10% in poultry and beef cattle feed mix while 20% mix as swine feed (MDA 2012). These attributes make Camelina an attractive crop for commercially important downstream products.

11.3.1 Genetic Engineering Approaches for Oil Content Improvement in Camelina

The efforts over the past decade are advancing our understanding for oil content and quality in Camelina seed. Approaches including overexpression, chemical mutagenesis and suppression via RNA interference (RNAi) are arguably successful (Jiang et al. 2017; Kang et al. 2011; Morineau et al. 2017; Nguyen et al. 2013). Recent advances in genome editing can also be used to achieve improved oil content by creating specific knockouts, leading to stable mutation, by the expression of nucleases like Zinc-Finger Nucleases, TALEN or by clustered regularly interspaced short palindromic repeats (CRISPR-Cas9) (Morineau et al. 2017). However, except for a few recent cases Camelina improvement was still dominated by classic genetic engineering methods such as the introduction of the novel enzymatic step or by the manipulation (over expression or down regulation) of any existing step which is crucial in seed-oil synthesis (Sainger et al. 2017). In the following sections, we discuss recent published reports using different genetic engineering approaches for improvement of fatty acid content in Camelina seeds.

11.3.1.1 Overexpression of Heterologous Genes

Camelina is a plant with small seeds, thus, oil yield increase could be complimented by increasing the seed size or seed number per plant. The effective genetic modification approaches to improve the overall yield should improve oil content in transformed plants in comparison to the non-transformed ones. For example, the expression of CO_{2}/HCO^{-} transporters with CaMV35S promoter enhanced CO_2 assimilation by 20–30% in leaf tissues and thus, resulted in increased seed yield by 20–44% (Paulose et al. 2014). Diacylglycerol acyltransferase1 (DGAT1) enzyme was a key regulating factor which was mainly responsible for the oil quantity and fatty acid composition of seed in the various crops including Camelina (Taylor et al. 2009; Xu et al. 2008).Overexpression of the oil biosynthetic genes under the control of tissue specific or stage specific promoters, results in enhanced oil accumulation in seeds. For instance, Chhikara et al. (2018) reported enhancement of oil content (13%) and seed yield per plant (46–86%) was obtained in *C. sativa (L.)* through the engineering by combining DGAT1 and glycerol-3-phosphate enzyme (Gly3PDH) (Chhikara et al. 2018). Similarly, overexpression of *DGAT1* homeologs (*CsDGAT1A*, *CsDGAT1B*, and *CsDGAT1C*) from developing Camelina seeds enhanced total seed oils by ~24% compared with non-transgenic lines (Kim et al. 2016). Additionally, overexpression of *pPLAIIIδ* gene, which hydrolyzes phosphatidylcholine, important for synthesis and modification of triacylglycerol to free-fatty acids and lysophosphatidyl choline with CaMV35S promoter or a seed specific promoter, resulted in the increased seed-oil content with or without affecting plant growth and seed production (Li et al. 2015). Overexpression of some other genes and their effects are discussed in Table 11.2.

The nucleotides from other heterologous systems has also been used for Camelina oil improvement. Integration of foreign genes to bypass the host process proved their

Table 11.2 The complied summary of genetic engineering approaches adopted in Camelina for fatty acid profile improvement.

Gene	Approach	Trait	References
pPLAIIIδ	Overexpression	increasing seed oil content	Li et al. (2015)
CsDGAT1B		Oleic acid and linoleic acids were 3 and 4 mol% higher, respectively	Kim et al. (2016)
AGG3		Increased seed number per plant by 15–40% with higher oil content	Roy Choudhury et al. (2014)
E. coli glycolate dehydrogenase (GDH) for partial bypass; GDH glyoxylate carboxyligase (GCL), tartronic semialdehyde reductase (TSR) for full bypass	Introduction of foreign gene	partial and full bypass increased seed yield by 50–57% & 57–73% respectively	Dalal et al. (2015)
CsFAD2	CRISPR-Cas9 based gene knockout	~ 50% significantly higher Oleic acid levels, decrease in linoleic and linolenic	Jiang et al. (2017)
CsDGAT1 or CsPDAT		reduced oil content and altered fatty acid composition but also increased the amount of VLCFAs	Aznar-Moreno and Durrett (2017)
CsCRUC		No change in total oil content but increase in relative abundance of all saturated fatty acids	Lyzenga et al. (2019)
3-keto-acyl-ACP synthase II and the FatB 16:0-ACP thioesterase	Seed specific downregulation	Increase accumulation of omega- 7 fatty acid to 60–65%	Nguyen et al. (2015)
AtWRI1-Arabidopsis WRINKLED1	seed specific expression of transcription factor regulating fatty acid synthesis	enhanced seed mass and total seed oil content by 14%	An and Suh (2015)
FAE1-B gene	EMS Inducing mutation	Reduction of 60% undesired C20-C24 VLCFAs	Ozseyhan et al. (2018)
CsFAD2 and *CsFAE1*	Gene silencing	Up to 71% increase oleic acid	Li et al. (2015)

efficiency in achieving desirable traits. For instance, about 50% more seeds per plant were obtained with better plant growth than wild type when Arabidopsis *Purple acid phosphate (AtPAP2)* was expressed, which improved the carbon flow and sink-source activities (Zhang et al. 2012). More recently, *Arabidopsis G-protein γ subunit 3 (AGG3)* was expressed under promoter CaMV35S and seed-specific soybean glycinin showed enlarged seed size, seed mass and increase in seed number per plant by 15–40% (Roy Choudhury et al. 2014). Another major approach to increase oil yield by overexpression of master regulators of fatty acid biosynthesis. *WRINKLED1* is an AP2/EREBP-type transcription factor, that works upstream of lipid biosynthesis pathways. The *AtWRI1* expression with seed-specific SiW6 promoter enhanced seed mass and total seed oil content by 14% compared to wild type seeds (An and Suh 2015).

Some reports emphasized the use of genes combination to ultimately meet the requirements. As done by the introduction of 3 *E. coli* genes, encoding for glycolate dehydrogenase for partial bypass or glycolate dehydrogenase with glyoxylate carboligase and tartonic semialdehyde reductase for full bypass to reduce carbon loss from photorespiration which resulted in significant increase in yield, with partial bypass increased the yield by 50–57% and in case of expression of full bypass it was increased by 57–73% (Dalal et al. 2015).

11.3.1.2 Gene Knockout/Silencing Approaches

Gene silencing is a very commonly used tool to reduce the transcript level of target genes post-transcriptionally. The knockdown of negative regulators of lipid biosynthesis also proved its worth in proposing successful approaches to increase fatty acid/TAG content in Camelina seeds. Bansal et al. (2018) silenced *CsDGAT1* gene while expressing *Euonymus alatus* acetyltransferase *EaDAcT*. This co-expression/silencing approach resulted in increased levels of acetyl-TAG together with medium-chain fatty acids (MCFAs) (Bansal et al. 2018), which was consistent with previous reports (Liu et al. 2015a, 2015b). The seed-specific expression of RNAi cassette targeting to *CsFAD2* resulted in generation of mid-oleic acid lines around 50% higher compared to wild-type seeds (Kang et al. 2011), while co-silencing of both *CsFAD2* and *CsFAE1* further increased the levels of oleic acid to 70% in the best performing Camelina lines (Nguyen et al. 2013). An alternative approach reported 67% increase of 16 : 1Δ9 and 18 : 1Δ11 in Arabidopsis seeds by seed-specific expression of fungal 16 : 0–/18 : 0-CoA desaturases (targeted to the cytosol) and RNAi silencing of FAE1 *(FATTY ACID ENLONGASE 1)* gene, encoding the 3-ketoacyl-CoA synthase responsible for the elongation of C18 and C20 fatty acids (Liu et al. 2015b). There is general belief that silencing the key genes coding for fatty acid elongation enzyme should be equally successful in Camelina. Hence, it is considerably validated by seed silencing of *3-keto-acyl-ACP synthase II* and the *FatB 16 : 0-ACP thioesterase* genes in seed. The omega-7 fatty acid increases up to 60–65% of the total fatty acids in seeds using the co-silencing approach. This increase results from a higher level of substrate 16 : 0-ACP for the Δ9-acyl-ACP desaturase and blocked C18 fatty acid elongation (Nguyen et al. 2010). Further, in a classic approach like ethyl methane sulfonate (EMS) induced non-functional mutation in Camelina *FAE1-B* gene also significantly changed fatty acid composition by reduction of 60% of undesired C20-C24 very long-chain fatty acids (VLCFAs) in seed (Nguyen et al. 2015). Thus, we could speculate that fine-tuning the fatty acid elongation step provides an excellent target for more recent approaches for metabolic engineering like precise genome editing by RNA-guided nucleases.

11.3.1.3 CRISPR-Cas9 Approaches for Camelina Improvement

Plant metabolic engineering was never as exciting as the current time of advances in precise genome editing. The crop improvement sphere has significantly widened with these new breeding technologies. Nevertheless, the lipid biosynthesis and oil content improvement approaches in crops like Camelina must also be improvised. There are some recent reports on Camelina by those who employed the CRISPR-Cas9-based genome editing. For example, using CRISPR-Cas9 seed-storage protein CRUCIFERIN C (CsCRUC) was knockout. This changed fatty acid profile differentially by enriching all saturated fatty acids with relative abundance, however, there was no change in oil content for these knockout seeds (Ozseyhan et al. 2018). Additionally, to improve the Camelina seed-oil composition, *FAD2* genes were targeted in germlines and somatic cells, resulting in increased oleic content from 16% to over 50% in total fatty acid composition and a decrease in the level of less desirable polyunsaturated fatty acids, i.e. linoleic and linolenic (Jiang et al. 2017). The efficiency of CRISPR-Cas9 to target multiple copy of genes in polyploid plants has always been of great interest to the plant scientist. A recent report validated CRISPR-Cas9 targeting three copies of delta-12-desaturase (FAD2) which negatively impacted the polyunsaturated fatty acids while increasing oleic acid from 10 to 62% in the seed oil (Morineau et al. 2017). Other important work proved mutating homeologs of *CsDGAT1* or *CsPDAT1* improved TAG synthesis in developing seeds. Though, these knockout mutations reduced oil content and altered fatty acid composition in the seed oil of *C. sativa* in T1 and T2 generation (Lyzenga et al. 2019). The qualitative improvement in Camelina seed fatty acid content was also achieved by knockout of three *FAE1* alleles to obtain homozygous in a single generation. In the *FAE1*knockout lines, they reported high oleic acid/α-linolenic acid in seed, which is a desirable trait for industrial or food/feed uses compared to wild-type lines (Ozseyhan et al. 2018).

11.4 Conclusion

Lipids are an indispensable part of living cells; in plants they serve structural, signaling and energy purposes. The *de novo* fatty acid biosynthesis pathway components are now known in several plant/crop species. During last decade, genetic engineering-based strategies were developed for qualitative and quantitative improvement of oil content in several crops of commercial importance. In this chapter, we compiled the biotechnological approaches used in two different oil crops for quantitative improvement of oil/fatty acid content in seed. These approaches included overexpression of indigenous gene or gene from heterologous system such as yeast, bacteria or other plants. It is also conceivable in both crops viz. Soybean and Camelina that the right choice of promoter brings more advantages in trait achievement. The seed-specific promoters are most effective. However, using insights from *de novo* lipid biosynthesis during the course of seed development might be more worthy for choosing the right stage-specific active promoters at intersection of seed development and oil biosynthesis. Further, it is also evident that silencing genes coding for enzymes involved in fatty acid elongation also have a dramatic effect in overall fatty acid profile of seeds. Recent advances in genome editing and post-translational protein function regulation could be promising tools for precise fine-tuning of lipid metabolism in crops (Bhati et al., 2020; Soda et al. 2017). In future, these novel approaches should bring more industrial and economic values to oil crops considering the emerging challenges of environmental cues affecting productivity and product quality.

Acknowledgments

KKB would like to acknowledge FNRS (Fonds de la recherche scientifique), Belgium for CHARGÉ DE RECHERCHES- Fellowship for supporting the research stay at UCLouvain. RR is thanking Bilateral state scholarship 2019/2020 (AK-00210-002/2019). AS is grateful to funding from Ben Gurion University of Negev. The authors would like to apologize to colleagues those work could not be cited in this book chapter due to space limitations.

References

Abdullah, H.M., Akbari, P., Paulose, B. et al. (2016). Transcriptome profiling of Camelina sativa to identify genes involved in triacylglycerol biosynthesis and accumulation in the developing seeds. *Biotechnology for Biofuels* 9: 136.

Alt, J.L., Fehr, W.R., Welke, G.A., and Sandhu, D. (2005). Phenotypic and molecular analysis of oleate content in the mutant soybean line M23. *Crop Science* 45: 1997–2000.

American Soybean Association (2012). SoyStats. http://soystats.com (accessed 17 July 2020).

al Amin, N., Ahmad, N., Wu, N. et al. (2019). CRISPR-Cas9 mediated targeted disruption of FAD2–2 microsomal omega-6 desaturase in soybean (*Glycine max. L*). *BMC Biotechnology* 19: 9.

An, D. and Suh, M.C. (2015). Overexpression of Arabidopsis WRI1 enhanced seed mass and storage oil content in Camelina sativa. *Plant Biotechnology Reports* 9: 137–148. https://doi.org/10.1007/s11816-015-0351-x.

An, D. and Suh, M.C. (2015). Overexpression of ArabidopsisWRI1 enhanced seed mass and storage oil content in Camelina sativa. *Plant Biotechnology Reports* 9: 137–148.

Anai, T., Yamada, T., Hideshima, R. et al. (2008). Two high-oleic-acid soybean mutants, *M23* and *KK21*, have disrupted microsomal omega-6 fatty acid desaturase, encoded by GmFAD2-1a. *Breeding Science* 58: 447–452.

Aznar-Moreno, J.A. and Durrett, T.P. (2017). Simultaneous targeting of multiple gene homeologs to alter seed oil production in Camelina sativa. *Plant and Cell Physiology* 58: 1260–1267.

Bansal, S., Kim, H.J., Na, G. et al. (2018). Towards the synthetic design of camelina oil enriched in tailored acetyl-triacylglycerols with medium-chain fatty acids. *Journal of Experimental Botany* 69: 4395–4402.

Bellou, S., Baeshen, M.N., Elazzazy, A.M. et al. (2014). Microalgal lipids biochemistry and biotechnological perspectives. *Biotechnology Advances* 32: 1476–1493.

Berhow, M.A., Vaughn, S.F., Moser, B.R. et al. (2014). Evaluating the phytochemical potential of Camelina: An emerging new crop of old-world origin. In: *Phytochemicals–Biosynthesis, Function and Application* (ed. R. Jetter), 129–148. Cham: Springer.

Bhati, K. K., Kruusvee, V,. Straub, D. et al. (2020). Global Analysis of Cereal microProteins Suggests Diverse Roles in Crop Development and Environmental Adaptation. G3 (Bethesda)10(10): 3709–3717.

Bilyeu, K.D., Palavalli, L., Sleper, D.A., and Beuselinck, P.R. (2003). Three microsomal omega-3 fatty-acid desaturase genes contribute to soybean linolenic acid levels. *Crop Science* 43: 1833–1838.

Bilyeu, K., Palavalli, L., Sleper, D., and Beuselinck, P. (2005). Mutations in soybean microsomal omega-3 fatty acid desaturase genes reduce linolenic acid concentration in soybean seeds. *Crop Science* 45: 1830–1836.

Bilyeu, K., Palavalli, L., Sleper, D.A., and Beuselinck, P. (2006). Molecular genetic resources for development of 1% linolenic acid soybeans. *Crop Science* 46: 1913–1918.

Bilyeu, K., Gillman, J.D., and LeRoy, A.R. (2011). Novel FAD3 mutant allele combinations produce soybeans containing 1% linolenic acid in the seed oil. *Crop Science* 51: 259–264.

Boersma, J.G., Gillman, J.D., Bilyeu, K.D. et al. (2012). New mutations in a delta-9-stearoyl-acyl carrier protein desaturase gene associated with enhanced stearic acid levels in soybean seed. *Crop Science* 52: 1736–1742.

Briggs, M.A., Petersen, K.S., and Kris-Etherton, P.M. (2017). Saturated fatty acids and cardiovascular disease: replacements for saturated fat to reduce cardiovascular risk. *Healthcare (Basel)* 5: 29.

Buhr, T., Sato, S., Ebrahim, F. et al. (2002). Ribozyme termination of RNA transcripts down-regulate seed fatty acid genes in transgenic soybean. *The Plant Journal* 30: 155–163.

Burton, J.W., Wilson, R.F., Rebetzke, G.J., and Pantalone, V.R. (2006). Registration of N98-4445A mid-oleic soybean germplasm line. *Crop Science* 46: 1010–1012.

Byfield, G.E., Xue, H., and Upchurch, R.G. (2006). Two genes from soybean encoding soluble D9 stearoyl-ACP desaturases. *Crop Science* 46: 840–846.

Cagliari, A., Margis, R., dos Santos Maraschin, F. et al. (2011). Biosynthesis of triacylglycerols (TAGs) in plants and algae. *International Journal of Plant Biology* 2: 10.

Chapman, K.D. and Ohlrogge, J.B. (2012). Compartmentation of triacylglycerol accumulation in plants. *Journal of Biological Chemistry* 287: 2288–2294.

Chappell, A.S. and Bilyeu, K.D. (2006). A *GmFAD3A* mutation in the low linolenic acid soybean mutant C1640. *Plant Breeding* 125: 535–536.

Chen, J.E. and Smith, A.G. (2012). A look at diacylglycerol acyltransferases (DGATs) in algae. *Journal of Biotechnology* 162: 28–39.

Chhikara, S., Abdullah, H.M., Akbari, P. et al. (2018). Engineering *Camelina sativa (L.)* Crantz for enhanced oil and seed yields by combining diacylglycerol acyltransferase1 and glycerol-3-phosphate dehydrogenase expression. *Plant Biotechnology Journal* 16: 1034–1045.

Clemente, T.E. and Cahoon, E.B. (2009). Soybean oil: genetic approaches for modification of functionality and total content. *Plant Physiology* 51: 1030–1040.

Dalal, J., Lopez, H., Vasani, N.B. et al. (2015). A photorespiratory bypass increases plant growth and seed yield in biofuel crop Camelina sativa. *Biotechnology for Biofuels* 8: 175.

Demorest, Z.L., Coffman, A., Baltes, N.J. et al. (2016). Direct stacking of sequence-specific nuclease-induced mutations to produce high oleic and low linolenic soybean oil. *BMC Plant Biology* 16: 225.

Devi, A. and Khatkar, B.S. (2016). Physicochemical, rheological and functional properties of fats and oils in relation to cookie quality: a review. *Journal of food science and technology* 53: 3633–3641.

Dierking, E.C. and Bilyeu, K.D. (2009). New sources of soybean seed meal and oil composition traits identified through TILLING. *BMC Plant Biology* 9: 89.

Do, P.T., Nguyen, C.X., Bui, H.T. et al. (2019). Demonstration of highly efficient dual gRNA CRISPR/Cas9 editing of the homeologous *GmFAD2*-1A and *GmFAD2*-1B genes to yield a high oleic, low linoleic and α-linolenic acid phenotype in soybean. *BMC Plant Biology* 19: 311.

Fehr, W.R., Welke, G.A., Hammond, E.G. et al. (1992). Inheritance of reduced linolenic acid content in soybean genotypes A16 and A17. *Crop Science* 32: 903–906.

Gillman, J.D., Stacey, M.G., Cui, Y. et al. (2014). Deletions of the SACPD-C locus elevate seed stearic acid levels but also result in fatty acid and morphological alterations in nitrogen fixing nodules. *BMC Plant Biology* 14: 143.

Hajduch, M., Hearne, L.B., Miernyk, J.A. et al. (2010). Systems analysis of seed filling in Arabidopsis: using general linear modeling to assess concordance of transcript and protein expression. *Plant Physiology* 152: 2078–2087.

Haun, W., Coffman, A., Clasen, B.M. et al. (2014). Improved soybean oil quality by targeted mutagenesis of the fatty acid desaturase 2 gene family. *Plant Biotechnology Journal* 12: 934–940.

He, X., Chen, G.Q., Lin, J.T., and McKeon, T.A. (2004). Regulation of diacylglycerol acyltransferase in developing seeds of castor. *Lipids* 39: 865–871.

Held, J.P., Carrero-Colón, M., and Hudson, K.A. (2019). Combination of novel mutation in FAD3C and FAD3A for low linolenic acid soybean. *Agrosystems, Geosciences & Environment* 2: 1–4.

Hoshino, T., Takagi, Y., and Anai, T. (2010). Novel GmFAD2-1b mutant alleles created by reverse genetics induce marked elevation of oleic acid content in soybean seeds in combination with GmFAD2-1a mutant alleles. *Breeding science* 60: 419–425.

Hoshino, T., Watanabe, S., Takagi, Y., and Anai, T. (2014). A novel GmFAD3-2a mutant allele developed through TILLING reduces α-linolenic acid content in soybean seed oil. *Breeding Science* 64: 371–377.

Huang, A.H.C. (2018). Plant lipid droplets and their associated proteins: potential for rapid advances. *Plant Physiology* 176: 1894–1918.

Jeong, J.E., Kulkarni, K.P., Chang, J.H. et al. (2018). A novel allele of GmSACPD-C associated with high seed stearic acid concentration in an EMS-induced mutant PE980 in soybean. *Crop Science* 58: 192–203.

Jiang, W.Z., Henry, I.M., Lynagh, P.G. et al. (2017). Significant enhancement of fatty acid composition in seeds of the allohexaploid, *Camelina sativa*, using CRISPR/Cas9 gene editing. *Plant Biotechnology Journal* 15: 648–657.

Kang, J., Snapp, A.R., and Lu, C. (2011). Identification of three genes encoding microsomal oleate desaturases (FAD2) from the oilseed crop Camelina sativa. *Plant Physiology and Biochemistry* 49: 223–229.

Kim, H., Park, J.H., Kim, A.Y., and Suh, M.C. (2016). Functional analysis of diacylglycerol acyltransferase1 genes from Camelina sativa and effects of CsDGAT1B overexpression on seed mass and storage oil content in C. sativa. *Plant Biotechnology Reports* 10: 141–153.

Kim, S.H., Ryu, J., Kim, W.J. et al. (2019). Identification of a new GmSACPD-C allele in high stearic acid mutant Hfa180 derived from gamma-ray irradiation. *Molecular Breeding* 39: 17.

Klonoff, D.C. (2007). Replacements for trans fats-will there be an oil shortage? *Journal of Diabetes Science and Technology* 1: 415–422.

Kurasiak-Popowska, D., Tomkowiak, A., Członińska, M. et al. (2018). Analysis of yield and genetic similarity of polish and Ukrainian Camelina sativa genotypes. *Industrial Crops and Products* 123: 667–675.

Lacey, D.J. and Hills, M.J. (1996). Heterogeneity of endoplasmic reticulum with respect to lipid synthesis in developing seeds of Brassica napusL. *Planta* 199: 545–551.

Lakhssassi, N., Zhou, Z., Liu, S. et al. (2017). Characterization of the *FAD2* gene family in soybean reveals the limitations of gel-based TILLING in genes with high copy number. *Frontiers in Plant Science* 8: 324.

Li, M., Wei, F., Tawfall, A. et al. (2015). Overexpression of patatin-related phospholipase AIII δ altered plant growth and increased seed oil content in camelina. *Plant Biotechnology Journal* 13: 766–778.

Liu, J., Rice, A., McGlew, K. et al. (2015a). Metabolic engineering of oilseed crops to produce high levels of novel acetyl glyceride oils with reduced viscosity, freezing point and calorific value. *Plant Biotechnology Journal* 13: 858–865.

Liu, J., Tjellström, H., McGlew, K. et al. (2015b). Field production, purification and analysis of high-oleic acetyl-triacylglycerols from transgenic Camelina sativa. *Industrial Crops and Products* 65: 259–268.

Lyzenga, W.J., Harrington, M., Bekkaoui, D. et al. (2019). CRISPR/Cas9 editing of three CRUCIFERIN C homoeologues alters the seed protein profile in Camelina sativa. *BMC Plant Biology* 19: 292.

MDA (2012). Approved uses of camelina meal in feed. Helena Mont.: Montana Department of Agriculture. https://agr.mt.gov/Commodities (accessed 17 July 2020).

Morineau, C., Bellec, Y., Tellier, F. et al. (2017). Selective gene dosage by CRISPR-Cas9 genome editing in hexaploid Camelina sativa. *Plant Biotechnology Journal* 15: 729–739.

Moser, B.R. (2010). Camelina (*Camelina sativa L.*) oil as a biofuels feedstock: Golden opportunity or false hope? *Lipid Technology* 22: 270–273.

Natarajan, S., Luthria, D., Bae, H. et al. (2013). Transgenic soybeans and soybean protein analysis: an overview. *Journal of Agricultural and Food Chemistry* 61: 11736–11743.

Nguyen, H.T., Mishra, G., Whittle, E. et al. (2010). Metabolic engineering of seeds can achieve levels of ω-7 fatty acids comparable with the highest levels found in natural plant sources. *Plant Physiology* 154: 1897–1904.

Nguyen, H.T., Silva, J.E., Podicheti, R. et al. (2013). Camelina seed transcriptome: a tool for meal and oil improvement and translational research. *Plant Biotechnology Journal* 11: 759–769.

Nguyen, H.T., Park, H., Koster, K.L. et al. (2015). Redirection of metabolic flux for high levels of omega-7 monounsaturated fatty acid accumulation in camelina seeds. *Plant biotechnology journal* 13: 38–50.

Ozseyhan, M.E., Kang, J., Mu, X., and Lu, C. (2018). Mutagenesis of the FAE1 genes significantly changes fatty acid composition in seeds of Camelina sativa. *Plant Physiology and Biochemistry* 123: 1–7.

Ozseyhan, M., Kang, J., Mu, X., and Lu, C. (2018). Mutagenesis of the *FAE1* genes significantly changes fatty acid composition in seeds of Camelina sativa. *Plant Physiology and Biochemistry* 123: 1–7.

Pantalone, V.R., Wilson, R.F., Novitzky, W.P., and Burton, J.W. (2002). Genetic regulation of elevated stearic acid concentration in soybean oil. *Journal of the American Oil Chemists' Society* 79: 549–553.

Park, H., Graef, G., Xu, Y. et al. (2014). Stacking of a stearoyl-ACP thioesterase with a dual-silenced palmitoyl-ACP thioesterase and Δ 12 fatty acid desaturase in transgenic soybean. *Plant Biotechnology Journal* 12: 1035–1043.

Paulose, B., DaCosta, M., Normanly, J.et al. (2014). Approaches to increase carbon capture and redistribution to fuel molecules and co-products in *Camelina sativa*. *Proceedings of the 53rd Annual meeting of the Phytochemical Society of North America*, Raleigh, USA (9–13 August 2014). North Carolina, USA: PSNA.

Pham, A.T., Lee, J.D., Shannon, J.G., and Bilyeu, K.D. (2010). Mutant alleles of FAD2-1A and FAD2-1B combine to produce soybeans with the high oleic acid seed oil trait. *BMC Plant Biology* 10: 195.

Pham, A.T., Lee, J.D., Shannon, J.G., and Bilyeu, K.D. (2011). A novel FAD2-1 A allele in a soybean plant introduction offers an alternate means to produce soybean seed oil with 85% oleic acid content. *Theoretical and Applied Genetics* 123: 793–802.

Philip, D.B., Sten, S., and John, O. (2013). Biochemical pathways in seed oil synthesis. *Current Opinion in Plant Biology* 16: 358–364.

Pullen, J. and Saeed, K. (2012). An overview of biodiesel oxidation stability. *Renewable & Sustainable Energy Reviews* 16: 5924–5950.

Pyc, M., Cai, Y., Greer, M.S. et al. (2017). Turning over a new leaf in lipid droplet biology. *Trends In Plant Science* 22: 596–609.

Rahman, S.M., Takagi, Y., Miyamoto, K., and Kawakita, T. (1995). High stearic acid soybean mutant induced by X-ray irradiation. *Bioscience, Biotechnology, and Biochemistry* 59: 922–923. https://doi.org/10.1271/bbb.59.922.

Rahman, S.M., Takagi, Y., and Kinoshita, T. (1997). Genetic control of high stearic acid content in seed oil of two soybean mutants. *Theoretical and Applied Genetics* 95: 772–776.

Reinprecht, Y. and Pauls, K.P. (2016). Microsomal Omega-3 fatty acid desaturase genes in low linolenic acid soybean line RG10 and validation of major linolenic acid QTL. *Frontiers in Genetics* 7: 38.

Ross, A.J., Fehr, W.R., Welke, G.A., and Cianzio, S.R. (2000). Agronomic and seed traits of 1%-linolenate soybean genotypes. *Crop Science* 40: 383–386.

Roy Choudhury, S., Riesselman, A.J., and Pandey, S. (2014). Constitutive or seed-specific overexpression of Arabidopsis G-protein γ subunit 3 (AGG 3) results in increased seed and oil production and improved stress tolerance in Camelina sativa. *Plant Biotechnology Journal* 12: 49–59.

Ruddle, P., Whetten, R., Cardinal, A. et al. (2013). Effect of a novel mutation in a Δ9-stearoyl-ACP-desaturase on soybean seed oil composition. *Theoretical and Applied Genetics* 126: 241–249.

Ruddle, P., Whetten, R., Cardinal, A. et al. (2014). Effect of Δ9-stearoyl-ACP-desaturase-C mutants in a high oleic background on soybean seed oil composition. *Theoretical and Applied Genetics* 127: 349–358.

Sainger, M., Jaiwal, A., Sainger, P.A. et al. (2017). Advances in genetic improvement of *Camelina sativa* for biofuel and industrial bio-products. *Renewable and Sustainable Energy Reviews* 68: 623–637.

Sandhu, D., Alt, J.L., Scherder, C.W. et al. (2007). Enhanced oleic acid content in the soybean mutant *M23* is associated with the deletion in the *Fad2*-1a gene encoding a fatty acid desaturase. *Journal of the American Oil Chemists' Society* 84: 229–235.

Schlueter, J.A., Vasylenko-Sanders, I.F., Deshpande, S. et al. (2017). The FAD2 gene family of soybean: insights into the structural and functional divergence of a paleopolyploid genome. *Plant Genome.* 47: 14–26.

Schmidt, M.A. and Herman, E.M. (2008). Suppression of soybean oleosin produces micro-oil bodies that aggregate into oil body/ER complexes. *Molecular Plant* 1: 910–924.

Shimada, T.L., Hayashi, M., and Hara-Nishimura, I. (2018). Membrane dynamics and multiple functions of oil bodies in seeds and leaves. *Plant Physiology* 176: 199–207.

Shockey, J.M., Gidda, S.K., Chapital, D.C. et al. (2006). Tung tree DGAT1 and DGAT2 have nonredundant functions in triacylglycerol biosynthesis and are localized to different subdomains of the endoplasmic reticulum. *The Plant Cell* 18: 2294–2313.

Soda, N., Verma, L., and Giri, J. (2017). CRISPR-Cas9-based plant genome editing: significance, opportunities and recent advances. *Plant Physiology and Biochemistry* 131: 2–11.

Song, Y., Wang, X., and Rose, R.J. (2017). Oil body biogenesis and biotechnology in legume seeds. *Plant Cell Reports* 36: 1519–1532.

Sweeney, D.W., Carrero-Colón, M., and Hudson, K.A. (2017). Characterization of new allelic combinations for high-oleic soybeans. *Crop Science* 57: 611–616.

Tang, G.Q., Novitzky, W.P., Carol Griffin, H. et al. (2005). Oleate desaturase enzymes of soybean: evidence of regulation through differential stability and phosphorylation. *The Plant Journal* 44: 433–446.

Taylor, D.C., Zhang, Y., Kumar, A. et al. (2009). Molecular modification of triacylglycerol accumulation by over-expression of DGAT1 to produce canola with increased seed oil content under field conditions. *Botany* 87: 533–543.

Tzen, J.T.C. (2012). Integral proteins in plant oil bodies. *ISRN Botany* 2012: 16.

Wagner, N., Mroczka, A., Roberts, P.D. et al. (2011). RNAi trigger fragment truncation attenuates soybean *FAD2*-1 transcript suppression and yields intermediate oil phenotypes. *Plant Biotechnology Journal* 9: 723–728.

Xu, C. and Shanklin, J. (2016). Triacylglycerol metabolism, function, and accumulation in plant vegetative tissues. *Annual Review of Plant Biology* 67: 179–206.

Xu, J., Francis, T., Mietkiewska, E. et al. (2008). Cloning and characterization of an acyl-CoA-dependent diacylglycerol acyltransferase 1 (DGAT1) gene from Tropaeolum majus, and a study of the functional motifs of the DGAT protein using site-directed mutagenesis to modify enzyme activity and oil content. *Plant Biotechnology Journal* 6: 799–818.

Yang, J., Xing, G., Niu, L. et al. (2018). Improved oil quality in transgenic soybean seeds by RNAi-mediated knockdown of GmFAD2-1B. *Transgenic Research* 27: 155–166.

Zanetti, F., Eynck, C., Christou, M. et al. (2017). Agronomic performance and seed quality attributes of Camelina (*Camelina sativa L. crantz*) in multi-environment trials across Europe and Canada. *Industrial Crops and Products* 107: 602–608.

Zhang, P., Burton, J.W., Upchurch, R.G. et al. (2008). Mutations in a Δ 9–stearoyl-ACP-desaturase gene are associated with enhanced stearic acid levels in soybean seeds. *Crop Science* 48: 2305–2313.

Zhang, Y., Yu, L., Yung, K.F. et al. (2012). Over-expression of AtPAP2 in *Camelina sativa* leads to faster plant growth and higher seed yield. *Biotechnology for Biofuels* 5: 19.

Zhang, L., Yang, X.D., Zhang, Y.Y. et al. (2014). Changes in oleic acid content of transgenic soybeans by antisense RNA mediated posttranscriptional gene silencing. *International Journal of Genomics* 2014: 921–950.

12

Genome-Editing Mediated Improvement of Biotic Tolerance in Crop Plants

Krishan Mohan Rai[1] *and Harpal Singh*[2]

[1] *Department of Plant and Microbial Biology, University of Minnesota, Saint Paul, MN, USA*
[2] *Department of Nematology, University of California, Riverside, CA, USA*

CHAPTER MENU

12.1 Introduction

Plant pathogens such as bacteria, fungi and viruses are extremely diverse, have high multiplication rates and can readily spread in many ways. Distinct plant pathogens present a variety of challenges in combating infections and become the major limiting factor in crop productivity causing 20–40% yield penalty globally (Savary et al. 2012). Pathogens are equipped with several structural components and elicitors that facilitate establishment of infection and survival in the plant system. Plants have developed various defense mechanism to combat against pathogen attack. They harbor an array of genes capable of recognizing and responding against specific pathogen. The plant immune system is similar to innate immune system in animals, but plants have a significantly wider recognition repertoire so as to compensate for the unavailability of an adaptive immune system (Ausubel 2005).

Genome Engineering for Crop Improvement, First Edition. Edited by Santosh Kumar Upadhyay.

The enduring research for the identification of potential of plant immune system and pathogen biology facilitates our awareness of plant pathogen interactions. Better perception of underlying mechanisms aids the development of new approaches for crop protection (Dodds and Rathjen 2010). Plant breeders have been utilizing the knowledge of plant resistance genes for the development of resistant crop varieties for over 100 years. But the recent introduction of advanced genome-editing technologies such as CRISPR/Cas9, Transcription activator-like effector nucleases (TALENs), zinc-finger nucleases (ZFNs) etc. have revolutionized the development of plant varieties with desired potential. Genome-editing technologies facilitate precise editing of genomic DNA of an organism. These techniques are relatively simple, less time consuming, and offer exceptional efficiency as well as lesser risk of off-target effects. In recent years, several economically important crops have been edited for improved resistance against different pathogens or for the generation of basic knowledge. These approaches are mainly utilized to target specific host susceptibility genes, cleavage of pathogen DNA or for the inhibition of pathogen proliferation.

12.2 Plant Defense Response

Plants have developed a variety of defense pathways for containment of damage caused by different pathogens. The first line of defense in plants constitutes in their ability to recognize pathogens at cellular level and then triggering the underlying signaling pathways (Jones and Dangl 2006; Andolfo and Ercolano 2015). Pathogens display certain signals that can be recognized by plants known as Danger-associated molecular patterns or Pathogen-associated molecular patterns (PAMPs) that are recognized by Pattern recognition receptors (PRRs). PRRs act as immunity activation components that eventually trigger a complex pathway known as PAMP-triggered immunity (PTI) (Dodds and Rathjen 2010). The second level of immunity activation involves recognition of pathogen virulence molecules known as effectors, by intracellular receptors. This recognition triggers Effector triggered immunity (ETI). PTI is innate immunity of the plant defense system whereas ETI is adapted immunity against specific pathogens. Though PTI and ETI induce similar types of responses against pathogen infection, ETI is faster and has a stronger response usually accompanied by a localized cell death also known as the Hypersensitive response (HR) (Dodds and Rathjen 2010).

Phytohormones such as salicylic acid, jasmonic acid, ethylene, etc. play a crucial role in signaling and regulation of plant immune response (Bari and Jones 2009). Salicylic acid and jasmonic acid are among the major defense hormones. Distinct hormonal signaling pathways undergo antagonistic and synergistic interactions that act as an additional regulatory level also known as Hormone cross-talk (Jaillais and Chory 2010). This facilitates the plant to finely regulate the immune response against the different invading pathogens with a cost-efficient approach (Pieterse et al. 2012). The salicylic acid response is usually effective against biotrophic microbes, during both PTI and ETI. Pathogen invasion affects the Ca^{2+} levels and is an important trigger for initial signaling events upstream of salicylic acid biosynthesis (Du et al. 2009). Successively, *Enhanced Disease Susceptibility1 (EDS1) and Phytoalexin Deficient4 (PAD4)* induce the salicylic acid biosynthesis pathways (Wiermer et al. 2005). On the other hand, jasmonic acid response is more prominent against herbivores and necrotrophs (Verhage et al. 2010). SA and JA work in an antagonistic fashion and SA-JA crosstalk can be a strong process to decide the priority of induced defense pathways depending upon the sequence and kind of attacker involved.

12.3 Genome Engineering Tools for Engineering Disease Resistance

Genome engineering tools offer a superior biotechnological approach that facilitates specific and efficient gene targeting in the organism of interest. Genome-editing systems have been used in a variety of plant species for gene characterization and improvement of agricultural traits. This book chapter focuses on the applications of genome editing for the improvement of plant resistance and resilience against different pathogens.

12.3.1 Clustered Regularly Interspaced Short Palindromic Repeats (CRISPR)

The applications of the CRISPR/Cas9 system has transformed the face of life science over the past few years. Bacteria and archaea have CRISPR as a part of their adaptive immune system implied to generate DNA breaks to provide immunity against invading viruses (Rai et al. 2019; Zhang et al. 2019). This system was modified for implication in modification of organisms. CRISPR/Cas9 recognizes the target sequence with the help of a short-guide RNA (gRNA). The gRNA is designed specifically for the targeted gene and is usually accompanied by a protospacer adjacent motif (PAM) (Vats et al. 2019).

12.3.2 CRISPR/Cas9 for Resistance against Bacterial Pathogens

Bacterial blight susceptibility genes, *OsSWEET14* and *OsSWEET11* were one of the earliest plant immunity related genes targeted using CRISPR/Cas9, supporting the successful utilization of CRISPR/Cas9 system for agriculture applications (Jiang et al. 2013). Biotrophic and necrotrophic pathogens have different lifestyles, therefore, defense against one pathogen can increase the susceptibility toward another. To overcome this issue, Ortigosa et al. tried to uncouple the SA-JA antagonism in Arabidopsis stomata to achieve resistance against *P. syringae* pv. tomato (Pto) (causal agent of tomato bacterial speck disease) without compromising resistance against *Botrytis cinereal* (causal agent of tomato gray mold). Pto induces bacterial leaf colonization in stomata by production of coronatine (COR). Ortigosa et al. edited a major co-receptor of COR, SlJAZ2. The edited protein lost the C-terminal Jas domain and prevented stomatal opening by COR (Ortigosa et al. 2019). Direct delivery of purified CRISPR/Cas9 ribonucleoproteins for *DIPM-1, DIPM-2*, and *DIPM-4* in apple (Golden delicious variety) protoplasts facilitated the generation of resistance against an enterobacterial phytopathogen *Erwinia amylovora*, the causal agent of fire blight disease (Malnoy et al. 2016).

Xanthomonas citri causes Citrus canker disease compromising yield worldwide. *X. citri* activates two alleles of canker susceptibility lateral organ boundaries 1 (*CsLOB1*) gene in the host. Targeting Type I *CsLOB1* promoter region did not help the susceptibility against *X. citri* (Jia et al. 2016). Apparently, the activation of either of the alleles is capable of provoking disease symptoms. CRISPR/Cas9 targeted mutations in both the *CsLOB1* alleles developed resistant varieties for Citrus canker in Duncan grapefruit (Jia et al. 2016). In a similar study on *Citrus sinensis* Osbeck (Wanjincheng orange), *CsLOB1* promoter was targeted to enhance resistance for citrus canker. Wanjincheng orange *CsLOB1* also possesses two alleles *CsLOB1G* and *CsLOB1*$^-$. Peng et al. targeted promoter region of only *CsLOB1G* and observed significant improvement against citrus canker (Peng et al. 2017).

12.3.3 CRISPR/Cas9 for Resistance against Fungal Pathogens

A soybean fungal pathogen *Phytophthora sojae* is considered as a model organism for study of oomycete genetics, physiology and pathology. *P. sojae Avr4/6* gene encodes effector protein RXLR. CRISPR/Cas9 system was utilized to explore the involvement of RXLR effector in pathogenicity of *P. sojae*. Targeting *Avr4/6* confirmed the recognition of RXLR by soybean R genes, *Rps6* and *Rps4* (Fang and Tyler 2016; Mushtaq et al. 2019). This study raises possibilities for the study of pathogenomics of other important pathogens. Rice is one of the most important cereal crops worldwide. Infection by filamentous fungus *Magnaporthe oryzae* causes rice-blast disease, severely compromising yield. *Ptr* is a broad-spectrum blast resistance gene that functions in an NLR-mediated pathway. Zhao et al. created a two-base pair deletion line in the rice *Ptr* gene using CRISPR/Cas9. The mutated line M2354 confers susceptibility for filamentous fungus *M. oryzae* suggesting its importance in blast resistance (Zhao et al. 2018). Further, targeting *OsERF922*, an ethylene responsive factor (ERF) facilitates enhanced resistance against rice blast (Wang et al. 2016).

Downy mildew related gene, *Downy mildew resistance 6* (*SlDMR6-1*) gene in tomato when mutated confers enhanced resistance in transgenic tomato plants against *Phytophthora capsici*, *Xanthomonas* spp and *Pseudomonas syringae pv. tomato* without any significant compromise in the plant growth and development (de Toledo Thomazella et al. 2016). Editing of *SlMlo1* through CRISPR/Cas9 facilitated generation of Tomelo, a non-transgenic tomato variety that is resistant against the powdery mildew causing pathogen *Oidium neolycopersici* (Nekrasov et al. 2017). Similar to the work on apple for developing resistance against *E. amylovora*, Malnoy et al. also targeted *MLO-7* for direct delivery of purified CRISPR/Cas9 ribonucleoproteins to Chardonnay (Grape cultivar) protoplasts for generation of a powdery mildew resistant grape variety (Malnoy et al. 2016).

One of the most devastating diseases affecting rice production worldwide is rice-blast disease caused by a filamentous ascomycete fungus *M. oryzae*. Multiple targeting of *OsERF922* gene in japonica rice generated lines resistant toward *M. oryzae* without affecting any of the agronomic traits such as, flag leaf length, flag leaf width, number and length of panicles, rate of seed setting, plant height, seed weight, etc. (Wang et al. 2016). Fungal pathogen *Ustilago maydis* is the causal agent of corn smut disease in maize. Expression of codon optimized *U. maydis* cas9 gene with appropriate sgRNA, under the control of U6 snRNA promoter adequately induced genome editing. Approximately 70% of the single transformant progenies showed mutation in the targeted gene (Schuster et al. 2016).

Fister et al. reported for the first-time utilization of CRISPR/Cas9 system in *Theobroma cacao*. They targeted a known suppressor of defense response, *Non-Expressor of Pathogenesis-Related 3 (TcNPR3)* gene. The detached leaf transformation and pathogen bioassay showed enhanced resistance against *Phytophthora tropicalis* and increased expression of underlying defense-associated genes (Fister et al. 2018). *Arabidopsis thaliana Enhanced disease resistance1 (*EDR1*)* mutants are less susceptible to powdery mildew. *EDR1* has three homoeologs in wheat. CRISPR/Cas9 targeting for all three homoeologs of *TaEDR1* generate wheat plants resistant toward powdery mildew (Zhang et al. 2017).

12.3.4 CRISPR/Cas9 for Resistance against Viral Pathogens

12.3.4.1 DNA Viruses

Plant viruses compromise up to 10–15% of the crop yield worldwide (van Regenmortel and Mahy 2009). Gemini viruses are ssDNA viruses that exist in an intermediate dsDNA during viral replication. Utilizing CRISPR/Cas9 to specifically target the dsDNA can facilitate disease resistance. sgRNA/Cas9 constructs targeting Beet severe curly top virus (BSCTV) in transient expression assays in *Nicotiana benthamiana* inhibits the accumulation of viral particles. Transgenic *N. benthamiana* and Arabidopsis plants expressing sgRNA/Cas9 construct against BSCTV are highly resistant to virus infection (Ji et al. 2015). Expression of specific sgRNAs targeting Tomato yellow leaf curl virus (TYLCV) and Bean yellow dwarf virus (BeYDV) in *N. benthamiana* also impart resistance to the transgenic plants (Ali et al. 2015; Baltes et al. 2015). In another similar study Cotton leaf curl Kokhran virus (CLCuKoV) was successfully targeted and cleaved. Also, targeting conserved nano-nucleotide sequence from multiple begomo viruses developed transgenic *N. benthamiana* plants showing resistance for wide range of gemini viruses such as TYLCV, CLCuKov, TYLCSV, BCTV-Worland, MeMV, and BCTV-Logan (Zaidi et al. 2016).

Ali et al. targeted non-coding sequences of three gemini viruses viz. Merremia mosaic virus (MeMV), TYLCV and beet curly top virus (BCTV). The *N. benthamiana* plants infiltrated with the constructs showed decreased virus count as the viruses were unfit for replication and eventually degraded (Ali et al. 2016). Another plant virus Wheat dwarf virus (WDV) infects two economically important crops, wheat and barley. CRISPR/Cas9 system with 4 different sgRNAs against WDV show effective resistance in barley (Kis et al. 2019). The interaction between rice tungro bacilliform virus and rice tungro spherical virus (RTSV) causes rice tungro disease (RTD). RTD imposes serious yield penalty in rice. RTSV relies on rice translation initiation factor 4 gamma gene (*eIF4G*). CRISPR/Cas9 induced mutation in *eIF4G* in RTSV-susceptible variety IR64 conferred resistance against RTD (Macovei et al. 2018).

CRISPR/Cas9 system utilized for gemini viruses mostly targeted the coding regions which lead to generation of viral variants that could be capable of replication and movement, thus escaping the CRISPR/Cas9 system. Further reports suggest that the targeting of non-coding intergenic sequences furnish viral interference activity and limits creation of viral variants. This strategy proves to be more promising for long-term control of viruses (Ali et al. 2016). Another recent study utilized coat protein and replicase as potential targets for successful generation of resistant transgenic *N. benthamiana* and tomato plants against TYLCV (Tashkandi et al. 2018).

12.3.4.2 RNA Viruses

RNA viruses depend on host factors for their survival. Thus, targeting the host genes essentially required for survival of virus particle can help in developing RNA virus resistance. Targeting *eIF4E*, eukaryotic translation initiation factor in Cucumber confers resistance to several members from Potyviridae family [Cucumber vein yellowing virus (CVYV), Zucchini yellow mosaic virus (ZYMV) and Papaya ringspot mosaic virus-W (PRSV-W)] (Chandrasekaran et al. 2016). Pyott et al. used a similar technique for successful engineering of Arabidopsis resulting in complete resistance against *Turnip mosaic virus* (TuMV) (Pyott et al. 2016).

Another strategy for combating against RNA viruses can be by directly targeting the specific RNA sequences. The CRISPR/Cas9 system was originally developed for identification of DNA targets. However, Cas9 protein, supplemented with PAM-presenting oligonucleotides, can also recognize specific ssRNA (O'Connell et al. 2014). Modified Cas9 derived from *Francisella novicida* along with Cucumber mosaic virus (CMV)/Tobacco mosaic virus (TMV) specific sgRNA was successfully utilized for attenuation of viral accumulation and infection symptoms in transgenic Arabidopsis and *N. benthamiana* plants (Zhang et al. 2018). The efficient use of CRISPR/Cas9 system to develop resistance for plant viruses is a promising approach for the management of viral diseases in crops.

12.3.5 Transcription Activator-Like Effector Nucleases (TALENs)

TALENs are sequence-specific nucleases that were first discovered in Xanthomonas bacterium, that modulate the host defense mechanism by injecting a virulence protein. TALENs have a highly conserved and customizable DNA-binding domain derived from transcription activator-like effectors (TALEs) fused to a non-specific *FokI* nuclease enzyme (Joung and Sander 2013).

12.3.5.1 TALENs for Resistance against Bacterial Pathogens

Bacterial blight (causal agent is *Xanthomonas oryzae* pv. *oryzae*) is a dreadful disease that makes a huge economic impact on rice production worldwide. *Xanthomonas* encodes Transcription activator-like (TAL) effectors that specifically activate rice disease-susceptibility (*S*) genes facilitating pathogen virulence. The *S* gene, *Os11N3* or *OsSWEET14* was targeted in this study. The promoter of *OsSWEET14* has two effector-binding elements recognized by *Xanthomonas* AvrXa7 and PthXo3 TAL effectors, facilitating divergence of sugar influx for pathogen proliferation. TALENs designed for specifically editing these two effector-binding elements enabled generation of rice plants resistant to bacterial blight without compromising the developmental function of *OsSWEET14* (Li et al. 2012).

12.3.5.2 TALENs for Resistance against Viral Pathogens

TALENs have been used to confer resistance against multiple begomo viruses. TALENs were designed to target conserved sequences in begomo virus genome, such as replication associated protein encoded by AC1 gene and inverted repeat genes. Transgenic *N. benthamiana* plants thus generated were tested against three different begomo viruses, TbCSV, TYLCCNV and Tomato leaf curl Yunnan virus (TLCYnV). TALEs conferred resistance against TbCSV and TYLCCNV but were only partially resistant for TLCYnV infection (Cheng et al. 2015).

12.3.6 Zinc-Finger Nucleases (ZFNs)

ZFNs are chimeric fusion proteins comprising of a zinc-finger DNA-binding domain and a nuclease domain *FokI* (Kim et al. 1996; Bibikova et al. 2001). The zinc fingers contain cysteine and histidine repeats designed for specific recognition of DNA (Federici et al. 2012).

12.3.6.1 ZFNs for Resistance against Viral Pathogens

ZFNs have not been very well utilized for the improvement of disease resistance in crops. There are a few reports employing the ZFNs for targeting viral replication as a measure to enhance tolerance for viral infection. Sera used an artificial zinc-finger protein (AZP) for targeting origin of replication of BSCTV. Transgenic Arabidopsis expressing the AZP showed complete suppression of BSCTV replication and were highly resistant against virus infection (Sera 2005). In a similar study, specific AZP was successfully shown to target replication origin of Tomato yellow curl virus (TYLCV) in *N. benthamiana* and inhibit viral replication (Takenaka et al. 2007). Chen et al. designed an AZP targeting a conserved sequence motif of begomo viruses and tested against two different viruses Tomato yellow leaf curl china virus TYLCCNV, and Tobacco curly shoot virus (TbCSV), through transient expression in *N. benthamiana*. The results showed effective termination of virus replication and a successful strategy for developing resistance against multiple begomo viruses (Chen et al. 2014).

12.3.7 Additional Genome-Editing Targets for Disease Resistance in Crop Plants

Comprehensive understanding of plant immune signaling will facilitate the construction of plant-immune response networks, enabling identification of potential targets for genome editing.

Broad-spectrum resistance can be achieved by targeting PRRs. Andolfo et al. identified candidate PRRs in *Solanum lycopersicum* (Andolfo et al. 2013). Tang et al. report LRR-RLKs from four different species *A. thaliana*, rice, poplar and grapevine (Tang et al. 2010). A similar effort was made for the identification of "NLRome" across 65 diverse accessions of *A. thaliana* (Van de Weyer et al. 2019). It is well understood that the plant pathogens secrete different effector proteins that suppress plant-immune response and facilitate pathogen proliferation (Irieda et al. 2019). Though, most of the effector proteins known so far pose negative effect on other traits and thus are not promising effector targets. Identification of more effector genes needs to be carried out to discover more favorable targets (Gawehns et al. 2013). Genome-editing techniques can be used to modulate the effector and target sequences to alter their interaction during immune response. Dong et al. reported that a single amino-acid alteration in *Phytophthora infestans* effector EPIC1, *Phytophthora mirabilis* effector PmEPIC and their corresponding targets proteins tomato PLCP and potato RCR3 respectively, impairs their mutual interaction and alters the resistance phenotype (Dong et al. 2014; Niks et al. 2015).

RIN4 is an important component of the immune system and is a known target for effectors, therefore, it can be a promising target for genome editing. Luo et al. reported that Arabidopsis *rin4* mutant shows enhanced resistance against the bacteria *P. syringae* and oomycete *Peronospora parasitica* infections (Luo et al. 2016). Several effectors suppress the plant-immune response by modulating the synthesis and regulation of hormonal pathways. Genome-engineering techniques can be used to target the negative regulators of hormonal pathways, such as MAP kinase *MPK4*, a negative regulator of salicylic acid response. Arabidopsis *mpk4* mutant shows increase in resistance for *P. syringae* and *P. parasitica* (Petersen et al. 2000).

Ubiquitination is another important pathway involved in plant immunity and is a target of several pathogen effectors. Mutants for PUB22, PUB23, and PUB24 ubiquitin ligases in Arabidopsis show improved resistance against oomycete *Hyaloperonospora arabidopsidis* and bacteria *P. syringe* (Trujillo et al. 2008). Ubiquitin ligase knockouts in potato also show resistance to *P. infestans* (Bos et al. 2010).

Genome-editing techniques can help in the development of resistant varieties by modulation of immunity players and lead to a more environmentally sustainable agriculture approach. Genome-editing technologies provide additional advantage by editing effectors by specific mutations despite gene deletion eventually avoiding harmful pleiotropic effects.

References

Ali, Z., Abulfaraj, A., Idris, A. et al. (2015). CRISPR/Cas9-mediated viral interference in plants. *Genome Biology* 16 (1): 238.

Ali, Z., Ali, S., Tashkandi, M. et al. (2016). CRISPR/Cas9-mediated immunity to geminiviruses: differential interference and evasion. *Scientific Reports* 6 (1): 1–13.

Andolfo, G. and Ercolano, M.R. (2015). Plant innate immunity multicomponent model. *Frontiers in Plant Science* 6: 987.

Andolfo, G., Sanseverino, W., Rombauts, S. et al. (2013). Overview of tomato (Solanum lycopersicum) candidate pathogen recognition genes reveals important Solanum R locus dynamics. *New Phytologist* 197 (1): 223–237.

Ausubel, F.M. (2005). Are innate immune signaling pathways in plants and animals conserved? *Nature Immunology* 6 (10): 973–979.

Baltes, N.J., Hummel, A.W., Konecna, E. et al. (2015). Conferring resistance to geminiviruses with the CRISPR–Cas prokaryotic immune system. *Nature Plants* 1 (10): 1–4.

Bari, R. and Jones, J.D. (2009). Role of plant hormones in plant defence responses. *Plant Molecular Biology* 69 (4): 473–488.

Bibikova, M., Carroll, D., Segal, D.J. et al. (2001). Stimulation of homologous recombination through targeted cleavage by chimeric nucleases. *Molecular and Cellular Biology* 21 (1): 289–297.

Bos, J.I., Armstrong, M.R., Gilroy, E.M. et al. (2010). Phytophthora infestans effector AVR3a is essential for virulence and manipulates plant immunity by stabilizing host E3 ligase CMPG1. *Proceedings of the National Academy of Sciences of the United States of America* 107 (21): 9909–9914.

Chandrasekaran, J., Brumin, M., Wolf, D. et al. (2016). Development of broad virus resistance in non-transgenic cucumber using CRISPR/Cas9 technology. *Molecular Plant Pathology* 17 (7): 1140–1153.

Chen, W., Qian, Y., Wu, X. et al. (2014). Inhibiting replication of begomoviruses using artificial zinc finger nucleases that target viral-conserved nucleotide motif. *Virus Genes* 48 (3): 494–501.

Cheng, X., Li, F., Cai, J. et al. (2015). Artificial TALE as a convenient protein platform for engineering broad-spectrum resistance to begomoviruses. *Viruses* 7 (8): 4772–4782.

Dodds, P.N. and Rathjen, J.P. (2010). Plant immunity: towards an integrated view of plant–pathogen interactions. *Nature Reviews Genetics* 11 (8): 539–548.

Dong, S., Stam, R., Cano, L.M. et al. (2014). Effector specialization in a lineage of the Irish potato famine pathogen. *Science* 343 (6170): 552–555.

Du, L., Ali, G.S., Simons, K.A. et al. (2009). Ca 2+/calmodulin regulates salicylic-acid-mediated plant immunity. *Nature* 457 (7233): 1154–1158.

Fang, Y. and Tyler, B.M. (2016). Efficient disruption and replacement of an effector gene in the oomycete P hytophthora sojae using CRISPR/C as9. *Molecular Plant Pathology* 17 (1): 127–139.

Federici, F., Dupuy, L., Laplaze, L. et al. (2012). Integrated genetic and computation methods for in planta cytometry. *Nature Methods* 9 (5): 483–485.

Fister, A.S., Landherr, L., Maximova, S.N., and Guiltinan, M.J. (2018). Transient expression of CRISPR/Cas9 machinery targeting TcNPR3 enhances defense response in Theobroma cacao. *Frontiers in Plant Science* 9: 268.

Gawehns, F., Cornelissen, B.J., and Takken, F.L. (2013). The potential of effector-target genes in breeding for plant innate immunity. *Microbial Biotechnology* 6 (3): 223–229.

Irieda, H., Inoue, Y., Mori, M. et al. (2019). Conserved fungal effector suppresses PAMP-triggered immunity by targeting plant immune kinases. *Proceedings of the National Academy of Sciences of the United States of America* 116 (2): 496–505.

Jaillais, Y. and Chory, J. (2010). Unraveling the paradoxes of plant hormone signaling integration. *Nature Structural & Molecular Biology* 17 (6): 642–645.

Ji, X., Zhang, H., Zhang, Y. et al. (2015). Establishing a CRISPR–Cas-like immune system conferring DNA virus resistance in plants. *Nature Plants* 1 (10): 1–4.

Jia, H., Orbovic, V., Jones, J.B., and Wang, N. (2016). Modification of the PthA4 effector binding elements in Type I Cs LOB 1 promoter using Cas9/sg RNA to produce transgenic Duncan grapefruit alleviating XccΔpthA4: dCs LOB 1.3 infection. *Plant Biotechnology Journal* 14 (5): 1291–1301.

Jiang, W., Zhou, H., Bi, H. et al. (2013). Demonstration of CRISPR/Cas9/sgRNA-mediated targeted gene modification in arabidopsis, tobacco, sorghum and rice. *Nucleic Acids Research* 41 (20): e188–e188.

Jones, J.D. and Dangl, J.L. (2006). The plant immune system. *Nature* 444 (7117): 323–329.

Joung, J.K. and Sander, J.D. (2013). TALENs: a widely applicable technology for targeted genome editing. *Nature Reviews Molecular Cell Biology* 14 (1): 49–55.

Kim, Y.-G., Cha, J., and Chandrasegaran, S. (1996). Hybrid restriction enzymes: zinc finger fusions to Fok I cleavage domain. *Proceedings of the National Academy of Sciences of the United States of America* 93 (3): 1156–1160.

Kis, A., Hamar, É., Tholt, G. et al. (2019). Creating highly efficient resistance against wheat dwarf virus in barley by employing CRISPR/Cas9 system. *Plant Biotechnology Journal* 17 (6): 1004.

Li, T., Liu, B., Spalding, M.H. et al. (2012). High-efficiency TALEN-based gene editing produces disease-resistant rice. *Nature Biotechnology* 30 (5): 390.

Luo, M., Gilbert, B., and Ayliffe, M. (2016). Applications of CRISPR/Cas9 technology for targeted mutagenesis, gene replacement and stacking of genes in higher plants. *Plant Cell Reports* 35 (7): 1439–1450.

Macovei, A., Sevilla, N.R., Cantos, C. et al. (2018). Novel alleles of rice eIF4G generated by CRISPR/Cas9-targeted mutagenesis confer resistance to Rice tungro spherical virus. *Plant Biotechnology Journal* 16 (11): 1918–1927.

Malnoy, M., Viola, R., Jung, M.-H. et al. (2016). DNA-free genetically edited grapevine and apple protoplast using CRISPR/Cas9 ribonucleoproteins. *Frontiers in Plant Science* 7: 1904.

Mushtaq, M., Sakina, A., Wani, S.H. et al. (2019). Harnessing genome editing techniques to engineer disease resistance in plants. *Frontiers in Plant Science* 10.

Nekrasov, V., Wang, C., Win, J. et al. (2017). Rapid generation of a transgene-free powdery mildew resistant tomato by genome deletion. *Scientific Reports* 7 (1): 1–6.

Niks, R.E., Qi, X., and Marcel, T.C. (2015). Quantitative resistance to biotrophic filamentous plant pathogens: concepts, misconceptions, and mechanisms. *Annual Review of Phytopathology* 53: 445–470.

O'Connell, M.R., Oakes, B.L., Sternberg, S.H. et al. (2014). Programmable RNA recognition and cleavage by CRISPR/Cas9. *Nature* 516 (7530): 263–266.

Ortigosa, A., Gimenez-Ibanez, S., Leonhardt, N., and Solano, R. (2019). Design of a bacterial speck resistant tomato by CRISPR/Cas9-mediated editing of Sl JAZ 2. *Plant Biotechnology Journal* 17 (3): 665–673.

Peng, A., Chen, S., Lei, T. et al. (2017). Engineering canker-resistant plants through CRISPR/Cas9-targeted editing of the susceptibility gene Cs LOB 1 promoter in citrus. *Plant Biotechnology Journal* 15 (12): 1509–1519.

Petersen, M., Brodersen, P., Naested, H. et al. (2000). Arabidopsis MAP kinase 4 negatively regulates systemic acquired resistance. *Cell* 103 (7): 1111–1120.

Pieterse, C.M., Van der Does, D., Zamioudis, C. et al. (2012). Hormonal modulation of plant immunity. *Annual Review of Cell and Developmental Biology* 28: 489–521.

Pyott, D.E., Sheehan, E., and Molnar, A. (2016). Engineering of CRISPR/Cas9-mediated potyvirus resistance in transgene-free Arabidopsis plants. *Molecular Plant Pathology* 17 (8): 1276–1288.

Rai, K.M., Ghose, K., Rai, A. et al. (2019). Genome engineering tools in plant synthetic biology. In: *Current Developments in Biotechnology and Bioengineering*, 47–73. Elsevier.

van Regenmortel, M.H. and Mahy, B.W. (2009). *Desk Encyclopedia of Plant and Fungal Virology*. Academic Press.

Savary, S., Ficke, A., Aubertot, J.-N., and Hollier, C. (2012). *Crop Losses Due to Diseases and Their Implications for Global Food Production Losses and Food Security*. Springer.

Schuster, M., Schweizer, G., Reissmann, S., and Kahmann, R. (2016). Genome editing in Ustilago maydis using the CRISPR–Cas system. *Fungal Genetics and Biology* 89: 3–9.

Sera, T. (2005). Inhibition of virus DNA replication by artificial zinc finger proteins. *Journal of Virology* 79 (4): 2614–2619.

Takenaka, K., Koshino-Kimura, Y., Aoyama, Y., and Sera, T. (2007). Inhibition of tomato yellow leaf curl virus replication by artificial zinc-finger proteins. In: *Nucleic Acids Symposium Series*, 429–430. Oxford: Oxford University Press.

Tang, P., Zhang, Y., Sun, X. et al. (2010). Disease resistance signature of the leucine-rich repeat receptor-like kinase genes in four plant species. *Plant Science* 179 (4): 399–406.

Tashkandi, M., Ali, Z., Aljedaani, F. et al. (2018). Engineering resistance against tomato yellow leaf curl virus via the CRISPR/Cas9 system in tomato. *Plant Signaling & Behavior* 13 (10): e1525996.

de Toledo Thomazella, D.P., Brail, Q., Dahlbeck, D., and Staskawicz, B. (2016). *CRISPR-Cas9 Mediated Mutagenesis of a DMR6 Ortholog in Tomato Confers Broad-Spectrum Disease Resistance*, 064824. BioRxiv.

Trujillo, M., Ichimura, K., Casais, C., and Shirasu, K. (2008). Negative regulation of PAMP-triggered immunity by an E3 ubiquitin ligase triplet in Arabidopsis. *Current Biology* 18 (18): 1396–1401.

Van de Weyer, A.-L., Monteiro, F., Furzer, O.J. et al. (2019). *The Arabidopsis thaliana pan-NLRome*, 537001. BioRxiv.

Vats, S., Kumawat, S., Kumar, V. et al. (2019). Genome editing in plants: exploration of technological advancements and challenges. *Cell* 8 (11): 1386.

Verhage, A., van Wees, S.C., and Pieterse, C.M. (2010). Plant immunity: it's the hormones talking, but what do they say? *Plant Physiology* 154 (2): 536–540.

Wang, F., Wang, C., Liu, P. et al. (2016). Enhanced rice blast resistance by CRISPR/Cas9-targeted mutagenesis of the ERF transcription factor gene OsERF922. *PLoS One* 11 (4).

Wiermer, M., Feys, B.J., and Parker, J.E. (2005). Plant immunity: the EDS1 regulatory node. *Current Opinion in Plant Biology* 8 (4): 383–389.

Zaidi, SS.-e.-A., Tashkandi, M., Mansoor, S., and Mahfouz, M.M. (2016). Engineering plant immunity: using CRISPR/Cas9 to generate virus resistance. *Frontiers in Plant Science* 7: 1673.

Zhang, Y., Bai, Y., Wu, G. et al. (2017). Simultaneous modification of three homoeologs of Ta EDR 1 by genome editing enhances powdery mildew resistance in wheat. *The Plant Journal* 91 (4): 714–724.

Zhang, T., Zheng, Q., Yi, X. et al. (2018). Establishing RNA virus resistance in plants by harnessing CRISPR immune system. *Plant Biotechnology Journal* 16 (8): 1415–1423.

Zhang, Y., Malzahn, A.A., Sretenovic, S., and Qi, Y. (2019). The emerging and uncultivated potential of CRISPR technology in plant science. *Nature Plants* 5 (8): 778–794.

Zhao, H., Wang, X., Jia, Y. et al. (2018). The rice blast resistance gene Ptr encodes an atypical protein required for broad-spectrum disease resistance. *Nature Communications* 9 (1): 1–12.

13

Genome Engineering and Essential Mineral Enrichment of Crops

Erum Shoeb, Uzma Badar, Srividhya Venkataraman, Ghyda Murad Hashim, and Kathleen Hefferon

Cell and Systems Biology, University of Toronto, Toronto, Ontario, Canada

CHAPTER MENU

13.1 Introduction

Plants provide macro and micronutrients such as proteins, minerals and vitamins as essential components of the human diet. These must be readily bioavailable and of sufficient quantity within the edible parts of crops. At present, more than 850 million of our global population are chronically undernourished (Global 2018). During the Green Revolution in the 1960s, the major focus was on enhancing crop yield to improve food security. This is no longer a concern for today's world populace; instead, the quality/nutritional value of food consumed has become the greatest priority in food programs worldwide. Nutritional insufficiency in food grains is the major cause of malnutrition. This is compounded by other factors such as population increase, unpredictable changes in climate, diminishing arable land, elevated food prices and other environmental parameters. The health impacts of poor food nutrition value include retarded physical growth, obesity, metabolic diseases such as diabetes, enhanced risk of infections and even death (FAO 2017).

In urban areas, strategies such as dietary diversification, nutraceutical supplementation and food fortification have proven to be effective (Ruel and Alderman 2013). On the other hand, biofortification of food has been found to be the most practical means of providing nutrition and nutritional security to poor, malnourished populations in rural areas with

Genome Engineering for Crop Improvement, First Edition. Edited by Santosh Kumar Upadhyay.

restricted access to micronutrients (Bouis and Welch 2010). The concentration of nutrients in grains or edible parts of plants could be enhanced by adopting conventional selective plant-breeding techniques in addition to integrated agricultural practices and high throughput "Omics" technologies.

Initially, crop-breeding programs focused on the selection of allelic variants, resulting in high yield while ignoring allelic forms determining nutritional quality factors. Crops improved in nutrient quality can be developed by changing the physical and chemical content of the soil used in their cultivation, in addition to selecting appropriate genotypes that are high-yield and nutrient-rich as donors in plant-breeding programs. Moreover, the genes and proteins involved in nutrient homeostasis can be suitably manipulated toward obtaining nutrient-rich varieties. To achieve this, in-depth insight into the molecular metabolism's behind nutrient and mineral regulatory pathways is required. Currently, such molecular discernment is available only for a few minerals such as selenium in maize, iron and zinc in barley, rice, wheat and maize as well as calcium in finger millet (Bauer and Hell 2006; Malagoli et al. 2015; Sharma et al. 2016). Additionally, in most cases, nutrient bioavailability is not well characterized. In addition to this, the sequestering of nutrients to different parts of the edible portions of plants can be highly variable.

There exists a highly complex physiochemical interaction between the soil, the microflora and plant genotypes. Our agricultural land worldwide has diminished drastically due to modern industrialization, urbanization and environmental issues. Also, our population is growing at an unprecedented rate endangering our food security. This emphasizes the need for enhanced agricultural productivity within the nearest timeframe. Using conventional breeding techniques to address this concern, intensive agricultural practices can be adopted which require more arable land, augmented mechanization of farms and wide-ranging use of agrochemicals toward improving crop production. However, this approach is not very effective and does not achieve food security in the long term. Issues such as soil saturation (Dumka and Bindraban 2016) and deterioration of soil system sustainability can impact agricultural growth and yield. In addition to this, nutrient quality is compromised, and bioavailability of nutrients becomes diminished, resulting in malnutrition. Therefore, there exists an urgent need for augmented production of nutrient-enriched crops without compromising soil health. Biofortification through both agronomic (microbial assisted) breeding and genetic engineering (genomic-assisted breeding) poses a viable solution toward achieving food and nutrient security.

Phytoremediation is an emerging ecological engineering technology that involves the application of green plants toward treatment and control of chemical wastes and noxious pollutants in diverse environments such as soil, water and air. Metal-enriched plant biomass could be used to generate energy and metal recovery through phyto-mining and biofortification. Phytoremediation affords a low-cost, sustainable alternative to improve the economies of developing countries. The dual problems of malnutrition and environmental pollution can be overcome by combining biofortification and phytoremediation. Thus, polluted soils can be recovered for agricultural purposes and this forms a vital link between farmers and biotechnology researchers. The application of phytoremediation is based on underlying knowledge of plant molecular processes and metabolic mechanisms as well as the tolerance of plants to saline and noxious substances. The use of phytoremediation has become vitally important for ensuring food and nutrient security. Novel genetic engineering technologies such as genome editing using clustered regularly interspaced short palindromic repeat (CRISPR)/Cas molecular systems have become increasingly adopted for achieving phytoremediation.

This review focuses on current technologies available for enhancing biofortification and nutrient security using mineral-enrichment strategies. In addition to this, the review describes the use of genetically engineered plants for phytoremediation and how this technology can be combined with biofortification to ensure both food security and nutrient bioavailability. The technology of rhizosphere engineering is also discussed.

13.2 Root Engineering of Cereals: A Promising Strategy to Improve Nutrient Efficiency, Biofortification, and Drought Tolerance

Millions of people worldwide are improperly nourished, not only because of insufficient quantity of food, but also due to deficient nutritional contents. Lack of bioavailable micronutrients create severe health complications (Kumssa et al. 2015). Besides several examples of genetically modified crops (Tian et al. 2012), there are very few attempts reported to improve nutrient efficiency (Chen and Liao 2017). Crop plants complete all the vital functions through their root systems (Ramireddy et al. 2018). Factors such as drought and scarcity of soil nutrients affect crop productivity and effectiveness (Lynch 1995; Hochholdinger 2016). One of the most appropriate solutions to these challenges is to improve the root systems of crop plants (Lynch and Brown 2012; Comas et al. 2013; White et al. 2013; Rogers and Benfey 2015; Koevoets et al. 2016).

The plant root system is a complex polygenic trait (Lynch and Brown 2012) and is thus difficult to improve through traditional breeding practices (Ramireddy et al. 2018). In the field, roots face multiple factors that interact with plant genetics to produce root system architecture. Bioavailability, distribution and mobility are variable for different nutrients and unfortunately there is no standard root architecture for all (Chen and Liao 2017). For nitrogen, sulfur and boron, soil mobility and bioavailability are high and have a tendency to accumulate in deep soil. In order to absorb these nutrients, the root system must be able to penetrate deep soil (Lynch 2013; White et al. 2013). This root penetrance system can also be beneficial in drought affected areas for plants to exhibit drought-tolerance.

Genetic studies related to root regulation systems and correlated quantitative trait loci (QTLs) have been widely reported in Arabidopsis (Liang et al. 2014; Li et al. 2016). The plant hormone cytokinin (CK) has been recognized as a major negative regulator of root development. It has been proven that degradation of CK leads to formation of an improved root system, triggers accumulation of nutrients in the plant and enhances chances of survival in drought conditions (Werner et al. 2010). In 2018, (Ramireddy et al. 2018) Genetic engineering of CK-degrading (CK oxidase/dehydrogenase CKX) genes resulted in barley plants that have an enhanced root system without any effect on shoot development. As compared to wild type, transformed plants showed increased accumulation of soil nutrients and drought tolerance.

Yan et al. (2014) related overexpression of the microRNA (miR444a) with nitrate accumulation and reduced length of lateral roots coordinated with nitrate-dependent elevated growth of primary and adventitious roots. The MADS-box transcription factor in Arabidopsis plays a major role in the nitrate-signaling pathway through the ANR1 gene (Yu et al. 2015). In rice, an ANR1 homologous gene OsMADS25 was reported as a positive regulator controlling lateral and primary root development (Zhang et al. 2018). Another gene found in rice, OsNAR2.1 has also been reported to repress lateral root formation in a nitrate-dependent manner (Huang et al. 2015). In wheat, overexpression of TaNAC2-5A improves root growth and the rate of nitrate intake, which improves wheat yield (He et al. 2015). GmNMHC5 is a sucrose-regulatory MADS-box transcription factor in soybean, and was reported to stimulate nodulation and root development (Liu et al. 2015).

It has been reported that in crops such as rice, maize, wheat and soybean, phosphorus deficiency can reduce the growth of roots and root hair (López-Arredondo et al. 2014). In rice OsMYB2P-1 and in maize ZmPTF1 transcription factors act as positive monitors of changes in root architecture under low phosphorous conditions (Li et al. 2011; Dai et al. 2012). Another gene GmEXPB2 enhances plant growth under low phosphorous levels in the soil in Arabidopsis and soybean (Guo et al. 2011; Xie et al. 2015).

In 2011 (Weber and Bodanese-Zanettini 2011), an attempt was made for the induction of hairy roots through *Agrobacterium tumefaciens*-mediated transformation in soybean (*Glycine max* L.). The hairy root phenotype consists of highly branched roots that grow rapidly and plagiotropically in the absence of exogenous plant-growth regulators (Collier et al. 2005). The gene-editing technology CRISPR/Cas (CRISPR-associated) system has been extensively used for genome engineering in different organisms including plants. Genome editing was applied to target soybean (*Glycine max* [L.] Merrill.) hairy root genes (GmFEI2 and GmSHR). The targeted DNA mutations were detected in soybean hairy roots. The results demonstrated that this customized CRISPR/Cas9 system provides a powerful tool for root-specific functional genomics studies (Cai et al. 2015).

13.3 Use of Genome Edited Plants in Phytoremediation

Heavy metal contamination in the environment is mainly due to anthropogenic activities and/or natural processes and is a serious, widespread problem. Heavy metals occur in various forms in the soil, which differ greatly with respect to their solubility/bioavailability (Shahid et al. 2015). Bioaccumulation and biomagnification of lethal pesticides are the main cause of the loss of plants, microbes and animal biodiversity (Boudh and Singh 2019); (Vera-Reyes et al. 2019). Phytoremediation is the process of cleaning up of environmental pollutants using plants, but plants are unable to detoxify the pollutants very rapidly, resulting in accumulation of toxic compounds (Van Aken 2008). Gene-editing tools are now available for bioremediation of persistent organic pollutants, heavy metal, petroleum and all the pollutants in pesticidal compounds (Jaiswal et al. 2019).

Successes over the past few years suggest that genetic engineering can be effective in producing plants for phytoremediation of soils and waters contaminated with pollutants. These plants benefit from the genes that enhance the capability of plants to mobilize, stabilize and/or accumulate metals (Fasani et al. 2018). Phytoextraction of heavy metals such as cadmium, lead and copper from soil is possible due to their accumulation in the plants through metal transporters, improved production of enzymes of sulfur metabolism, production of metal-detoxifying chelators and activation of various antioxidants (Kotrba et al. 2009). A large number of genes for increased metal tolerance, metal uptake and hyperaccumulation have already been identified (Basharat et al. 2018). Transgenic plants with an enhanced phytoremediation capacity are usually produced by the introduction of a gene that is not native to the plant (Gunarathne et al. 2019). Genes of interest including degradative enzymes to achieve complete mineralization of organic pollutants (Eapen et al. 2007).

For phytoremediation, transgenic plants were initially produced to improve heavy metal tolerance. Cadmium tolerance gene for yeast metallothionein expressed in tobacco plants (*Nicotiana tabacum*) (Misra and Gedamu 1989), and mercuric ion reductase gene overexpressed in *Arabidopsis thaliana* (Rugh et al. 1996) for tolerance to mercury are a few of the examples. Phytoremediation of organic compounds was also examined, and tobacco plants were transformed to target explosives and halogenated organic compounds (French et al. 1999; Doty et al. 2000). The first transgenic poplars for phytoremediation were developed by Gellner (Gullner et al. 2001) to treat chloroacetanilide herbicides by the overexpression of an enzyme gamma-glutamyl cysteine synthetase.

A large enzyme superfamily cytochrome P450s (CYPs) is involved in the clearance and breakdown of toxic compounds. Human CYP was expressed in transgenic tobacco plants and was capable of metabolizing trichloroethylene (TCE) much more than wild-type plants (Doty et al. 2000). Later, faster metabolism of TCE in transgenic cultures of *Atropa belladonna* was reported using rabbit CYP (Banerjee et al. 2002). In 2007 (Doty et al. 2007), hybrid poplar plants transformed with cytochrome P450 2E1 (CYP2E1) were capable of metabolism of five environmental pollutants: trichloroethylene, a heavily used industrial degreaser and major soil pollutant; vinyl chloride, a proven human carcinogen that is produced as a result of microbial biodegradation of TCE; carbon tetrachloride, a common toxic soil pollutant; chloroform, a toxin formed during disinfection of water by chlorination; and benzene, a carcinogenic compound associated with petroleum pollution.

The chloroplast is the main target of organic mercury in plants as it restricts electron transport and photosynthesis. Ruiz et al. (2003) modified chloroplast genetic composition by inserting an operon containing mercuric ion reductase (merA) and organomercurial lyase (merB) genes. merAB operon increased tolerance of transgenic tobacco (*N. tabacum*) plants against phenylmercuric acetate (PMA) as compared to the wild types.

Mutant lines of rice (*Oryza sativa* L.), OsLCT1 and OsNramp5, by CRISPR/Cas9-mediated modifications were grown in cadmium-polluted fields (Songmei et al. 2019). These lines showed less accumulation of cadmium, whereas the growth was reduced, which could be utilized to produce crop safe for human use in cadmium-contaminated soils.

Genomes of classic phytoremediators, *Arabidopsis helleri* (Cd and Zn hyperaccumulator) (Briskine et al. 2017), *Pteris vittata* (As hyperaccumulator) (Xie et al. 2009), *Noccaea caerulescens* (zinc, cadmium, and nickel hyperaccumulator) (Mandáková et al. 2015). *Hirschfeldia incana* (tolerate and accumulate Pb) (Auguy et al. 2016) and several other species, have been sequenced. It is well established that transgenic plants are capable of contributing to the phytoremediation process. However, the studies on these transgenic plants must be further investigated to provide a thorough understanding. An environmental risk assessment of those transgenic plants for phytoremediation require long-term field testing.

13.4 Genetic Engineering and Crop Biofortification

The world's population faces various health conditions linked to nutrition-related disorders (Stover 2006). Early researchers mainly focused on various conventional breeding methods for biofortification to improve the micronutrients in crops and also agronomical characters such as yield and pest resistance. If required genotypic variation is not available, suggested alternatives to traditional breeding techniques include genetic manipulation approaches in order to insert the desired trait in the selected crops.

Nutrigenomics is the new scientific field that links essential molecules involved in nutrition with genomics. It is the action of single nutrients on different molecular targets which causes the alteration of nutrient-gene expression within a biological system. This transgenic technology has been extensively used for improving various nutritionally important traits such as starch content in waxy sorghum to make it more digestible. It is controlled by the waxy allele at the GBSS I locus (the granule-bound starch synthase I) (McIntyre et al. 2008), waxy gene in foxtail millets (Van et al. 2008; Kim et al. 2009); grains with waxy endosperm wheat (Sestili et al. 2010); quality proteins in maize (Gibbon and Larkins 2005).

A second major group of required micronutrients in humans is vitamins, required for growth and development and also to provide tolerance against number of diseases. β-carotene is the precursor of pro-vitamin A, which is found in very low concentrations in crops such as rice (Van Eck et al. 2007). Transgenics offer a sustainable approach to concentrate the vitamin content of edible parts of the plants. Classic examples of transgenic based biofortification technology are the crops with higher levels of provitamin A, known as "golden" crops (Giuliano 2017) including "Golden Rice" (Potrykus 2012) and sorghum (Lipkie et al. 2013).

The bacterial phytoene synthase gene, *CrtB* was used to increase levels of provitamin A in crops (Giuliano et al. 2008). Overexpression of *CrtB* was found to be associated with a higher accumulation of β-carotene, lutein and violaxanthin in potato (Ducreux et al. 2004). Expression of the daffodil phytoene *PSY* gene with bacterial *crtI* increased carotenoid content in rice (Ye et al. 2000). Similarly, when bacterial *crtB* and *crtI* genes were overexpressed, increased levels of β-carotene in maize was observed (Aluru et al. 2008). Genetically engineered potato, canola, tomato, carrot, and cauliflower were also reported for higher β-carotene content (Shahriari et al. 2013). Carotenoid are readily degradable in cereal crops due to the presence of the lipoxygenase enzyme (Carrera et al. 2007; Che et al. 2016). Genetic engineering has been showed to decrease the regulation of lipoxygenase which, in turn, increases stability of carotenoids during storage (Gayen et al. 2015).

13.5 Genome Engineering Technology and Its Use in Essential Mineral Enrichment of Crop Plants

Major edible cash crops are considered for major biofortification projects to increase micronutrient deficiencies (Brinch-Pedersen et al. 2007). Various studies were carried out to increase the Iron (Fe) and Zinc (Zn) content acquired from the environment in plants (Morrissey and Guerinot 2009).

Iron (Fe) is stored in plants in the form of ferritin. It is a localized bioavailable protein within plant plastids and is released whenever needed for any metabolic activity (Darbani et al. 2013). Improved accumulation of Fe by expression of *ferritin* gene is a popular strategy for Fe biofortification. Various studies have been completed on the overexpression of ferritin demonstrating increased Fe content as well as the bioavailability of Fe (Aluru et al. 2011; Borg et al. 2012). Goto et al. (1999), generated brown rice transformants to increase Fe accumulation in endosperm by using the *SoyferH1* gene from soybean, under the endosperm specific GluB1 rice promoter. This resulted in a transformant strain that exhibited a three times higher level of Fe content in seeds when compared with original strain. Likewise, ferritin cDNA of soybean was inserted in wheat and rice controlled by maize promoter (Kumar et al. 2019). In this case, the transformants demonstrated a high Fe content in leaves rather than in seeds. Qu et al. (2005) introduced soybean ferritin into rice with an endosperm specific promoter of globulin which caused an increase of expression of ferritin protein much more than that reported by Goto et al. (1999). These experiments for Fe biofortification in rice provide a very good example of genetic engineering in cereal crops.

Very few examples of Fe biofortification in crops other than rice are available to report. Wheat ferritin genes *TaFer1-A* and *TaFer 2* were cloned and analyzed by Borg et al. (2012) under the control of HMW glutein 1DX5 promoter and demonstrated the overexpression potential of ferritin in the endosperm of wheat to enhance the Fe content. Therefore, the endosperm targeted intragenic overexpressing of the *TaFer1-A* gene results in a 50–85% higher iron content in the transgenic grains.

Phytic acid is a potent inhibitor of Fe and Zn absorption. Ascorbic acid and nicotianamine (NA) are enhancers of Fe absorption (Hunt 2005; Zheng et al. 2010). Phytic acid chelates iron and zinc ions that plays an important role in transporting these metals within the plant. Johnson et al. (2011) produced biofortified rice lines with enhanced Fe and Zn concentrations in polished grain using constitutive overexpression of the OsNAS2 gene. It has been suggested that the NA overproduction enhances the translocation of Fe and Zn to rice grains.

Similarly, the chelation of Fe and Zn ions with phytic acid has a strong adverse effect on absorption of these minerals in humans and other monogastric animals that largely lack the phytase enzyme to degrade phytate. Therefore, it is important to reduce phytate levels which, in turn, increase Fe and Zn bioavailability. Mutants of barley and maize with reduced phytate levels (up to 95%) have been produced by Raboy (2002). Mutations in phytic acid biosynthesis in cereals can be an effective strategy to reduce phytic acid (Holm et al. 2002). The transformation of the phytase gene from *Aspergillus fumigatus* into rice endosperm, shows increased levels of phytase up to 130-fold (Lucca et al. 2001). In the case of maize, this phytase gene was expressed along with the soybean ferritin gene, which significantly increased Fe and phytase up to 20–70% (Drakakaki et al. 2005).

Besides iron, zinc is an essential element and plays an important role as a co-factor for number of enzymes and transcription factors (Palmgren et al. 2008). As far as Zn content of cereals is concerned, there is a very low degree of diversity reported. Consequently, anticipating genetic manipulations to increase the cereal grains, Zn content must be crucial for advancing human diet and metabolic uptake. Strong correlations between content of protein and Fe and Zn content was reported by Ozturk et al. (2006). In wheat, the Grain protein content B1 (*Gpc-B1*) is a quantitative trait locus found to be associated with improved protein contents which also increased the Zn and Fe content of grain (Uauy et al. 2006). Role of Gpc-B1 was observed in leaves to grains mobilization of proteins, Zn, Fe and Mn in wild species of tetraploid wheat *Triticum turgidum* (Distelfeld et al. 2007).

13.6 Conclusion

Plant breeding is a process that helps to improve plant traits (Ramireddy et al. 2018). The main principle of breeding is combining different desirable traits to produce more productive varieties. Breeding techniques include pedigree breeding, idiotype breeding, breeding based on observed variations, and hybrid breeding (Breseghello and Coelho 2013).

Pedigree breeding is mostly used in breeding self-pollinated plants. The method requires the evaluation of a variety of parents which are later hybridized to produce plants with more desirable qualities. The method is useful in breeding qualitative traits such as resistance to drought and diseases, the height and color of plants, among other classifiable features. The downside of this method is that it relies on observable traits, and the outcome can only be determined after the breeding process (Breseghello and Coelho 2013). Idiotype breeding basically relies on plant traits. The technique involves selecting the variants from their natural habitats and cultivating them on the field. However, this method could also provide poor quality due to poor selection practice. The resultant offspring portray the traits of both parents, thus can be used to produce plants with a variety of characteristics from different parents. The offspring from a hybrid are more resilient than the parents. However, the offspring are not viable; hence cannot be used for further breeding (Breseghello and Coelho 2013).

Genome engineering is the breeding process that uses the most advanced technology. Some of genome engineering include CRISPR-Cas9 and rhizoengineering among other techniques (Lopes et al. 2011). The main focus of the methods is altering plants' DNA, which helps in the production of plants with desired qualities. The technology is used for increasing plant immunity, improving resistance to drought and pathogens and producing plants enriched with different mineral content such as zinc and iron. Scientists can alter plant DNA that controls various traits and processes in a plant. The technology can be used to modify plant leaves and root cells to help in the uptake of a specific type of mineral in plants or even improve the mineral generation process (Nehnevajova et al. 2019). Therefore, CRISPR-Cas9 technology can be used to modify the hormones that control root development in plants. CK hormones regulate the elongation of the roots; therefore, increasing the hormone in a plant will help in increasing the rate of root elongation (Chandra et al. 2020). Consequently, plants with high cytokines have a high response rate (Nehnevajova et al. 2019). The technology is also applied in plant shoots to reduce the rate of transpiration by regulating the osmotic processes in leaves. Therefore, the approach helps produce drought-resistant crops. Technology such as rhizoengineering assists in solving issues related to nutrient deficiency in plants (Chandra et al. 2019).

Both modern and traditional methods are convenient for improving crop productivity. However, advanced biotechnology is more suitable for plant breeding; the technique is efficient and helps to modify targeted plant genes to produce plants with improved quality.

The technology can be used to control plant diseases and adaptation, hence increasing the survival rate of the plant under adverse conditions. Genome-engineering technology can be used to increase root size and architecture by increasing cytokine concentration and production rate. Also, the technology helps repair DNA strands responsible for root development to enhance the root physiological processes. Generating crops with high mineral content will improve human and animal nutrition, which, in turn, provides a positive impact on both health and the economy.

References

Aluru, M., Xu, Y., Guo, R. et al. (2008). Generation of transgenic maize with enhanced provitamin a content. *Journal of Experimental Botany* 59 (13): 3551–3562.

Aluru, M.R., Rodermel, S.R., and Reddy, M.B. (2011). Genetic modification of low phytic acid 1-1 maize to enhance iron content and bioavailability. *Journal of Agricultural and Food Chemistry* 59 (24): 12954–12962.

Auguy, F., Fahr, M., Moulin, P. et al. (2016). Transcriptome changes in Hirschfeldia incana in response to lead exposure. *Frontiers in Plant Science* 6: 1231.

Banerjee, S., Shang, T.Q., Wilson, A.M. et al. (2002). Expression of functional mammalian P450 2E1 in hairy root cultures. *Biotechnology and Bioengineering* 77 (4): 462–466.

Basharat, Z., Novo, L., and Yasmin, A. (2018). Genome editing weds CRISPR: what is in it for phytoremediation? *Plants* 7.

Bauer, P. and Hell, R. (2006). *Translocation of Iron in Plant Tissues*, Iron Nutrition in Plants and Rhizospheric Microorganisms. Dordrecht: Springer.

Borg, S., Brinch-Pedersen, H., Tauris, B. et al. (2012). Wheat ferritins: improving the iron content of the wheat grain. *Journal of Cereal Science* 56 (2): 204–213.

Boudh, S. and Singh, J.S. (2019). Pesticide contamination: environmental problems and remediation strategies. In: *Emerging and Eco-Friendly Approaches for Waste Management* (eds. R.N. Bharagava and P. Chowdhary), 245–269. Singapore: Springer.

Bouis, H. and Welch, R. (2010). Biofortification--a sustainable agricultural strategy for reducing micronutrient malnutrition in the global south. *Crop Science* 50: S20–S32.

Breseghello, F. and Coelho, A.S. (2013). Traditional and modern plant breeding methods with examples in rice (Oryza sativa L.). *Journal of Agricultural and Food Chemistry* 61 (35): 8277–8286.

Brinch-Pedersen, H., Borg, S., Tauris, B. et al. (2007). Molecular genetic approaches to increasing mineral availability and vitamin content of cereals. *Journal of Cereal Science* 46 (3): 308–326.

Briskine, R.V., Paape, T., Shimizu-Inatsugi, R. et al. (2017). Genome assembly and annotation of Arabidopsis halleri, a model for heavy metal hyperaccumulation and evolutionary ecology. *Molecular Ecology Resources* 17 (5): 1025–1036.

Cai, Y., Chen, L., Liu, X. et al. (2015). CRISPR/Cas9-mediated genome editing in soybean hairy roots. *PLoS One* 10 (8): e0136064.

Carrera, A., Echenique, V., Zhang, W. et al. (2007). A deletion at the Lpx-B1 locus is associated with low lipoxygenase activity and improved pasta color in durum wheat (Triticum turgidum ssp. durum). *Journal of Cereal Science* 45 (1): 67–77.

Chandra, A.K., Kumar, A., Bharati, A. et al. (2019). Microbial-assisted and genomic-assisted breeding: a two way approach for the improvement of nutritional quality traits in agricultural crops. *3 Biotech* 10 (1): 2.

Che, P. et al. (2016). β-Carotene accumulation and stability in sorghum. *Proceedings of the National Academy of Sciences* 113 (39): 11040–11045.

Chen, L. and Liao, H. (2017). Engineering crop nutrient efficiency for sustainable agriculture. *Journal of Integrative Plant Biology* 59 (10): 710–735.

Collier, R., Fuchs, B., Walter, N. et al. (2005). Ex vitro composite plants: an inexpensive, rapid method for root biology. *The Plant Journal* 43 (3): 449–457.

Comas, L., Becker, S.R., Cruz, V.M.V. et al. (2013). Root traits contributing to plant productivity under drought. *Frontiers in Plant Science* 4: 442.

Dai, X., Wang, Y., Yang, A. et al. (2012). OsMYB2P-1, an R2R3 MYB transcription factor, is involved in the regulation of phosphate-starvation responses and root architecture in rice. *Plant Physiology* 159 (1): 169–183.

Darbani, B., Briat, J.F., Holm, P.B. et al. (2013). Dissecting plant iron homeostasis under short and long-term iron fluctuations. *Biotechnology Advances* 31 (8): 1292–1307.

Dimkpa, C.O. and Bindraban, P.S. (2016). Fortification of micronutrients for efficient agronomic production: a review. *Agronomy for Sustainable Development* 36 (1): 7.

Distelfeld, A., Cakmak, I., Peleg, Z. et al. (2007). Multiple QTL-effects of wheat Gpc-B1 locus on grain protein and micronutrient concentrations. *Physiologia Plantarum* 129 (3): 635–643.

Doty, S.L., Shang, T.Q., Wilson, A.M. et al. (2000). Enhanced metabolism of halogenated hydrocarbons in transgenic plants containing mammalian cytochrome P450 2E1. *Proceedings of the National Academy of Sciences* 97 (12): 6287.

Doty, S.L., James, C.A., Moore, A.L. et al. (2007). Enhanced phytoremediation of volatile environmental pollutants with transgenic trees. *Proceedings of the National Academy of Sciences of the United States of America* 104 (43): 16816–16821.

Drakakaki, G., Marcel, S., Glahn, R.P. et al. (2005). Endosperm-specific co-expression of recombinant soybean ferritin and Aspergillus phytase in maize results in significant increases in the levels of bioavailable iron. *Plant Molecular Biology* 59 (6): 869–880.

Ducreux, L.J.M., Moris, W.L., Hedley, P.E. et al. (2004). Metabolic engineering of high carotenoid potato tubers containing enhanced levels of β-carotene and lutein. *Journal of Experimental Botany* 56 (409): 81–89.

Eapen, S., Singh, S., and D'Souza, S.F. (2007). Advances in development of transgenic plants for remediation of xenobiotic pollutants. *Biotechnology Advances* 25 (5): 442–451.

FAO (2017).The state of food insecurity in the World–Meeting the 2016. International Hunger Targets: taking stock of unevenprogress. http://www.fao.org/3/a-i4646e.pdf (accessed 20 July 2020).

Fasani, E., Manara, A., Martini, F. et al. (2018). The potential of genetic engineering of plants for the remediation of soils contaminated with heavy metals. *Plant, Cell & Environment* 41 (5): 1201–1232.

French, C.E., Rosser, S.J., Davies, G.J. et al. (1999). Biodegradation of explosives by transgenic plants expressing pentaerythritol tetranitrate reductase. *Nature Biotechnology* 17 (5): 491–494.

Gayen, D., Ali, N., Sarkar, S.N. et al. (2015). Down-regulation of lipoxygenase gene reduces degradation of carotenoids of golden rice during storage. *Planta* 242 (1): 353–363.

Gibbon, B.C. and Larkins, B.A. (2005). Molecular genetic approaches to developing quality protein maize. *Trends in Genetics* 21 (4): 227–233.

Giuliano, G. (2017). Provitamin a biofortification of crop plants: a gold rush with many miners. *Current Opinion in Biotechnology* 44: 169–180.

Giuliano, G., Tavazza, R., Diretto, G. et al. (2008). Metabolic engineering of carotenoid biosynthesis in plants. *Trends in Biotechnology* 26 (3): 139–145.

Global, H.I.(2018). http://www.ifpri.org/publication/2018-global-hunger-index-inequalities-hunger.

Goto, F., Yoshihara, T., Shigemoto, N. et al. (1999). Iron fortification of rice seed by the soybean ferritin gene. *Nature Biotechnology* 17 (3): 282–286.

Gullner, G., Komives, T., and Rennenberg, H. (2001). Enhanced tolerance of transgenic poplar plants overexpressing gamma-glutamylcysteine synthetase towards chloroacetanilide herbicides. *Journal of Experimental Botany* 52 (358): 971–979.

Gunarathne, V., Mayakaduwa, S., Ashiq, A. et al. (2019). *Transgenic Plants: Benefits, Applications, and Potential Risks in Phytoremediation*, 89–102. Academic Press.

Guo, W., Zhao, J., Li, Z. et al. (2011). A soybean beta-expansin gene GmEXPB2 intrinsically involved in root system architecture responses to abiotic stresses. *The Plant Journal* 66 (3): 541–552.

He, X., Qu, B., Li, W. et al. (2015). The nitrate-inducible NAC transcription factor TaNAC2-5A controls nitrate response and increases wheat yield. *Plant Physiology* 169 (3): 1991–2005.

Hochholdinger, F. (2016). Untapping root system architecture for crop improvement. *Journal of Experimental Botany* 67 (15): 4431–4433.

Holm, P.B., Kristiansen, K.N., and Pedersen, H.B. (2002). Transgenic approaches in commonly consumed cereals to improve iron and zinc content and bioavailability. *The Journal of Nutrition* 132 (3): 514s–516s.

Huang, S., Chen, S., Liang, Z. et al. (2015). Knockdown of the partner protein OsNAR2.1 for high-affinity nitrate transport represses lateral root formation in a nitrate-dependent manner. *Scientific Reports* 5: 18192.

Hunt, J.R. (2005). Dietary and physiological factors that affect the absorption and bioavailability of iron. *International Journal for Vitamin and Nutrition Research* 75 (6): 375–384.

Jaiswal, S., Singh, D.K., and Shukla, P. (2019). Gene editing and systems biology tools for pesticide bioremediation: a review. *Frontiers in Microbiology* 10: 87.

Johnson, A.A., Kyriacou, B., Callahan, D.L. et al. (2011). Constitutive overexpression of the OsNAS gene family reveals single-gene strategies for effective iron- and zinc-biofortification of rice endosperm. *PLoS One* 6 (9): e24476.

Kim, S.-K., Sohn, E.-Y., and Lee, I.-J. (2009). Starch properties of native foxtail millet, Setaria italica Beauv. *Journal of Crop Science and Biotechnology* 12 (1): 59–62.

Koevoets, I.T., Venema, J.H., Elzenga, J.T.M., and Testerink, C. (2016). Roots withstanding their environment: exploiting root system architecture responses to abiotic stress to improve crop tolerance. *Frontiers in Plant Science* 7: 1335.

Kotrba, P., Najmanova, J., Macek, T. et al. (2009). Genetically modified plants in phytoremediation of heavy metal and metalloid soil and sediment pollution. *Biotechnology Advances* 27 (6): 799–810.

Kumar, S., Palve, A., Joshi, C. et al. (2019). Crop biofortification for iron (Fe), zinc (Zn) and vitamin a with transgenic approaches. *Heliyon* 5 (6): e01914.

Kumssa, D.B. et al. (2015). Dietary calcium and zinc deficiency risks are decreasing but remain prevalent. *Scientific Reports* 5: 10974.

Li, Z., Gao, Q., Liu, Y. et al. (2011). Overexpression of transcription factor ZmPTF1 improves low phosphate tolerance of maize by regulating carbon metabolism and root growth. *Planta* 233 (6): 1129–1143.

Li, B., Zhang, Y., Mohammadi, S.A. et al. (2016). An integrative genetic study of Rice metabolism, growth and stochastic variation reveals potential C/N partitioning loci. *Scientific Reports* 6: 30143.

Liang, C., Wang, J., Zhao, J. et al. (2014). Control of phosphate homeostasis through gene regulation in crops. *Current Opinion in Plant Biology* 21: 59–66.

Lipkie, T.E., De Moura, F.F., Zhao, Z.-Y. et al. (2013). Bioaccessibility of carotenoids from transgenic Provitamin a biofortified Sorghum. *Journal of Agricultural and Food Chemistry* 61 (24): 5764–5771.

Liu, W., Han, X., Zhan, G. et al. (2015). A novel sucrose-regulatory MADS-box transcription factor GmNMHC5 promotes root development and nodulation in soybean (Glycine max [L.] Merr.). *International Journal of Molecular Sciences* 16 (9): 20657–20673.

Lopes, M.S., Araus, J.L., van Heerden, P.D.R. et al. (2011). Enhancing drought tolerance in C(4) crops. *Journal of Experimental Botany* 62 (9): 3135–3153.

López-Arredondo, D., Levya-González, M.A., and González-Morales, S.I. (2014). Phosphate nutrition: improving low-phosphate tolerance in crops. *Annual Review of Plant Biology* 65: 95–123.

Lucca, P., Hurrell, R., and Potrykus, I. (2001). Approaches to improving the bioavailability and level of iron in rice seeds. *Journal of the Science of Food and Agriculture* 81: 828–834.

Lynch, J. (1995). Root architecture and plant productivity. *Plant Physiology* 109 (1): 7–13.

Lynch, J.P. (2013). Steep, cheap and deep: an ideotype to optimize water and N acquisition by maize root systems. *Annals of Botany* 112 (2): 347–357.

Lynch, J.P. and Brown, K.M. (2012). New roots for agriculture: exploiting the root phenome. *Philosophical Transactions of the Royal Society, B: Biological Sciences* 367 (1595): 1598–1604.

Malagoli, M., Schiavon, M., dall'Acqua, S. et al. (2015). Effects of selenium biofortification on crop nutritional quality. *Frontiers in Plant Science* 6: 280.

Mandáková, T., Singh, V., Krämer, U., and Lysak, M.A. (2015). Genome structure of the heavy metal Hyperaccumulator Noccaea caerulescens and its stability on metalliferous and nonmetalliferous soils. *Plant Physiology* 169: 674–689.

McIntyre, C.L., Gonzalez, J.D.N., Henzell, R.G., and Jordan, D.R. (2008). Molecular characterization of the waxy locus in sorghum. *Genome* 51 (7): 524–533.

Misra, S. and Gedamu, L. (1989). Heavy metal tolerant transgenic Brassica napus L. and Nicotiana tabacum L. plants. *Theoretical and Applied Genetics* 78 (2): 161–168.

Morrissey, J. and Guerinot, M.L. (2009). Iron uptake and transport in plants: the good, the bad, and the Ionome. *Chemical Reviews* 109 (10): 4553–4567.

Nehnevajova, E., Ramireddy, E., Stolz, A. et al. (2019). Root enhancement in cytokinin-deficient oilseed rape causes leaf mineral enrichment, increases the chlorophyll concentration under nutrient limitation and enhances the phytoremediation capacity. *BMC Plant Biology* 19 (1): 83.

Ozturk, L., Yazici, M.A., Yucel, C. et al. (2006). Concentration and localization of zinc during seed development and germination in wheat. *Physiologia Plantarum* 128 (1): 144–152.

Palmgren, M.G., Clemens, S., William, L.E. et al. (2008). Zinc biofortification of cereals: problems and solutions. *Trends in Plant Science* 13 (9): 464–473.

Potrykus, I. (2012). “Golden Rice”, a GMO-product for public good, and the consequences of GE-regulation. *Journal of Plant Biochemistry and Biotechnology* 21 (1): 68–75.

Qu, L.Q., Yoshihara, T., Ooyama, A. et al. (2005). Iron accumulation does not parallel the high expression level of ferritin in transgenic rice seeds. *Planta* 222 (2): 225–233.

Raboy, V. (2002). Progress in breeding low phytate crops. *The Journal of Nutrition* 132 (3): 503S–505S.

Ramireddy, E., Hosseini, S.A., Eggert, K. et al. (2018). Root engineering in barley: increasing Cytokinin degradation produces a larger root system, mineral enrichment in the shoot and improved drought tolerance. *Plant Physiology* 177 (3): 1078–1095.

Rogers, E.D. and Benfey, P.N. (2015). Regulation of plant root system architecture: implications for crop advancement. *Current Opinion in Biotechnology* 32: 93–98.

Ruel, M.T. and Alderman, H. (2013). Nutrition-sensitive interventions and programmes: how can they help to accelerate progress in improving maternal and child nutrition? *Lancet* 382 (9891): 536–551.

Rugh, C.L., Wilde, H.D., Stack, N.M. et al. (1996). Mercuric ion reduction and resistance in transgenic Arabidopsis thaliana plants expressing a modified bacterial merA gene. *Proceedings of the National Academy of Sciences of the United States of America* 93 (8): 3182–3187.

Ruiz, O.N., Hussein, H.S., Terry, N., and Daniell, H. (2003). Phytoremediation of organomercurial compounds via chloroplast genetic engineering. *Plant Physiology* 132 (3): 1344.

Sestili, F., Janni, M., Doherty, A. et al. (2010). Increasing the amylose content of durum wheat through silencing of the SBEIIagenes. *BMC Plant Biology* 10 (1): 144.

Shahid, M., Khalid, S., Abbas, G. et al. (2015). *Heavy Metal Stress and Crop Productivity*, Crop Production and Global Environmental Issues. Cham: Springer.

Shahriari, Z., Heidari, B., Cheraghi, M., and Shahriari, A.G. (2013). Biofortification of staple food crops: engineering the metabolic pathways. *International Research Journal of Applied and Basic Sciences*: 287–290.

Sharma, D., Jamra, G., Singh, U.M. et al. (2016). Calcium biofortification: three pronged molecular approaches for dissecting complex trait of calcium nutrition in finger millet (Eleusine coracana) for devising strategies of enrichment of food crops. *Frontiers in Plant Science* 7: 2028.

Songmei, L., Jie, J., Yang, L. et al. (2019). Characterization and evaluation of OsLCT1 and OsNramp5 mutants generated through CRISPR/Cas9-mediated mutagenesis for breeding low cd Rice. *Rice Science* 26 (2): 88–97.

Stover, P.J. (2006). Influence of human genetic variation on nutritional requirements. *The American Journal of Clinical Nutrition* 83 (2): 436s–442s.

Tian, Z., Zhao, M., She, M. et al. (2012). Genome-wide characterization of nonreference transposons reveals evolutionary propensities of transposons in soybean. *The Plant Cell* 24 (11): 4422.

Uauy, C., Brevis, J.C., and Dubcovsky, J. (2006). The high grain protein content gene Gpc-B1 accelerates senescence and has pleiotropic effects on protein content in wheat. *Journal of Experimental Botany* 57 (11): 2785–2794.

Van Aken, B. (2008). Transgenic plants for phytoremediation: helping nature to clean up environmental pollution. *Trends in Biotechnology* 26 (5): 225–227.

Van Eck, J., Conlin, B., Garvin, D.F. et al. (2007). Enhancing beta-carotene content in potato by rnai-mediated silencing of the beta-carotene hydroxylase gene. *American Journal of Potato Research* 84 (4): 331.

Van, K., Onoda, S., Kim, M.Y. et al. (2008). Allelic variation of the waxy gene in foxtail millet [Setaria italica (L.) P. Beauv.] by single nucleotide polymorphisms. *Molecular Genetics and Genomics* 279 (3): 255–266.

Vera-Reyes, I., Esparza-Arrendondo, I.J.E., Lira-Saldivar, R.H. et al. (2019). In vitro antimicrobial effect of metallic nanoparticles on phytopathogenic strains of crop plants. *Journal of Phytopathology* 167 (7–8): 461–469.

Weber, R. and Bodanese-Zanettini, M. (2011). Induction of transgenic hairy roots in soybean genotypes by agrobacterium rhizogenes-mediated transformation. *Pesquisa Agropecuária Brasileira* 46: 1070–1075.

Werner, T., Nehnevajova, E., Köllmer, I. et al. (2010). Root-specific reduction of cytokinin causes enhanced root growth, drought tolerance, and leaf mineral enrichment in Arabidopsis and tobacco. *Plant Cell* 22 (12): 3905–3920.

White, P.J., George, T.S., Gregory, P.J. et al. (2013). Matching roots to their environment. *Annals of Botany* 112 (2): 207–222.

White, P., George, T.S., Dupuy, L.X. et al. (2013). Root traits for infertile soils. *Frontiers in Plant Science* 4: 193.

Xie, Q.-E. et al. (2009). The arsenic Hyperaccumulator Fern Pteris vittata L. *Environmental Science & Technology* 43 (22): 8488–8495.

Xie, J., Zhou, J., Wang, X., and Liao, H. (2015). Effects of transgenic soybean on growth and phosphorus acquisition in mixed culture system. *Journal of Integrative Plant Biology* 57 (5): 477–485.

Yan, Y., Wang, H., Hamera, S. et al. (2014). miR444a has multiple functions in the rice nitrate-signaling pathway. *The Plant Journal* 78 (1): 44–55.

Ye, X. et al. (2000). *Engineering the Provitamin A (β -Carotene) Biosynthetic Pathway into (Carotenoid-Free) Rice Endosperm, Science* 287 (5451): 303–305 DOI: 10.1126/science.287.5451.303.

Yu, C., Liu, Y., Zhang, A. et al. (2015). MADS-box transcription factor OsMADS25 regulates root development through affection of nitrate accumulation in rice. *PLoS One* 10 (8): e0135196.

Zhang, G., Xu, N., Chen, H. et al. (2018). OsMADS25 regulates root system development via auxin signalling in rice. *The Plant Journal* 95 (6): 1004–1022.

Zheng, L., Cheng, Z., Ai, C. et al. (2010). Nicotianamine, a novel enhancer of Rice Iron bioavailability to humans. *PLoS One* 5 (4): e10190.

14

Genome Editing to Develop Disease Resistance in Crops

Kashaf Zafar[1,2], Azka Noureen[1,3], Muhammad Jawad Akbar Awan[1], Naveed Anjum[1,4], Muhammad Qasim Aslam[1], Muhammad Zuhaib Khan[1], Imran Amin[1], and Shahid Mansoor[1]

[1] *Agricultural Biotechnology Division, National Institute for Biotechnology and Genetic Engineering (NIBGE), Constituent College of Pakistan Institute of Engineering and Applied Sciences, Faisalabad, Pakistan*
[2] *Department of Biotechnology, Balochistan University of Information Technology, Engineering and Management Sciences (BUITEMS), Quetta, Pakistan*
[3] *University Institute of Biochemistry and Biotechnology (UIBB), Pir Mehar Ali Shah Arid Agriculture University, Rawalpindi, Pakistan*
[4] *Department of Biotechnology, University of Sialkot, Sialkot, Punjab, Pakistan*

CHAPTER MENU

14.1 Introduction

Modern-day agriculture is under tremendous pressure to ensure food security because of the estimated increase of 2.4 billion in the world's population by 2050 (Wheeler and Von Braun 2013). While, in terms of agricultural production, an approximate 60% increase in crop productivity is required to fulfill global demand (Alexandratos and Bruinsma 2012). In the presence of overwhelming plant pathogens such as bacteria, fungi and viruses, this goal will be more challenging to achieve. It is stated that 11–30% of losses to the world's production are due to pests and diseases (Oerke and Dehne 2004; Savary et al. 2019).

Genome Engineering for Crop Improvement, First Edition. Edited by Santosh Kumar Upadhyay.

Researchers are manipulating the genetic composition of plants to enhance their disease-resistance ability against pathogens.

Plants interact with pathogens using their innate immunity in two ways (Figure 14.1). One way is by directly recognizing pathogens-associated molecular patterns (PAMP) through cell-surface receptors (receptor-like kinases or receptor-like proteins). The second way is by directly or indirectly interacting with pathogens secreted by virulence effectors from inside the cell through cytoplasmic receptors like nucleotide-binding leucine-rich repeat proteins (NLR). These immune receptors in plants are largely encoded by resistance (R) or R-like genes which detect the extra and intracellular danger signals and activate plant defenses (Van der Burgh and Joosten 2019). On the other hand, over the course of evolution, plant pathogens have learned to evade the host immunity by hijacking or misdirecting the host immune response. In this way, they maintain host susceptibility for their own good. For instance, the crosstalk between phytohormones like salicylic acid and jasmonic acid production is crucial for effective activation of plant defense response against certain pathogens (De Vleesschauwer et al. 2014; He et al. 2004). In case of *Coronatine-insensitive protein* 1 (*COI*1) jasmonic acid receptor in *Arabidopsis thaliana*, some bacterial pathogens takeover this crosstalk and suppress the effective salicylic acid defense pathway by deliberately inducing jasmonic acid-based defenses (He et al. 2004). Likewise, *Pyrenophora tritici*-repentis and *Stagonospora nodorum* are two important necrotrophic fungal pathogens of wheat and related cereals which, by activating an NLR-based Tsn1 receptor, trigger an inappropriate cell-death response in host plants for their benefits (Faris et al. 2010). This understanding helps us to manipulate the plant immunity with better pathogen perception abilities. These include the introduction of novel receptors from other

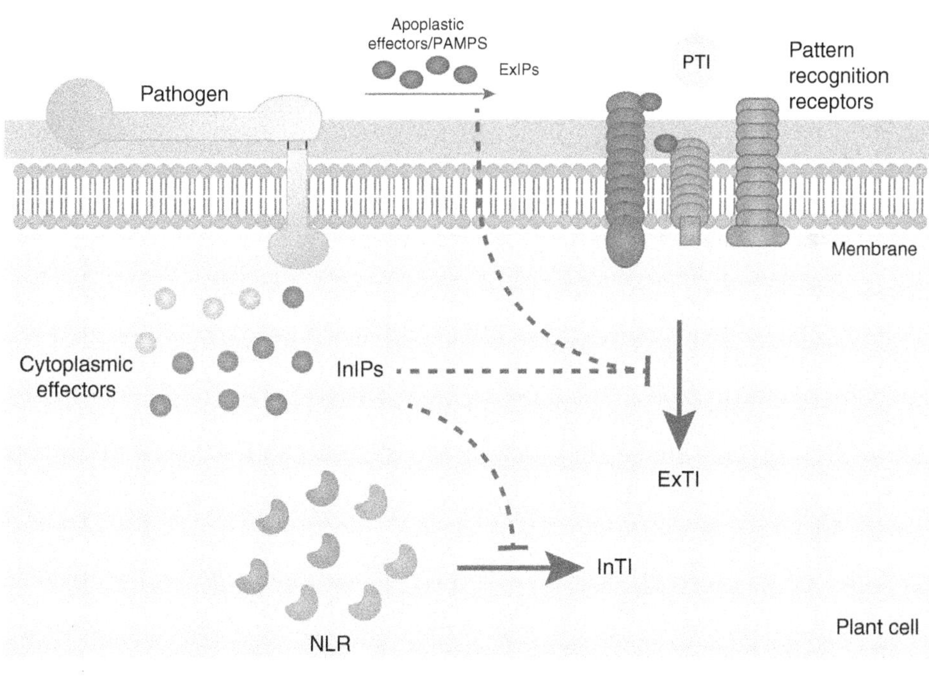

Figure 14.1 Schematic picture of the basal plant immunity.

plant species (Albert et al. 2015; Ghislain et al. 2019), transfer of diverse helper NLRs to reactivate the host plant NLR genes attacked by pathogen-derived effectors (Wu et al. 2017) or alterations of pathogen targeted NLR domains (Maqbool et al. 2015) that enhance the pathogen recognition capability of subjected crop plants. A growing understanding of susceptibility genes is pulling our attention due to their useful nature in offering plant immunity and modifications provide more promising and long-lasting recessive resistance than dominant R genes (Gust et al. 2010). These susceptibility genes are the potential vulnerable host targets, exposed during effector interaction and are necessary for successful pathogen infection. These opportunities have been successfully exploited in different crops using genome-editing technologies (Zaidi et al. 2018).

14.2 Traditional Approaches to Develop Disease Resistance in Crops

Practices of plant breeding in human culture has very deep-rooted links, tracing back to the early times of agriculture (Gross and Olsen 2010; Meyer and Purugganan 2013). Prior intentions of plant breeding were to improve crop yield, disease resistance, stress tolerance and nutritional quality enhancement. Though conventional breeding approaches have widely proven successful in developing elite germplasm with novel agronomic traits. This technique still relies on a considerable amount of plant-population evaluation and empirical selection. The marker-assisted breeding (MAB) brought a great revolution in conventional breeding methods offering greater efficiency in selection with increased practical and accurate evaluation, even though, it is still time consuming and labor-intensive (Jiang 2013). Moreover, in conventional breeding, it is challenging to make a selection against multiple diseases at a time while upholding the best performance of elite varieties such as selection of wheat against newly emerging wheat-blast disease is an issue for breeders when they also have to maintain resistance against the devastating stem-rust disease (Islam et al. 2016).

Genetic engineering has the potential to make direct modifications in a genome (Christou 2013), hence can be an effective approach in developing resistance in crops against various challenging pathogens. In comparison with conventional breeding, it allows modification, fine-tuning and introduction of novel disease-resistance genes into elite cultivars without imposing considerable changes to the rest of their genome. Thus, in less time, the best combination of agronomic traits can be obtained in a few generations. Moreover, there is no limitation on inter- or intraspecies selection of target genes. Thus, the need for genetic material required for developing transgenic organisms is not confined to the genes available within the species.

In conventional transgenic methods, genes responsible for certain agronomic characters are inserted into the plant genome at random sites (Lorence and Verpoorte 2004) and the plants produced through this procedure are called transgenic plants. In terms of disease-resistance development in crop plants, most success stories of conventional genetic engineering involve the single pathway of resistance based on a dominant R gene. For instance, currently, a late blight-resistant variety of potato is available in the market in the USA (www.isaaa.org) which is expressing an NLR receptor introduced from its wild relative.

Though it is an important advancement, it must be considered that resistance, based on a single-dominant gene can be easily overcome by plant pathogens which have a shorter lifecycle with surprising adaptation ability as compared to their hosts (Van Esse et al. 2020).

As an alternative to single-gene induction, gene stacking or molecular stacking offers an assembly of multiple R gene cassettes onto one plasmid followed by transformation into a plant cell at a single locus through plant transformation methods (Dafny-Yelin and Tzfira 2007; Que et al. 2010). In this way, the R-genes cluster will remain intact and positioned at a single genetic locus during the process of inheritance and it will further simplify the selection process. Some significant achievements have been made by adopting this strategy against late blight disease of potato. In two different studies, double-gene stack of potato late blight R genes (*Rpi-vnt1.1* and *Rpi-sto1*) and triple-gene stack of three potato late blight R genes (*Rpisto1*, *Rpi-vnt1.1*, and *Rpi-blb3*) have been deployed in potato crop using *Agrobacterium*-mediated transformation (Jo et al. 2014; Zhu et al. 2012). Both studies have reported that a broad-spectrum resistance against late blight disease was witnessed in transgenic lines of potato. It is assumed that growing of these late blight-resistant transgenic plants can minimize the use of fungicide by over 80% (Que et al. 2010).

RNA interference (RNAi) is an innate immune response of plants against non-self-nucleotides sequences triggered by a double-stranded RNA molecule. Essentially, plants deploy this interference against viral pathogens and cleave viral RNA to stop translation (Rosa et al. 2018). By adopting transgenic overexpression of viral components (Galvez et al. 2014), the introduction of a hairpin structure through cloning the viral sequences in a self-complementary way (Bonfim et al. 2007) and engineering artificial micro RNA sequences (Zhang et al. 2011), RNAi mechanism has been successfully exploited in several crops against pathogens. On the other hand, in the evolutionary arms race almost all plant viruses produce specialized suppressers to counteract plant RNAi-based defenses (Zamore 2004) thus, in this invisible warfare, some of the viruses eventually defeat host RNAi machinery by counter silencing host genes which ultimately leads to viral pathogenicity (Wang et al. 2012). Still, even with these challenges, RNAi technology is a very strong tool in developing resistance in crop plants against plant pathogens. The integration of the RNAi mechanism with R-gene stacking would be another powerful approach to tackle plant-infecting pathogens. In the past, this approach has proven extremely effective against plant viruses. Particularly, in the 1990s, the introduction of genetically modified papaya named "Rainbow papaya" against the epidemic of papaya ringspot virus has prevented the collapse of the Hawaiian papaya industry (Gonsalves et al. 2004). A similar strategy was used in transgenic squash (Tricoll et al. 1995) and both have been cultivated in the USA for more than 20 years.

In contrast, the face of modern-day genetic engineering involves the genome-editing tools which allow changes in the genome of plants at desired sites (Barrangou and Doudna 2016) and transgene-free edited plants can be generated. In some parts of the world, including the USA, Brazil and Argentina, genome-edited plants without foreign DNA are not liable to additional regulatory measures as applicable to transgenic plants (Orozco 2018).

Extracellular immunogenic patterns (ExIPs) are pathogen-derived danger signals, sensed by plasma membrane affiliated cell-surface receptors of the plant. This recognition activates extracellularly triggered immunity (ExTI) which is also known as PAMPs triggered immunity (PTI). On the other hand, pathogens release effector molecules named as intracellular immunogenic patterns (InIPs), inside the cell to suppress the PTI-based plant defenses. Pathogen associated InIPs are encountered by cytoplasmic receptors of the plant defense system, mostly NLRs. This InIPs-NLRs interaction produces an intracellular defense response in the plant, named as intracellularly triggered immunity (InTI). Both ExTI and InTI helps the plant to cope with the diverse race of plant pathogens.

14.3 Genome Editing-a New Way Forward

Though traditional approaches of developing disease-resistant crops are in use but the advent of genome-editing technologies has revolutionized plant research. These technologies empower researchers to modify genetic functions using endogenous genes of different organisms (Barrangou and Horvath 2017; Doudna and Charpentier 2014; Gaj et al. 2013; Hsu et al. 2014; Joung and Sander 2013; Kim and Kim 2014; Sander and Joung 2014; Voytas 2013). Genome editing was started when researchers came to know that sequence-specific nucleases (SSNs) are reprogrammable and can be used to create double-stranded breaks (DSBs) at the desired location. There are four major classes of DNA-binding proteins which have been used so far to accomplish genome editing. These four classes comprise of meganucleases (Pâques and Duchateau 2007; Smith et al. 2006), Zinc-Finger Nucleases (ZFNs) (Maeder et al. 2008), Transcription-Activator-Like Effector Nucleases (TALENs) (Bogdanove and Voytas 2011; Miller et al. 2011), and numerous derivatives from CRISPR-Cas endonucleases (Abudayyeh et al. 2016, 2017; Jinek et al. 2012, 2013; Shmakov et al. 2015; Zetsche et al. 2017). Meganucleases have the natural ability to target DNA and can be modified to recognize desired location. The most well-characterized meganucleases for use in genome editing are I-SceI meganucleases, which are coded by a mitochondrial ribosomal RNA (Puchta and Fauser 2014; Voytas 2013). ZFNs consist of DNA-binding domain which is used for target site identification. The DNA-binding domain is further attached to FokI endonuclease to cleave DNA at the desired location (Urnov et al. 2010). One zinc-finger binding domain usually identify three nucleotides in the DNA (Gaj et al. 2013; Urnov et al. 2010; Voytas 2013). TALENs also consist of a DNA-binding domain fused to the catalytic domain of the FokI endonuclease. The central domain contains almost identical repeats and only positions 12 and 13 are exceptional. Thus, these positions are used to form a code to define specificity for DNA binding (Boch et al. 2009; Boch and Bonas 2010; Christian et al. 2010; Moscou and Bogdanove 2009; Schornack et al. 2013). These three are protein-dependent systems and extensive protein engineering is involved for their customization whereas CRISPR-Cas endonucleases are easily programmable RNA- or DNA-cleavage systems in which DSBs are introduced at RNA-guided position.

CRISPR-Cas9 was initially discovered in prokaryotes for providing adaptive immunity against any invading phages or nucleic acids. There are various forms of CRISPR-Cas systems present in bacteria and archaea. Among all the CRISPR-associated (Cas) proteins, Cas9 and its variants are well characterized and widely exploited for crop improvement, i.e. Cas9-nickase and dead Cas9 (dCas9) (Chatterjee et al. 2018; Jensen et al. 2017; Jiang and Doudna 2017; Nishimasu et al. 2014; Qi et al. 2013). CRISPR-Cas9 introduces DSBs at the target site which are repaired either by non-homologous end joining or homology directed repair (Aird et al. 2018; Gil-Humanes et al. 2017; Lin et al. 2014; Salsman and Dellaire 2017;

Suzuki et al. 2016). Another variant includes base editing, which is an efficient way of creating desired point mutation. Two types of base editors, cytosine base editors and adenosine base editors are available for various applications of crop improvement (Gaudelli et al. 2017; Komor et al. 2017; Li et al. 2018a, 2018b; Ma et al. 2016 Negishi et al. 2019; Shimatani et al. 2017; Yan et al. 2018; Zong et al. 2018). The genome-editing toolbox was further diversified by the addition of another system CRISPR-Cpf1 also known as CRISPR-Cas12a which is able to perform editing at T-rich PAM sequence. There are many examples of exploiting this tool for crop improvement (Gao et al. 2017; Kim et al. 2016, 2017; Kleinstiver et al. 2016; Liu et al. 2017; Malzahn et al. 2019; Moon et al. 2018; Moreno-Mateos et al. 2017; Tóth et al. 2016; Wang et al. 2017; Yan et al. 2017; Zaidi et al. 2017; Zetsche et al. 2015). The genome-editing tools can be used to target multiple locations in the genome which is possible by multiplexed genome editing (MGE) approach. It involves more than one gRNA to target multiple regions in the genome, simultaneously (Čermák et al. 2017; Li et al. 2013; Ma et al. 2015; Wang et al. 2018; Xie et al. 2015; Xing et al. 2014; Zhou et al. 2014).

14.4 Genome Editing Examples to Develop Disease Resistance in Crops

Various genome-editing tools have been used to develop disease resistance in many crops (Figure 14.2). Here, we enlist some examples (Table 14.1) of genome-editing technologies to develop disease resistance in several crops.

Disruption or base editing of S gene/s or regulatory element/s in the targeted variety/plant through GETs like Cas9, Cpf1 and their variants, with a high level of precision and

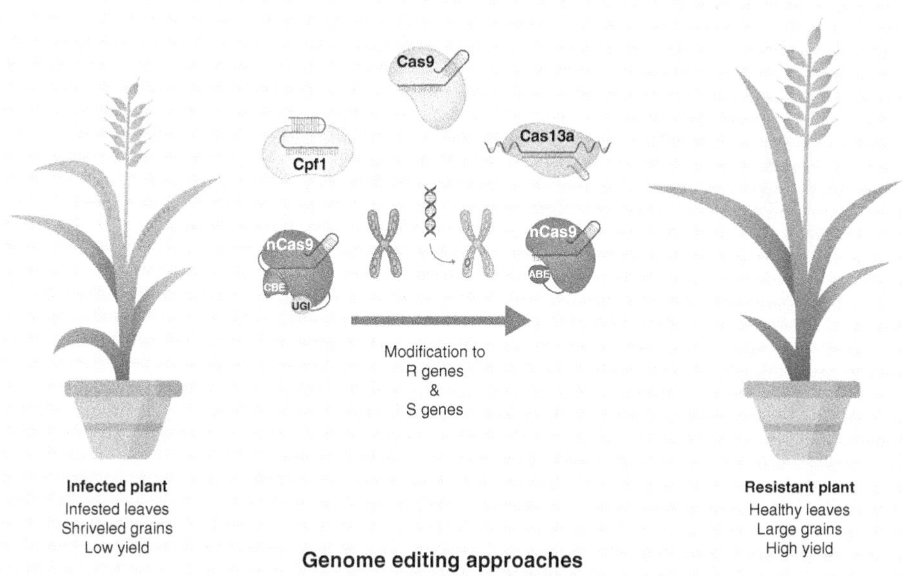

Figure 14.2 CRISPR/Cas-based genome-editing tools (GETs) for the development of disease-resistant plants with improved traits.

Table 14.1 Engineered nucleases applications for disease-resistance development.

Type	Pathogen						
	Name	Plant	Targeted gene	Gene function	Engineered nuclease	Result	References
Fungi	*Verticillium dahliae*	*Gossypium hirsutum*	*Gh14-3-3d*	Negative regulators of BR signaling	CRISPR-Cas9	Enhanced resistance to disease	Zhang et al. (2018)
	Blumeria graminis f. sp	*Triticum aestivum*	*TaMLO*	Susceptibility (S) gene involved in powdery mildew disease	CRISPR-Cas9 and TALEN	Enhanced tolerance to disease	Wang et al. (2014), Wang et al. (2018)
	Fusarium graminearum	*Triticum aestivum*	*TaLpx-1*	Immunity regulator	CRISPR-Cas9	Enhanced resistance to disease	Wang et al. (2018)
	Blumeria graminis f. sp	*Triticum aestivum*	*TaEDR1*	Immunity regulator	CRISPR-Cas9	Enhanced resistance to disease	Zhang et al. (2017)
	Botrytis cinerea (Gray mold)	*Vitis vinifera*	*WRKY52*	A transcription factor from WRKY family involved in response to biotic stress	CRISPR-Cas9	Enhanced resistance to disease	Wang et al. (2018)
	Erysiphe necator (Powdery mildew)	*Vitis vinifera*	*MLO-7*	Susceptibility (S) gene involved in powdery mildew disease	CRISPR-Cas9	Resistance to disease developed	Malnoy et al. (2016)
	Phytophthora tropicalis (Black pod disease)	*Theobroma cacao*	*NPR3*	Immunity regulator	CRISPR-Cas9	Enhanced resistance to disease	Fister et al. (2018)
	Oidium neolycopersici (Powdery Mildew)	*Solanum lycopersicum*	*SlMlo1*/cds	the SlMlo1 is the susceptible allele for escalating Powdery mildew	CRISPR-Cas9	Enhanced to diseases resistance	Nekrasov et al. (2017)

	Oidium neolycopersici (Powdery Mildew)	*Solanum lycopersicum*	*PMR4/cds*	PMR4 belongs to Glucan-synthase-like (GSL) and act as negative regulator for Salicyclic acid (SA) pathway	CRISPR-Cas9	Enhanced to diseases resistance	Koseoglou (2017)
	Fusarium oxysporum f.sp. lycopersici, (Fusarium wilt)	*Solnum lycopersicum*	*Solyc08g075770/ genomic*	gene is predicted to encode an exopolysaccharide production negative regulator	CRISPR-Cas9	Fusarium wilt resistance	Prihatna et al. (2018)
	Botrytis cinerea (Gray mold)	*Solanum lycopersicum*	*SlMAPK3*	*activated protein kinase 3* (*MAPK3*) regulating various pathways	CRISPR-Cas9	Enhanced to diseases resistance	Zhang et al. (2018)
	Magnaporthe oryzae.	*Oryza sativa*	*OsERF922*	Encode ERF family transcription factor	CRISPR-Cas9	Edited lines showed considerably reduced blast lesions	Wang et al. (2016)
Bacteria	*Xanthomonas citri* subspecies *citric* (Citrus canker)	*Citrus paradisi*	*LOB1* (cds/promoter region)	Susceptibility (S) gene boosting pathogen growth and pustule development	CRISPR-Cas9	Enhanced resistance to disease	Jia et al. (2016), Jia et al. (2017)
	Xanthomonas citri subspecies *citric* (Citrus canker)	*Citrus sinensis Osbeck*	*LOB1* (promoter)	Susceptibility (S) gene boosting pathogen growth and pustule development	CRISPR-Cas9	Enhanced resistance to disease	Peng et al. (2017)
	Erwinia amylovora (Fire blight)	*Malus domestica*	*DIPM-1 DIPM-2 DIPM-4*	Susceptibility factor/gene play role in fire blight disease	CRISPR-Cas9	Disease resistance not confirmed	Malnoy et al. (2016)

(*Continued*)

Table 14.1 (Continued)

	Pathogen						
Type	**Name**	**Plant**	**Targeted gene**	**Gene function**	**Engineered nuclease**	**Result**	**References**
	Pseudomonas syringae	*Solanum Lycopersicum*	*SIDMR6–1*/cds	Susceptible gene and facilitate the pathogen's spreading	CRISPR-Cas9	Enhanced diseases resistance against Bacterial speck	De Toledo Thomazella et al. (2016)
	Phytophthora capsici	*Solanum Lycopersicum*	*SIDMR6–1*/cds	Its paralogue to plants immunity, involve in boosting pathogen growth	CRISPR-Cas9	Enhanced disease resistance	De Toledo Thomazella et al. (2016)
	Xanthomonas spp.	*Solanum lycopersicum*	*SIDMR6–1*/cds	Susceptible gene and facilitate the pathogen's spreading	CRISPR-Cas9	Enhanced resistance against Bacterial spot	De Toledo Thomazella et al. (2016)
	Pseudomonas syringae pv. tomato (Pto) DC3000	*Solanum Lycoperscum*	*SlJAZ2/cds*	repressor proteins facilitate COR for interaction by suppressing the JA defense pathway	CRISPR-Cas9	Bacterial speck CRISPR-Cas9 n.d. Enhanced disease resistance, defense trade-off solved	Ortigosa et al. (2019)
	Xanthomonas oryzae pv. *oryzae*	*Oryza sativa*	*OsSWEET14*	Encode SWEET sugar transporter	TALENS	Enhanced resistance against Bacterial blight	Blanvillain-Baufumé et al. (2017)
	Xanthomonas oryzae pv. *oryzae*	*Oryza sativa*	*OsSWEET11, OsSWEET13, OsSWEET14*	Encode SWEET sugar transporter	CRISPR-Cas9	Broad spectrum resistance bacterial blight	Oliva et al. (2019), Xu et al. (2019)
	Xanthomonas oryzae pv. *oryzae*	*Oryza sativa*	*OsSWEET14*	Encode SWEET sugar transporter	CRISPR-Cas9	Reduced lesion length	Zafar et al. (2020)

Virus	eBSV	Gonja Manjaya (*Musa* spp. AAB)	eBSV sequence	Disruption of integrated endogenous *banana streak virus* (eBSV) in B genome of plantain	CRISPR-Cas9	Resistance developed	Tripathi et al. (2019)
	CVYV, PRSV-W and ZYMV	*Cucumis sativus* L.	*eIF4E*	Host factor for RNA viruses	CRISPR-Cas9	Resistance to CVYV, PRSV-W and ZYMV	Chandrasekaran et al. (2016)
	CBSV	*Cassava*	*nCBP*-1and*nCBP-2*	Host factor for RNA viruses	CRISPR-Cas9	Suppressed symptom and viral load	Gomez et al. (2019)
	WDV	*Hordeum vulgare*L.	MP, CP, Rep/RepA, IR	Directly targeting the virus genome to develop resistance	CRISPR-Cas9	Directly digesting the virus genome gave asymptomatic plants and no virus presence	Kis et al. (2019)
	BSCTV	Tobacco and *Arabidopsis*	IR, CP, Rep	Directly targeting the virus genome to develop resistance	CRISPR-Cas9	Resistance developed in both	Ji et al. (2015)
	CMV	Tobacco and *Arabidopsis*	ORF1a, ORFCP and 3'-UTR	Directly targeting the virus genome to develop resistance	CRISPR-Cas9	Suppressed symptom and viral load (no cleavage)	Zhang et al. (2018)
	TuMV	Tobacco	GFP, Hc-pro and CP	Directly targeting the virus genome to develop resistance	CRISPR-Cas13a	Suppressed symptom and viral load	Aman et al. (2018)

(*Continued*)

Table 14.1 (Continued)

Pathogen							
Type	Name	Plant	Targeted gene	Gene function	Engineered nuclease	Result	References
	PVY	*Solanum tuberosum*	*P3*, *CI*, *NIb* and *CP*	Targeting the virus genome to develop resistance	CRISPR-Cas13a	Development of resistance	Zhan et al. (2019)
	ToMV and PVX	*Solanum lycopersicum*	DCL2b	Targeting the virus genome to develop resistance	CRISPR-Cas9	Suppressed symptom and viral load	Wang et al. (2018)
	RTBV and RTSV	*Oryza sativa*	*eIF4G*	Eukaryotic translation initiation factor 4G (eIF4G) is a protein involved in eukaryotic translation initiation	CRISPR-Cas9	Resistance against Rice Tungro disease	Macovei et al. (2018)

eBSV; endogenous *banana-streak virus*; CVYV, cucumber vein-yellowing virus; PRSV-W, papaya ring spot mosaic virus-W; ZYMV, zucchini yellow-mosaic virus; Cassava brown-streak virus, CBSV; WDV, *Wheat dwarf virus*; *Beet severe curly top virus*, BSCTV; *cucumber mosaic virus*, CMV; *Turnip mosaic virus*, TuMV; Potato virus Y, PVY; Tomato mosaic virus, ToMV; Potato virus X, PVX; *Rice tungro bacilliform virus*, RTBV; *Rice tungro spherical virus*, RTSV.

flexibility to develop disease resistance against particular pathogen/s. Tools like Cas9, Cpf1, and Cas13 could also be used to directly target the pathogen genome in the host plant for disease-resistance development. Further, precise modification of R gene/s could be done to gain lost resistance or develop broad-spectrum resistance.

14.4.1 Rice

Rice (*Oryza sativa*) is one of the crops which are required to feed billion of new mouths (Gill et al. 2014). It is an important cereal and source of calories to human population as it provides 23% calories globally (Hossain et al. 2003). About 50% of the human population consume rice as primary food. Among food crops, rice production should be increased from 25 to 40% to fulfill the nutritional demands of the people (Clarke and Zhang 2013; Datta 2004; Maclean et al. 2002). To address major issues which include growing population, environmental changes and food scarcity (Clarke and Zhang 2013), there is a need to develop rice varieties tolerant to abiotic and biotic stresses, with higher yield and adaptability to changing climate.

There are different biotic and abiotic stresses responsible for decreased rice production. Different pathogens including bacteria, fungus and virus attack rice plants for causing various diseases which ultimately result in yield loss. Many strategies had been applied to reduce disease pressure in the field (Heinrichs and Muniappan 2017). There is a significant contribution of genome-editing technologies for developing disease-resistant rice crop.

14.4.1.1 Bacterial Blight Resistant Rice

Bacterial diseases are very hard to control, especially after the widespread occurrence of infection. A bacterial disease named “rice bacterial blight” caused by *Xanthomonas oryzae* pv. *oryzae* (*Xoo*) is one the major diseases of rice crop. It is a vascular bundle disease which causes up to 20% yield loss. In the favorable conditions of infection (i.e. high humidity), this disease can cause 50% to complete yield loss (Mew et al. 1993). *Xoo* infect rice by secreting transcription activator-like effectors TALEs into the host through type III secretion system (Makino et al. 2006). For a successful infection, these TALEs target susceptibility genes to activate their expression. Thus, expression of these genes helps *Xoo* to establish infection (Boch and Bonas 2010; Doyle et al. 2013). One example of such TALE protein is *AvrXa7* which has its effector-binding element (EBE) in the promoter of *OsSWEET14*. The binding of *AvrXa7* with its respective EBE activates the expression of *OsSWEET14*. This gene codes for sucrose-efflux transporter, which is hijacked by *Xoo* for transporting sugars from rice cells to the pathogen to promote infection (Chen et al. 2012). One strategy could be to knock out this susceptibility gene but as *OsSWEET14* also has important plant-development roles. So, this strategy may not be feasible. Therefore, such a strategy can be adopted which will disrupt the TALE binding with its respective EBE. Genome editing has this potential because, using these tools, the EBEs can be edited, thus interrupting TALE binding without any disruption in normal functions of *OsSWEET14*. This type of editing was done using a pair of TALENs to develop resistance against bacterial blight (Li et al. 2012). Geologically different *Xoo* strains target *OsSWEET14* making it a major target. Out of six known TALEs four (*AvrXa7, PthXo3, TalF*, and *TalC*) are present in *OsSWEET14*. Knowing the importance of this very significant susceptibility gene its role was studied extensively by doing different types of editing in its promoter using TALEN as a genome-editing tool. Their results indicated that all these four TALEs (*AvrXa7, PthXo3, TalF*, and *TalC*) target *OsSWEET14*. However, *TalC* also targets an EBE in a gene other than *SWEET14* (Blanvillain-Baufumé et al. 2017). *Xoo* strains also possess another protein for disease development that is *PthXo2* (Ochiai et al. 2005; Yang and White 2004). Its binding promotes the expression of *OsSWEET13*. CRISPR-Cas9-based genome editing was used to knockout *OsSWEET13* by targeting its coding region. This loss of function mutant showed increased resistance and did not cause any developmental defects in plants (Zhou et al. 2015). Another attempt was made to develop bacterial blight resistance using CRISPR-Cas9 to target *OsSWEET11* and *OsSWEET14* which results in symptom reduction (Jiang et al. 2013). The same technology was then used to create a new germplasm with broad-spectrum resistance against a diversity of *Xoo* strains by targeting *SWEET* genes (Xu et al. 2019). CRISPR-Cas9 was also used to develop edited lines of two mega-rice varieties (IR64 and Ciherang-Sub1) having broad-spectrum resistance against bacterial blight. The resistance was created by targeting promoter of three *SWEET* genes (Oliva et al. 2019). Recently, super basmati was also edited for resistance against bacterial blight (Zafar et al. 2020). These success stories show the potential of genome editing to deal with bacterial diseases of rice.

14.4.1.2 Virus-Resistant Rice

Viral diseases are one of the constraints in rice production. Rice tungro disease is an example of viral disease which severely affected rice yield in Asia. This disease is caused by *Rice tungro bacilliform virus* (RTBV) and *Rice tungro spherical virus* (RTSV). The candidate gene *eIF4G* (*TRANSLATION INITIATION FACTOR 4 GAMMA*) to develop resistance against RTSV was identified following extensive studies. There is a specific region in the*eIF4G* gene which interacts with RTSV to cause disease (Lee et al. 2010). This region of *eIF4G* can be edited in virus-susceptible rice varieties to develop resistance against rice tungro disease. CRISPR-Cas9 was used for this purpose to target *eIF4G* in rice as an attempt to create an RTSV-resistant variety. The CRISPR edited plants were no longer susceptible to RTSV and mutations were successfully transferred to next generations. This study exhibited the importance of CRISPR-Cas9 to develop resistance against rice tungro disease (Macovei et al. 2018).

14.4.1.3 Fungus-Resistant Rice

Rice blast is a very damaging fungal disease of rice. The causative agent is filamentous fungus *Magnaporthe oryzae*. Worldwide rice production is threatened by this disease (Liu et al. 2014). The development of blast-resistant varieties is the most cost-effective way to deal with this disease (Miah et al. 2013). The *OsERF922* was targeted through CRISPR-Cas9 to develop resistance against rice blast. This approach was successful in generating mutations with a 42% mutation frequency. The edited lines showed considerably reduced blast lesions as compared to the control plants (Wang et al. 2016). Such examples highlight the worth of CRISPR-Cas9 for improving rice crop against fungal diseases.

14.4.2 Wheat

Bread wheat (*Triticum aestivum*) is being consumed as one of the substantial staple crops around the globe. Its nutritional contents fulfill an ample amount of the caloric intake of humans. This defines the intensifying demand of consumers and also highlights the economic impact of wheat. Breeders desire to enhance the yield of wheat crop without disrupting the other essential traits of elite cultivars (Atlin et al. 2017). Among these significant traits such as high yield and quality, the capability to sustain and adapt under biotic and abiotic stresses is inevitable to achieve enhanced wheat crop production. Over the year, the biotic factors have limited profitable wheat production. For instance, a 15–50% decrease in yield has been reported under the leaf-rust epidemic in wheat, depending upon the growth stages at which infection occurs (Huerta-Espino et al. 2011; Roelfs and Bushnell 1985). Traditional breeding approaches maximize the chances of linkage drag and it takes years for an arduous screening of the large population, even with the facilitation of MAB.

The CRISPR-Cas9 system has paved the way for efficient and timely genome editing. Further, this system defies all the constraints that can be presented by other approaches. In wheat, the CRISPR-Cas9 system has been deployed very effectively to acquire resistance against certain pathogens. In several cases, susceptibility (S) genes or negative regulators have been knocked out to attain resistance. The study suggests that the Mildew resistance locus (MLO) gene makes the plants susceptible to fungal powdery mildew pathogen known as *Blumeria graminis* f. sp. A seven-transmembrane domain protein localized at the plasma membrane is encoded by the *MLO* gene. The loss-of-function mutant alleles have shown resistance against powdery mildew in tomato, pea and *A. thaliana* (Acevedo-Garcia et al. 2014). Wang et al. downregulated *MLO* gene in wheat to confer resistance against *B. graminisf. sp. tritici* by TALEN-induced targeted-mutations in all three homeologs of *TaMLO* and single *TaMLO-A1* homeolog through CRISPR-Cas9 (Wang et al. 2014). Similarly, a previously confirmed negative regulator termed as enhanced-disease resistance1 (*TaEDR1*) gene demonstrated resistance against powdery mildew in wheat when targeted with virus-induced gene silencing (VIGS), RNAi and CRISPR-Cas9 (Zhang et al. 2017).

Recently, Wang and the team generated a wheat crop by targeting multiple agronomic traits. They used multiplexed gene editing (MGE)-based CRISPR-Cas9 system to knockout *TaGW2*, *TaLpx-1* and *TaMLO* genes, simultaneously (Wang et al. 2018). All three genes are reported to be negative regulators (Hong et al. 2014; Nalam et al. 2015; Wang et al. 2014). Silencing all three homeologs of *TaGW2* at conserved region increased thousand-grain weight (TGW) and silencing two homeologs of *TaLpx-1* at conserved region conferred resistance to *Fusarium graminearum*. While knocking out the *TaMLO* gene at one homeolog rendered wheat resistant to powdery mildew (Wang et al. 2018). First, the MGE using tRNA as spacers in between multiple gRNAs was achieved in rice. The efficiency was reported to be as high as 100% in stable transgenic rice (Xie et al. 2015).

14.4.3 Tomato

The development of broad-spectrum resistance of crops against necrotrophs can be a sustainable approach to fulfill the food demand of a bourgeoning population. CRISPR-Cas9 and their variants have been applied to improve tomato and potato crops. Tomato (*Solanum lycopersicum*), is the second most consumed vegetable (Schwartz et al. 2015), which is affected by various plant pathogens.

Bacterial pathogens cause upregulation of *SlDMR6-1* gene. So, it can be a potential target for genome editing. CRISPR-Cas9 was used to edit this gene, resulting in small deletion that ultimately leads toward frameshift mutations causing the formation of truncated protein. This truncated protein is responsible for deterioration of interaction pattern of *Phytophthora* spp., *Pseudomonas syringae*, *Xanthomon* as spp. with tomato plants. Such mutations exhibit resistance in tomato against these pathogens (De Toledo Thomazella et al. 2016).

Bacterial speck disease in tomato has been caused by *Pseudomonas syringae* (PTO). The bacterial spots on tomato has reduced fruit weight and affected up to 75% growth rate at the early stages (Yunis et al. 1980). Plants attain immunity by plethora of hormones like jasmonate (JA) and salicylate (SA) defense pathways. In tomato, PTO strain DC3000 produce coronatine (COR), that initiate the opening of stomata and assist bacterial colonization in leaf. Further, Jasmonate Zim Domain (JAZ) repressor proteins facilitate COR for interaction with plant receptors by suppressing the JA defense pathway. The *SlJAZ2* edited by CRISPR-Cas9 to generate JAZ2 repressors lacking the C-terminal JAZ-domain (*SlJAZ2Djas*) that prevent the reopening of stomata by COR and provide resistance against PTO-DC3000 (Ortigosa et al. 2019).

In tomato, there are 16 *MILDEW RESISTANT LOCUS O* (*Mlo*), the *SlMlo1* is the susceptible allele for escalating powdery mildew. The loss-of-function mutation obtained through CRISPR-Cas9 in *SlMol* provides resistance in tomato against *Oidium neolycopersici* (Nekrasov et al. 2017). Koseoglou et al. developed resistance against pathogen *Erysiphe cichoracearum* by knock-down of the Powdery Mildew Resistance 4 (PMR4) gene. The *PMR4* belongs to Glucan-synthase-like (GSL) and acts as a negative regulator for Salicylic acid (SA) pathway. CRISPR-Cas9 edit PMR4 and it activates the SA pathway that provides tolerance against PM (Koseoglou 2017). Further manipulation of *activated protein kinase 3* (*MAPK3*) (Zhang et al. 2018) and *Solyc08g075770* (Prihatna et al. 2018) through CRISPR-Cas9 provides immunity against gray mold and Fusarium wilt diseases, respectively. The CRISPR-Cas9 has promising potential of editing the genome to obtain desirable traits and it can be exploited to develop new cultivars against various pathogens.

14.4.4 Miscellaneous Stories of Disease Resistance Through Genome Editing

The targeted genome-edited system has paved the way for efficient and timely genome editing. Further, this system defies all the constraints that can be presented by other approaches. The genome-editing system has been deployed very effectively to acquire resistance against certain pathogens. In several cases, susceptibility (S) genes or negative regulators have been knocked out to attain resistance. The study suggests that the *MLO* gene makes the plants susceptible to fungal powdery mildew pathogen known as *B. graminis* f. sp. A seven-transmembrane domain protein localized at the plasma membrane is encoded by the *MLO* gene. The loss-of-function mutant alleles have shown resistance against powdery mildew in tomato, pea and *A. thaliana* (Acevedo-Garcia et al. 2014).

Gossypium hirsutum (cotton) is an allotetraploid species and cotton yield is severely affected by disease and insect-pest complex. Zhang et al. developed transgene-free *G. hirsutum* that has enhanced the level of resistance against *Verticillium dahliae* infestation by disrupting the *Gh14-3-3d* gene through indels. The *Gh14-3-3d* is a negative regulator of BR signaling and is involved in the biotic stress regulation (Zhang et al. 2018).

The knock-out of *VvWRKY52* gene in grape (*Vitis vinifera*) increase Botrytis *cinerea* resistance (Vriet et al. 2012; Wang et al. 2018). Malnoy et al., mutated the *MLO*-7 (a susceptible gene) and *DIPM-1, DIPM-2, DIPM-4* gene/s to develop resistance against powdery mildew in the grape and fire-blight disease in the apple respectively (Malnoy et al. 2016). The *Theobroma cacao* affected by a variety of disease which is the source of cocoa. Fister et al. targeted the *Non-Expressor of Pathogenesis-Related3* (*TcNPR3*) gene, which can mask the defense response. The mutated/edited plant showed an enhanced level of resistance against the infection of *Phytophthora tropicalis* pathogen. This study also confirmed the function of *TcNPR3* as an immune suppressor because it also increases the downstream expression of immune genes (Fister et al. 2018).

The *Xanthomonas citri* subspecies *citri* (Xcc) cause citrus canker disease in most of the citrus cultivars. *Citrus sinensis* Lateral Organ Boundaries (*CsLOB1*) is a susceptibility gene for citrus canker which is induced by bind of the pathogenicity factor PthA4 to its promoter element. Jia et al., disrupt the promoter element of *CsLOB1* gene and improved the canker-resistance in grapefruit (Jia et al. 2016). In another study, disruption of the coding region of both the allele of *CsLOB1* gene gave canker-resistant grapefruit (Jia et al. 2017). Peng et al. mutate the PthA4-binding element in the promoter of *CsLOB1* gene and homozygous mutant gave an enhanced level of canker-resistance in Wanjincheng orange (*Citrus sinensis* Osbeck) compared to wild type (Peng et al. 2017).

Banana streak virus (BSV) is affects production of banana (*Musa* spp.) and it can integrate in the plant genome. The BSV integrated in genome of banana is called endogenous BSV (eBSV). CRISPR-Cas9-based inactivation of eBSV in the B genome of plantain "Gonja Manjaya" gave asymptomatic plants as compared to wild type under stress conditions (Harper et al. 1999; Tripathi et al. 2019). The Cas9 nuclease has been deployed to develop resistance in Cucumber (*Cucumis sativus* L.) against *cucumber vein yellowing virus* (*Ipomovirus*), *Papaya ringspot mosaic virus-W* and potyviruses *Zucchini yellow mosaic virus* by mutating the recessive *eIF4E* (*eukaryotic translation initiation factor 4E*) gene (Chandrasekaran et al. 2016). Cassava brown-streak disease (CBSD) damaging the cassava crop caused by two species of positive-sense RNA viruses *Cassava brown streak virus* (CBSV) and *Ugandan cassava brown streak virus* (UCBSV). For disease development, there is required interaction between viral genome-linked (VPg) and eIF4E isoforms, related novel cap-binding protein-1 (nCBP-1), and nCBP-2 have interactions with VPg. CRISPR-Cas9-based genome editing of ncbp-1/ncbp-2 showed reduced disease severity (Gomez et al. 2019). The *Wheat dwarf virus* (WDV) damage both wheat and barley crop. Directly digestion of WDV genome in the plant by CRISPR-Cas9 system in barley gave an enhanced level of resistance and the approach of directly targeting virus has also been used in other crops (Ji et al. 2015; Kis et al. 2019; Zhang et al. 2018). CRISPR-Cas13a (discussed below in detail) can be used against RNA viruses by a mechanism of RNA-guided RNAi. It gave promising results against *turnip mosaic virus* (TuMV) in tobacco (Aman et al. 2018).

14.5 Recent Trends in Genome Editing

Scientists are continuously expanding the toolkit for genome editing to address limitations present in already available tools. The expansion in tools is a result of either discovering new variants of Cas proteins or due to fusion of already available proteins. There are some recently developed tools which can be used for creating disease resistance in plants. One of them is type-VI CRISPR-Cas13 effector (~900–1300aa), that is an RNA-degrading protein. (Cox et al. 2017; East-Seletsky et al. 2016; Garcia-Doval and Jinek 2017; Liu et al. 2017; Marraffini and Sontheimer 2010; O'Connell 2019; Wolter and Puchta 2018; Yan et al. 2018). The CRISPR-Cas13 can be used for detection of various pathogens and degradation of the newly formed transcript. This system can be used against ssRNA viruses by targeting multiple genes in virus-genome to attain resistance against multiple viruses (Aman et al. 2018; Khan et al. 2018; Matsoukas 2018; Yang et al. 2019; Zhan et al. 2019; Zhang et al. 2018). Another system named CRISPR-Cas14is highly compact and has the potential to target and cleave single-stranded DNA (ssDNA)(Harrington et al. 2018). It is almost half the size of Cas9 protein and does not require any PAM sequence near target site for its functioning (Harrington et al. 2018; Marraffini and Sontheimer 2010). It can be used for creating disease resistance against ssDNA viruses, i.e. from families Nanoviridae and Geminiviridae (Ali et al. 2016; Khan et al. 2019).

Genome modifications including insertions, deletions and substitutions are required for crop improvement, i.e. creating disease-resistant crops (Wang et al. 2018). There are some problems with previous genome-editing tools to achieve desired mutations through HDR like low efficiency and difficulty in delivering donor template (Chen et al. 2019; Puchta and Fauser 2013; Ran et al. 2017). Base editors are also a good alternative, but they are unable to install transversions and Indels (Gaudelli et al. 2017; Kim et al. 2017; Komor et al. 2016). These limitations move toward usage of a recently developed advance tool entitled "Prime Editor"(Anzalone et al. 2019). It works in a "search-and-replace" manner and holds great potential for performing precise genome editing (Anzalone et al. 2019; Cohen 2019; Li et al. 2020a; Lin et al. 2020; Marzec et al. 2020; Tang et al. 2020).

The novel base editor Saturated targeted endogenous mutagenesis editors (STEMs) was developed recently by fusion of adenosine deaminase with cytidine deaminase. It has the ability to produce *de novo* mutations and also facilitates directed evolution of genes (Li et al. 2020b). These examples clearly showed that genome-editing tools are in continuous development to solve previous limitations and can be used in future crop improvement with much precision and accuracy.

14.6 Conclusion and Prospects

Genome editing holds tremendous potential for solving problems associated with conventional breeding methods to develop disease resistance. It can directly modify target genes; thus, a mating process is not required. Another benefit of genome editing is that changes are not introduced beyond target sites and there is no need for genetic crosses and segregant progeny selection. Together with these benefits, simplicity and efficiency of using genome-editing tools make them an excellent choice for developing disease resistance in crops.

For a successful disease predisposition, three main factors are required: a lethal pathogen, to infect a vulnerable host, in a promising environment. The presence of three factors at the same time cause disease initiation. Genome editing can be applied on any of these factors for disturbing host–pathogen interactions in order to control disease.

Typically, some pathogens of plant infecting viruses essentially demand a vector agent to infect their plant host. Such diseases are usually transmitted through insect-pests, ultimately leading to major yield losses. The predominant agrochemical-based methods to control insects are costly, dangerous for the environment and also kill non-target insects. These methods gradually become inefficient due to the development of resistance against insecticide. A novel approach of CRISPR-Cas-based gene drives can selectively propagate edited genes in the whole population leading to sterility or lethality of target-insect species. This type of insect control or developing insect populations carrying virus resistance genes are promising approaches to control insects and their vectored diseases (Hammond et al. 2017; Unckless et al. 2017).

The application of genome editing for developing disease resistance is not only limited to gene disruption or replacement, but gene regulation can also be used for this purpose. There are some examples where enhanced expression of a certain gene is associated with disease resistance. One such example is increased expression of *Xa3* gene in japonica rice as compared to Indica making them resistant against *Xoo* (Cao et al. 2007). Similarly, *Hm2* gene's enhanced expression level in maize also provides resistance against ear mold disease and leaf spot (Chintamanani et al. 2008). In contrast to these examples, the down regulation of *GhLMMD* gene in cotton make them resistant to Verticillium dahlia infection (Chai et al. 2017). These examples suggest that the newly established approaches CRISPR interference (CRISPRi) or CRISPR activation (CRISPRa) can be used in the future to regulate expression of certain genes to make crops resistant against different diseases. Additionally, mRNA translation can also be modified by editing open-reading frames of target genes (Zhang et al. 2018).

Genome editing can also play a role for crop improvement against diseases by analyzing genes and engineering different pathways to increase molecular immunity in crops. There is a tremendous advancement in next-generation sequencing (NGS) platforms, due to which total number of sequences from different plants and pathogens are rapidly increasing, and are available in databases (Buermans and Den Dunnen 2014). This sequencing information can be used to design CRISPR-Cas experiments to engineer disease resistance in crops. Further, the availability of whole genome sequences will also reduce off-targets.

Bioinformatic tools are developing continuously to provide ease in genome editing. There are many tools available to design and analyze CRISPR-Cas-associated experiments. These tools are used for designing gRNA and finding potential off-target effects. One of the tools mostly used for identifying off targets is Cas-OFFinder (http://www.rgenome.net/cas-offinder). There is also a recently developed suite of software tools, CRISPRitz, available for this purpose (Cancellieri et al. 2020). Another tool MultiGuideScan is available for designing CRISPR guide RNA libraries (Li et al. 2020c). The tools are also available for predicting mutations resulting from CRIPSR-Cas9-induced DSB. The FORECasT (https://partslab.sanger.ac.uk/FORECasT) is available for this purpose; you just have to enter sequence of target site and PAM location and this tool will predict resulting mutations at the target site (Allen et al. 2019). These continuously developed bioinformatic tools and software can be utilized for designing and predicting CRISPR-Cas-mediated experiments for developing disease-resistant crops with no or minimum off-targets.

As well as all the competent tools, there is also a need for an efficient delivery system to transfer genome-editing components with maximum editing efficiency and minimum off-target effects. There is still a need to explore more systems to deliver change. Though a considerable advancement has been made to reduce any off targeting, as a result of genome editing by keenly choosing a target site, exploring new variants of Cas and creating fusion proteins, but still more work is needed to deliver these systems into the plants. Recently, a new genome-editing technology "Cas-CLOVER" was developed by Hera BioLabs and presented on April 14, 2020 in a webinar held by Genetic Engineering and Biotechnology News (GEN). This latest tool is a fusion of Cas9 having inactivated nuclease activity and Clo51 endonuclease. It has higher fidelity compared to CRISPR-Cas9 with efficient editing and no detectable off-targets (Xianghong et al. 2018). This system can be exploited for crop improvement in future. Similarly, ExpressEdit is another emerging approach, having great potential of editing with reduced off-targets and no need for tissue culturing. Genome editors are also trying to develop such systems where transgene-free genome-edited plants can be obtained bypassing the need for tissue culturing (Ji et al. 2020; Li et al. 2020d).

Further, the achievement of transgene-free genome-edited plants is needed to escape the biosafety and regulatory concerns. Scientists have made successful leaps to achieve transgene-free disease-resistant crops (Li et al. 2020). However, in order to effectively use genome-editing technologies for crop improvement, scientists must address several biosafety and social concerns about it. There is also a need to educate the general public about genome editing. The re-evaluation of the regulations on genome-edited plants are also required to release such crops in the field.

Overall, use of genome-editing technologies might be a promising way to reduce losses due to various diseases. The advances in functional genomics, synthetic biology, NGS, genome-editing technologies and several other related techniques, will allow scientists to develop new varieties of different crops with significantly upgraded traits including plant fitness.

References

Abudayyeh, O.O., Gootenberg, J.S., Konermann, S. et al. (2016). C2c2 is a single-component programmable RNA-guided RNA-targeting CRISPR effector. *Science* 353 (6299): aaf5573.

Abudayyeh, O.O., Gootenberg, J.S., Essletzbichler, P. et al. (2017). RNA targeting with CRISPR–Cas13. *Nature* 550 (7675): 280–284.

Acevedo-Garcia, J., Kusch, S., and Panstruga, R. (2014). Magical mystery tour: MLO proteins in plant immunity and beyond. *New Phytologist* 204 (2): 273–281.

Aird, E.J., Lovendahl, K.N., Martin, A.S. et al. (2018). Increasing Cas9-mediated homology-directed repair efficiency through covalent tethering of DNA repair template. *Communications Biology* 1 (1): 1–6.

Albert, I., Böhm, H., Albert, M. et al. (2015). An RLP23–SOBIR1–BAK1 complex mediates NLP-triggered immunity. *Nature Plants* 1 (10): 1–9.

Alexandratos, N. and Bruinsma, J. (2012).World agriculture towards 2030/2050: the 2012 revision. ESA Working Paper No. 12-03.

Ali, Z., Ali, S., Tashkandi, M. et al. (2016). CRISPR/Cas9-mediated immunity to gemini viruses: differential interference and evasion. *Scientific Reports* 6 (1): 1–13.

Allen, F., Crepaldi, L., Alsinet, C. et al. (2019). Predicting the mutations generated by repair of Cas9-induced double-strand breaks. *Nature Biotechnology* 37 (1): 64–72.

Aman, R., Ali, Z., Butt, H. et al. (2018). RNA virus interference via CRISPR/Cas13a system in plants. *Genome Biology* 19 (1): 1.

Aman, R., Mahas, A., Butt, H. et al. (2018). Engineering RNA virus interference via the CRISPR/Cas13 machinery in Arabidopsis. *Viruses* 10 (12): 732.

Anzalone, A.V., Randolph, P.B., Davis, J.R. et al. (2019). Search-and-replace genome editing without double-strand breaks or donor DNA. *Nature* 576 (7785): 149–157.

Atlin, G.N., Cairns, J.E., and Das, B. (2017). Rapid breeding and varietal replacement are critical to adaptation of cropping systems in the developing world to climate change. *Global Food Security* 12: 31–37.

Barrangou, R. and Doudna, J.A. (2016). Applications of CRISPR technologies in research and beyond. *Nature Biotechnology* 34 (9): 933.

Barrangou, R. and Horvath, P. (2017). A decade of discovery: CRISPR functions and applications. *Nature Microbiology* 2 (7): 1–9.

Blanvillain-Baufumé, S., Reschke, M., Solé, M. et al. (2017). Targeted promoter editing for rice resistance to Xanthomonas oryzae pv. Oryzae reveals differential activities for SWEET 14-inducing TAL effectors. *Plant Biotechnology Journal* 15 (3): 306–317.

Boch, J. and Bonas, U. (2010). Xanthomonas AvrBs3 family-type III effectors: discovery and function. *Annual Review of Phytopathology* 48: 419–436.

Boch, J., Scholze, H., Schornack, S. et al. (2009). Breaking the code of DNA binding specificity of TAL-type III effectors. *Science* 326 (5959): 1509–1512.

Bogdanove, A.J. and Voytas, D.F. (2011). TAL effectors: customizable proteins for DNA targeting. *Science* 333 (6051): 1843–1846.

Bonfim, K., Faria, J.C., Nogueira, E.O. et al. (2007). RNAi-mediated resistance to bean golden mosaic virus in genetically engineered common bean (Phaseolus vulgaris). *Molecular Plant-Microbe Interactions* 20 (6): 717–726.

Buermans, H. and Den Dunnen, J. (2014). Next generation sequencing technology: advances and applications. *Biochimica et Biophysica Acta (BBA)-Molecular Basis of Disease* 1842 (10): 1932–1941.

Cancellieri, S., Canver, M.C., Bombieri, N. et al. (2020). CRISPRitz: rapid, high-throughput and variant-aware in silico off-target site identification for CRISPR genome editing. *Bioinformatics* 36 (7): 2001–2008.

Cao, Y., Ding, X., Cai, M. et al. (2007). The expression pattern of a rice disease resistance gene Xa3/Xa26 is differentially regulated by the genetic backgrounds and developmental stages that influence its function. *Genetics* 177 (1): 523–533.

Čermák, T., Curtin, S.J., Gil-Humanes, J. et al. (2017). A multipurpose toolkit to enable advanced genome engineering in plants. *The Plant Cell* 29 (6): 1196–1217.

Chai, Q., Shang, X., Wu, S. et al. (2017). 5-Aminolevulinic acid dehydratase gene dosage affects programmed cell death and immunity. *Plant Physiology* 175 (1): 511–528.

Chandrasekaran, J., Brumin, M., Wolf, D. et al. (2016). Development of broad virus resistance in non-transgenic cucumber using CRISPR/Cas9 technology. *Molecular Plant Pathology* 17 (7): 1140–1153.

Chatterjee, P., Jakimo, N., and Jacobson, J.M. (2018). Divergent PAM specificity of a highly-similar SpCas9 ortholog. *BioRxiv*: 258939. https://doi.org/10.1101/258939.

Chen, L.-Q., Qu, X.-Q., Hou, B.-H. et al. (2012). Sucrose efflux mediated by SWEET proteins as a key step for phloem transport. *Science* 335 (6065): 207–211.

Chen, K., Wang, Y., Zhang, R. et al. (2019). CRISPR/Cas genome editing and precision plant breeding in agriculture. *Annual Review of Plant Biology* 70: 667–697.

Chintamanani, S., Multani, D.S., Ruess, H., and Johal, G.S. (2008). Distinct mechanisms govern the dosage-dependent and developmentally regulated resistance conferred by the maize Hm2 gene. *Molecular Plant-Microbe Interactions* 21 (1): 79–86.

Christian, M., Cermak, T., Doyle, E.L. et al. (2010). Targeting DNA double-strand breaks with TAL effector nucleases. *Genetics* 186 (2): 757–761.

Christou, P. (2013). Plant genetic engineering and agricultural biotechnology 1983–2013. *Trends in Biotechnology* 31 (3): 125–127.

Clarke, J.L. and Zhang, P. (2013). Plant biotechnology for food security and bioeconomy. *Plant Molecular Biology* 83 (1): 1–3.

Cohen, J. (2019). Prime editing promises to be a cut above CRISPR. *American Association for the Advancement of Science* 366 (6464): 406.

Cox, D.B., Gootenberg, J.S., Abudayyeh, O.O. et al. (2017). RNA editing with CRISPR-Cas13. *Science* 358 (6366): 1019–1027.

Dafny-Yelin, M. and Tzfira, T. (2007). Delivery of multiple transgenes to plant cells. *Plant Physiology* 145 (4): 1118–1128.

Datta, S.K. (2004). Rice Biotechnology: A Need for Developing Countries. *AgBio Forum* 7 (1): 31–35.

De Toledo Thomazella, D.P., Brail, Q., Dahlbeck, D., and Staskawicz, B. (2016). CRISPR-Cas9 mediated mutagenesis of a DMR6 ortholog in tomato confers broad-spectrum disease resistance. *BioRxiv*: 064824. https://doi.org/10.1101/064824.

De Vleesschauwer, D., Xu, J., and Höfte, M. (2014). Making sense of hormone-mediated defense networking: from rice to Arabidopsis. *Frontiers in Plant Science* 5: 611.

Doudna, J.A. and Charpentier, E. (2014). The new frontier of genome engineering with CRISPR-Cas9. *Science* 346 (6213): 1258096.

Doyle, E.L., Stoddard, B.L., Voytas, D.F., and Bogdanove, A.J. (2013). TAL effectors: highly adaptable phytobacterial virulence factors and readily engineered DNA-targeting proteins. *Trends in Cell Biology* 23 (8): 390–398.

East-Seletsky, A., O'Connell, M.R., Knight, S.C. et al. (2016). Two distinct RNase activities of CRISPR-C2c2 enable guide-RNA processing and RNA detection. *Nature* 538 (7624): 270–273.

Faris, J.D., Zhang, Z., Lu, H. et al. (2010). A unique wheat disease resistance-like gene governs effector-triggered susceptibility to necrotrophic pathogens. *Proceedings of the National Academy of Sciences* 107 (30): 13544–13549.

Fister, A.S., Landherr, L., Maximova, S.N., and Guiltinan, M.J. (2018). Transient expression of CRISPR/Cas9 machinery targeting TcNPR3 enhances defense response in Theobroma cacao. *Frontiers in Plant Science* 9: 268.

Gaj, T., Gersbach, C.A., and Barbas, C.F. III (2013). ZFN, TALEN, and CRISPR/Cas-based methods for genome engineering. *Trends in Biotechnology* 31 (7): 397–405.

Galvez, L.C., Banerjee, J., Pinar, H., and Mitra, A. (2014). Engineered plant virus resistance. *Plant Science* 228: 11–25.

Gao, L., Cox, D.B., Yan, W.X. et al. (2017). Engineered Cpf1 variants with altered PAM specificities. *Nature Biotechnology* 35 (8): 789.

Garcia-Doval, C. and Jinek, M. (2017). Molecular architectures and mechanisms of class 2 CRISPR-associated nucleases. *Current Opinion in Structural Biology* 47: 157–166.

Gaudelli, N.M., Komor, A.C., Rees, H.A. et al. (2017). Programmable base editing of A• T to G• C in genomic DNA without DNA cleavage. *Nature* 551 (7681): 464.

Ghislain, M., Byarugaba, A.A., Magembe, E. et al. (2019). Stacking three late blight resistance genes from wild species directly into African highland potato varieties confers complete field resistance to local blight races. *Plant Biotechnology Journal* 17 (6): 1119–1129.

Gil-Humanes, J., Wang, Y., Liang, Z. et al. (2017). High-efficiency gene targeting in hexaploid wheat using DNA replicons and CRISPR/Cas9. *The Plant Journal* 89 (6): 1251–1262.

Gill, S.S., Gill, R., Tuteja, R., and Tuteja, N. (2014). Genetic engineering of crops: a ray of hope for enhanced food security. *Plant Signaling &Behavior* 9 (3): e28545.

Gomez, M.A., Lin, Z.D., Moll, T. et al. (2019). Simultaneous CRISPR/Cas9-mediated editing of cassava eIF 4E isoforms nCBP-1 and nCBP-2 reduces cassava brown streak disease symptom severity and incidence. *Plant Biotechnology Journal* 17 (2): 421–434.

Gonsalves, C., Lee, D.R. and Gonsalves, D. (2004) Transgenic virus-resistant papaya: the Hawaiian 'Rainbow' was rapidly adopted by farmers and is of major importance in Hawaii today. APSnet Features. https://www.apsnet.org/edcenter/apsnetfeatures/Pages/PapayaHawaiianRainbow.aspx (Accessed 29 July 2020)

Gross, B.L. and Olsen, K.M. (2010). Genetic perspectives on crop domestication. *Trends in Plant Science* 15 (9): 529–537.

Gust, A.A., Brunner, F., and Nürnberger, T. (2010). Biotechnological concepts for improving plant innate immunity. *Current Opinion in Biotechnology* 21 (2): 204–210.

Hammond, A.M., Kyrou, K., Bruttini, M. et al. (2017). The creation and selection of mutations resistant to a gene drive over multiple generations in the malaria mosquito. *PLoS Genetics* 13 (10): e1007039.

Harper, G., Osuji, J.O., Heslop-Harrison, J., and Hull, R. (1999). Integration of banana streak badna virus into the Musa genome: molecular and cytogenetic evidence. *Virology-New York* 255: 207–213.

Harrington, L.B., Burstein, D., Chen, J.S. et al. (2018). Programmed DNA destruction by miniature CRISPR-Cas14 enzymes. *Science* 362 (6416): 839–842.

He, P., Chintamanani, S., Chen, Z. et al. (2004). Activation of a COI1-dependent pathway in Arabidopsis by Pseudomonas syringae type III effectors and coronatine. *The Plant Journal* 37 (4): 589–602.

Heinrichs, E. and Muniappan, R. (2017). IPM for tropical crops: rice. *CAB Reviews* 12 (30): 1–30.

Hong, Y., Chen, L., Du, L.-p. et al. (2014). Transcript suppression of TaGW2 increased grain width and weight in bread wheat. *Functional &Integrative Genomics* 14 (2): 341–349.

Hossain, M., Gollin, D., Cabanilla, V. et al. (2003). International research and genetic improvement in rice: evidence from Asia and Latin America. In: *Crop Variety Improvement and its Effect on Productivity: The Impact of International Agricultural Research* (eds. D. Gollin and R.E. Evenson), 71–108. Cambridge, MA: CABI Publishing.

Hsu, P.D., Lander, E.S., and Zhang, F. (2014). Development and applications of CRISPR-Cas9 for genome engineering. *Cell* 157 (6): 1262–1278.

Huerta-Espino, J., Singh, R., German, S. et al. (2011). Global status of wheat leaf rust caused by Puccinia triticina. *Euphytica* 179 (1): 143–160.

Islam, M.T., Croll, D., Gladieux, P. et al. (2016). Emergence of wheat blast in Bangladesh was caused by a south American lineage of Magnaporthe oryzae. *BMC Biology* 14 (1): 84.

Jensen, K.T., Fløe, L., Petersen, T.S. et al. (2017). Chromatin accessibility and guide sequence secondary structure affect CRISPR-Cas9 gene editing efficiency. *FEBS letters* 591 (13): 1892–1901.

Ji, X., Zhang, H., Zhang, Y. et al. (2015). Establishing a CRISPR–Cas-like immune system conferring DNA virus resistance in plants. *Nature Plants* 1 (10): 1–4.

Ji, X., Yang, B., and Wang, D. (2020). Achieving plant genome editing while bypassing tissue culture. *Trends in Plant Science* 25 (5): 427–429.

Jia, H., Orbovic, V., Jones, J.B., and Wang, N. (2016). Modification of the PthA4 effector binding elements in type I Cs LOB 1 promoter using Cas9/sg RNA to produce transgenic Duncan grapefruit alleviating XccΔpthA4: dCs LOB 1.3 infection. *Plant Biotechnology Journal* 14 (5): 1291–1301.

Jia, H., Zhang, Y., Orbović, V. et al. (2017). Genome editing of the disease susceptibility gene Cs LOB 1 in citrus confers resistance to citrus canker. *Plant Biotechnology Journal* 15 (7): 817–823.

Jiang, G.-L. (2013). Plant marker-assisted breeding and conventional breeding: challenges and perspectives. *Advancesin Crop Science and Technology* 1: e106.

Jiang, F. and Doudna, J.A. (2017). CRISPR–Cas9 structures and mechanisms. *Annual Review of Biophysics* 46: 505–529.

Jiang, W., Zhou, H., Bi, H. et al. (2013). Demonstration of CRISPR/Cas9/sgRNA-mediated targeted gene modification in Arabidopsis, tobacco, sorghum and rice. *Nucleic Acids Research* 41 (20): e188–e188.

Jinek, M., Chylinski, K., Fonfara, I. et al. (2012). A programmable dual-RNA–guided DNA endonuclease in adaptive bacterial immunity. *Science* 337 (6096): 816–821.

Jinek, M., East, A., Cheng, A. et al. (2013). RNA-programmed genome editing in human cells. *elife* 2: e00471.

Jo, K.-R., Kim, C.-J., Kim, S.-J. et al. (2014). Development of late blight resistant potatoes by cisgene stacking. *BMC Biotechnology* 14 (1): 50.

Joung, J.K. and Sander, J.D. (2013). TALENs: a widely applicable technology for targeted genome editing. *Nature reviews Molecular Cell Biology* 14 (1): 49–55.

Khan, M.Z., Amin, I., Hameed, A., and Mansoor, S. (2018). CRISPR–Cas13a: prospects for plant virus resistance. *Trends in Biotechnology* 36 (12): 1207–1210.

Khan, M.Z., Haider, S., Mansoor, S., and Amin, I. (2019). Targeting plant ssDNA viruses with engineered miniature CRISPR-Cas14a. *Trends in Biotechnology* 37 (8): 800–804.

Kim, H. and Kim, J.-S. (2014). A guide to genome engineering with programmable nucleases. *Nature Reviews Genetics* 15 (5): 321–334.

Kim, D., Kim, J., Hur, J.K. et al. (2016). Genome-wide analysis reveals specificities of Cpf1 endonucleases in human cells. *Nature Biotechnology* 34 (8): 863.

Kim, H.K., Song, M., Lee, J. et al. (2017). In vivo high-throughput profiling of CRISPR–Cpf1 activity. *Nature Methods* 14 (2): 153.

Kim, Y.B., Komor, A.C., Levy, J.M. et al. (2017). Increasing the genome-targeting scope and precision of base editing with engineered Cas9-cytidine deaminase fusions. *Nature Biotechnology* 35 (4): 371.

Kis, A., Hamar, É., Tholt, G. et al. (2019). Creating highly efficient resistance against wheat dwarf virus in barley by employing CRISPR/Cas9 system. *Plant Biotechnology Journal* 17 (6): 1004.

Kleinstiver, B.P., Tsai, S.Q., Prew, M.S. et al. (2016). Genome-wide specificities of CRISPR-Cas Cpf1 nucleases in human cells. *Nature Biotechnology* 34 (8): 869.

Komor, A.C., Kim, Y.B., Packer, M.S. et al. (2016). Programmable editing of a target base in genomic DNA without double-stranded DNA cleavage. *Nature* 533 (7603): 420.

Komor, A.C., Zhao, K.T., Packer, M.S. et al. (2017). Improved base excision repair inhibition and bacteriophage mu gam protein yields C: G-to-T: a base editors with higher efficiency and product purity. *Science Advances* 3 (8): 4774.

Koseoglou, E.(2017).The study of SlPMR4 CRISPR/Cas9-mediated tomato allelic series for resistance against powdery mildew. Master thesis. Wageningen University and Research.

Lee, J.-H., Muhsin, M., Atienza, G.A. et al. (2010). Single nucleotide polymorphisms in a gene for translation initiation factor (eIF4G) of rice (Oryza sativa) associated with resistance to Rice tungro spherical virus. *Molecular Plant Microbe Interacteractions* 23 (1): 29–38.

Li, T., Liu, B., Spalding, M.H. et al. (2012). High-efficiency TALEN-based gene editing produces disease-resistant rice. *Nature Biotechnology* 30 (5): 390.

Li, J.-F., Norville, J.E., Aach, J. et al. (2013). Multiplex and homologous recombination–mediated genome editing in Arabidopsis and Nicotiana benthamiana using guide RNA and Cas9. *Nature Biotechnology* 31 (8): 688.

Li, Z., Xiong, X., and Jian-Feng, L. (2018a). New cytosine base editor for plant genome editing. *Science China. Life Sciences* 61 (12): 1602.

Li, C., Zong, Y., Wang, Y. et al. (2018b). Expanded base editing in rice and wheat using a Cas9-adenosine deaminase fusion. *Genome Biology* 19 (1): 59.

Li, H., Jingying, L., Chen, J. et al. (2020a). Precise modifications of both exogenous and endogenous genes in rice by prime editing. *Molecular Plant* 13 (5): 671–674.

Li, C., Zhang, R., Meng, X. et al. (2020b). Targeted, random mutagenesis of plant genes with dual cytosine and adenine base editors. *Nature Biotechnology* 38 (7): 875–882.

Li, T., Wang, S., Luo, F. et al. (2020c). MultiGuideScan: a multi-processing tool for designing CRISPR guide RNA libraries. *Bioinformatics* 36 (3): 920–921.

Li, C., Li, W., Zhou, Z. et al. (2020d). A new rice breeding method: CRISPR/Cas9 system editing of the Xa13 promoter to cultivate transgene-free bacterial blight-resistant rice. *Plant Biotechnology Journal* 18 (2): 313–315.

Lin, S., Staahl, B.T., Alla, R.K., and Doudna, J.A. (2014). Enhanced homology-directed human genome engineering by controlled timing of CRISPR/Cas9 delivery. *Elife* 3: e04766.

Lin, Q., Zong, Y., Xue, C. et al. (2020). Prime genome editing in rice and wheat. *Nature Biotechnology* 38: 582–585.

Liu, W., Liu, J., Triplett, L. et al. (2014). Novel insights into rice innate immunity against bacterial and fungal pathogens. *Annual Reviews of Phytopathology* 52: 213–241.

Liu, Y., Han, J., Chen, Z. et al. (2017). Engineering cell signaling using tunable CRISPR–Cpf1-based transcription factors. *Nature Communications* 8 (1): 1–8.

Liu, L., Li, X., Wang, J. et al. (2017). Two distant catalytic sites are responsible for C2c2 RNase activities. *Cell* 168 (1–2): 121–134. e12.

Lorence, A. and Verpoorte, R. (2004). Gene transfer and expression in plants. In: *Recombinant Gene Expression* (eds. P. Balbas and A. Lorence), 329–350. Springer.

Ma, X., Zhang, Q., Zhu, Q. et al. (2015). A robust CRISPR/Cas9 system for convenient, high-efficiency multiplex genome editing in monocot and dicot plants. *Molecular Plant* 8 (8): 1274–1284.

Ma, Y., Zhang, J., Yin, W. et al. (2016). Targeted AID-mediated mutagenesis (TAM) enables efficient genomic diversification in mammalian cells. *Nature Methods* 13 (12): 1029–1035.

Maclean, J., Dawe, D., Hardy, B. and Hettel, G.(2002).*Rice Almanac*. Los Baños:International Rice Research Institute (IRRI).

Macovei, A., Sevilla, N.R., Cantos, C. et al. (2018). Novel alleles of rice eIF4G generated by CRISPR/Cas9-targeted mutagenesis confer resistance to Rice tungro spherical virus. *Plant Biotechnology Journal.* 16 (11): 1918–1927.

Maeder, M.L., Thibodeau-Beganny, S., Osiak, A. et al. (2008). Rapid "open-source" engineering of customized zinc-finger nucleases for highly efficient gene modification. *Molecular Cell* 31 (2): 294–301.

Makino, S., Sugio, A., White, F., and Bogdanove, A.J. (2006). Inhibition of resistance gene-mediated defense in rice by Xanthomonas oryzae pv. Oryzicola. *Molecular Plant-Microbe Interactions* 19 (3): 240–249.

Malnoy, M., Viola, R., Jung, M.-H. et al. (2016). DNA-free genetically edited grapevine and apple protoplast using CRISPR/Cas9 ribonucleoproteins. *Frontiers in Plant Science* 7: 1904.

Malzahn, A.A., Tang, X., Lee, K. et al. (2019). Application of CRISPR-Cas12a temperature sensitivity for improved genome editing in rice, maize, and Arabidopsis. *BMC Biology* 17 (1): 9.

Maqbool, A., Saitoh, H., Franceschetti, M. et al. (2015). Structural basis of pathogen recognition by an integrated HMA domain in a plant NLR immune receptor. *Elife* 4: e08709.

Marraffini, L.A. and Sontheimer, E.J. (2010). Self versus non-self-discrimination during CRISPR RNA-directed immunity. *Nature* 463 (7280): 568–571.

Marraffini, L.A. and Sontheimer, E.J. (2010). CRISPR interference: RNA-directed adaptive immunity in bacteria and archaea. *Nature Reviews Genetics* 11 (3): 181–190.

Marzec, M., Brąszewska-Zalewska, A., and Hensel, G. (2020). Prime editing: a new way for genome editing. *Trends in Cell Biology* 30 (4): 257–259.

Matsoukas, I.G. (2018). Commentary: RNA editing with CRISPR-Cas13. *Frontiers in Genetics* 9: 134.

Mew, T.W., Alvarez, A.M., Leach, J., and Swings, J. (1993). Focus on bacterial blight of rice. *Plant Disease* 77 (1): 5–12.

Meyer, R.S. and Purugganan, M.D. (2013). Evolution of crop species: genetics of domestication and diversification. *Nature Reviews Genetics* 14 (12): 840–852.

Miah, G., Rafii, M.Y., Ismail, M.R. et al. (2013). Blast resistance in rice: a review of conventional breeding to molecular approaches. *Molecular Biology Reports* 40 (3): 2369–2388.

Miller, J.C., Tan, S., Qiao, G. et al. (2011). A TALE nuclease architecture for efficient genome editing. *Nature Biotechnology* 29 (2): 143.

Moon, S.B., Lee, J.M., Kang, J.G. et al. (2018). Highly efficient genome editing by CRISPR-Cpf1 using CRISPR RNA with a uridinylate-rich 3′-overhang. *Nature Communications* 9 (1): 1–11.

Moreno-Mateos, M.A., Fernandez, J.P., Rouet, R. et al. (2017). CRISPR-Cpf1 mediates efficient homology-directed repair and temperature-controlled genome editing. *Nature Communications* 8 (1): 1–9.

Moscou, M.J. and Bogdanove, A.J. (2009). A simple cipher governs DNA recognition by TAL effectors. *Science* 326 (5959): 1501–1501.

Nalam, V.J., Alam, S., Keereetaweep, J. et al. (2015). Facilitation of Fusarium graminearum infection by 9-lipoxygenases in Arabidopsis and wheat. *Molecular Plant-Microbe Interactions* 28 (10): 1142–1152.

Negishi, K., Kaya, H., Abe, K. et al. (2019). An adenine base editor with expanded targeting scope using SpCas9-NG v1 in rice. *Plant Biotechnology Journal* 17 (8): 1476–1478.

Nekrasov, V., Wang, C., Win, J. et al. (2017). Rapid generation of a transgene-free powdery mildew resistant tomato by genome deletion. *Scientific Reports* 7 (1): 1–6.

Nishimasu, H., Ran, F.A., Hsu, P.D. et al. (2014). Crystal structure of Cas9 in complex with guide RNA and target DNA. *Cell* 156 (5): 935–949.

Ochiai, H., Inoue, Y., Takeya, M. et al. (2005). Genome sequence of *Xanthomonas oryzae pv. oryzae* suggests contribution of large numbers of effector genes and insertion sequences to its race diversity. *Japan Agricultural Research Quarterly: JARQ* 39 (4): 275–287.

O'Connell, M.R. (2019). Molecular mechanisms of RNA targeting by Cas13-containing type VI CRISPR–Cas systems. *Journal of Molecular Biology* 431 (1): 66–87.

Oerke, E.-C. and Dehne, H.-W. (2004). Safeguarding production – losses in major crops and the role of crop protection. *Crop Protection* 23 (4): 275–285.

Oliva, R., Ji, C., Atienza-Grande, G. et al. (2019). Broad-spectrum resistance to bacterial blight in rice using genome editing. *Nature Biotechnology* 37 (11): 1344–1350.

Oliva, R., Ji, C., Atienza-Grande, G. et al. (2019). Broad-spectrum resistance to bacterial blight in rice using genome editing. *Nature Biotechnology* 37: 1344–1350.

Orozco, P. (2018) Argentina and Brazil merge law and science to regulate new breeding techniques. https://allianceforscience.cornell.edu/blog/2018/01/argentina-and-brazil-merge-law-and-science-to-regulate-new-breeding-techniques/(accessed 20 July 2020).

Ortigosa, A., Gimenez-Ibanez, S., Leonhardt, N., and Solano, R. (2019). Design of a bacterial speck resistant tomato by CRISPR/Cas9-mediated editing of Sl JAZ 2. *Plant Biotechnology Journal* 17 (3): 665–673.

Pâques, F. and Duchateau, P. (2007). Meganucleases and DNA double-strand break-induced recombination: perspectives for gene therapy. *Current Gene Therapy* 7 (1): 49–66.

Peng, A., Chen, S., Lei, T. et al. (2017). Engineering canker-resistant plants through CRISPR/Cas9-targeted editing of the susceptibility gene Cs LOB 1 promoter in citrus. *Plant Biotechnology Journal* 15 (12): 1509–1519.

Prihatna, C., Barbetti, M.J., and Barker, S.J. (2018). A novel tomato fusarium wilt tolerance gene. *Frontiers in Microbiology* 9: 1226.

Puchta, H. and Fauser, F. (2013). Gene targeting in plants: 25 years later. *International Journal of Developmental Biology* 57 (6-7-8): 629–637.

Puchta, H. and Fauser, F. (2014). Synthetic nucleases for genome engineering in plants: prospects for a bright future. *The Plant Journal* 78 (5): 727–741.

Qi, L.S., Larson, M.H., Gilbert, L.A. et al. (2013). Repurposing CRISPR as an RNA-guided platform for sequence-specific control of gene expression. *Cell* 152 (5): 1173–1183.

Que, Q., Chilton, M.-D.M., de Fontes, C.M. et al. (2010). Trait stacking in transgenic crops: challenges and opportunities. *GM Crops* 1 (4): 220–229.

Ran, Y., Liang, Z., and Gao, C. (2017). Current and future editing reagent delivery systems for plant genome editing. *Science China Life Sciences* 60 (5): 490–505.

Roelfs, A.P. and Bushnell, W.R. (1985). *The Cereal Rusts*. Orlando, FL: Academic Press.

Rosa, C., Kuo, Y.-W., Wuriyanghan, H., and Falk, B.W. (2018). RNA interference mechanisms and applications in plant pathology. *Annual Review of Phytopathology* 56: 581–610.

Salsman, J. and Dellaire, G. (2017). Precision genome editing in the CRISPR era. *Biochemistry and Cell Biology* 95 (2): 187–201.

Sander, J.D. and Joung, J.K. (2014). CRISPR-Cas systems for editing, regulating and targeting genomes. *Nature Biotechnology* 32 (4): 347.

Savary, S., Willocquet, L., Pethybridge, S.J. et al. (2019). The global burden of pathogens and pests on major food crops. *Nature Ecology &Evolution* 3 (3): 430–439.

Schornack, S., Moscou, M.J., Ward, E.R., and Horvath, D.M. (2013). Engineering plant disease resistance based on TAL effectors. *Annual Review of Phytopathology* 51: 383–406.

Schwartz, A.R., Potnis, N., Timilsina, S. et al. (2015). Phylogenomics of Xanthomonas field strains infecting pepper and tomato reveals diversity in effector repertoires and identifies determinants of host specificity. *Frontiers in Microbiology* 6: 535.

Shimatani, Z., Kashojiya, S., Takayama, M. et al. (2017). Targeted base editing in rice and tomato using a CRISPR-Cas9 cytidine deaminase fusion. *Nature Biotechnology* 35 (5): 441–443.

Shmakov, S., Abudayyeh, O.O., Makarova, K.S. et al. (2015). Discovery and functional characterization of diverse class 2 CRISPR-Cas systems. *Molecular Cell* 60 (3): 385–397.

Smith, J., Grizot, S., Arnould, S. et al. (2006). A combinatorial approach to create artificial homing endonucleases cleaving chosen sequences. *Nucleic Acids Research* 34 (22): e149–e149.

Suzuki, K., Tsunekawa, Y., Hernandez-Benitez, R. et al. (2016). In vivo genome editing via CRISPR/Cas9 mediated homology-independent targeted integration. *Nature* 540 (7631): 144–149.

Tang, X., Sretenovic, S., Ren, Q. et al. (2020). Plant prime editors enable precise gene editing in rice cells. *Molecular Plant* 13 (5): 667–670.

Tóth, E., Weinhardt, N., Bencsura, P. et al. (2016). Cpf1 nucleases demonstrate robust activity to induce DNA modification by exploiting homology directed repair pathways in mammalian cells. *Biology Direct* 11 (1): 46.

Tricoll, D.M., Carney, K.J., Russell, P.F. et al. (1995). Field evaluation of transgenic squash containing single or multiple virus coat protein gene constructs for resistance to cucumber mosaic virus, watermelon mosaic virus 2, and zucchini yellow mosaic virus. *Biotechnology* 13 (12): 1458–1465.

Tripathi, J.N., Ntui, V.O., Ron, M. et al. (2019). CRISPR/Cas9 editing of endogenous banana streak virus in the B genome of Musa spp. overcomes a major challenge in banana breeding. *Communications Biology* 2 (1): 1–11.

Unckless, R.L., Clark, A.G., and Messer, P.W. (2017). Evolution of resistance against CRISPR/Cas9 gene drive. *Genetics* 205 (2): 827–841.

Urnov, F.D., Rebar, E.J., Holmes, M.C. et al. (2010). Genome editing with engineered zinc finger nucleases. *Nature Reviews Genetics* 11 (9): 636–646.

Van der Burgh, A.M. and Joosten, M.H. (2019). Plant immunity: thinking outside and inside the box. *Trends in Plant Science* 24 (7): 587–601.

Van Esse, H.P., Reuber, T.L., and van der Does, D. (2020). Genetic modification to improve disease resistance in crops. *New Phytologist* 225 (1): 70–86.

Voytas, D.F. (2013). Plant genome engineering with sequence-specific nucleases. *Annual Review of Plant Biology* 64: 327–350.

Vriet, C., Russinova, E., and Reuzeau, C. (2012). Boosting crop yields with plant steroids. *The Plant Cell* 24 (3): 842–857.

Wang, M.-B., Masuta, C., Smith, N.A., and Shimura, H. (2012). RNA silencing and plant viral diseases. *Molecular Plant-Microbe Interactions* 25 (10): 1275–1285.

Wang, Y., Cheng, X., Shan, Q. et al. (2014). Simultaneous editing of three homoeoalleles in hexaploid bread wheat confers heritable resistance to powdery mildew. *Nature Biotechnology* 32 (9): 947.

Wang, F., Wang, C., Liu, P. et al. (2016). Enhanced rice blast resistance by CRISPR/Cas9-targeted mutagenesis of the ERF transcription factor gene OsERF922. *PLoS One* 11 (4): e0154027.

Wang, M., Mao, Y., Lu, Y. et al. (2017). Multiplex gene editing in rice using the CRISPR-Cpf1 system. *Molecular Plant* 10 (7): 1011–1013.

Wang, W., Pan, Q., He, F. et al. (2018). Transgenerational CRISPR-Cas9 activity facilitates multiplex gene editing in allopolyploid wheat. *The CRISPR Journal* 1 (1): 65–74.

Wang, X., Tu, M., Wang, D. et al. (2018). CRISPR/Cas9-mediated efficient targeted mutagenesis in grape in the first generation. *Plant Biotechnology Journal* 16 (4): 844–855.

Wang, W., Mauleon, R., Hu, Z. et al. (2018). Genomic variation in 3,010 diverse accessions of Asian cultivated rice. *Nature* 557 (7703): 43–49.

Wang, T., Deng, Z., Zhang, X. et al. (2018). Tomato DCL2b is required for the biosynthesis of 22-nt small RNAs, the resulting secondary siRNAs, and the host defense against ToMV. *Horticulture Research* 5 (1): 1–14.

Wheeler, T. and Von Braun, J. (2013). Climate change impacts on global food security. *Science* 341 (6145): 508–513.

Wolter, F. and Puchta, H. (2018). The CRISPR/Cas revolution reaches the RNA world: Cas13, a new Swiss Army knife for plant biologists. *The Plant Journal* 94 (5): 767–775.

Wu, C.-H., Abd-El-Haliem, A., Bozkurt, T.O. et al. (2017). NLR network mediates immunity to diverse plant pathogens. *Proceedings of the National Academy of Sciences* 114 (30): 8113–8118.

Xianghong, L., Xinxin, W., Min T., et al. (2018) Cas-CLOVER™: A High-Fidelity Genome Editing System for Safe and Efficient Modification of Cells for Immunotherapy. *2018 Precision CRISPR Congress Poster Presentation, Boston, MA*. https://www.herabiolabs.com/cas-clover/ (Accessed 10 August 2020).

Xie, K., Minkenberg, B., and Yang, Y. (2015). Boosting CRISPR/Cas9 multiplex editing capability with the endogenous tRNA-processing system. *Proceedings of the National Academy of Sciences* 112 (11): 3570–3575.

Xing, H.-L., Dong, L., Wang, Z.-P. et al. (2014). A CRISPR/Cas9 toolkit for multiplex genome editing in plants. *BMC Plant Biology* 14 (1): 327.

Xu, Z., Xu, X., Gong, Q. et al. (2019). Engineering broad-spectrum bacterial blight resistance by simultaneously disrupting variable TALE-binding elements of multiple susceptibility genes in rice. *Molecular Plant* 12 (11): 1434–1446.

Yan, W.X., Mirzazadeh, R., Garnerone, S. et al. (2017). BLISS is a versatile and quantitative method for genome-wide profiling of DNA double-strand breaks. *Nature Communications* 8 (1): 1–9.

Yan, F., Kuang, Y., Ren, B. et al. (2018). Highly efficient A· T to G· C base editing by Cas9n-guided tRNA adenosine deaminase in rice. *Molecular Plant* 11 (4): 631–634.

Yan, W.X., Chong, S., Zhang, H. et al. (2018). Cas13d is a compact RNA-targeting type VI CRISPR effector positively modulated by a WYL-domain-containing accessory protein. *Molecular Cell* 70 (2): 327–339. e5.

Yang, B. and White, F.F. (2004). Diverse members of the AvrBs3/PthA family of type III effectors are major virulence determinants in bacterial blight disease of rice. *Molecular Plant-Microbe Interactions* 17 (11): 1192–1200.

Yang, L.-Z., Wang, Y., Li, S.-Q. et al. (2019). Dynamic imaging of RNA in living cells by CRISPR-Cas13 systems. *Molecular Cell* 76 (6): 981–997. e7.

Yunis, H., Bashan, Y., Okon, Y., and Henis, Y. (1980). Weather dependence, yield losses, and control of bacterial speck of tomato caused by pseudomonas tomato. *Plant Diseases* 64: 937–939.

Zafar, K., Khan, M.Z., Amin, I. et al. (2020). Precise CRISPR-Cas9 mediated genome editing in super basmati rice for resistance against bacterial blight by targeting the major susceptibility gene. *Frontiers in Plant Science* 11: 575.

Zaidi, S.S.-e.-A., Mahfouz, M.M., and Mansoor, S. (2017). CRISPR-Cpf1: a new tool for plant genome editing. *Trends in Plant Science* 22 (7): 550–553.

Zaidi, S.S.-e.-A., Mukhtar, M.S., and Mansoor, S. (2018). Genome editing: targeting susceptibility genes for plant disease resistance. *Trends Biotechnol* 36 (9): 898–906.

Zamore, P.D. (2004). Plant RNAi: how a Viral silencing suppressor inactivates siRNA. *Current Biology* 14 (5): R198–R200.

Zetsche, B., Gootenberg, J.S., Abudayyeh, O.O. et al. (2015). Cpf1 is a single RNA-guided endonuclease of a class 2 CRISPR-Cas system. *Cell* 163 (3): 759–771.

Zetsche, B., Heidenreich, M., Mohanraju, P. et al. (2017). Multiplex gene editing by CRISPR–Cpf1 using a single crRNA array. *Nature Biotechnology* 35 (1): 31.

Zhan, X., Zhang, F., Zhong, Z. et al. (2019). Generation of virus-resistant potato plants by RNA genome targeting. *Plant Biotechnology Journal* 17 (9): 1814–1822.

Zhang, X., Li, H., Zhang, J. et al. (2011). Expression of artificial microRNAs in tomato confers efficient and stable virus resistance in a cell-autonomous manner. *Transgenic Research* 20 (3): 569–581.

Zhang, Y., Bai, Y., Wu, G. et al. (2017). Simultaneous modification of three homoeologs of ta EDR 1 by genome editing enhances powdery mildew resistance in wheat. *The Plant Journal* 91 (4): 714–724.

Zhang, S., Wang, L., Zhao, R. et al. (2018). Knockout of SlMAPK3 reduced disease resistance to Botrytis cinerea in tomato plants. *Journal of Agricultural and Food Chemistry* 66 (34): 8949–8956.

Zhang, Z., Ge, X., Luo, X. et al. (2018). Simultaneous editing of two copies of Gh14-3-3d confers enhanced transgene-clean plant defense against Verticillium dahliae in allotetraploid upland cotton. *Frontiers in Plant Science* 9: 842.

Zhang, T., Zheng, Q., Yi, X. et al. (2018). Establishing RNA virus resistance in plants by harnessing CRISPR immune system. *Plant Biotechnology Journal* 16 (8): 1415–1423.

Zhang, C., Konermann, S., Brideau, N.J. et al. (2018). Structural basis for the RNA-guided ribonuclease activity of CRISPR-Cas13d. *Cell* 175 (1): 212–223. e17.

Zhang, H., Si, X., Ji, X. et al. (2018). Genome editing of upstream open reading frames enables translational control in plants. *Nature Biotechnology* 36 (9): 894–898.

Zhou, H., Liu, B., Weeks, D.P. et al. (2014). Large chromosomal deletions and heritable small genetic changes induced by CRISPR/Cas9 in rice. *Nucleic Acids Research* 42 (17): 10903–10914.

Zhou, J., Peng, Z., Long, J. et al. (2015). Gene targeting by the TAL effector PthXo2 reveals cryptic resistance gene for bacterial blight of rice. *The Plant Journal* 82 (4): 632–643.

Zhu, S., Li, Y., Vossen, J.H. et al. (2012). Functional stacking of three resistance genes against Phytophthora infestans in potato. *Transgenic Research* 21 (1): 89–99.

Zong, Y., Song, Q., Li, C. et al. (2018). Efficient C-to-T base editing in plants using a fusion of nCas9 and human APOBEC3A. *Nature Biotechnology* 36 (10): 950–953.

15

Biotechnological Approaches for Nutritional Improvement in Potato (*Solanum tuberosum* L.)

Chandrama Prakash Upadhyaya[1] *and Deepak Singh Bagri*[1,2]

[1] *Laboratory of Plant Molecular Biology, Department of Biotechnology, Dr Harisingh Gour Central University, Sagar, Madhya Pradesh, India*
[2] *Department of Botany, Government Degree College Amanganj, Panna, Madhya Pradesh, India*

CHAPTER MENU

15.1 Introduction

Plants are a major source of the human nutrition. A huge number of the human population, particularly in developing countries, rely on staple crops, such as rice, maize, wheat and potato which do not provide the full complement of essential nutrients. Potato (*Solanum tuberosum* L.) is grown for its tubers which is a staple food crop and potentially valuable for health benefits. The cultivation of potato has been reported from over 100 countries throughout the world and billions of people depend on the potato for daily food (Bagri et al. 2018). The potato tubers are the morphological transformation of the underground stem into stolon which leads to the formation of tubers having axillary dormant buds. Potato is grown directly from tuber seeds by vegetative propagation containing eyes or dormant buds. These buds grow under favorable conditions. Moreover, the nutritional value of potato is conserved because of its vegetative mode of propagation (Navarre et al. 2016). The potato is an herbaceous plant, 0.5–1.2-meter height. The leaves are alternate and irregularly pinnately compound. Inflorescence consists of several flowers that form pentamerous, actinomorphic, and perfect with sympetalous colored corollas. Fruits

Genome Engineering for Crop Improvement, First Edition. Edited by Santosh Kumar Upadhyay.

are berries, absent in many cultivars. Potato tuber is a starchy, edible part that is cultivated throughout the world and is consumed as the third largest staple crop after rice and legumes. Potato has the potential to combat hunger and malnutrition in the developing world in an era of shrinking cultivated land due to its exceptionally high yield per acre when compared to other grain crops (Burton 1989). Potato also has a variety of uses such as in the food-processing industry; livestock feed and other industrial uses (Feustel 1987).Among the major food crops, potato is superior in producing highest dry matter, protein and other nutrients per unit/time as it is a short-duration crop and can be harvested within only a 60-day period. Potato is a low-calorie food containing protein with biological value of 75% of egg and milk protein and is also rich in minerals, vitamins B and C. With the increasing population of the globe and decreasing resources; the potato has great potential in supplementing traditional cereal crops that exist today. India is placed as the second-largest producer of potato after the Republic of China in the world. It produces around 45 million metric tons of potatoes that contribute to approximately 7.75% of the world's total produce and its production is constantly growing with the increase in cultivated area. (Breithaupt and Bamedi 2002; Ensminger et al. 1983; Fortin 1996; Liu et al. 2003; Parr et al. 2005; Tudela et al. 2002; Wood and Rebecca 1988).

Potato is a good source of fiber, largely found in the flesh, however; eating the potato skin can help increase overall intake. 100 g serving of a baked potato with the skin contains 2.6 g of fiber, and without the skin itis 1.8 g. The daily requirement of fibers for women is 21–25 g of fiber a day and for men it is 30–38 g a day. Potatoes are a good source of potassium; 100 g serving of baked potato with the flesh contains 652 mg of potassium, meeting 18% of human's daily value (Table 15.1). Potassium is a mineral that aids in fluid and electrolyte balance. It may also help to improve blood pressure and in treating muscle weakness in humans. Potatoes are high in vitamins; 100 g serving of a baked potato with the skin contains 11.8 mg of vitamin C, 0.25 mg of vitamin B6 meeting 20 and 10%respectively of daily requirements for humans. Vitamins play a role as important antioxidants, protecting cells from oxidation and reducing the risk of heart disease and cancer. The potato is an important source of energy. Most of the calories in the potato come from its carbohydrate content, and they are a preferred source of energy. In fact, healthy diet guidelines recommend 45–65% of calories come from carbohydrates to support daily energy needs (Breithaupt and Bamedi 2002; Ensminger et al. 1983; Fortin 1996; Liu et al. 2003; Parr et al. 2005; Tudela et al. 2002; Wood and Rebecca 1988). In addition to helping people lose weight, potatoes have the function of curing wounds, increasing the discharge of urine and releasing spasm. Potatoes prevent and cure ecchymosis, neuralgia, arthritis, coronary heart disease and eye pain (Fortin 1996; Liu et al. 2003). As a result, to eat potatoes regularly not only makes healthy body, but also maintains youth and keeps a person smart. Various nutritional constituents of the potato are shown below in Table 15.1

15.2 Genetic Transformation of Potato

A large number of transgenic crops have been developed through genetic-engineering technologies and released for commercial cultivation which mainly include fiber crops such as cotton, oil crops such as canola, vegetable crops, cereals and millets (Chandra and Pental 2003;

Table 15.1 Various nutritional constituents of the potato.

Nutrient	Amount /100 g
Total Fat	0.1 g
Protein	2 g
Total Carbohydrate	17 g
Total Energy	77 kcal
Potassium	421 mg
Sodium	6 mg
Phosphorus	57 mg
Magnesium	23 g
Calcium	12 mg
Iron	0.8 mg
Zinc	0.3 mg
Vitamin C	19.7 mg
Vitamin B4	12.1 mg
Vitamin B3	1 mg
Vitamin B5	0.3 mg
Vitamin B6	0.3 mg
Vitamin B1	0.08 mg
Vitamin B2	0.032 mg
Vitamin B9	16 µg
Vitamin E	0.01 mg
Vitamin K	1.9 µg

Maximum volume of the edible part of potato is carbohydrate followed by the proteins and minerals. Vitamins and other constituents are scanty.
Source: Modified from Potatoes USA. Retrieved from: www.potatogoodness.com.

Dale et al. 1993; Dunwell et al. 2000). A number of transgenic crops have been moved from the laboratory to farm land areas around the world. Intensive efforts have been made to produce transgenic plants of important crops using the genes of different agronomic traits.

Like any other plant, for the transformation of potato, it is essential to have a target of totipotent cells for introduction of DNA and meristematic cells which can serve as a source of totipotent cells in potato. With the development of super-virulent strains of *Agrobacterium* and improvement in other methods of DNA transfer, it is now possible to increase the success rate of transgenic plant development in crops (Dita et al. 2006).For introduction of transgenes into the potato, either *Agrobacterium*-mediated or particle gun bombardment-mediated methods of gene transfer have been used. Some of the other direct methods of gene transfer necessitate isolation of protoplasts capable of regenerating plants. For potato, the selectable marker gene normally used is hygromycin phosphotransferase *(hpt).*

A number of protocols for the introduction of foreign genes into potato genome via *Agrobacterium-mediated* potato transformation are available (Beaujean et al. 1998; De Block 1988; Mitten et al. 1990; Sheerman and Bevan 1988; Stiekema et al. 1988; Tavazza et al. 1988; Trujillo et al. 2001; Wenzler et al. 1989). The table below summarizes major reports on genetic transformation of potato for enhanced nutritional value.

Table 15.2 Gene(s) used to enhance nutritional value in transgenic potato (*Solanum tuberosum L*).

S.N.	Gene(s) description	Comments	References
1	Bacterial phytoene synthase	Increased carotenoids up to sevenfold in tubers.	Laurence et al. (2005)
2	Steroidal *glycosyl transferases*	Antisense and sense expression in transgenic and the gene modulated the expression of the Gas	McCue et al. (2007)
3	Lycopene epsilon cyclase. (*Lcy-e*)	Significant increases in beta-carotenoid levels	Gianfranco et al. (2006)
4	*SnRK1* (sucrose non-fermenting-1-related protein kinase-1)	Causes a significant increase in starch content and a decrease in glucose levels.	Rowan et al. (2006)
5	*GalUR* and (GLOase)	Over-accumulation of vitamin C.	Hemavathi et al. (2009), Hemavathi et al. (2011)
6	*CrtB, CrtI*, and *Crty* genes	Enhance the amount of beta-carotene and total carotenoids	Gianfranco et al. (2006)
7	*AmA1* (*Amaranth* Albumin 1)	Tuber-specific expression of a seed protein enhanced nutritive value.	Chakraborty et al. (2010)
8	*At-HPPD* and *At-HPT*	Increase in alpha-tocopherol.	Crowell et al. (2008)
9	*DHARs*	Increased DHAR activities and AsA contents	Qin et al. (2011)
10	*GGP* (*Geranyl-geranyl pyrophosphate*)	Increase in tuber ascorbate	Bulley et al. (2012)
11	*HMG1* (3-Hydroxy-3-methylglutaryl coenzyme A reductase 1)	Transgenic producing high SGA (steroidal glycoalkaloids).	Idit et al. (2012)
12	cauliflower Orange (*Or*)	Promoted retention of β-carotene level	Li et al. (2012)
13	Dihydroxy phenylalanine (*DOPA*), dioxygenases (*DODs*) and feeding of *DOD* substrate (L-*DOPA*)	Induce betalain production	Harris et al. (2012)
14	*I bLCY-ε* by RNAi	Higher accumulation of β-carotene content	Kim et al. (2013)
15	β-*CAROTENE HYDROXYLASE* 2 (*CrtRb2*) and *PHYTOENE SYNTHASE* 2 (*PSY2*)	Carotenoid levels significantly increased	Pasare et al. (2013)

Table 15.2 (Continued)

S.N.	Gene(s) description	Comments	References
16	Dihydroflavonol-4-reductase (*DFRi*)	Dramatically reduced anthocyanin accumulation in young leaves	Wang et al. (2013)
17	Dihydroflavonol reductase gene (*DFR*).	Produced transgenic potato plants served as a nutrition source	Kostyn et al. (2013)
18	*StAN11*, a WD40-repeat gene	Accumulation of anthocyanin	Yastreb et al. (2016)
19	*IbOr-Ins*	Ib r plants exhibited increased carotenoid levels	Park et al. (2015)
20	*Ib Or*	Level of total carotenoid was elevated	Goo et al. (2015)
21	Δ12 acyl-lipid desaturase gene (*desA*)	Accumulation of photosynthetic pigments and phenolic compounds	Zagoskina et al. (2014)
22	*Or* gene	Enhanced astaxanthin content	Campbell et al. (2015)
23	4′ β-oxygenase (*crtW*) and 3-3′ β-hydroxylase (*crtZ*)	Dramatic increase in the carotenoid content of plastoglobuli	Mortimer et al. (2015)
24	*Stearoyl-acyl* carrier protein desaturase (*ScoSAD*, *SaSAD*, *ScaSAD*, and *StSAD*)	Increased linoleic acid content	Li et al. (2015)
25	PDX II gene	Increased the vitamin B6 content	Bagri et al. (2018)
26	crtO b-carotene ketolase gene Potato	Accumulation of echinenone, 39-hydroxyechinenone, and 4-ketozeaxanthin in leaves, as well as 39-hydroxyechinenone, 4-ketozeaxanthin together with astaxanthin in the tuber	Tanja and Gerhard (2006)
27	DnaJ cysteine-rich domain Potato	Overexpression of the cauliflower *Or* gene Or containing protein that mediates high levels of beta-carotene accumulation.	Lu et al. (2006)
28	*CrtB*, *CrtI*, and *Crty* genes Potato	Enhance the amount of beta-carotene and total carotenoids	Gianfranco et al. (2010)
29	*galUR* gene. *A. thaliana*	Ascorbic acid increase by two to three times	Agius et al. (2003)
30	DHAR gene, Potato	Increased AsA content in the tubers	Young et al. (2008)

(*Continued*)

Table 15.2 (Continued)

S.N.	Gene(s) description	Comments	References
31	Sterol methyl transferase. Potato	Overexpression of this gene in and the modulation in the content of GAs	Arnqvist et al. (2003)
32	Steroidal glycosyl transferases Potato	Antisense and sense expression in transgenic and the gene modulated the expression of the GAs	McCue et al. (2005), McCue et al. (2007)
33	*ubi*C Potato	4HB glucoside content reached4.0% of dry weight	Köhle et al. (2003)
34	chalcone synthase (CHS), chalcone isomerase (CHI), and dihydroflavonol reductase (DFR) Potato	Significant increase of measured phenolic acids and anthocyanins	Lukaszewicz et al. (2004)
35	*HMG1* (3-Hydroxy-3-methylglutaryl coenzyme A reductase 1) Potato	Transgenic producing high SGA (steroidal glycoalkaloids)	Idit et al. (2012)
36	*Fructosyl transferase* genes. Potato	Increased non-structural carbohydrate content in the leaves	Lngrid et al. (1994)
37	phosphofructokinase Potato	Degradation of simple sugars via glycolytic pathway	Navratil et al. (1998)
38	SnRK1 (sucrose non-fermenting-1-related protein kinase-1) Potato	Causes a significant increase in starch content and a decrease in glucose levels	Rowan et al. (2006)
39	14-3-3 protein derived from *Cucurbita pepo* Potato	Overexpression showed changes in lipid content and composition	Anna et al. (2001)
40	acetyl- CoA Potato	Transgenic tubers led to an increase in fatty acid synthesis	Klaus et al. (2004)
41	*DesA* gene and thermostable lichenase (LicB) Potato	Lipid content and the unsaturation level of their fatty acid acyl moieties significantly increased	Reza et al. (2007)
42	β-CAROTENE HYDROXYLASE 2 (CrtRb2) and PHYTOENE SYNTHASE 2 (PSY2) Potato	Carotenoid levels significantly increased	Pasare et al. (2013)
43	Silencing of StMGL1 Potato	Increased methionine to isoleucine ratio and change in the free amino acid profile	Huang et al. (2014)
44	Silence the vacuolar invertase gene (VInv) Potato	Plants had undetectable levels of reducing sugars	Clasen et al. (2015)

Table 15.2 (Continued)

S.N.	Gene(s) description	Comments	References
45	PrLeg polypeptide Potato	Methionine content was elevated	Goo et al. (2013)
46	Polyprotein (DEFp) glycolate dehydrogenase (GlcDH) Potato	Transgenic lines with the highest DEFp levels and GlcDH activity produced significantly higher levels of glucose (5.8-fold), fructose (3.8-fold), sucrose (1.6-fold) and transitory starch (threefold)	Nölke et al. (2014)
47	phosphatase 2 of Arabidopsis (AtPAP2) Potato	There was a significant change in sucrose, glucose and fructose levels in leaves	Zhang et al. (2014)
48	SGT1, SGT2 and SGT3 potato	Modifying glycoalkaloid content	Shepherd et al. (2015)
49	cysteine *StPI 143* and *StPI 146*	lowering protease activities and tyrosine along with total free amino acid (FAA) contents	Dong et al. (2020)
50	*AtCYP21-4* and *OsCYP21-4* a	20% higher in mannosidic glycoproteins	Park et al. (2017)
51	cystathionine γ-synthase (AtCGS)	increases tuber methionine levels	Kumar and Jander (2017)
52	*Wrinkled1, Diacylglycerol acyl transferase 1 and oleosin*	30-fold increase in triacylglycerols in the leaves	Xu et al. (2020)
53	downregulation of *ADP-glucose pyrophosphorylase* (AGPase) and *Sugar-dependent1* (SDP1) by RNAi	increase of triacylglycerol along with other lipids in transgenic potato tubers	Xu et al. (2019)
54	down regulation of pyrophosphatase (psPPase) of and an ATP/ADP translocator (NTT)	resulted in reduction in starch and amylose content	Andersson et al. (2018)
55	chimeric RNAi ISA1, ISA2 and ISA3	resulted in lower tuber starch	Ferreira et al. (2017)
56	*Wrinkled1, Diacylglycerol acyl transferase 1 and oleosin*	100-fold increase in TAG accumulation	Liu et al. (2017)
57	Phytochrome B (*PHYB*)	formation of more defective or more ordered starch structures	Wasserman et al. (2015)
58	*D-galacturonic acid reductase* (*GalUR*) gene	enhanced level of cellular L-ascorbate (L-AsA)	Chung et al. (2019)
59	cytosolic DHAR	resulted in increased level of AsA content in leaves	Qin et al. (2011)
60	*sCAX1*	deficiency symptoms of calcium	Gao et al. (2020)
61	AtLAS	enhanced SGA level and lower phytosterol level in leaves	Kumar et al. (2017)

Various transgenic potato with enhanced nutritional content are summarized in the following sections.

15.3 Protein Engineering of Potato for Protein Content

Humans require a diverse and balanced nutritional diet to maintain optimum health which depends essentially on plants for their daily nutritional needs. The amino acid and proteins are the primary components of a balanced diet. However, it is very rare in nature to find all of the essential amino acids in a single food crop. Protein malnutrition is essentially caused by poor quality diets with low-quality protein in terms of amino-acid composition. Protein deficiency lowers the immune system, delays physical growth and development. Out of 20 amino acid, 10 "essential amino acids" cannot be synthesized by the animals (including the humans) required for protein synthesis, therefore, these must be obtained from the diet source (Chakraborty et al. 2010). There have been several attempts through breeding and genetic engineering of crop plants via altering the expression of the major genes encoding amino-acid biosynthesis pathway enzymes to enhance the free essential amino acids but with little achievement (Matthews and Hughes 1993).The potato is gaining greater importance in agriculture due to both nutritional and industrial needs. To supply a good quality potato with enhanced protein is of prime importance and therefore, the genetic-engineering approach has been undertaken to develop a genetically engineered potato.

The quantity of protein in potato ranges from 1.1–1.5% of tuber fresh weight which is much less compared to other staple food sources and hence (Mary et al. 2009; Ortiz-Medina 2007), in order to understand the *denovo* amino acid biosynthesis in potato, the gene expression studies have been done which showed that amino-acid biosynthesis in potato tubers depend upon the sucrose level in the plant. The photo-assimilate in particular, the carbohydrate, is transported from source to sinks regulating plant growth and development. The sucrose coordinates the amino-acid biosynthesis within the tuber, most probably at the level of transcription of the key enzymes for the pathways concerned (Tunali et al. 2003).There are quite a few reports on the improvement of the amino-acid content the in potato using genetic-engineering approaches. The transgenic potato with enhanced methionine and systeine amino acid was reported via the expression of individual genes involved in cysteine and methionine metabolism as their biosynthetic regulation is different in the potato as reported in other plants such as *Arabidopsis* (Nikiforova et al. 2002). It has been already reported that the cystathionine-beta-lyase (CBL) is a rate-limiting enzyme in essential amino-acid biosynthesis in the plants. Maimann et al. (2000) reported that the regulation of the CBL through antisense inhibition in transgenic potato resulted in increased levels of the upstream metabolic amino acid such as cysteine, cystathionine, and homoserine, while threonine remained constant and methionine was only slightly reduced. It was also established that the expression of the sunflower albumin or an amaranth seed albumin *(AmA1)* in potato tubers initiated 5.0–7.0 folds' increase in total methionine (Chakraborty et al. 2010). Analysis of these transgenic tubers confirmed an enhanced content of total protein (up to 60%) in comparison to the untransformed potato tubers. In addition, the concentrations of several essential amino acids were also significantly enhanced in these transgenic tubers, which are otherwise limited in potato. This was done in seven genotypic backgrounds suitable for cultivation in different agro-climatic regions. Goo et al. (2013) reported development of transgenic potato lines with three-fold increase in the methionine content by overexpressing the *PrLeg* cDNA (isolated from *Perilla* seed) driven by *patatin* promoter, without any significant changes in other amino acids. The transgenic approaches to elevate methionine content in potato is focused on increasing the methionine biosynthesis which also result in higher isoleucine accumulation suggesting that the potatoes compensate for increased methionine biosynthesis with enhanced catabolism via methionine gamma-lyase (MGL), thereby producing 2-ketylbutyrate for isoleucine biosynthesis. Huang et al. (2014) reported that potato StMGL1 encodes a functional MGL in potato tubers and its silencing via RNAi technique resulted in an increased methionine to isoleucine ratio of free amino-acid profile in potato tubers and, in some transgenic lines, elevated accumulation of free methionine. The ratio of methionine to isoleucine is negatively correlated with the level of StMGL1 transcript. The report on development of transgenic potato overexpressing the gene encoding beta-casein (1177 bp) of human milk driven by the auxin-inducible (bidirectional mannopine synthase) promoter resulting in 0.01% increase in the beta–casein protein in the tubers of potato is also published (Chong et al. 1997).This experiment also established that the edible food crop can also be fortified with human milk, the beta-casein which might replace the addition of bovine protein in baby foods as a nutritional source. Similarly, the development of transgenic potato was also reported by overexpressing low-weight glutenin (*LMW-GS-MB1*) isolated from wheat seed (*Triticumaestivum* variety Chinese spring) which contain this protein as one of the essential food stock proteins (Benmoussa et al. 2004). This was done with the aim of developing efficient baking quality of flour made of potato. It was also observed that *LMW-GS-MB1* units accumulated in transgenic potato tubers function as a polymer and interconnect them either mutually or with other constituents present in tubers resulting in good-quality potato flour dough with better thickness.

Very recently, Dong et al. (2020) reported overexpression of protease inhibitors cysteine *StPI 143* and *StPI 146* in potato which resulted in lowering protease activities and lower levels of tyrosine along with total free amino acid (FAA) contents. In another study, the transgenic potato overexpressing *AtCYP21-4* and *OsCYP21-4*, a cyclophilin protein resulted in an approximately 20% increase in mannosidic-glycoproteins which also lead to significantly higher tuber number and weight (Park et al. 2017). Methionine deficiency is common in potatoes which are essentially required for tuber development, overexpression of cystathionine γ-synthase (AtCGS) from *Arabidopsis thaliana* in potato also increases tuber methionine levels and better yield (Kumar and Jander 2017). Progress in understanding the structure, function, folding and topology of proteins allow the design and synthesis of a gene encoding artificial protein with desirable essential amino-acid composition. The high essential amino-acid encoding synthetic gene (*HEAAE- DNA*) was inserted into the chloramphenicol acetyltransferase (CAT) coding sequence to generate a CAT-HEAAE fusion protein (Jaynes et al. 1986; Yang et al. 1989). The transgenic study indicated that CAT-HEAAE protein was accumulated at 0.02–0.35% of total tuber protein in transgenic potato.

15.4 Genetic Engineering of Potato for Starch Modification

Major plant photosynthetic carbohydrates are starch which are universally constituted of amylopectin (75%) and amylose (25%). It is synthesized in the plastids, and assemble into a semi-crystalline granule. Starch, being of major commercial importance, has been isolated from corn, wheat, rice, sorghum, roots and tubers such as potato, sweet potato, cassava, and pith such as sago. Starch granules play an important role in commercial application, therefore, the granule alternation via chemical modification is of major concern for industrial application. Potato starch has been recognized as an important source for industrial application due to greater granule size, long amylopectin chains, higher molecular weight amylose and smooth surface via higher covalently linked phosphate chain which make the starch granule resistant to enzymatic degradation by hydrolytic enzymes such as amyloglucosidases and α-amylases (Sun et al. 2006). Approximately 3–5% of the crop produced is used for the production of starch. The bulk of it is treated with high-fructose syrup or dextrose and approximately one third of the remaining used for the industrial application. The industrial application of starch comprises fabrication of high-quality paper and synthesis of viscous hydrocolloid systems for food-processing industries (Blennow et al. 2003; Wiesenborn et al. 1994). Several plants have been genetically modified through their starch biosynthetic pathways in order to modify the starch composition and structure suitable for higher yields and industrial applications. The most commonly denoted objective for starch improvement is the modification in the starch amylose-amylopectin ratio. All genetically modified technology used for starch modification is discussed below. Proposed pathway for starch biosynthesis in plants is presented in Figure 15.1.

15.4.1 Starch Synthase Enzyme Engineering in Potato

It has been established that the starch is metabolically synthesized by complex enzymes including starch synthases, starch branching and debranching enzymes (Figure 15.1). The potato tuber is known to store the starch in the plastids called amyloplasts. The sucrose photo-assimilate is delivered to the tuber cell via the phloem which convert metabolically into the hexose-phosphates and triose-phosphates that lead to the synthesis of the starch. The starch is made up of glucose monomers joined together by α1-4 orα1-6 glycosidic bonds. Out of two dominant forms of starch, the amylose starch formed by unbranched chains of glucose monomers (onlyα1-4 linkages), whereas amylopectin is a branched polysaccharide (α1-6 linkages at the branch points). At the time of starch synthesis, the transfer of the glucosyl moieties from ADP-glucose to the non-reducing end of an α-1,4-glucan is catalyzed by the enzyme Starch Synthase (SS; ADP Glucose:α-1,4 glucosyl transferase). The last enzyme in starch biosynthesis, the starch-branching enzyme (EC 2.4.1.11), that introduces α-1-6 branch points in the α-1-4 glucan chain thus forms the amylopectin. The sucrose synthase (SuSy) is considered an extremely ordered cytosolic enzyme that catalyzes the conversion of sucrose and a nucleoside diphosphate into the corresponding nucleoside diphosphate glucose and fructose leading to biosynthesis of the amylopectin. The transgenic potato was developed overexpressing the SuSy. The SuSy-overexpressing tubers showed a significant rise in starch, UDP-glucose, and ADP-glucose content when compared to untransformed potato. Other morphological characters such as tuber dry weight, starch content per plant and total tubers yield were also enhanced considerably (Edurne et al. 2009). Wang et al. (2017) reported the cloning of the starch synthase from sweet potato (IbSS1) and overexpressed in sweet potato. Molecular analysis revealed high expression of *IbSSI* in the leaves and storage roots of the sweet potato which significantly enhanced the content and granule size of starch and the proportion of amylopectin by

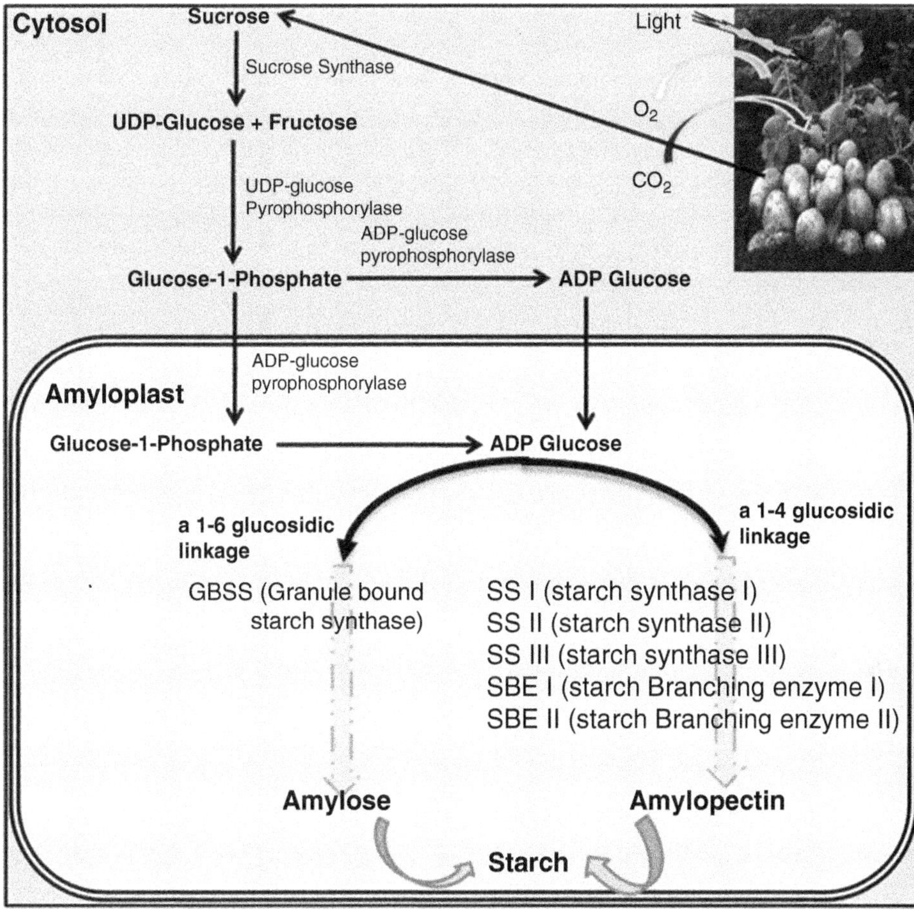

Figure 15.1 Schematic representation for the starch biosynthesis and modification pathway in potato (*Solanum tuberosum L.*). The sucrose is converted to UDP-glucose and fructose in tuber cells by SuSy in cytosol, in which a large amount is transported to amyloplast for starch biosynthesis. ADP glucose are converted to Amylopectin by various isoforms of the Su Syvia and 1-4 glycosidic linkage while its conversion to amylose is done by GBSS via 1-6 glycosidic linkage.

up-regulating starch biosynthetic genes in the transgenic plants compared with untransformed controlled plants. The transgenic potato was also developed by overexpressing the maize sucrose-phosphate synthase (SPS) gene which is the key enzyme in sucrose biosynthesis (Ishimaru et al. 2008). The transgenic potato showed improvement in the transport of photosynthetic products from source (leaves) to sink (tubers). Additionally, leaf senescence of the transgenic potato plants was delayed relative to that of untransformed control potato. The average tuber weight and total yield of transgenic plants were at least 20% higher, and the tuber sucrose content, as well as starch content was also higher.

A promising study on the function of potato starch branching enzyme (SBEII) using overexpression in potato tubers was done by Emes and Neuhaus (1997). The transgenic potato lines with SBEII overexpression was generated. Compared with wild-type, starch from these tubers possessed an increased degree of amylopectin branching, with more short chains of degree of polymerization (DP) 6–12 and particularly of DP6. Further, the increased ratio of short to long amylopectin branches facilitated gelatinization, which occurred at a reduced temperature (by up to 3 °C) or lower urea concentration. The initiation of starch biosynthesis in storage organs inevitably requires the mobilization of sucrose into glucose-6-phosphate (G6P), import of G6P into the amyloplast through inorganic phosphate (Pi) exchange, and subsequent conversion of G6P into glucose-1-phosphate (G1P) by plastidial phosphoglucomutase. The first committed step to starch synthesis is the formation of ADP-glucose (ADP-Glc) through ATP activation of G1P, catalyzed by ADPG pyrophosphorylase (AGPase). The AGPases enzymes are reported to be allosterically regulated with 3-phosphoglyceric acid (3PGA) being the main activator and Pi the main inhibitor (Emes and Neuhaus 1997; Neuhaus and Emes 2000). The transgenic potato lines were also developed via inhibiting the expression of the ADP-glucose pyrophosphorylase (AGPase) by introducing a chimeric gene containing the coding region of one of the subunits of the AGPase linked in an antisense orientation to the *CaMV 35S* promoter. This enzyme catalyzes the first committed step of starch biosynthesis by transferring the Pi from ATP to glucose for the formation of ADP-glucose (ADP-Glc). Molecular analysis of the transgenic plant revealed partial inhibition of the AGPase enzyme in leaves and complete inhibition in tubers resulting in no synthesis of starch in tubers, thus proving that AGPase has an exclusive function in starch biosynthesis in plants. Biochemical analysis also showed 30% increase in the sucrose and 8% increase in the glucose with the transgenic tubers. (Lloyd et al. 1999; Müller-Röber et al. 1992). Rowan et al. (2006) also reported the overexpression of *SnRK1* (sucrose non-fermenting-1-related protein kinase-1) gene under the control of a *patatin* (tuber-specific) promoter in potato tubers which causes a significant increase in starch content and a decrease in glucose levels. Biochemical analysis of the tubers revealed in increase in the starch content and lowering the glucose content up to 20–30% in comparison to the untransformed control potato tubers.

15.4.2 Genetic Engineering of Potato to Modify the Starch Granule Size

Starch granule size is an important feature that determines the suitability for many of the food and non food uses. Starch granule size is highly species-specific and starch granules can appear in many different forms. The large size of starch granules in potato tubers are advantageous in applications demanding high viscosity but limits its suitability for noodle making. In order to alter the granule size using genetic engineering approaches, the transgenic potato lines were developed by overexpression of the *cyclodextrin glycosyltransferase* gene isolated from *Bacillus circulans* bacterial strain which is a multiple tandem starch-binding domain (SBD). Biochemical analysis of transgenic potato tuber starch revealed a large number of small starch granules without affecting the starch yield (Ji et al. 2004; Nazarian-Firouzabadi et al. 2012).

Two starch phosphorylases in plants, plastidic phosphorylase A (Pho 1a) and cytosolic phosphorylase (Pho 2), catalyze reversible transfer of glucose from glucose-1-phosphate to a-glucan chain-releasing phosphorus (pi). A gene was cloned which translated protein counterpart involved in starch metabolism identified by its ability to bind the potato starch granules (Lorberth et al. 1998). The gene was introduced as an RNAi construct in the potato using the *Agrobacterium*-mediated transformation method. Biochemical analysis revealed that the reduction in the protein level of transgenic potato augmented with reduction in the phosphate content of the starch. The complementary result is obtained when the same gene was expressed in *E coli*, as this leads to an increased phosphate content of the glycogen. It was assumed that this protein might be responsible for the incorporation of phosphate into starch-like glucans, a process that is not understood at the biochemical level. The reduced phosphate content in potato starch has some secondary effects on its degradability, as the respective plants show a starch excess phenotype in leaves and a reduction in cold-sweetening in tubers. In a mapping study in potato, Pho 1a emerged as a candidate gene linked to starch gelling and starch granule size (Werij et al. 2012). In another study, it was found that the antisense inhibition of isoamylase in potato induces massive numbers of small granules in tubers, suggesting that the debranching activity is necessary to prevent excessive granule initiation (Bustos et al. 2004).

The post-harvest shelf life of potato is extended by the storage at cold temperature that reduces potato sprouting. Low temperature encourages the increase of reducing sugars in potato tubers that reacts with free amino acids during the food processing, resulting in dark golden browning, bitter-tasting products and elevated levels of acrylamide which is a potential carcinogen. Because RNAi often results in incomplete gene silencing and requires the plant to be transgenic, a study was done where the transcription activator-like effector nucleases (TALENs) were used to knockout *VInv* within the commercial potato variety (Clasen et al. 2015). Tubers from full *VInv*-knockout plants had undetectable levels of reducing sugars, and processed chips contained reduced levels of acrylamide and were lightly colored. These results provide a framework for using TALENs to quickly improve traits in commercially relevant auto-tetraploid potato lines. The down regulation of pyrophosphatase (psPPase) and an ATP/ADP translocator (NTT) in potato also demonstrated that pyrophosphate (PPi) hydrolysis by PPase is crucial for functional starch accumulation in potato (Andersson et al. 2018). In another study, the Phytochrome B (*PHYB*) gene from *A. thaliana* was overexpressed in potato which resulted in the formation of more ordered starch structures in transgenic potato as compared with the wild type (Wasserman et al. 2015).

15.5 Lipids Biosynthesis Engineering in Potato

Potato tubers are rich in starch but the lipids quantity is scanty. Gene expression analysis has been done to investigate the lipid biosynthesis in different parts of potato as well as the factors limiting the lipid synthesis in potato tubers.

The gene *acetyl-CoA carboxylase* isolated from *Arabidopsis* was overexpressed in potato amyloplasts of tubers which resulted in an increase in fatty acid synthesis and also enhanced the triacylglycerol up to five fold than the untransformed controls (Klaus et al. 2004). However, the changes in lipid content achieved in this study represent only the initial steps toward the attempt to redirect a considerable volume of carbon from starch to lipid. The overexpression of 14-3-3 protein gene isolated from *Cucurbitapepo* in potato was also reported (Anna et al. 2001). The biochemical analysis of transgenic line revealed the changes in lipid content in the tubers with 69% more total fat and enhanced linoleic acid content compared to the untransformed control potato. Further analysis confirmed that 14-3-3 protein gene Lorberth et al. 1998 affected the carbohydrate metabolism in plant via regulation of catecholamine synthesis involved in the lipid metabolism in potato. Reza et al. (2007) reported the overexpression of the *Cyanobacterium* acyl-lipid Δ12-desaturase (*Des-A*) gene fused with a thermostable lichenase (*LicB*) gene of *Clostridium thermocellum*, both driven by the *35SCaMV* promoter. Biochemical analysis confirmed the increase in the unsaturated fatty acid acyl moieties and fatty acid composition in leaves of transgenic potato as compared to control potato plants. Report also existed on the isolation and cloning of four new full-length cDNA of Stearoyl-Acyl carrier protein Desaturase (SAD) gene from four different *Solanum* species including *Solanum commersonii*, *Solanum acaule*, *Solanum cardiophyllum*, and *S. tuberosum*. The genes were cloned under the control of the patatin promoter in four different binary vectors and overexpressed in potato independently. The analysis of the transgenic potato lines confirmed that the gene isolated from *Solanum commersonii* (*Sco SAD*) showed significantly enhanced freeze tolerance and increased linoleic acid content, suggesting that linoleic acid is likely to play an important role for better freeze tolerance in potato plants (Li et al. 2015). In a recent finding, the transgenic potato was raised with a 30-fold increase in triacylglycerols in the leaves as compared to wild type by the overexpression of three genes, first the *At-Wrinkled1* driven by a senescence-inducible promoter, second the *Diacylglycerol acyl transferase 1* driven by the constitutive *35S* promoter, and third gene the *oleosin* from *Sesamumindicum* driven by a small subunit of rubisco promoter (Xu et al. 2020). This increase in triacylglycerols was accommodated with reduction in starch content and simultaneously increases in total soluble sugars. In another study, downregulation of *ADP-glucose pyrophosphorylase* (AGPase) and *Sugar-dependent1* (SDP1) by RNAi resulted in the increase of triacylglycerol along with other lipids in transgenic potato tubers and the increase in triacylglycerol was 16-fold higher in comparison to that of the wild-type potato, which is about 7% of tuber dry weight. This also changed the sugar accumulation, and starch content in tuber and leaf tissues. It also changed the lipid-droplet distribution and starch granule morphology as well (Liu et al. 2017).

15.6 Vitamins Genetic Engineering in Potato

Vitamins are a class of organic compounds, categorically vital for upholding the healthy life processes. Though they are required in small amounts, their nutrition capability and their ability to perform biochemical functions is remarkable. Vitamins are reported to play avital role in the blood cells synthesis, hormones and genetic materials. The vitamins are reported to regulate the metabolism via supporting various biochemical processes in the body. The vitamins have been grouped into water- and fat-soluble vitamins based on the solubility properties. Fat-soluble vitamins are A, D, E, and K and the remaining are water soluble. Most of the vitamins have been found to act as co-enzymes in metabolic reactions, some act as growth regulators and most of them as antioxidants (Jube and Borthakur 2006). The important groups of vitamins such as Vit A, C, and E must be obtained through vegetarian diets, therefore, inadequate vitamin intake may cause several health-related complications. Since, two-thirds of the world population depend on food crop containing the rice, potato, and cereals, therefore, it is important to fortify these crops with the various vitamins through the biofortification. The plant breeding and genetic engineering techniques has been utilized for biosynthesis of additional vitamins in the several edible crops, including the potato.

ß-carotene is known as the precursor for vitamin A (Vit-A), which is an important component for the visual pigments of rod and cone cells in the retina. In some cases, deficiency of Vit-A causes night blindness to total blindness. Plant foods such as carrots and many other vegetables contain ß-carotene and related compounds which is oxidatively cleaved into the intestine to produce two molecules of retinal, that finally reduce to form retinol or Vit-A. Cultivated potato is extremely poor in carotenoids, comprised primarily of the xanthophylls lutein, xanthophyll esters, antheraxanthin, and violaxanthin, lacking the provitamin A activity (Duroy et al. 2009). Carotenoids as isoprenoids also play an important biological role in plants and animals. A generalized overview of carotenoid biosynthesis in plants is presented in Figure 15.2. Laurence et al. (2005) reported the cloning and overexpression of the bacterial *phytoene synthase* in potato cultivar Desiree, resulting in an over seven fold increase in the carotenoids content compared to the untransformed control

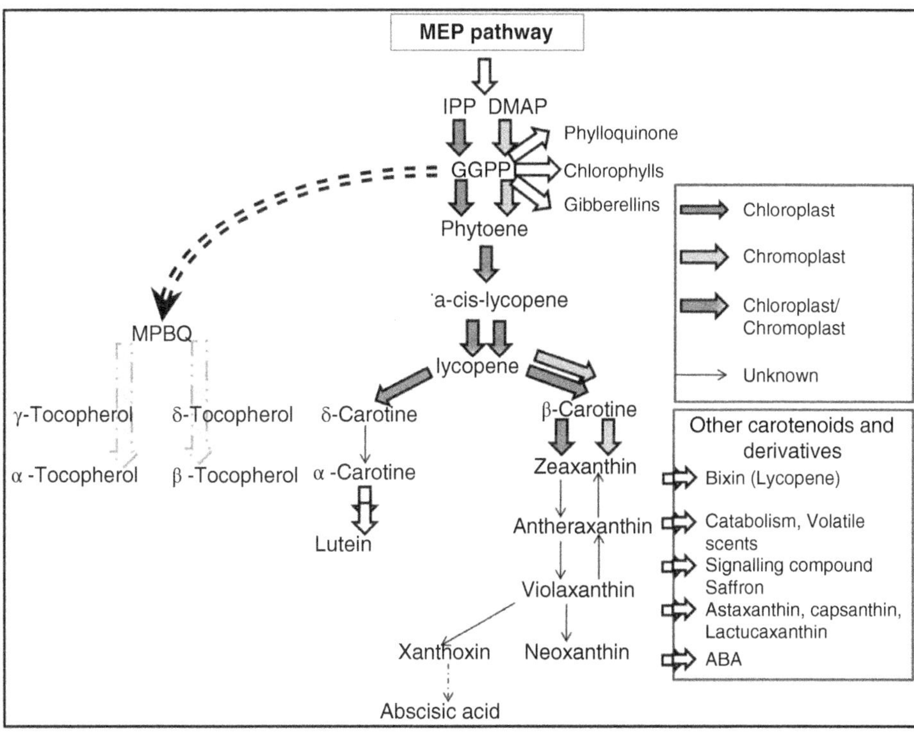

Figure 15.2 Schematic representation of various carotenoids and tocopherol biosynthesis along with other carotenoid derivatives in different cell organelles of plants. The MEP (2-C-methyl-d-erythritol-4-phosphate) pathway provides isopetenyl pyro phosphate (IPP) and its isomer dimethylallyl diphosphate (DMAP) for synthesis geranylgeranyl diphosphate (GGDP), the central intermediate. This GGDP can be used for synthesis of phytoene which give rise to lycopene, chlorophylls and tocotrienols. Lycopene is also the branching intermediate from which all carotenoids are formed, including tocopherol family.

potato. The transgenic lines also exhibited variations in the proportions of distinct carotenoids where the carotene content increased from trace amounts to 11 mg/g DW and lutein content amplified up to 19-fold than the untransformed control potato. Similarly, a report published by Gianfranco and his coworkers for the development of transgenic potato via the RNAi-mediated silencing of the *lycopene epsilon cyclase* (*Lcy-e*) gene driven by the *patatin* promoter, which regulate the first committed step in the beta-epsilon branch of carotenoid biosynthesis. Molecular analysis of the transgenic lines showed substantial rises in beta-carotene level up to 14-fold and total carotenoids content up to 2.5-fold in comparison to the untransformed control potato (Gianfranco et al. 2006). The changes in beta-carotene and total carotenoids did not affect the level of lutein, suggesting that *Lcy-e* is not rate-limiting for lutein accumulation in the potato tubers. The development of transgenic potato was also reported by the overexpression of*β-Carotene Hydroxalase 2* (*CrtRb2*) and *Phytoene Synthase 2 (PSY2)* genes, fused to red fluorescent protein (RFP) exhibited higher levels of carotenoids in the potato tubers confirming functionality of the fluorescently tagged proteins (Pasare et al. 2013). Similarly, the overexpression of the *4, 4′ β-oxygenase (crtW) and 3, 3′ β-hydroxylase (crtZ)* genes from *Brevundimonas* species in the potato driven by the constitutive promoter also resulted in enhanced hydroxylated and ketolated carotenoid derivatives (Mortimer et al. 2015). The astaxanthin and zeaxanthin have been recognized as commercially important keto-carotenoids as a feed supplement in poultry farming and aquaculture. Consumption of ketocarotenoids, most notably astaxanthin, is also associated with several human health benefits. The development of a genetically engineered potato variety containing the zeaxanthin lines was also reported by overexpression of the *crtO b-carotene ketolase* gene isolated from the cyanobacterium *Synechocystis*, driven by constitutive promoter (*CaMV35S*) which resulted in a several fold increase in all the keto-carotenoids including the zeaxanthin (Tanja and Gerhard 2006). To produce commercially valuable ketocarotenoids in potato, Harris et al. (2012) reported the biolistic introduction of DNA constructs for transient overexpression of two different *dihydroxyphenylalanine* (*DOPA*), *dioxygenases* (*DOD*s) genes, and feeding of DOD substrate (L-DOPA) which was sufficient to induce betalains formation in cell cultures of potato. High performance liquid chromatography (HPLC) analysis confirmed the presence of both betaxanthins and the betacyanins in the cell lines. Campbell et al. (2015) also showed that the selection of appropriate parental genotype for transgenic approaches and stacking carotenoid biosynthetic pathway genes with the *cauliflower orange (Or)* gene resulted in enhanced astaxanthin content (up to six fold) in tuber tissues in comparison to the wild- type control potato.

The vitamin E (Vit-E) is another vital fat-soluble vitamin which has also been amplified in plants via genetic engineering. A group of tocopherols including α-tocopherol, β-tocopherol, γ-tocopherol and δ-tocopherol is collectively known as vitamin E and recognized as a powerful antioxidant. Biosynthetic pathways involved in tocopherols (Vit-E) biosynthesis and proposed interrelation with vitamin B6 is presented **in** Figure 15.3. Though the importance of vitamin E for human health was recognized in literature in 1922, the plant genetic engineered model plants were developed in early 1990s. The first report on the genetic engineering of potato with enhanced tocopherols in tuber tissues was reported in 2008 (Crowell et al. 2008). Two major Vitamin-E biosynthetic genes, the *Arabidopsis thaliana* isolate p-*hydroxyphenylpyruvate dioxygenase (At-HPPD)* and *homogentis atphytyl transferase (At-HPT)* were cloned in two different binary vectors driven by the constitutive

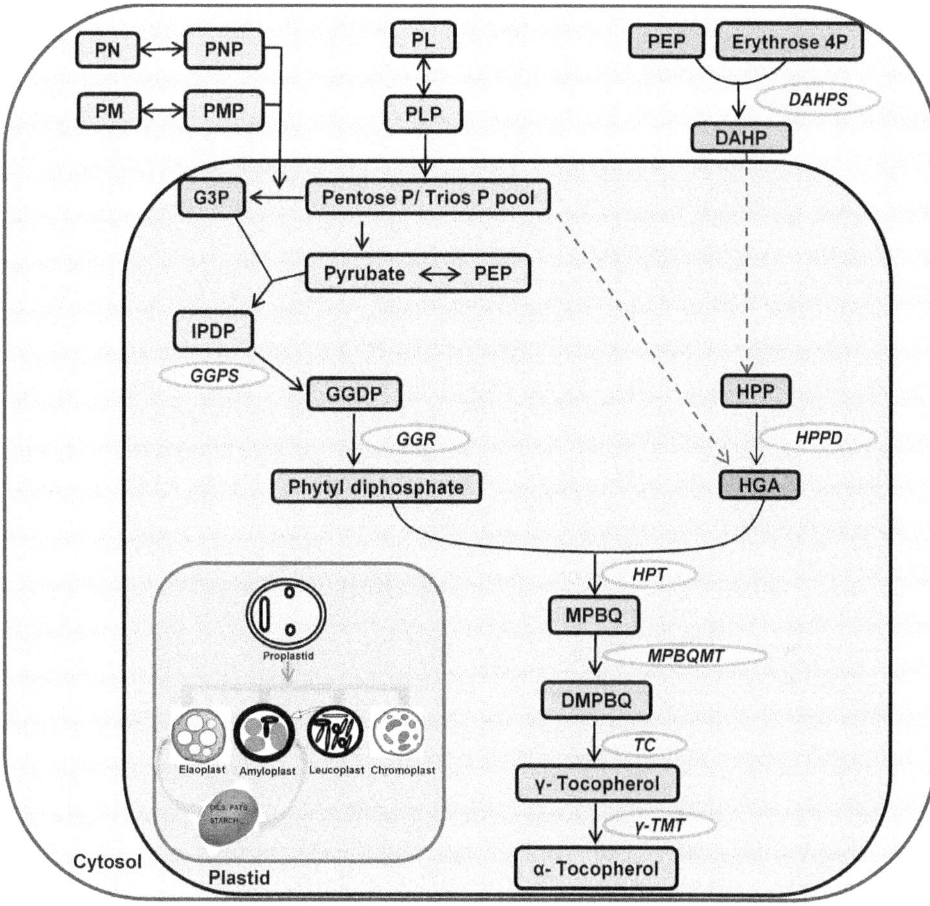

Figure 15.3 Biosynthetic pathways involved in tocopherols biosynthesis at gene level in plants and proposed interrelation with vitamin B6. Biosynthetic pathways forming the tocopherol precursors homogentisic acid (HGA) in the cytosol is derived from MEP pathway which ultimately comes from Pentose P/ Trios P pool. The phytyl diphosphate obtained from 1-deoxy-D-xylulose-5phosphate (DXP) pathway by reduction of GGDP which leads to the synthesis of the different vitamers of vitamin E (α-, β-, γ-, and δ-forms of tocopherols and tocotrienols).The biosynthesis of vitamin E is also facilitated by vitamers of the vitamin B6 pathway: PNP (pyridoxin 5-phosphate), PMP (pyridoxamine 5-phosphate) and (PLP) pyridoxal 5-phosphate that give rise to G3P (glyceraldehyde 3-phosphate) that ultimately increases Phytyl diphosphate.

promoter (CaMV35S). The *Agrobacterium* mediated genetic engineering was done with two independent transformation events in potato. The tocopherols (Vit-E) levels in the transgenic plants were determined by HPLC. The HPLC analysis of tocopherols (Vit-E) in the potato tubers revealed that over-expression of At-HPPD resulted in a maximum 266% increase in α-tocopherol, and overexpression of At-HPT resulted in over 106% increase of in α-tocopherol in potato tubers. However, tubers from transgenic plants still accumulated approximately 10- and 100-fold lesserα-tocopherol than leaves or real potato seeds, respectively. The results showed that physiological and regulatory constrictions may be limiting factors for the accumulation of tocopherols in the potato tubers.

Vitamin B (Vit-B) is also a vital metabolite needed for living organisms as a cofactor in several metabolic biochemical reactions and also acknowledged as an antioxidant molecule. It has been well-known that the microorganisms and plants can synthesize the B6 *de novo*, therefore, all animals, including humans, must obtain it from the plant dietary resources. However, the bioavailability of the vitamin in the edible portions of the commonly consumed plants is insufficient to meet the daily recommended doses. Genetic-engineering techniques have proven successful in increasing the vitamin B6 content in the model plants. The development of transgenic potato (*S. tuberosum* L. *cv.* Kufri chipsona) overexpressing key vitamin B6 pathway gene, the *PDXII* isolated from *A. thaliana* under the control of *CaMV 35S* constitutive promoter (Bagri et al. 2018). Transgenic tubers exhibited considerably enhanced B6 accumulation (up to 107–150%) in comparison to the untransformed control potato. The B6 increase in the transgenic potato also led to the capability to survive under abiotic stresses. The Figure 15.3 shown above represented the interdependence of the Vit-E and Vit-B biosynthesis. Our group is engaged in exploring the interdependent pathway gene of Vit-E and Vit-B synthesis for genetic engineering of potato so that a common pathway gene could enhance both vitamins in the potato by imposing minimum gene load.

Another essential component in human nutrition with antioxidant properties is the vitamin C (L-ascorbic acid; Vit-C). The Vit-C has been recognized as an important factor for human health which help in immuno-boosting, improved cardiovascular functions, prevents diseases linked to the connective tissue, and is also essential for the iron assimilation in the metabolism (Davey et al. 2000; Hallber et al. 1989). Besides, the ascorbate is a cofactor for several enzymes and imitates the expression of genes involved in plant protection and hormone signaling pathways (Pastori et al. 2003). During the process of evolution, humans have lost the capability to produce Vit-C by a transmutation in gene encoding the L-gulono-g-lactone oxidase enzyme liable for Vit-C biosynthesis (Davey et al. 2000). Therefore, diet is only source for the Vit-C in human. A proposed pathway for ascorbic acid biosynthesis in plants in shown in Figure 15.4. At first, the model plant *A. thaliana* was genetically engineered to show the proof-of-concept for enhanced Vit-C content in the plants (Agius et al. 2003). The research group isolated a major pathway gene, the *D-galacturonic acid reductase (galUR)* from strawberry and cloned in binary vector under the constitutive promoter. The overexpression of the gene in *Arabidopsis* resulted in a 2–3-fold increase in ascorbic acid in comparison to the untransformed control plans. The result proved that the transgenic crops, including the potato, can also be developed with enhanced Vit-C content. Thereafter, Hemavathi et al. (2009, 2011) reported the isolation and cloning of strawberry *D-galacturonic acid reductase (GalUR)* and mouse *L-gulono-c-lactone oxidase (GLOase)* in two cloning vectors driven by the constitutive promoter. The independent transformation event of the potato leads to development of transgenic potato with enhanced Vit-C content. The overexpression of *GalUR* and *GLOase* resulted respectively in 1.6 and 3-fold increase in Vit-C content in transgenic potato. Besides, the transgenic potato also showed tolerance against various abiotic stresses as compared to untransformed control plants. In a recent study, similar findings were reported by the overexpression of the *D-galacturonic acid reductase* (*GalUR*) gene driven by tissue-specific patatin promoter in potato which resulted in an enhanced level of cellular L-ascorbate (L-AsA) in the tuber tissues (Chung et al. 2019).

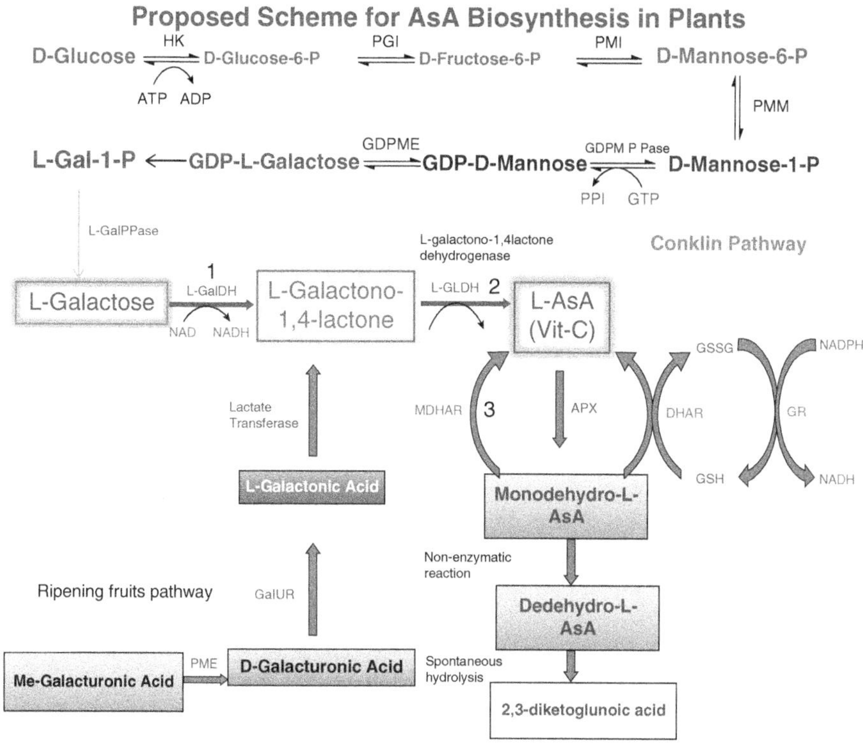

Figure 15.4 schematic representation of the ascorbic acid (Vit-C) biosynthesis and its recycling pathways scheme in plants.

The recycling of ascorbic acid is used in the cellular ROS scavenging process as a prosthetic group of various enzymes. The alternative approach for synthesis of ascorbic acid in plants is recycling of the used ascorbate lying in the cells. For recycling, the first step is the generation of the monodehydroascorbate (MDHA) radical upon oxidation of ascorbic acid, which readily convert back into ascorbic acid by the enzyme monodehydroascorbate reductase (MDHAR), which further oxidized non-enzymatically forming dehydroascorbate (DHA). The DHA can then undergo irreversible hydrolysis to synthesize the ascorbic acid (AsA) by the enzyme dehydroascorbate reductase (DHAR) with the help of a reducing organic agent called glutathione (GSH) (Smirnoff et al. 2001; Washko et al. 1992; Wheeler et al. 1998). Similarly, researchers showed that the overexpression of cytosolic DHAR in potato also significantly increased the activity of DHAR resulting in increased level of AsA (Vit-C) content in leaves (Chen et al. 2003; Qin et al. 2011). Young et al. (2008) reported development of transgenic potato by overexpressing *DHAR* gene driven by the either constitutive *CaMV35S* promoter and a potato tuber-specific promoter, the *patatin*. The ascorbic acid content in the transgenic tubers increased up to 1.3-fold as compared with that of untransformed control plants. However, the transgenic leaves showed less accumulation of AsA than the tubers when the tissue-specific promoter patatin was used. Besides, the potato transgenic with the construct *CaMV35*S promoter showed Vit-C accumulations in both tubers and the leaves.

15.7 Metabolic Engineering of Potato for Enhanced Mineral Content

The requirement of various minerals in human nutrition is vital for maintaining better health (Welch 2002). The Iron (Fe) and zinc (Zn) have been recognized as a nutritionally important mineral in which deficiency may lead to disturbed immune function and other ailments of health conditions (Zimmermann and Hurrell 2002). The iron deficiency alone has been reported to affect one third of the world's population and probably 800 000 deaths worldwide each year (Masuda et al. 2012). Potato has been considered as an important source of the minerals such as Fe, Zn, potassium (K) and phosphorous (p), in the form of phytate ions, therefore, the genetic engineering for mineral biofortification of potato has also been reported in literature (White and Broadley 2005). All the plants acquire these minerals from soil as a nutrient. As a result, engineering of plant mineral content is quite different from modifications of compounds like vitamins that the plant itself synthesizes. Genetic engineering has been focused on alteration of the gene involved in the mineral absorption and translocation from the soil. Though most minerals are transported through xylem, some minerals are transported by phloem (Grusak 2002). The xylem moves minerals to those tissues/organs with the greatest water loss; mostly to leaves. In contrast, the phloem that preferentially transports compounds from source (photosynthesis) organs to sinks (storage) such as seeds, fruits, and tubers. Thus, depending on the method of transport and utilization of the plant mineral content, genetic-engineering technique is also focused on the modification of the plant's normal absorption and transport system.

There is scanty report on the development of genetically engineered potato with enhanced accumulation of iron content. Park et al. (2005) reported the isolation of the *Arabidopsis sCAX* (Cationic Exchanger 1) and H^+/Ca^{2+} transporter genes and cloned in a vector under the control of the tissue-specific promoter. The overexpression of the gene in potato resulted in up to threefold more calcium content compared to the untransformed control potato plants. Similarly, a report on development of transgenic potato overexpressing a chimeric, N-terminus truncated *Arabidopsis* cation transporter (*CAX2B*) that contains a domain from *CAX1* for increased substrate specificity was also published (Kim et al. 2006). The transgenic lines showed 50–65% improved tuber calcium content compared to the control plants without any adverse effects on plant growth or development. This result opens the possibility of development of transgenic potato lines to the increase of other minerals content in tuber tissues.

15.8 Pathway Engineering for the Functional Secondary Metabolites

Potato steroidal glycol-alkaloids (SGAs) are toxic secondary metabolites of which total content in tubers must be regulated. The SGAs are biosynthesized by the sterol branch of the mevalonic acid/isoprenoid pathway (Heftmann 1983; Johnson et al. 1963). The SGA metabolites provide important components for plant defense. In case of potato, high content of SGA is reported to be present in the leaves rather than in tubers. Higher plants produce many secondary metabolites which are also commercially important. Genetic engineering opens a new route to increase the content of secondary metabolites in the easily cultivatable host plant (Broun and Somerville 2001; Della 2001; Moeh et al. 1997; Stapleton et al. 1991; Verpoorte et al. 1999; Zimowski 1991). Recently, Idit et al. (2012) developed transgenic potato with gene construct containing the coding region of either *HMG1* (3-Hydroxy-3-methylglutaryl CoA reductase-1) or *SQS1* (squalene synthase 1), which were reported to be a high *SGA* (steroidal glycoalkaloids) producer. The *SGA* levels in transgenic were greater than untransformed controls as determined by HPLC. This proved the correlation between SGA levels and transcript coding for *HMG-CoA reductase 1* (HMG1) and *squalene synthase 1* (SQS1) and also the key enzyme for the biosynthesis of the SGA in potato. Identification and modifications of the glycoalkaloid (GA) biosynthetic genes has also enabled transgenic approaches to increase or decrease GA content in potato. A steroidal *glycosyl transferases* gene was also overexpressed in transgenic potato that also altered the cellular GAs content in potato (McCue et al. 2005).

A report was also published for higher accumulation of 4HB (4-Hydroxybenzoic acid) glucosides as new, artificial secondary metabolites in the transgenic potato, by overexpression of the bacterial gene *ubiC* which encodes chorismate pyruvate-lyase (CPL) that catalyzes the conversion of chorismate to 4-hydroxybenzoate (4HB). The *ubi*C gene was cloned in a binary vector driven by *Ocs* constitutive promotor and Agrobacterium-mediated transformation of potato was done. Biochemical analysis of the transgenic potato revealed an increase in the 4HB glucoside content in the leaves of potato shoots rather than the tubers (Köhle et al. 2003). Moreover, another report published on the development of transgenic potato lines by overexpression of flavonoid biosynthetic pathway gene (Lukaszewicz et al. 2004). For this, a vector was designed carrying multigene construct and several transgenic plants were generated with overexpressing key biosynthesis pathway enzymes. The single-gene overexpression or simultaneous expression of genes encoding *chalcone synthase* (CHS), *chalcone isomerase* (CHI), and *dihydroflavonol reductase* (DFR) resulted in a significant increase of the phenolic acids and anthocyanins. The increase in phenolic compounds synthesis is accompanied by decrease in starch and glucose levels in transgenic plants. The flavonoids-enriched plants also showed improved antioxidant capacity.

15.9 Future Prospective and Conclusions

Biotechnology-based tools are now widely used to enhance and expand the traditional remit of potato in food production. By modifying its functionality, the capacity of the potato to produce, for example, therapeutic or industrial compounds is now a reality, and its ability to resist disease can also be radically improved. Recent advances in the field of structural and functional potato genomics and the ability to integrate genes of interest into the potato genome have revolutionized potato research. Efforts to maximize the nutritional potential of potatoes are in their very early stages. Allied with emerging high-throughput techniques and the excellent germplasm resources currently available, the sum of large-scale genomics-based potato research should have a radical impact on our ability to link genes with traits in potato over the next decade.

References

Agius, F., Gonzales-Lamothe, R., Caballero, J. et al. (2003). Engineering increased vitamin C levels in plants by over-expression of a D-galacturonic acid reductase. *Nat. Biotechnol.* 21: 177–181.

Andersson, M., Turesson, H., Arrivault, S. et al. (2018). Inhibition of plastid PPase and NTT leads to major changes in starch and tuber formation in potato. *J. Exp. Bot.* 69 (8): 1913–1924.

Anna, P., Anna, S., Jadwiga, B., and Jan, S. (2001). Increase in lipid content in potato tubers modified by 14-3-3 gene overexpression. *J. Agric. Food Chem.* 49: 3638–3643.

Arnqvist, L., Dutta, P.C., Jonsson, L., and Sitbon, F. (2003). Reduction of cholesterol and glycoalkaloid levels in transgenic potato plants by overexpression of a type 1 sterol methyl transferase cDNA. *Plant Physiol.* 131 (4): 1792–1799.

Bagri, D.S., Upadhyaya, D.C., Kumar, A., and Upadhyaya, C.P. (2018). Overexpression of PDX-II gene in potato (Solanum tuberosum L.) leads to the enhanced accumulation of vitamin B6 in tuber tissues and tolerance to abiotic stresses. *Plant Sci.* 272: 267–275.

Beaujean, A., Sangwan, R.S., Lecardonnel, A., and Sangwan-Norreel, B.S. (1998). Agrobacterium-mediated transformation of three economically important potato cultivars using sliced internodal explants: an efficient protocol of transformation. *J. Exp. Bot.* 49: 1589–1595.

Benmoussa, M., Vezina, L.P., Page, M. et al. (2004). Potato flour viscosity improvement is associated with the expression of a wheat LMW-glutenin gene. *Biotechnol. Bioeng.* 87: 495–500.

Blennow, A., Bay-Smidt, A.M., Leonhardt, P. et al. (2003). Starch paste stickiness is a relevant native starch selection criterion for wet-end paper manufacturing. *Starch* 55: 381–389.

Breithaupt, D.E. and Bamedi, A. (2002). Carotenoids and carotenoid esters in potatoes (*Solanum tuberosum L.*): new insights into an ancient vegetable. *J. Agric. Food Chem.* 50 (24): 7175–7181.

Broun, P. and Somerville, C. (2001). Progress in plant metabolic engineering. *PNAS* 98: 8925–8927.

Bulley, S., Wright, M., Rommans, C. et al. (2012). Enhancing ascorbate in fruits and tubers through over-expression of the L-galactose pathway gene GDP-L-galactose phosphorylase. *Plant Biotechnol. J.* 10: 390–397.

Burton, W.G. (1989). The Potato. New York, USA: Longmann.

Bustos, R., Fahy, B., Hylton, C.M. et al. (2004). Starch granule initiation is controlled by a heteromultimeric isoamylase in potato tubers. *PNAS* 101: 2215–2220.

Campbell, R., Morris, W.L., Mortimer, C.L. et al. (2015). Optimising ketocarotenoid production in potato tubers: effect of genetic background, transgene combinations and environment. *Plant Sci, Int J. Exp Plant Biol* 234: 24–37.

Chakraborty, S., Chakraborty, N., Agrawal, L. et al. (2010). Next-generation protein-rich potato expressing the seed protein gene *AmA1* is a result of proteome rebalancing in transgenic tuber. *Proc. Natl. Acad. Sci. U. S. A.* 107 (41): 17533–17538.

Chandra, P. and Pental, D. (2003). Regeneration and genetic transformation of grain legumes: An overview. *Curr. Sci.* 84: 381–387.

Chen, Z., Young, T.E., Ling, J. et al. (2003). Increasing vitamin C content of plants through enhanced ascorbate recycling. *PNAS* 100 (6): 3525–3530.

Chong, D.K., Roberts, W., Arakawa, T. et al. (1997). Expression of the human milk protein beta-casein in transgenic potato plants. *Transgenic Res.* 6 (4): 289–296.

Chung, I.M., Venkidasamy, B., Upadhyaya, C.P. et al. (2019). Alleviation of *Phytophthora infestans* mediated necrotic stress in the transgenic potato (*Solanum tuberosum* L.) with enhanced ascorbic acid accumulation. *Plants (Basel)* 23 (8): 10.

Clasen, B.M., Stoddard, T.J., Luo, S. et al. (2015). Improving cold storage and processing traits in potato through targeted gene knockout. *Plant Biotechnol. J.* 13: 1–8.

Crowell, J., Elizabeth, F., Mitchell, M. et al. (2008). Accumulation of vitamin E in potato (*Solanum tuberosum*) tubers. *Transgenic Res.* 17 (2): 205–217.

Dale, P.J., Irwin, J.A., and Scheffer, J.A. (1993). Experimental and commercial release of transgenic crop plants. *Plant Breed.* 11: 11–22.

Davey, M.W., Van, M.M., Sanmatin, M. et al. (2000). Plant L-ascorbic acid: chemistry, function, metabolism, bio-availability and effects of processing. *J. Sci. Food Agric.* 80: 825–860.

De Block, M. (1988). Genotype-independent leaf disc transformation of potato (Solanum tuberosum) using Agrobacterium tumefaciens. *Theor. Appl. Genet.* 76: 767–774.

Della, P.D. (2001). Plant metabolic engineering. *Plant Physiol.* 125: 160–163.

Dita, E., Rispail, N., Prats, E. et al. (2006). Biotechnology approaches to overcome biotic and abiotic stress constraints in legumes. *Euphytica* 147: 1–24.

Dong, T., Cao, Y., Jiang, C.Z. et al. (2020). Cysteine protease inhibitors reduce enzymatic browning of potato by lowering the accumulation of free amino acids. *J. Agric. Food Chem.* 68 (8): 2467–2476.

Dunwell, J.M., Khuri, S., and Gane, P.J. (2000). Microbial relatives of the seed storage proteins of higher plants: conservation of structure and diversification of function during evolution of the cupin superfamily. *Microbiol. Mol. Biol. Rev.* 64: 153–117.

Duroy, A., Aymeric, G., and Roshani, S. (2009). Nutritional value of potatoes: vitamin, phytonutrient, and mineral content. In: Advances in Potato Chemistry and Technology (eds. J. Singh and L. Kaur), 395–424. Academic Press.

Edurne, B.F., Francisco, J.M., Manuel, M. et al. (2009). Enhancing sucrose synthase activity in transgenic potato (*Solanum tuberosum* L.) tubers results in increased levels of starch, ADP glucose and UDP glucose and total yield. *Plant Cell Physiol.* 50 (9): 1651–1662.

Emes, M.J. and Neuhaus, H.E. (1997). Metabolism and transport in non-photosynthetic plastids. *J. Exp. Bot.* 48: 1995–2005.

Ensminger, A.H., Ensminger, M.E., Kondale, J.E. et al. (1983). Robson, Foods & Nutrition Encyclopedia. Clovis, California: Pegus Press.

Ferreira, S.J., Senning, M., Fischer-Stettler, M. et al. (2017). Simultaneous silencing of isoamylases ISA1, ISA2 and ISA3 by multi-target RNAi in potato tubers leads to decreased starch content and an early sprouting phenotype. *PLoS One* 12 (7): e0181444.

Feustel, I.C. (1987). Miscellaneous products from potatoes. In: potato processing, 4e (eds. W.F. Talburt and O. Smith), 727–746. New York: Van Nostand.

Fortin (1996). Francois, Editorial Director, the Visual Foods Encyclopedia. New York: Macmillan.

Gao, H., Wu, X., Zorrilla, C. et al. (2020). Fractionating of calcium in tuber and leaf tissues explains the calcium deficiency symptoms in potato plant overexpressing CAX1. *Front. Plant Sci.* 10 (31): 1793–1800.

Gianfranco, D., Raffaela, T., Ralf, W. et al. (2006). Metabolic engineering of potato tuber carotenoids through tuber-specific silencing of lycopene epsilon cyclase. *BMC Plant Biol.* 6: 13.

Gianfranco, D., Salim, A., Raffaela, T. et al. (2010). Transcriptional-metabolic networks in beta carotene enriched potato tubers the long and widening road to the golden phenotype. *Plant Physiol.* 154 (2): 899–912.

Goo, Y.M., Kim, T.W., Lee, M.K., and Lee, S.W. (2013). Accumulation of PrLeg, a Perilla legumin protein in potato tuber results in enhanced level of Sulphur-containing amino acids. *C. R. Biol.* 336 (9): 433–439.

Goo, Y.M., Han, E.H., Jeong, J.C. et al. (2015). Overexpression of the sweet potato IbOr gene results in the increased accumulation of carotenoid and confers tolerance to environmental stresses in transgenic potato. *C R Biol* 338 (1): 12–20.

Grusak, M.A. (2002). Genetic assisted plant improvement to benefit human nutrition and health benefits. *Curr. Opin. Biotechnol.* 13: 508–511.

Hallber, L., Brune, M., and Rossander, L. (1989). Iron absorption in man: ascorbic acid and dose-dependent inhibition by phytate. *Am. J. Clin. Nutr.* 49 (1): 140–144.

Harris, N.N., Javellana, J., Davies, K.M. et al. (2012). Betalain production is possible in anthocyanin-producing plant species given the presence of DOPA-dioxygenase and L-DOPA. *BMC Plant Biol.* 12: 34.

Heftmann, E. (1983). Biogenesis of steroids in Solanaceae. *Phytochemistrv* 22 (9): 1843–1860.

Hemavathi, H., Upadhyaya, C.P., Young, K.E. et al. (2009). Over-expression of strawberry D-galacturonic acid reductase in potato leads to accumulation of vitamin C with enhanced abiotic stress tolerance. *Plant Sci.* 177: 659–667.

Hemavathi, H., Upadhyaya, C.P., Nookaraju, A. et al. (2011). Biochemical analysis of enhanced tolerance in transgenic potato plants overexpressing D-galacturonic acid reductase gene in response to various abiotic stresses. *Mol. Breed.* 28 (1): 105–115.

Huang, T., Joshi, V., and Jander, G. (2014). The catabolic enzyme methionine gamma-lyase limits methionine accumulation in potato tubers. *Plant Biotechnol. J.* 12 (7): 883–893.

Idit, G., Muddarangappa, T., Edna, F. et al. (2012). Induction of potato steroidal glycoalkaloid biosynthetic pathway by overexpression of cDNA encoding primary metabolism HMG-CoA reductase and squalene synthase. *Planta* 235: 1341–1353.

Ishimaru, K., Hirotsu, N., Kashiwagi, T. et al. (2008). Overexpression of maize SPS gene improves yield character of potato under field condition. *Plant Prot. Sci.* 1: 104–107.

Jaynes, J.M., Yang, M.S., Espinoz, N., and Dodds, J.H. (1986). Plant protein improvement by genetic engineering- use of synthetic genes. *Trends Biotechnol.* 4: 314–320.

Ji, Q., Oomen, R.J.F.J., Vincken, J.P. et al. (2004). Reduction of starch granule size by expression of an engineered tandem starch-binding domain in potato plants. *Plant Biotechnol. J.* 2 (3): 251–260.

Johnson, D.F., Bennett, R.D., and Heftmann, E. (1963). Cholesterol in higher plants. *Science* 140: 198–199.

Jube, S. and Borthakur, D. (2006). Recent advances in food biotechnology research. In: Food Biochemistry and Food Bioprocessing (eds. Y.H. Hui, W.-K. Nip and B.K. Simpson), 35–70. Oxford, U.K: Blackwell Publishing.

Kim, C.K., Han, J.S., Lee, H.S. et al. (2006). Expression of an *Arabidopsis CAX2* variant in potato tubers increases calcium levels with no accumulation of manganese. *Plant Cell Rep.* 25: 1226–1232.

Kim, S.H., Ahn, Y.O., Ahn, M.J. et al. (2013). Cloning and characterization of an Orange gene that increases carotenoid accumulation and salt stress tolerance in transgenic sweet potato cultures. *Plant Physiol. Biochem.* 70: 445–454.

Klaus, D., Ohlrogge, J.B., Neuhaus, H.E., and Dormann, P. (2004). Increased fatty acid production in potato by engineering of acetyl-CoA carboxylase. *Planta* 219: 389–396.

Köhle, A., Sommer, S., Li, S.M. et al. (2003). Secondary metabolites in transgenic tobacco and potato, high accumulation of 4-hydroxybenzoic acid glucosides results from high expression of the bacterial gene ubiC. *Mol. Breed.* 11: 15–24.

Kostyn, K., Szatkowski, M., Kulma, A. et al. (2013). Transgenic potato plants with overexpression of dihydroflavonol reductase can serve as efficient nutrition sources. *J. Agric. Food Chem.* 61 (27): 6743–6753.

Kumar, P. and Jander, G. (2017). Concurrent overexpression of Arabidopsis thaliana cystathionine γ-synthase and silencing of endogenous methionine γ-Lyase enhance tuber methionine content in Solanum tuberosum. *J. Agric. Food Chem.* 65 (13): 2737–2742.

Kumar, A., Fogelman, E., Weissberg, M. et al. (2017). Lanosterol synthase-like is involved with differential accumulation of steroidal glycoalkaloids in potato. *Planta* 246 (6): 1189–1202.

Laurence, J.M.D., Wayne, L.M., Peter, E.H. et al. (2005). Metabolic engineering of high carotenoid potato tubers containing enhanced levels of b-carotene and lutein. *J. Exp. Bot.* 56 (409): 81–89.

Li, L., Yang, Y., Xu, Q. et al. (2012). The or gene enhances carotenoid accumulation and stability during post-harvest storage of potato tubers. *Mol. Plant* 5 (2): 339–352.

Li, F., Bian, C.S., Xu, J.F. et al. (2015). Cloning and functional characterization of SAD genes in potato. *PLoS One* 10 (3): 122036–122047.

Li, F., Bian, C.S., Xu, J.F. et al. (2015). Cloning and functional characterization of SAD genes in potato. *PLoS One* 10 (3): 122–136.

Liu, Y.W., Han, C.H., and Lee, M.H. (2003). Patatin, the tuber storage protein of potato (*Solanum tuberosum L.*), exhibits antioxidant activity in vitro. *J. Agric. Food Chem.* 51 (15): 4389–4393.

Liu, Q., Guo, Q., Akbar, S. et al. (2017). Genetic enhancement of oil content in potato tuber (*Solanum tuberosum* L.) through an integrated metabolic engineering strategy. *Plant Biotechnol. J.* 15 (1): 56–67.

Lngrid, M., Vander, M., Michel, J.M. et al. (1994). Smeekens fructan as a new carbohydrate sink in transgenic potato plants. *Plant Cell* 6: 561–570.

Lorberth, R., Rite, G., Willmizer, L., and Kossmann, J. (1998). Inhibition of a starch-granule-bound protein leads to modified starch and repression of cold sweetening. *Nat. Biotechnol.* 16: 473–477.

Lu, S., Eck, V.J., Zhou, X. et al. (2006). Cauliflower *Or* gene encodes a DnaJ cysteine-rich domain-containing protein that mediates high levels of β-carotene accumulation. *Plant Cell* 18: 3594–3605.

Lukaszewicz, M., Matysika, K., Skala, J. et al. (2004). Antioxidant capacity manipulation in transgenic potato tuber by changes in phenolic compounds content. *J. Agric. Food Chem.* 52: 1526–1533.

Maimann, S., Wagner, C., Kreft, O. et al. (2000). Transgenic potato plants reveal the indispensable role of cystathionine beta-lyase in plant growth and development. *Plant J.* 23: 747–758.

Mary, E.C., Stan, K., and Danielle, J.D. (2009). Potatoes and human health. *Crit. Rev. Food Sci. Nutr.* 49: 823–840.

Masuda, H., Ishimaru, Y., Aung, M.S. et al. (2012). Iron biofortification in rice by the introduction of multiple genes involved in iron nutrition. *Sci. Rep.* 2: 543.

Matthews, B.F. and Hughes, C.A. (1993). Nutritional improvement of the aspartate family of amino acids in edible crop plants. *Amino Acids* 4: 21–34.

McCue, K.F., Shepherd, L.V.T., Allen, P.V. et al. (2005). Metabolic compensation of steroidal glycoalkaloid biosynthesis in transgenic potato tubers: using reverse genetics to confirm the in vivo enzyme function of a steroidal alkaloid galactosyltransferase. *Plant Sci.* 168: 267–273.

McCue, K.F., Allen, P.V., Shepherd, L.V.T. et al. (2007). Potato glycosterol-rhamnosyltransferase, the terminal step in triocside-chain biosynthesis. *Phytochemistry* 68: 327–334.

Mitten, D.H., Horn, M., Burrel, M.M., and Blundy, K.S. (1990). Strategies for potato transformation and regeneration. In: The Molecular and Cellular Biology of the Potato (eds. M.E. Vayda and W.D. Park), 181–191. Wallingford: CAB International.

Moeh, C.P., Allen, P.V., Friedman, M., and Belknap, W.R. (1997). Cloning and expression of solanidine UDP-glucose glucosyl transferase from potato. *Plant Journal: For Cell and Molecular Biology* 11: 227–236.

Mortimer, C.L., Misawa, N., Ducreux, L. et al. (2015). Product stability and sequestration mechanisms in Solanum tuberosum engineered to biosynthesize high value ketocarotenoids. *Plant Biotechnol. J.* 14: 140–152.

Navarre, D.A., Shakya, R., and Hellmann, H. (2016). Vitamins, phytonutrients, and minerals in potato. In: Advances in Potato Chemistry and Technology, 2e (eds. J. Singh and L. Kaur), 117–166. Academic Press U.K.

Navratil, O., Vojtechova, M., Fischer, L. et al. (1998). Characterization of transgenic potato plants with an additional bacterial gene coding for phosphofructokinase. *Chem. Pap.* 52: 598–598.

Nazarian-Firouzabadi, F., Trindade, L.M., and Visser, R.G.F. (2012). Production of small starch granules by expression of a tandem-repeat of a family 20 starch-binding domain (SBD3-SBD5) in an amylose-free potato genetic background. *Funct. Plant Biol.* 39: 146–155.

Neuhaus, H.E. and Emes, M.J. (2000). Nonphotosynthetic metaboism in plastids. *Annual rev of Plant Phisol and Plant Mol Biol.* 51: 111–140.

Nikiforova, V., Kempa, S., Zeh, M. et al. (2002). Engineering of cysteine and methionine biosynthesis in potato. *Amino Acids* 22: 259–278.

Nölke, G., Houdelet, M., Kreuzaler, F. et al. (2014). The expression of a recombinant glycolate dehydrogenase polyprotein in potato (*Solanum tuberosum*) plastids strongly enhances photosynthesis and tuber yield. *Plant Biotechnol. J.* 12 (6): 734–742.

Ortiz-Medina, E. (2007).Potato tuber protein and its manipulation by chimeral disassembly using specific tissue explantation for somatic embryogenesis. Ph.D. Dissertation. McGill University.

Park, S., Cheng, N.H., Pittman, J.K. et al. (2005). Increased calcium levels and prolonged shelf life in tomatoes expressing *Arabidopsis* H+/Ca2+ transporters. *Plant Physiol.* 139: 1194–1206.

Park, S.C., Kim, S.H., Park, S. et al. (2015). Enhanced accumulation of carotenoids in sweet potato plants overexpressing IbOr-Ins gene in purple-fleshed sweet potato cultivar. *Plant Physiol. Biochem.* 86: 82–90.

Park, H.J., Lee, A., Lee, S.S. et al. (2017). Overexpression of Golgi protein CYP21-4s improves crop productivity in potato and rice by increasing the abundance of Mannosidic glycoproteins. *Front. Plant Sci.* 20 (8): 1250–1261.

Parr, A., Mellon, F., Colquhoun, I., and Davies, H. (2005). Dihydro caffeoyl polyamines (Kukoamine and allies) in potato (*Solanum tuberosum*) tubers detected during metabolite profiling. *J. Agric. Food Chem.* 53: 5461–5466.

Pasare, S., Wright, K., Campbell, R. et al. (2013). The sub-cellular localisation of the potato (*Solanum tuberosum* L.) carotenoid biosynthetic enzymes, CrtRb2 and PSY2. *Protoplasma* 250 (6): 1381–1392.

Pastori, G.M., Kiddle, G., Antoniw, J. et al. (2003). Leaf vitamin C content modulates plant defence transcripts and regulate genes that control development through hormone signalling. *Plant Cell* 15: 939–951.

Qin, A., Shi, Q., and Yu, X. (2011). Ascorbic acid contents in transgenic potato plants overexpressing two dehydroascorbate reductase genes. *Mol. Biol. Rep.* 38 (3): 1557–1566.

Reza, M.A., Irina, V.G.P., Vasiliy, P.P. et al. (2007). Acyl-lipid Δ12-desaturase of the cyanobacterium increases the unsaturation degree in transgenic potato (Solanum tuberousum L). *Biologia* 53 (2): 4–7.

Rowan, S.M.K., Muttucumaru, N., Paul, M.J. et al. (2006). Production of high-starch, low-glucose potatoes through over-expression of the metabolic regulator SnRK1. *Plant Biotechnol. J.* 4 (4): 409–418.

Sheerman, S. and Bevan, M.W. (1988). A rapid transformation method for *Solanum tuberosum* using binary *Agrobacterium tumefaciens* vectors. *Plant Cell Rep.* 7: 13–15.

Shepherd, L.V.T., Hackett, C.A., Alexander, C.J. et al. (2015). Modifying glycoalkaloid content in transgenic potato, metabolome impacts. *Food Chem.* 187: 437–443.

Smirnoff, N., Conklin, P., and Loewus, F. (2001). Biosynthesis of ascorbic acid: a renaissance. *Annu. Rev. Plant Physiol. Plant Mol. Biol.* 52: 437–467.

Stapleton, A., Allen, P.V., Friedman, M., and Belknap, W.R. (1991). Purification and characterization of solanidine glycosyltransferase from the potato (*Solanum tuberosum*). *J. Agric. Food Chem.* 39: 1187–1193.

Stiekema, W.J., Heidekamp, F., Louwerse, J.D. et al. (1988). Introduction of foreign genes into potato cultivars Bintje and Desiree using an *Agrobacterium tumefaciens* binary vector. *Plant Cell Rep.* 7: 47–50.

Sun, T., Lærke, H.N., Jørgensen, H., and Bach Knudsen, K.E. (2006). The effect of heat processing of different starch sources on the in vitro and in vivo digestibility in growing pigs. *Animal Feed Sci. and Tech.* 131: 66–85.

Tanja, A. and Gerhard, S. (2006). Ketocarotenoid formation in transgenic potato. *J. Exp. Bot.* 57 (14): 639–3645.

Tanja, G. and Gerhard, S. (2006). Ketocarotenoid formation in transgenic potato. *J. Exp. Bot.* 57 (14): 3639–3645.

Tavazza, R., Tavazza, M., Ordas, R.J. et al. (1988). Genetic transformation of potato (*Solanum tuberosum*): an efficient method to obtain transgenic plants. *Plant Sci.* 59: 175–181.

Trujillo, C., Rodriguez-Arango, E., Jaramillo, S. et al. (2001). One step transformation of two Andean potato cultivars (*Solanum tuberosum L.* subsp. *and igena*). *Plant Cell Rep.* 20: 637–641.

Tudela, J.A., Cantos, E., and Espin, J.C. (2002). Induction of antioxidant flavonol biosynthesis in fresh-cut potatoes, effect of domestic cooking. *J. Agric. Food Chem.* 21: 5925–5931.

Tunali, U.R., Wochniak, E.U., Czechowski, T. et al. (2003). De novo amino acid biosynthesis in potato tubers is regulated by sucrose level. *Plant Physiol.* 133: 683–692.

Verpoorte, R., Vander, H.R., Hoopen, H.J.G., and Memelink, J. (1999). Metabolic engineering of plant secondary metabolite pathways for the production of fine chemicals. *Biotechnol. Lett.* 21: 467–479.

Wang, H., Fan, W., Li, H. et al. (2013). Functional characterization of dihydroflavonol-4-reductase in anthocyanin biosynthesis of purple sweet potato underlies the direct evidence of anthocyanins function against abiotic stresses. *PLoS One* 11: 78484.

Wang, Y., Li, Y., and Zhang, H. (2017). A soluble starch synthase I gene, *IbSSI*, alters the content, composition, granule size and structure of starch in transgenic sweet potato. *Sci. Rep.* 7: 2315.

Washko, P.W., Welch, R.W., Dhariwal, K.R. et al. (1992). Ascorbic acid and dehydro ascorbic acid analyses in biological samples. *Anal. Biochem.* 204: 1–14.

Wasserman, L.A., Sergeev, A.I., Vasil'ev, V.G. et al. (2015). Thermodynamic and structural properties of tuber starches from transgenic potato plants grown in vitro and in vivo. *Carbohydr. Polym.* 10 (125): 214–223.

Welch, R.M. (2002). The impact of mineral nutrients in food crops on global human health. *Plant Soil* 247: 83–90.

Wenzler, H., Mignery, G., May, G., and Park, W. (1989). A rapid and efficient transformation method for the production of large numbers of transgenic potato plants. *Plant Sci.* 63: 79–85.

Werij, J.S., Furrer, H., Van Eck, H.J. et al. (2012). A limited set of starch related genes explain several interrelated traits in potato. *Euphytica* 186: 501–516.

Wheeler, G., Jones, M., and Smirnoff, N. (1998). The biosynthetic pathway of vitamin C in higher plants. *Nature* 393: 365–369.

White, P.J. and Broadley, M.R. (2005). Biofortifying crops with essential mineral elements. *Trends Plant Sci.* 10: 586–593.

Wiesenborn, D.P., Orr, P.H., Casper, H.H., and Tacke, B.K. (1994). Potato starch paste behaviour as related to some physical/chemical properties. *J. Food Sci.* 59: 644–648.

Wood and Rebecca (1988). The Whole Foods Encyclopedia. New York, NY: Prentice-Hall Press, PMID 15220.

Xu, X., Vanhercke, T., Shrestha, P. et al. (2019). Upregulated lipid biosynthesis at the expense of starch production in potato (*Solanum tuberosum*) vegetative tissues *via* simultaneous downregulation of *ADP-Glucose Pyrophosphorylase* and *Sugar Dependent1* expressions. *Front. Plant Sci.* 12 (10): 1444.

Xu, X.Y., Akbar, S., Shrestha, P. et al. (2020). A synergistic genetic engineering strategy induced triacylglycerol accumulation in potato (*Solanum tuberosum*) leaf. *Front Plant Sci, Mar* 6 (11): 215–222.

Yang, M.S., Espinoza, N.O., Nagplala, P.G. et al. (1989). Expression of a synthetic gene for improved protein quality in transformed potato plants. *Plant Sci.* 64: 99–111.

Yastreb, T.O., Kolupaev, Y.E., Lugovaya, A.A. et al. (2016). Content of Osmolytes and flavonoids under salt stress in *Arabidopsis thaliana* plants defective in jasmonate signaling. *Appl. Biochem. Microbiol.* 52: 210–215.

Young, M.G., Hyun, J.C., Tae, W.K. et al. (2008). Expressional characterization of dehydroascorbate reductase C-DNA in transgenic potato plants. *J. Plant. Biol.* 51 (1): 35–41.

Zagoskina, N.V., Priadekhina, E.V., Lapshin, P.V. et al. (2014). Morphophysiological and biochemical characteristics of potato plants with various expression rates of the Δ12 acyl-lipid desaturase gene. *Nauk Ser. Biol.* (2): 142–149.

Zhang, Y., Sun, F., Fettke, J. et al. (2014). Heterologous expression of AtPAP2 in transgenic potato influences carbon metabolism and tuber development. *FEBS Lett.* 588 (20): 3726–3731.

Zimmermann, M.B. and Hurrell, R.F. (2002). Improving iron, zinc and vitamin a nutrition through plant biotechnology. *Curr. Opin. Biotechnol.* 13: 142–145.

Zimowski, J. (1991). Occurrence of a glucosyltransferase specific for solanidine in potato plants. *Phyrochem* 30 (6): 1827–1831.

16

Genome Engineering Strategies for Quality Improvement in Tomato

Tian Wang[1], *Hongyan Zhang*[1], *and Hongliang Zhu*[2]

[1] *Key Laboratory of Food Nutrition and Safety of Shandong Normal University, College of Life Science, Shandong Normal University, Jinan, China*
[2] *College of Food Science and Nutritional Engineering, China Agricultural University, Beijing, China*

CHAPTER MENU

16.1 Introduction

Fruits are the main sources of vitamins, and minerals (Giovannoni et al. 2017). They have been cultivated for food and nutritional use throughout the course of human history (Ravichandra 2014). Because of global warming, the growing human population, and natural disasters, an increase of approximately 25–70% above current production levels is needed to meet 2050 food demand (Hunter et al. 2017). With improvement in living standards, people nowadays pay more attention to the quality of fruits. Thus, it is of great need to enhance the yield and quality of fruit crops. To achieve this goal, humans have adopted diverse strategies such as utilization of chemical fertilizers, and hybrid breeding techniques (Yamamoto et al. 2018). However, the overuse of chemical fertilizers leads to salt accumulation and chemical fertilizer production requires a great deal of energy (Yamamoto et al. 2018). Conventional breeding for fruit crop improvement is a time-consuming practice, accompanied by lack of environmental fitness and genetic diversity (Meyer and Purugganan 2013). Therefore, it is of great value to generate genetic resources with diverse and desirable characteristics (Xu et al. 2019).

Genome Engineering for Crop Improvement, First Edition. Edited by Santosh Kumar Upadhyay.

Designer nucleases allow the generation of genetic resources by introducing precisely-targeted double-strand breaks (DSBs) that trigger cellular DNA repair mechanisms. DSBs enable efficient gene modifications by either non-homologous end joining (NHEJ) or homology directed repair (HDR) with the introduction of insertions, deletions and substitutions at pre-determined sites (Ran et al. 2017). The major nuclease technologies include meganucleases (MNs), zinc-finger nucleases (ZFNs), transcription activator-like effector nucleases (TALENs), and clustered regulatory interspaced short palindromic repeats (CRISPR)/Cas systems (e.g. Cas9 and Cpf1) (Zhu et al. 2017). These genome-editing tools provide the possibility of performing robust genetic engineering in many plant species (Araki and Ishii 2015). For fruit crops, they have, so far, been successfully applied to tomato, strawberry, banana, grape, apple, watermelon, and kiwifruit (Wang et al. 2019). From a scientific standpoint, the genome-edited fruit crops are different from classically bred crops and conventional transgenic plants. Therefore, they might be evaluated according to the new traits rather than the technique used to create the varieties. In other words, the crops should not be classified as genetically modified organisms (GMOs) or under the strict GMO regulation (Hartung and Schiemann 2014).

Tomato (*Solanum lycopersicum*) originated from the Andean region of South America and belongs to the Solanaceae family (Zhu et al. 2018). The easily achieved mutant populations, simple diploid inheritance, short growing period, and easy transformation methodology make tomato an ideal model system for fruit biology and these resources provide a comprehensive platform that aids genome engineering in tomato (Reem and Van Eck 2019). In addition, tomato is also one of the important fruit crops from an economic perspective and makes a great contribution to human nutrition. Nowadays, current demands for tomato improvement are better quality such as longer shelf-life and more attractive colors, and the higher content of health-beneficial components such as vitamin C and lycopene (Vos et al. 2011).

In this chapter, we first compare the development of three engineered nucleases that are used for genome editing. Next, we focus on their applications in tomato quality enhancement, including environmental adaptation and fruit quality. Finally, we discuss the challenges and opportunities of genome engineering for the improvement of tomato.

16.2 Genome Editing Systems in Plants

Targeted genome modification was first achieved by MNs (Smith et al. 2006). However, one MNs can only recognize one DNA target site and it is difficult to modify MNs without influencing their catalytic activity (Voytas 2013; Mao et al. 2017). Thus, MNs have not been widely applied in plants. Its limitation has been overcome by more powerful nucleases, including ZFNs, TALENs and CRISPRs.

(a) ZFNs consist of DNA binding domain (DBD) linked with *Fok*I nuclease. Each zinc finger recognizes a 3-bp target sequence. Each zinc-finger (ZF) array binds to one DNA strand to cause double-strand break (DSB) Source: Modified from Carroll (2011). (b) TALENs also consist of DBD fused to *Fok*I nuclease. Each TALE module recognizes and binds one nucleotide. Paired TALENs in dimer form cut the DNA strands to introduce sticky DSBs. Source: Modified from Mahfouz (2014). (c) The CRISPR/Cas9 system consists of an sgRNA and the Cas9 endonuclease. The Cas9 protein is guided by an sgRNA to bind to genomic DNA through a 20-nt sequence and further cuts the DNA strands at a location 3-nt upstream of the PAM motif, forming a blunt-end DSB. Abbreviations: PAM, protospacer adjacent motif; sgRNA, single guide RNA; TALENs, TAL effector nucleases; ZFNs, zinc-finger nucleases. Source: Modified from Knott et al. (2018).

16.2.1 ZFNs

More than 30 years ago, scientists identified a repeat motif within a protein from *Xenopus oocytes* which was later called ZF (Miller et al. 1985). After artificial modification and fusion with *Fok*I endonuclease, this amazing protein is able to cut DNA at preferred site (Kim et al. 1996). In general, ZFNs consist of two functional domains: a DNA-binding domain of ZF and a DNA-cleaving domain *Fok*I (Carroll 2011) (Figure 16.1a). The Cys2-His2 ZF domain is the most frequently occurring DNA-binding domain in higher and lower eukaryotes (Persikov et al. 2015). Each ZF is composed of approximately 30 amino acids which fold into

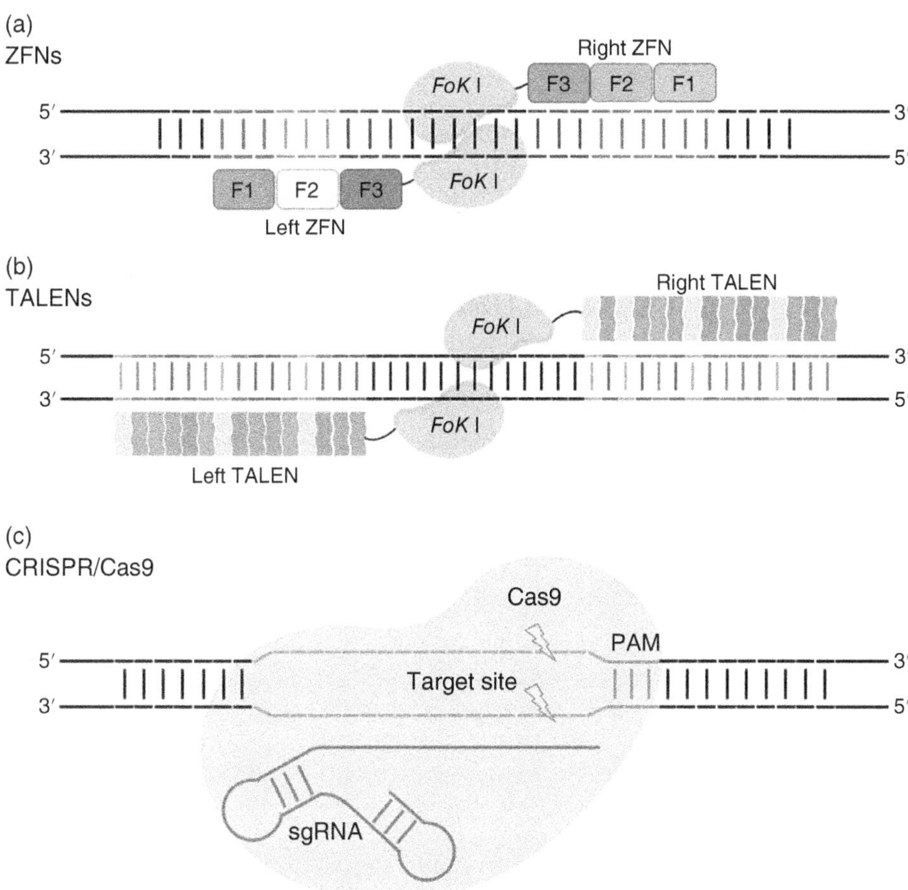

Figure 16.1 Schematic of the three nucleases used for genome engineering.

a conserved ββα configuration, stabilized by a single zinc ion (Saha et al. 2019). The ZF comprises three or four tandem arrays and each of them can recognize and bind to a specific 3-bp target sequence (Bhakta et al. 2013). *Fok*I restriction enzyme was isolated from *Flavobacterium okeanokoites* (Aggarwal et al. 1998). It links to the C-terminal of each ZF array to form ZFN. Two ZFNs usually function as a dimer to cleave the DNA sequence effectively, inducing DSB that stimulates the cellular DNA damage response pathways (Mc et al. 2007).

There are several methods to generate engineered ZFs: "modular assembly"(Mh et al. 2008), Compozer (Moore et al.2001), oligomerized pool engineering (OPEN) (Maeder et al. 2008), and context-dependent assembly (CoDA) (Thibodeau-Beganny et al. 2011). ZFNs have been employed to target plant genomes, including *Arabidopsis* (Reyon et al. 2010), tobacco (Townsend et al. 2009), and maize (Shukla et al. 2009). However, the application of ZFNs is seriously limited by its low efficiency and high cost.

16.2.2 TALENs

In 2007, German scientists discovered a transcription activator-like effector (TALE) in plant that delivered by a type III secretion system of *Xanthomonas*. TALE could mimic eukaryotic transcription factors to bind specific DNA sequences and activate the transcription of target genes (Kay et al. 2007). With deepening study of TALE, its DNA-binding mechanism was clearly illustrated and TALEN was regarded as a new promising tool for genome engineering (Boch et al. 2009; Moscou and Bogdanove 2009; Bedell et al. 2012). Similar to ZFNs, TALENs also consist of two components: TALE DBD with a *Fok*I catalytic domain linked to its C terminus (Mahfouz et al. 2014) (Figure 16.1b). TALEs are modular proteins that comprise N-terminal secretion and translocation signals, a central DBD, C-terminal nuclear localization signals (NLSs), and an acidic transcriptional activation domain (AD) (Gao et al. 2012). The central DBD is highly repetitive and typically contains 33–35 amino-acid sequence with divergent 12 and 13th positions which can recognize and bind to one specific nucleotide. These two positions, designated "Repeat Variable Di-residues" (RVDs), are highly variable and show a strong correlation with specific nucleotide recognition (Mahfouz et al. 2012). The recognition preferences of RVDs have been well established through experimental and computational methods: RVDs NI, NN, HD, and NG can recognize the DNA bases A, G, C, and T, respectively (Deng et al. 2012). Like ZFNs, the *Fok*I nuclease of TALENs need to dimerize to efficiently cleave desired DNA sequence (Catto et al. 2006). In practical application, TALENs have advantages over ZFNs since they are easily designed to recognize any DNA sequence (Zhang et al. 2013). Three TALEN assembly approaches have been developed, including standard cloning assembly methods (Khayter et al. 2011), Golden Gate assembly methods (Bogdanove et al. 2011), and solid-phase assembly methods (Foden et al. 2012). TALEN technology has made knockout mutations in many plant genomes such as *Arabidopsis* (Christian et al. 2010), tobacco (Luo et al. 2015), and rice (Chen et al. 2014).

16.2.3 CRISPRs

CRISPR was first found in *Escherichia coli* genome and proved to be an adaptive immune system in bacteria and Archaea (Sternberg et al. 2016; Ishino et al. 1987). A CRISPR/Cas locus typically involves several components, including a CRISPR array, an AT-rich leader sequence, and a *Cas* gene that encodes the Cas protein (Hille and Charpentier 2016). Based on the participating Cas proteins, scientists have identified more than 11 CRISPR systems which can be grouped into three types (Makarova et al. 2011). Type I and type III require multiple proteins to form a large functional Cas complex (Nam et al. 2012; Rouillon et al. 2013), while the type II CRISPR system relies on a single endonuclease (Gasiunas et al. 2012). The simplicity of the type II CRISPR system allowed it to be the most favorable genome-editing system (Meisam et al. 2018). CRISPR-associated protein 9 (Cas9), which belongs to the type II endonuclease, was the first Cas effector to be harnessed for genome engineering (Knott and Doudna 2018). The Cas9 protein contains two domains, RuvC and HNH. The HNH nuclease domain cleaves the complementary strand and the RuvC domain cleaves the non-complementary strand (Hiroshi et al. 2014).

The process of the CRISPR/Cas9 system can be usually be divided into three stages in response to invading foreign DNA or plasmid. An initial step is the acquisition stage where invading DNA is inserted into the host CRISPR array. The second is the expression stage in which the Cas9 protein is expressed and CRISPR RNA (crRNA) is generated. The third step is the interference stage; the crRNA guides the Cas9 protein to recognize and cleave the DNA target (Amitai and Sorek 2016; Wang et al. 2016). To exploit CRISPR/Cas9 system for genome editing, only a synthetic single-guide RNA (sgRNA) that encoded by a sequence of approximately 100-nt is required to construct the CRISPR/Cas9 expression cassette. So the Cas9 protein is guided to identify the NGG-type PAM and targets a specific sequence through complementary base pairing (Liu et al. 2017) (Figure 16.1c). Nowadays, the CRISPR/Cas9 system has been successfully applied in numerous plant species including the model plants (e.g. *Arabidopsis* (Akihiro et al. 2019) and tobacco (Hirohata et al. 2019)), the crop plants (e.g. rice (Saika et al. 2019) and wheat (Zhang et al. 2019)), and the horticultural plants (e.g. tomato (Ortigosa et al. 2019) and petunia (Xu et al. 2019)).

Much efforts have been made to modify CRISPR to become a more efficient and robust tool. Since sgRNA is one of the two key components of CRISPR/Cas9 system, scientists have tried to empower sgRNAs with multiplexing capability by assembling several sgRNA expression cassettes into one CRISPR/Cas9 vector (Ma et al. 2015). Maet al. develop a strategy to produce numerous gRNAs from a single polycistronic gene via the endogenous tRNA-processing system, which processes tRNA-gRNA architecture into sgRNAs to edit multiple targets (Cunfa et al. 2019). The Cas9 is the other key factor to be engineered, including the promoters used to drive *Cas* gene expression and the Cas protein itself. To increase production of heritable mutations in plants, researchers test different promoters and find the ones specifically expressed in germ-line cells are of high efficiency (Yan et al. 2015; Mao et al. 2016). The most commonly used Cas9 protein was isolated from *Streptococcus pyogenes* (SpCas9) and it recognizes the canonical NGG PAM (Long et al. 2018; Wang et al. 2018; Zhu et al. 2019). To overcome this constraint, SpCas9 and its derivatives are engineered to recognize alternative PAM sequences. For example, the VQR and VRER variants of the SpCas9 are able to identify NGA and NGCG PAMs (Hu et al. 2016). Cas9 orthologues derived from *Streptococcus thermophilus* (St1Cas9) and *Staphylococcus aureus* (SaCas9) are discovered to recognize six nucleotides ("NNAGAA" or "NNGGAA") and five to six nucleotides ("NNGGGT" or "NNGAA"), respectively, which may significantly reduce the off-target rate compared with SpCas9 (Steinert et al. 2015). Recently, Cpf1 and Cas13a have emerged as new type II components to mediate genome modification.

The PAM that Cpf1 recognizes is "TTTN," which helps Cpf1 to complement the SpCas9 system (G-rich) (Zaidi et al. 2017). Cas13a was identified to be the first RNA hunter in CRISPR/Cas system. With the guidance of a crRNA, Cas13a binds to a single-strand RNA and cleaves it (Aman et al. 2018).

16.3 Current Applications of Genome Editing in Tomato Improvement

16.3.1 Environmental Adaptation

Breeding efforts have focused on useful traits improvement in tomato plants. For example, adaptation to new growth conditions, and resistance to biotic and abiotic stresses (Rothan et al. 2019) (Table 16.1). For biotic stress, CRISPR/Cas9 system has been explored to evaluate the genes involved in defense against viruses, fungus, bacteria and parasitic weeds. The viral infections can significantly reduce plant yield and consequently cause economic losses (Alexander et al. 2017). *Dicer-like 2* (DCL2) genes are proved to play critical roles in defending against RNA viruses such as tomato mosaic virus, tobacco mosaic virus, and potato virus X (Wang et al. 2018a,b). The CRISPR/Cas9 system is also used for the delivery of mutations in the *MLO1* gene to generate tomato plant resistant to powdery mildew, which is caused by fungus *Oidiumneolycopersici* (Nekrasov et al. 2017). A similar study was accomplished by using four sgRNAs to target *Powdery mildew resistance 4* (PMR4), generating plants with resistance against *O. neolycopersici* (Koseoglou 2017). Another destructive disease that affects tomato production is downy mildew, a CRISPR/Cas9-generated tomato of *Downy mildew resistant 6 (DMR6)* showed stronger resistance to several pathogens such as *Pseudomonas syringae*, *Phytophthora capsica*, and *Xanthomonas* spp.(Thomazella et al. 2016). Moreover, *Mitogen-activated protein kinase 3 (MAPK3)*, and *Solyc08g075770* are identified to be responsible for the defense against *Fusarium oxysporum* and *Botrytis cinereal*, respectively, using the CRISPR/Cas9 technology (Prihatna et al. 2018; Zhang et al. 2018). *P. syringae*, living both on the surface and in the apoplast of the tomato plant, is the causative agent of the bacterial speck disease (Scalschi et al. 2020). Researchers use CRISPR/Cas9 strategy to generate *jaz2Δjas* mutants that are able to become resistant to *P. syringae* (Ortigosa et al. 2019). Parasitic weeds attack the host roots of economic crops and cause extreme damage to agricultural production. Bari et al. demonstrate a CRISPR/Cas9-mediated mutagenesis of the *Carotenoid cleavage dioxygenase 8 (CCD8)* gene in tomato. The mutated tomato lines can host resistance to the parasitic weed *Phelipancheaegyptiaca* (Bari et al. 2019).

In addition to microorganisms, crops usually confront other stress factors during growth, such as drought, heat, chilling, and herbicides stresses, which limit the survival chance of the crop in certain areas and restricts larger production. The CRISPR/Cas9 technology could be a solution to help crops adapt to the cultivation environment (Eş et al. 2019). *Brassinazole resistant 1* (BZR1), *C-repeat binding factor 1* (CBF1) are characterized to be related to heat and chilling tolerance, respectively (Li et al. 2018; Yin et al. 2018). *MAPK3* and *non expressor of pathogenesis-related gene 1* (NPR1) are involved in tomato drought tolerance (Wang et al. 2017; Rui et al. 2019). Herbicides are used to kill weeds in the field

Table 16.1 Current status of genome editing in tomato.

Targeted genes	Targeted traits	References
Environmental adaptation		
DCL2	Susceptibility to RNA viruses	Wang et al. (2018) and Wang et al. (2018)
MLO1	Resistance against powdery mildew	Nekrasov et al. (2017)
PMR4	Resistance against powdery mildew	Koseoglou (2017)
DMR6	Resistance against downy mildew	Thomazella et al. (2016)
Solyc08g075770	Susceptibility to *Fusarium* wilt disease	Prihatna et al. (2018)
MAPK3	Susceptibility to gray mold disease	Zhang et al. (2018)
JAZ2	Resistance against bacterial speck disease	Ortigosa et al. (2019)
CCD8	Resistance against parasitic weed	Bari et al. (2019)
BZR1	Decrease of heat stress tolerance	(Yin et al. 2018)
CBF1	Decrease of chilling stress tolerance	Li et al. (2018)
NPR1	Decrease of drought stress tolerance	Rui et al. (2019)
MAPK3	Decrease of drought stress tolerance	Wang et al. (2017)
ALS1	Enhanced herbicide resistance	Benoit et al. (2019)
ALS1, ALS2	Enhanced herbicide resistance	Florian et al. (2019)
MBP21	Generation of jointless pedicle	Roldan et al. (2017)
Solyc12g038510 *Solyc03g114840*	Generation of jointless pedicle	Soyk et al. (2017) and Soyk et al. (2019)
GAI	Generation of dwarf tomato plants	Tomlinson et al. (2019)
Fruit quality		
CLV3	Fruits with increasing locule numbers	Rodríguez-Leal et al. (2017)
PSY1	Generation of yellow-colored tomato	Filler Hayut et al. (2017)
MYB12	Generation of pink-colored tomato	Deng et al. (2018)
ANT2 (gene insertion)	Generation of purple-colored tomato	Cermák et al. (2015)
ALC	Generation of long shelf life tomato	Yu et al. (2017)
PL	Generation of long shelf life tomato	Duoduo et al. (2018) and Uluisik et al. (2016)
GAD2, GAD3	Increase of GABA content	Nonaka et al. (2017)
GABA-TP1, GABA-TP2, GABA-TP3, CAT9, SSADH	Increase of GABA content	Li et al. (2018)
SGR1, LCY-E, Blc, LCY-B1, LCY-B2	Increase of lycopene content	Li et al. (2018)
LCY	Increase of lycopene content	Zsögön et al. (2018)
GGP1	Increase of vitamin C content	Li et al. (2018)
AGL6	Production of parthenocarpic fruit	Klap et al. (2017)
IAA9	Production of parthenocarpic fruit	Ueta et al. (2017))
ARF7	Production of parthenocarpic fruit	Hu et al. (2018)

without significant damage to desirable crops. However, sometimes they can harm the plants due to low resistance against these chemicals (Eş et al. 2019). Increased herbicide-resistance in plant is a good solution to the problem. CRISPR/Cas9 genome editing is carried out to target *Acetolactate synthase 1 (ALS1)* and *ALS2*. The generated tomato mutants showed enhanced herbicide-resistance (Benoit et al. 2019; Florian et al. 2019).

The jointless pedicel eliminates the abscission zone on the stems of fruits and allows mechanical harvesting (Zsögön et al. 2017). So, it is a preferred trait by farmers and breeders. CRISPR/Cas9-induced deletions in a MADS-box gene, *MBP21*, can lead to jointless tomato phenotype (Roldan et al. 2017). Two studies by Soy ket al. provide evidence that the inactivation of *Solyc12g038510* and *Solyc03g114840* can also eliminate fruit-stem abscission and reduce fruit dropping (Soyk et al. 2017, 2019). Dwarf plants are more likely to survive in windy conditions. Scientists design a sgRNA to target the *gibberellic-acid insensitive (GAI)* gene and obtain a dominant dwarf mutation that carries one amino-acid deletion. These tomato plants will have more survival chances in windy environments (Tomlinson et al. 2019).

16.3.2 Fruit Quality

Apart from environmental adaptation, genome-editing technologies has proven to be an especially powerful tool for the elucidation of genes influencing tomato fruit size, color, firmness, nutrition values, and parthenocarpy (Miller et al. 1985). Fruit size largely depends on the number of locules. Scientists designed eight sgRNAs to target the promoter of *CLAVATA (CLV3)* gene, creating tomato fruits with more locules and bigger size (Rodríguez-Leal et al. 2017). Fruit color is one of the important commercial traits to attract consumers to pick. Now, with the availability of CRISPR/Cas9 approach, it is possible to cultivate yellow, pink, and purple tomatoes by engineering *Phytoene synthase 1 (PSY1)* (Filler Hayut et al. 2017), *MYB transcription factor 12 (MYB12)* (Deng et al. 2018), and *Anthocyanin 2 (ANT2)* genes (Cermák et al. 2015), respectively. For fresh market tomatoes, long shelf life is desired, which allows the producer to reach distant markets (Zsögön et al. 2017). This goal was achieved by the inactivation of *Ripening inhibitor* (RIN) that incomplete ripening fruits were obtained. However, these fruits lacked the ripening attributes such as flavor, color, and nutrients accumulation (Li et al. 2020). *Alcobaca* (ALC) is a naturally occurring long-shelf life allele in tomato (Zhou et al. 2020). CRISPR/Cas9 technology is used to knock out *ALC* and the generated fruits displayed increased storage performance without sacrificing fruit organoleptic quality (Yu et al. 2017). Similar experiments have conducted to verify the function of *pectate lyase* (PL), fruits with long shelf life are generated (Uluisik et al. 2016; Duoduo et al. 2018).

CRISPR/Cas9 technology has also been applied to improve the nutritional quality of fruits. By targeting the genes that involved in γ-aminobutyric acid (GABA) metabolic pathway, researchers produce high GABA content tomato fruits (Nonaka et al. 2017; Li et al. 2018). Lycopene is known as a bioactive component that can reduce the risk of cancer. Tomato is one of the dietary lycopene sources and numerous studies have focused on the enhancement of lycopene accumulation. Two research groups employ CRISPR/Cas9 technology to rewrite lycopene metabolic pathway, either by targeting five genes or one gene, increase the lycopene content significantly (Li et al. 2018; Zsögön et al. 2018). The upstream open reading frame (uORF) of *GGP1* that encode a vitamin C-biosynthetic enzyme is edited to increase vitamin C levels in tomato fruit (Li et al. 2018). Greenhouse tomato production is hampered by insufficient pollination and parthenocarpy is the best way to solve the problem of low fruit set (Ficcadenti et al. 1999). *Agamous-like* 6 (AGL6) gene is mutated to produce parthenocarpic fruits under heat stress conditions. It is worth noting that the CRISPR-edited fruits are the similar weight and shape compared with the wild-type fruits, making *AGL6* an ideal gene for parthenocarpy (Klap et al. 2017). Another two research groups targeted *Indole-3-acetic acid inducible 9* (IAA9) (Ueta et al. 2017) and *Auxin response factor 7* (ARF7) (Hu et al. 2018) to generated seedless fruits.

16.4 Challenges and Future of Genome Editing in Tomato

Tomato is both a model organism and a leading horticultural crop, which can contribute to basic research and applied research (Zsögön et al. 2017). Development and application of genome-editing technologies in tomato has successfully extended to other horticultural crop improvement. For example, the CRISPR-edited papaya (Gumtow et al. 2018), citrus (Peng et al. 2017), and apple (Malnoy et al. 2016) are able to confer immunity against plant diseases. Since gynoecious lines allow higher yield, and a more concentrated fruit set, researchers employed CRISPR/Cas9 system to target *WIP domain containing protein 1 (WIP1)* to inhibit development of cucumber carpel, generating mutants with gynoecious phenotype (Hu et al. 2017). However, some challenges still remain for the wider application of genome-editing tools in fruit crops. One of them is the insufficient genomic information which limits the primer design for target sequence. The second problem is the difficulties of transformation. Once the target gene is identified, a vector should be delivered into the plant cell to produce the edited mutant. Nowadays, the most commonly used method is tissue culture-based *Agrobacterium*-mediated gene transformation (Niazian et al. 2017). To date, more than 12 fruit crops have been successfully edited by this transformation approach (Xu et al. 2019). Other protocols should be established to broaden the application of CRISPR/Cas9 in fruit crops. Another problem is that tomato is a true diploid species, unlike many fruit crops, which are allopolyploid, or autopolyploid. Moreover, most examples presented here are recessive mutations with high impact on productivity. These complex genomic contexts demand fine-tuned gene editing for which current technology has not yet met the requirement (Park et al. 2014; Zsögön et al. 2017).

Genome-editing tools such as CRISPR/Cas9 can produce T-DNA-free mutants, either by high-throughput screening of transgene-free lines (Florian et al. 2019), or delivering editing reagents via ribonucleoproteins (Liang et al. 2017). Recently, the CRISPR-edited false flax, soybean, and waxy corn are given the green light by the Food and Drug Administration (FDA), indicating that these crops will not be subject to the same strict regulations as GMOs (Waltz 2018). However, the exploitation of CRISPR/Cas9 systems for breeding purposes is still under uncertain regulation in other countries. According to a decision made by the European Court of Justice (ECJ), the genome-edited crops should be subject to a law that was developed for older breeding techniques (Callaway 2018). In other words, the plant varieties generated by genome-editing technologies require a lengthy approval process before entering the market. Moreover, the public acceptance of genome-edited crops in European and Asian countries is a challenging issue (Karkute et al. 2017). Increasing their confidence in the safety of genome-edited crops will take a lot of efforts. Given the enormous potential of genome-editing technologies, these challenges will be overcome in the near future.

References

Aggarwal, A., Bitinaite, J., Schildkraut, I., and Wah, D. (1998). FokI dimerization is required for DNA cleavage. *Proceedings of the National Academy of Sciences of the United States of America* 95 (18): 10570–10575.

Akihiro, Y., Takashi, I., Mika, Y. et al. (2019). Developing heritable mutations in Arabidopsis thaliana using a modified CRISPR/Cas9 toolkit comprising PAM-altered Cas9 variants and gRNAs. *Plant & Cell Physiology* 60 (10): 2255–2262.

Alexander, H.M., Bruns, E., Schebor, H., and Malmstrom, C.M. (2017). Crop-associated virus infection in a native perennial grass: reduction in plant fitness and dynamic patterns of virus detection. *Journal of Ecology* 105 (4SI): 1021–1031.

Aman, R., Ali, Z., Butt, H. et al. (2018). RNA virus interference via CRISPR/Cas13a system in plants. *Genome Biology* 19 (1): 1.

Amitai, G. and Sorek, R. (2016). CRISPR–Cas adaptation: insights into the mechanism of action. *Nature Reviews Microbiology* 14 (2): 67–76.

Araki, M. and Ishii, T. (2015). Towards social acceptance of plant breeding by genome editing. *Trends in Plant Science* 20 (3): 145–149.

Bari, V.K., Nassar, J.A., Kheredin, S.M. et al. (2019). CRISPR/Cas9-mediated mutagenesis of CAROTENOID CLEAVAGE DIOXYGENASE 8 in tomato provides resistance against the parasitic weed Phelipancheaegyptiaca. *Scientific Reports* 9 (1): 11438.

Bedell, V.M., Wang, Y., Campbell, J.M. et al. (2012). In vivo genome editing using a high-efficiency TALEN system. *Nature* 491 (Nov.1 TN.7422): 114–118.

Benoit, D., Laura, P., Kostlend, M. et al. (2019). Efficient and transgene-free gene targeting using agrobacterium-mediated delivery of the CRISPR/Cas9 system in tomato. *Plant Cell Reports* 38 (4): 459–462.

Bhakta, M.S., Henry, I.M., Ousterout, D.G. et al. (2013). Highly active zinc finger nucleases by extended modular assembly. *Genome Research* 23 (3): 530–538.

Boch, J., Scholze, H., Schornack, S. et al. (2009). Breaking the code of DNA binding specificity of TAL-type III effectors. *Science* 326 (Dec.11 TN.5959): 1509.

Bogdanove, A.J., Schmidt, C., Christian, M. et al. (2011). Efficient design and assembly of custom TALEN and other TAL effector-based constructs for DNA targeting. *Nucleic Acids Research* 39 (12): e82.

Callaway, E. (2018). EU law deals blow to CRISPR crops. *Nature* 560 (Aug.2 TN.7716): 16.

Carroll, D. (2011). Genome engineering with zinc-finger nucleases. *Genetics: A Periodical Record of Investigations Bearing on Heredity and Variation* 188 (4): 773–782.

Catto, L.E., Ganguly, S., Milsom, S.E. et al. (2006). Protein assembly and DNA looping by the FokI restriction endonuclease. *Nucleic Acids Research* 34 (6): 1711–1720.

Cermák, T., Baltes, N.J., Cegan, R. et al. (2015). High-frequency, precise modification of the tomato genome. *Genome Biology* 16 (1): 232.

Chen, K., Shan, Q., and Gao, C. (2014). An efficient TALEN mutagenesis system in rice. *Methods: A Companion to Methods in Enzymology* 69 (1): 2–8.

Christian, M., Cermak, T., Doyle, E.L. et al. (2010). Targeting DNA double-strand breaks with TAL effector nucleases. *Genetics* 186 (2): 757–761.

Cunfa, M., Chenzeng, Z., Min, Z. et al. (2019). CRISPR/Cas9-mediated multiple gene editing in Brassica oleracea var. capitata using the endogenous tRNA-processing system. *Horticulture Research* 6 (20).

Deng, D., Yin, P., Yan, C. et al. (2012). Recognition of methylated DNA by TAL effectors. *Cell Research* 22 (10): 1502–1504.

Deng, L., Wang, H., Sun, C. et al. (2018). Efficient generation of pink-fruited tomatoes using CRISPR/Cas9 system. *Journal of Genetics and Genomics* 45 (1): 51–54.

Duoduo, W., Nurul, S., Cheng, Y. et al. (2018). Characterization of CRISPR mutants targeting genes modulating pectin degradation in ripening tomato. *Plant Physiology* 178 (2): 544–557.

Eş, I., Gavahian, M., Marti-Quijal, F.J. et al. (2019). The application of the CRISPR-Cas9 genome editing machinery in food and agricultural science: current status, future perspectives, and associated challenges. *Biotechnology Advances* 37 (3): 410–421.

Ficcadenti, N., Sestili, S., Pandolfini, T. et al. (1999). Genetic engineering of parthenocarpic fruit development in tomato. *Molecular Breeding* 5 (5): 463–470.

Filler Hayut, S., Melamed Bessudo, C., and Levy, A.A. (2017). Targeted recombination between homologous chromosomes for precise breeding in tomato. *Nature Communications* 8 (15605): 15605.

Florian, V., Laura, P., Laura, C. et al. (2019). Transgene-free genome editing in tomato and potato plants using agrobacterium-mediated delivery of a CRISPR/Cas9 cytidine base editor. *International Journal of Molecular Sciences* 20 (2): 402.

Foden, J.A., Khayter, C., Tsai, S.Q. et al. (2012). FLASH assembly of TALENs for high-throughput genome editing. *Nature Biotechnology* 30 (5): 460–465.

Gao, H., Wu, X., Chai, J., and Han, Z. (2012). Crystal structure of a TALE protein reveals an extended N-terminal DNA binding region. *Cell Research* 22 (12): 1716–1720.

Gasiunas, G., Barrangou, R., Horvath, P., and Siksnys, V. (2012). Cas9-crRNA ribonucleoprotein complex mediates specific DNA cleavage for adaptive immunity in bacteria. *Proceedings of the National Academy of Sciences* 109 (39): E2579–E2586.

Giovannoni, J., Nguyen, C., Ampofo, B. et al. (2017). The epigenome and transcriptional dynamics of fruit ripening. *Annual Review of Plant Biology* 68 (1): 61–84.

Gumtow, R., Wu, D., Uchida, J., and Tian, M. (2018). A Phytophthora palmivora extracellular cystatin-like protease inhibitor targets papain to contribute to virulence on papaya. *Molecular Plant-Microbe Interactions* 31 (3): 363–373.

Hartung, F. and Schiemann, J. (2014). Precise plant breeding using new genome editing techniques: opportunities, safety and regulation in the EU. *The Plant Journal* 78 (5): 742–752.

Hille, F. and Charpentier, E. (2016). CRISPR-Cas: biology, mechanisms and relevance. *Philosophical Transactions of the Royal Society, B: Biological Sciences* 371 (1707): 20150496.

Hirohata, A., Sato, I., Kaino, K. et al. (2019). CRISPR/Cas9-mediated homologous recombination in tobacco. *Plant Cell Reports* 38 (4): 463–473.

Hiroshi, N., Ann, R.F., D, H.P. et al. (2014). Crystal structure of Cas9 in complex with guide RNA and target DNA. *Cell* 156 (5): 935–949.

Hu, X., Wang, C., Fu, Y. et al. (2016). Expanding the range of CRISPR/Cas9 genome editing in rice. *Molecular Plant* 9 (6): 943–945.

Hu, B., Li, D., Liu, X. et al. (2017). Engineering non-transgenic gynoecious cucumber using an improved transformation protocol and optimized CRISPR/Cas9 system. *Molecular Plant* 10 (12): 1575–1578.

Hu, J., Israeli, A., Ori, N., and Sun, T. (2018). The interaction between DELLA and ARF/IAA mediates crosstalk between gibberellin and auxin signaling to control fruit initiation in tomato. *The Plant Cell* 30 (8): 1710–1728.

Hunter, M.C., Smith, R.G., Schipanski, M.E. et al. (2017). Agriculture in 2050: recalibrating targets for sustainable intensification. *BioScience* 67 (4): 386–391.

Ishino, Y., Shinagawa, H., Makino, K. et al. (1987). Nucleotide sequence of the iap gene, responsible for alkaline phosphatase isozyme conversion in Escherichia coli, and identification of the gene product. *Journal of Bacteriology* 169 (12): 5429–5433.

Karkute, S.G., Singh, A.K., Gupta, O.P. et al. (2017). CRISPR/Cas9 mediated genome engineering for improvement of horticultural crops. *Frontiers in Plant Science* 8: 1635.

Kay, S., Hahn, S., Marois, E. et al. (2007). A bacterial effector acts as a plant transcription factor and induces a cell size regulator. *Science* 318 (5850): 648–651.

Khayter, C., Reyon, D., Joung, J.K. et al. (2011). Targeted gene disruption in somatic zebrafish cells using engineered TALENs. *Nature Genetics* 43 (S.Nasture): S8–S9.

Kim, Y.G., Cha, J., and Chandrasegaran, S. (1996). Hybrid restriction enzymes: zinc finger fusions to Fok I cleavage domain. *Proceedings of the National Academy of Sciences of the United States of America* 93 (3): 1156–1160.

Klap, C., Yeshayahou, E., Bolger, A.M. et al. (2017). Tomato facultative parthenocarpy results from SlAGAMOUS-LIKE 6 loss of function. *Plant Biotechnology Journal* 15 (5): 634–647.

Knott, G.J. and Doudna, J.A. (2018). CRISPR-Cas guides the future of genetic engineering. *Science* 361 (Aug.31 TN.6405): 866–869.

Koseoglou, E. (2017). *The Study of SlPMR4 CRISPR/Cas9- Mediated Tomato Allelic Series for Resistance against Powdery Mildew*. Wageningen University.

Li, R., Zhang, L., Wang, L. et al. (2018). Reduction of tomato-plant chilling tolerance by CRISPR-Cas9-mediated SICBF1 mutagenesis. *Journal of Agricultural and Food Chemistry* 66 (34): 9042–9051.

Li, R., Li, R., Li, X. et al. (2018). Multiplexed CRISPR/Cas9-mediated metabolic engineering of γ-aminobutyric acid levels in Solanum lycopersicum. *Plant Biotechnology Journal* 16 (2): 415–427.

Li, X., Wang, Y., Chen, S. et al. (2018). Lycopene is enriched in tomato fruit by CRISPR/Cas9-mediated multiplex genome editing. *Frontiers in Plant Science* 9: 559.

Li, T., Yang, X., Yu, Y. et al. (2018). Domestication of wild tomato is accelerated by genome editing. *Nature Biotechnology* 36 (12): 1160–1163.

Li, S., Zhu, B., Pirrello, J. et al. (2020). Roles of RIN and ethylene in tomato fruit ripening and ripening-associated traits. *New Phytologist* 226 (2): 460–475.

Liang, Z., Chen, K., Li, T. et al. (2017). Efficient DNA-free genome editing of bread wheat using CRISPR/Cas9 ribonucleoprotein complexes. *Nature Communications* 8: 14261.

Liu, X., Xie, C., Si, H., and Yang, J. (2017). CRISPR/Cas9-mediated genome editing in plants. *Methods* 121-122: 94–102.

Long, L., Guo, D., Gao, W. et al. (2018). Optimization of CRISPR/Cas9 genome editing in cotton by improved sgRNA expression. *Plant Methods* 14: 85.

Luo, S., Li, J., Stoddard, T.J. et al. (2015). Non-transgenic plant genome editing using purified sequence-specific nucleases. *Molecular Plant* 8 (9): 1425–1427.

Ma, X., Zhang, Q., Zhu, Q. et al. (2015). A robust CRISPR/Cas9 system for convenient, high-efficiency multiplex genome editing in monocot and dicot plants. *Molecular Plant* 8 (8): 1274–1284.

Maeder, M.L., Thibodeau-Beganny, S., Osiak, A. et al. (2008). Rapid "open-source" engineering of customized zinc-finger nucleases for highly efficient gene modification. *Molecular Cell* 31 (2): 294–301.

Mahfouz, M., Zhu, J., Yan, C. et al. (2012). Structural basis for sequence-specific recognition of DNA by TAL effectors. *Science* 335 (Feb.10 TN.6069): 720–723.

Mahfouz, M.M., Piatek, A., and Stewart, C.N. (2014). Genome engineering via TALENs and CRISPR/Cas9 systems: challenges and perspectives. *Plant Biotechnology Journal* 12 (8): 1006–1014.

Makarova, K.S., Haft, D.H., Barrangou, R. et al. (2011). Evolution and classification of the CRISPR-Cas systems. *Nature Reviews. Microbiology* 9 (6): 467–477.

Malnoy, M., Viola, R., Jung, M. et al. (2016). DNA-free genetically edited grapevine and apple protoplast using CRISPR/Cas9 ribonucleoproteins. *Frontiers in Plant Science* 7: 1904.

Mao, Y., Zhang, Z., Feng, Z. et al. (2016). Development of germ-line-specific CRISPR-Cas9 systems to improve the production of heritable gene modifications in Arabidopsis. *Plant Biotechnology Journal* 14 (2): 519–532.

Mao, Y., Botella, J.R., and Zhu, J. (2017). Heritability of targeted gene modifications induced by plant-optimized CRISPR systems. *Cellular and Molecular Life Sciences* 74 (6): 1075–1093.

Mc, H., Ns, W., Ka, K. et al. (2007). An improved zinc-finger nuclease architecture for highly specific genome editing. *Nature Biotechnology* 25 (7): 778–785.

Meisam, N.K., Hejazi, S.S., Ezgi, E. et al. (2018). Clustered regularly interspaced short palindromic repeats/Cas9 gene editing technique in xenotransplantation. *Frontiers in Immunology* 9: 1711.

Meyer, R.S. and Purugganan, M.D. (2013). Evolution of crop species: genetics of domestication and diversification. *Nature Reviews Genetics* 14 (12): 840–852.

Mh, P., Jp, C., Jk, J. et al. (2008). Comparison of zinc finger nucleases for use in gene targeting in mammalian cells. *Molecular therapy: the Journal of the American Society of Gene Therapy* 16 (4): 707–717.

Miller, J., Mclachlan, A.D., and Klug, A. (1985). Repetitive zinc-binding domains in the protein transcription factor IIIA from Xenopus oocytes. *The EMBO Journal* 4 (6): 1609–1614.

Moore, M., Klug, A., and Choo, Y. (2001). Improved DNA binding specificity from polyzinc finger peptides by using strings of two-finger units. *Proceedings of the National Academy of Sciences of the United States of America* 98 (4): 1437–1441.

Moscou, M.J. and Bogdanove, A.J. (2009). A simple cipher governs DNA recognition by TAL effectors. *Science* 326 (Dec.11 TN.5959): 1501.

Nam, K.H., Haitjema, C., Liu, X. et al. (2012). Cas5d protein processes pre-crRNA and assembles into a Cascade-like interference complex in subtype I-C/Dvulg CRISPR-Cas system. *Structure* 20 (9): 1574–1584.

Nekrasov, V., Wang, C., Win, J. et al. (2017). Rapid generation of a transgene-free powdery mildew resistant tomato by genome deletion. *Scientific Reports* 7 (1): 482–486.

Niazian, M., Sadat Noori, S.A., Galuszka, P., and Mortazavian, S.M.M. (2017). Tissue culture-based agrobacterium-mediated and in planta transformation methods. *Czech Journal of Genetics and Plant Breeding* 53 (4): 133–143.

Nonaka, S., Arai, C., Takayama, M. et al. (2017). Efficient increase of gamma-aminobutyric acid (GABA) content in tomato fruits by targeted mutagenesis. *Scientific Reports* 7 (1): 7057.

Ortigosa, A., Gimenez-Ibanez, S., Leonhardt, N., and Solano, R. (2019). Design of a bacterial speck resistant tomato by CRISPR/Cas9-mediated editing of SlJAZ2. *Plant Biotechnology Journal* 17 (3): 665–673.

Park, S.J., Jiang, K., Tal, L. et al. (2014). Optimization of crop productivity in tomato using induced mutations in the florigen pathway. *Nature Genetics* 46 (12): 1337–1342.

Peng, A., Chen, S., Lei, T. et al. (2017). Engineering canker-resistant plants through CRISPR/Cas9-targeted editing of the susceptibility gene CsLOB1 promoter in citrus. *Plant Biotechnology Journal* 15 (12): 1509–1519.

Persikov, A.V., Wetzel, J.L., Rowland, E.F. et al. (2015). A systematic survey of the Cys(2) his(2) zinc finger DNA-binding landscape. *Nucleic Acids Research* 43 (3): 1965–1984.

Prihatna, C., Barbetti, M.J., and Barker, S.J. (2018). A novel tomato Fusarium wilt tolerance gene. *Frontiers in Microbiology* 9: 1226.

Ran, Y., Liang, Z., and Gao, C. (2017). Current and future editing reagent delivery systems for plant genome editing. *Science China. Life Sciences* 60 (5): 490–505.

Ravichandra, N.G. (2014). *Horticulture and its Role in the National Economies.* New Delhi: Horticultural Nematology, Springer.

Reem, N.T. and Van Eck, J. (2019). Application of CRISPR/Cas9-mediated gene editing in tomato. *Methods in Molecular Biology* 1917: 171–182.

Reyon, D., Peterson, T., Zhang, F. et al. (2010). High frequency targeted mutagenesis in Arabidopsis thaliana using zinc finger nucleases. *Proceedings of the National Academy of Sciences of the United States of America* 107 (26): 12028–12033.

Rodríguez-Leal, D., Lemmon, Z.H., Man, J. et al. (2017). Engineering quantitative trait variation for crop improvement by genome editing. *Cell* 171 (2): 470–480.

Roldan, M.V.G., Périlleux, C., Morin, H. et al. (2017). Natural and induced loss of function mutations in SlMBP21 MADS-box gene led to jointless-2 phenotype in tomato. *Scientific Reports* 7 (1): 4402.

Rothan, C., Diouf, I., and Causse, M. (2019). Trait discovery and editing in tomato. *The Plant Journal* 97 (1): 73–90.

Rouillon, C., Zhou, M., Zhang, J. et al. (2013). Structure of the CRISPR interference complex CSM reveals key similarities with cascade. *Molecular Cell* 52 (1): 124–134.

Rui, L., Chunxue, L., Ruirui, Z. et al. (2019). CRISPR/Cas9-mediated SlNPR1 mutagenesis reduces tomato plant drought tolerance. *BMC Plant Biology* 19 (1): 38.

Saha, S.K., Saikot, F.K., Rahman, S. et al. (2019). Programmable molecular scissors: applications of a new tool for genome editing in biotech. *Molecular Therapy-Nucleic Acids* 14: 212–238.

Saika, H., Mori, A., Endo, M., and Toki, S. (2019). Targeted deletion of rice retrotransposon Tos17 via CRISPR/Cas9. *Plant Cell Reports* 38 (4): 455–458.

Scalschi, L., Llorens, E., Garcia-Agustin, P., and Vicedo, B. (2020). Role of Jasmonic acid pathway in tomato plant-Pseudomonas syringae interaction. *Plants (Basel)* 9 (2).

Shukla, V.K., Doyon, Y., Miller, J.C. et al. (2009). Precise genome modification in the crop species Zea mays using zinc-finger nucleases. *Nature* 459 (TN.7245): 437–474.

Smith, J., Grizot, S., Arnould, S. et al. (2006). A combinatorial approach to create artificial homing endonucleases cleaving chosen sequences. *Nucleic Acids Research* 34 (e14922): e14922.

Soyk, S., Lemmon, Z.H., Oved, M. et al. (2017). Bypassing negative epistasis on yield in tomato imposed by a domestication gene. *Cell* 169 (6): 1142–1155.

Soyk, S., Lemmon, Z.H., Sedlazeck, F.J. et al. (2019). Duplication of a domestication locus neutralized a cryptic variant that caused a breeding barrier in tomato. *Nature Plants* 5 (5): 471–479.

Steinert, J., Schiml, S., Fauser, F., and Puchta, H. (2015). Highly efficient heritable plant genome engineering using Cas9 orthologues from Streptococcus thermophilus and Staphylococcus aureus. *The Plant Journal* 84 (6): 1295–1305.

Sternberg, S.H., Richter, H., Charpentier, E., and Qimron, U. (2016). Adaptation in CRISPR-Cas systems. *Molecular Cell* 61 (6): 797–808.

Thibodeau-Beganny, S., Reyon, D., Stupar, R.M. et al. (2011). Selection-free zinc-finger-nuclease engineering by context-dependent assembly (CoDA). *Nature Methods* 8 (1): 67–69.

Thomazella, D., Brail, Q., Dahlbeck, D., and Staskawicz, B. (2016). CRISPR-Cas9 mediated mutagenesis of a DMR6 ortholog in tomato confers broad-spectrum disease resistance. *BioRxiv* https://doi.org/10.1101/064824.

Tomlinson, L., Yang, Y., Emenecker, R. et al. (2019). Using CRISPR/Cas9 genome editing in tomato to create a gibberellin-responsive dominant dwarf DELLA allele. *Plant Biotechnology Journal* 17 (1): 132–140.

Townsend, J.A., Wright, D.A., Winfrey, R.J. et al. (2009). High-frequency modification of plant genes using engineered zinc-finger nucleases. *Nature* 459 (TN.7245): 442–474.

Ueta, R., Abe, C., Watanabe, T. et al. (2017). Rapid breeding of parthenocarpic tomato plants using CRISPR/Cas9. *Scientific Reports* 7: 507.

Uluisik, S., Chapman, N.H., Smith, R. et al. (2016). Genetic improvement of tomato by targeted control of fruit softening. *Nature Biotechnology* 34 (9): 950.

Vos, R.D., Hall, R.D., and Moing, A. (2011). *Metabolomics of a Model Fruit: Tomato*. Hoboken: Blackwell/Wiley.

Voytas, D.F. (2013). Plant genome engineering with sequence-specific nucleases. *Annual Review of Plant Biology* 64: 327–350.

Waltz, E. (2018). With a free pass, CRISPR-edited plants reach market in record time. *Nature Biotechnology* 36 (1): 6–7.

Wang, H., La Russa, M., and Qi, L.S. (2016). CRISPR/Cas9 in genome editing and beyond. *Annual Review of Biochemistry* 85 (1): 227–264.

Wang, L., Chen, L., Li, R. et al. (2017). Reduced drought tolerance by CRISPR/Cas9-mediated SlMAPK3 mutagenesis in tomato plants. *Journal of Agricultural and Food Chemistry* 65 (39): 8674–8682.

Wang, Z., Zheng, Z., Song, L., and Liu, D. (2018). Functional characterization of Arabidopsis PHL4 in plant response to phosphate starvation. *Frontiers in Plant Science* 9: 1432.

Wang, Z., Hardcastle, T.J., Canto Pastor, A. et al. (2018a). A novel DCL2-dependent miRNA pathway in tomato affects susceptibility to RNA viruses. *Genes & Development* 32 (17-18): 1155–1160.

Wang, T., Deng, Z., Zhang, X. et al. (2018b). Tomato DCL2b is required for the biosynthesis of 22-nt small RNAs, the resulting secondary siRNAs, and the host defense against ToMV. *Horticulture Research* 5 (1).

Wang, T., Zhang, H., and Zhu, H. (2019). CRISPR technology is revolutionizing the improvement of tomato and other fruit crops. *Horticulture Research* 6 (1): 77.

Xu, J., Hua, K., and Lang, Z. (2019). Genome editing for horticultural crop improvement. *Horticulture Research* 6 (113): 113.

Xu, J., Kang, B., Naing, A.H. et al. (2019). CRISPR/Cas9-mediated editing of 1-aminocyclopropane-1-carboxylate oxidase1 enhances Petunia flower longevity. *Plant Biotechnology Journal* 18: 287–297.

Yamamoto, T., Kashojiya, S., Kamimura, S. et al. (2018). Application and development of genome editing technologies to the Solanaceae plants. *Plant Physiology and Biochemistry* 131: 37–46.

Yan, L., Wei, S., Wu, Y. et al. (2015). High-efficiency genome editing in Arabidopsis using YAO promoter-driven CRISPR/Cas9 system. *Molecular Plant* 8 (12): 1820–1823.

Yin, Y., Qin, K., Song, X. et al. (2018). BZR1 transcription factor regulates heat stress tolerance through FERONIA receptor-like kinase-mediated reactive oxygen species signaling in tomato. *Plant and Cell Physiology* 59 (11): 2239–2254.

Yu, Q., Wang, B., Li, N. et al. (2017). CRISPR/Cas9-induced targeted mutagenesis and gene replacement to generate long-shelf life tomato lines. *Scientific Reports* 7 (1): 11874.

Zaidi, S.S., Mahfouz, M.M., and Mansoor, S. (2017). CRISPR-Cpf1: a new tool for plant genome editing. *Trends in Plant Science* 22 (7): 550–553.

Zhang, Y., Voytas, D.F., Starker, C.G. et al. (2013). Transcription activator-like effector nucleases enable efficient plant genome engineering. *Plant Physiology* 161 (1): 20–27.

Zhang, S., Wang, L., Zhao, R. et al. (2018). Knockout of SlMAPK3 reduced disease resistance to Botrytis cinerea in tomato plants. *Journal of Agricultural and Food Chemistry* 66 (34): 8949–8956.

Zhang, Z., Hua, L., Gupta, A. et al. (2019). Development of an agrobacterium-delivered CRISPR/Cas9 system for wheat genome editing. *Plant Biotechnology Journal* 17 (8): 1623–1635.

Zhou, J., Li, D., Wang, G. et al. (2020). Application and future perspective of CRISPR/Cas9 genome editing in fruit crops. *Journal of Integrative Plant Biology* 62 (3): 269–286.

Zhu, C., Bortesi, L., Baysal, C. et al. (2017). Characteristics of genome editing mutations in cereal crops. *Trends in Plant Science* 22 (1): 38–52.

Zhu, G., Wang, S., Huang, Z. et al. (2018). Rewiring of the fruit metabolome in tomato breeding. *Cell* 172 (1-2): 249–261.

Zhu, Z., Kang, X., Lor, V.S. et al. (2019). Characterization of a semidominant dwarfing PROCERA allele identified in a screen for CRISPR/Cas9-induced suppressors of loss-of-function alleles. *Plant Biotechnology Journal* 17 (2): 319–321.

Zsögön, A., Cermak, T., Voytas, D., and Peres, L.E.P. (2017). Genome editing as a tool to achieve the crop ideotype and de novo domestication of wild relatives: case study in tomato. *Plant Science* 256: 120–130.

Zsögön, A., Cermák, T., Naves, E.R. et al. (2018). De novo domestication of wild tomato using genome editing. *Nature Biotechnology* 36 (12): 1211–1216.

17

Genome Editing for Biofortification of Rice

Current Implications and Future Aspects

Suchismita Roy[1] *and Praveen Soni*[2]

[1] *Skaggs School of Pharmacy and Pharmaceutical Sciences, University of California, San Diego, CA, USA*
[2] *Department of Botany, University of Rajasthan, Jaipur, Rajasthan, India*

CHAPTER MENU

17.1 Introduction

Oryza sativa L. is the main crop of the world with the production of more than seven billion tonnes currently only in Asia, followed by other continents like America and Africa. However, due to harsh environmental constraints such as dry weather conditions, saline soils, extremely high or low temperatures, rice production is facing detrimental losses. The United Nations Food and Agriculture Organization (FAO) has anticipated a rise in agricultural demands with an increment of 70% by 2050. In the recent scenario, researchers all over the world have started to sort out means and methods to improve rice production and quality with the purpose of feeding billions of people. The generation of new rice varieties

Genome Engineering for Crop Improvement, First Edition. Edited by Santosh Kumar Upadhyay.

with high yields along with an increase in nutritional values is important for human health. Seed starch and amylose content are important traits that have gained wide interest. Cereal grains having a high content of amylose are decent providers of resistant starch which is resistant to digestion (Regina et al. 2006). Amylose is not absorbed in the stomach or the intestine. However, it feeds the friendly gut bacteria. It, in turn, helps in improving overall health (Regina et al. 2006). Different methods like mutagenesis, breeding, and transgenic approach have been employed to improve rice quality. Currently, several genome sequences have been elucidated and annotated (Michael and Jackson 2013). Moreover, several databases have been developed that store and provide genomic information of a horde of genes (Wingender et al. 2000). In-depth analysis of the dynamics of genomes, genes and their function have helped in the generation of tools that permit controlled alteration of genomic DNA sequences in various organisms including plants. Recently, genome-editing techniques have gained popularity. They have further advanced and replaced previous techniques as the older methods were time-consuming. Sequence-specific nucleases, for instance, zinc-finger nucleases (ZFNs), transcription activator-like effector nucleases (TALENs) as well as clustered regularly interspersed short palindromic repeats (CRISPR)/CRISPR-associated protein 9 nuclease (Cas9), are prevailing gears which have come into limelight for the discovery of gene function and evolving unique traits in plants. They have carved a niche as potential tools for molecular breeding and overall plant science (Mishra et al. 2018).

17.2 Genome Editing and its Tools

Genome editing is a powerful approach to investigate and bring about changes in gene functionality. The genome-editing process involves plant transformation and tissue-culture techniques. With the availability of optimized plant transformation protocols with high regeneration efficiency, the efficacy of editing the whole genome is undeniable. The tools of genome editing have been described in the following sections.

17.2.1 ZFNs

ZFNs are fusion proteins having a nuclease fused to a zinc-finger motif-containing protein. Nuclease is required for cleavage of DNA while the other one is required for DNA binding. ZFNs make DNA double-strand breaks (DSBs) at a precise base pair. It prompts the natural DNA repair processes in the cell and produces intriguing possibilities of base-pair modifications. ZFNs contain four to six zinc-finger motifs that each spot roughly three base pairs of DNA (Moore et al. 2001). These motifs effectively bind DNA in a sequence-specific manner. The trick in this technique lies in engineering and customizing zinc-finger motifs to bind to a specific sequence of the target DNA (Urnov et al. 2010). For clipping DNA, the enzyme FokI is used which falls under the category of type II restriction enzymes (Kim et al. 1996). FokI dimerizes to chop off target DNA. For the catalytic property to be active, two adjacent ZFNs must be aligned with proper spacing at specific target site. For the successful implementation of this technique, a longer recognition sequence (24–36 bp) is preferred which provides greater specificity. Several variants of FokI enzyme have been developed for heterodimer formation, enhancement of the specificity for the target sequence and dampening the chances of off-site cleavage (Miller et al. 2007). Site-specific transgene integration or repair mechanisms occur at the sites of the ZFN-induced DSBs via two methods (i) non-homologous end joining (NHEJ) and (ii) homology-directed repair (HDR). Genomes of numerous organisms have been edited through ZFNs (Beumer et al. 2006; Doyon et al. 2008; Urnov et al. 2010; Hauschild et al. 2011; Osiak et al. 2011; Straimer et al. 2012).

17.2.2 TALENs

TALENs have also been successfully utilized for editing genomes. They have been used expansively in plant bioengineering. The plus point of this technique is the low chance of off-targets and associated cytotoxic effects. It uses a simpler "DNA-protein code" in comparison to ZFNs. TALEN has two prominent features; first is a TALE DNA-binding domain. This domain has several tandem repeats extending from 1.5 to 33.5. Each repeat covers 34 amino acids with two different amino acid residues at 12 and 13th positions. They are called as "repeat variable diresidues" (RVDs). These RVDs can identify one nucleotide. Asparagine-isoleucine, histidine-aspartate, asparagine-asparagine and asparagine-glycine amino acid residues as RVDs can recognize A, C, G, and T nucleotides individually. The second feature of TALENs is the cleavage domain comprising of FokI restriction endonuclease. The dimer of this domain produces DSBs. The repairing of the DSBs takes place by either NHEJ or HDR similar to ZFNs. TALENs have generated huge enthusiasm in this field, owing to their ease and rapidity by which they can be designed by just using a lucid "protein-DNA code". Over the preceding years, TALENs have been used to modify genes in yeast, drosophila, rice and also human somatic and pluripotent stem cells (Hockemeyer et al. 2011; Joung and Sander 2013). Though ZFNs and TALENs have no straight comparison, the high efficiency of the cleavage and boundless choices of TALENs makes them suitable for scientific purposes. Though, one of the challenges is the generation of DNA-sequence encoding engineered repeat arrays for the TALE region. But now several platforms dealing with this issue have appeared to fulfill the increasing demands.

17.2.3 CRISPR/Cas9 System

In distinction to TALENs and ZFNs, which rely on protein-DNA-binding specificity, CRISPR/Cas9 system is a bit different as it depends upon an RNA-guided approach. CRISPR/Cas9 has arisen up as the most improved technique for genome editing. It came into play by an accidental discovery in *Escherichia coli* by Ishino et al., in the year 1987 (Ishino et al. 1987) and subsequently studied in organisms like *Shigella dysenteriae*, *Salmonella*, and *Mycobacterium* (Nakata et al. 1989; Hermans et al. 1991). With progress, this tool has been implemented for genome editing in mammalian cells, bacteria, fungi and plants (Jiang et al. 2013; Krappmann 2016). This tool has two components, first is the DNA endonuclease Cas9 protein and second is the customized solo guide RNA (sgRNA). Cas9 from *Streptococcus pyogenes* has been extensively used. Cas9 forms a complex with the sgRNA that, in turn, paves way for the enzyme to its specific-target DNA by matching conventional base pairing. Cas9 contains a bilobed structure having a larger recognition lobe (REC) and a smaller nuclease lobe (NUC). The latter has two nuclease domains RuvC and HNH. Each of the domains cleaves a specific strand of the DNA. The sgRNA is a petite non-coding RNA that is the combination of CRISPR RNA (crRNA) and trans-activating crRNA (tracrRNA) (Jinek et al. 2012). crRNA contains 18–20 nucleotide long sequence which acts as the guide and aims at recognition of a specific target site. tracrRNA, on the other hand, contains a greater span of nucleotides that generates a stem loop-like structure. It mainly interacts with Cas9 endonuclease. In bacteria, the CRISPR/Cas9 system plays a role similar to the adaptive immune system and destroys any foreign DNA. It has three phases, the first of which is adaptation that is the integration of foreign DNA. The immunological memory is accumulated in the form of spacer sequences from exogenous genomes that are put into arrays based on CRISPRs (Jinek et al. 2012). The spacer sequence and Cas9 protein work as a monitoring system to distinguish and destroy non-native DNA. A short stretch of a sequence having two to six nucleotides is found near the cleavage site of target DNA. It is known as the protospacer adjacent motif (PAM). It is vital for Cas9 for cutting the nucleic acid stretch. Cas9 from *S. pyogenes* recognizes 5′-NGG-3′ motif as PAM. Other Cas proteins such asCpf1/Cas12 recognize different PAM sequences that are 5′-TTTN-3′ or 5′-TTN-3′. Newly developed versions of Cas9 variants recognize other alternative PAMs (Anders et al. 2016). In the second phase, pre-crRNA is expressed from the CRISPR locus and subsequently, crRNA is formed after processing (Anders et al. 2016). In the third phase, called immunity or interference, a ribonucleoprotein complex with Cas proteins is formed with the crRNAs, along with tracrRNA. This super complex has the potential to simply recognize any external DNA to damage it (Bhaya et al. 2011). In terms of application, this tool is simple and quick to design. It only needs the generation of a short RNA to guide Cas9 to the specific target.

The CRISPR/Cas9 system was introduced by Shan et al., in wheat and rice in the year 2013 (Shan et al. 2013). Since then, there have been many success stories. One empowering feature of this tool is its capacity to act in-trans of the particular target of importance (Ricroch et al. 2017). In plants, it is essential to obtain stable genome-edited lines without any traces of CRISPR/Cas9 components in the genome of the plants of final generation. For this, several methods have been adopted such as transient expression of the CRISPR/Cas9 constituents, genetic segregation, and ribonucleoprotein complex mediated transformation (Pyott et al. 2016; Zhang et al. 2016; Liang et al. 2017. Though, genetic separation is the easiest, achievable and naturally occurring way.

Multiplex CRISPR/Cas9 system can target several genes at one go thereby dropping the price and time of making new traits (Wang et al. 2019). This approach has been successfully useful in rice (Hu et al. 2016). One drawback allied with this technique is the targeting of homologous sequences and homeologs that occur in polyploid crops and crops with larger genome sizes (Andersson et al. 2017; Jiang et al. 2013). For such experiments, a unique PAM site is required by SpCas9 (from *Streptococcus pyogenes*) (Kleinstiver et al. 2016). Nevertheless, several substitute PAM sequences like 5′-NAG-3′ and 5′-NGA-3′(Kleinstiver et al. 2016) and Cas9 variants, for instance, StCas9 (from *Streptococcus thermophilus*) and SaCas9 (from *Staphylococcus aureus*) have been highlighted in the latest studies (Kaya et al. 2016). The wild-type SpCas9 can recognize both 5′-NAG-3′ and 5′-NGG-3′ PAM sequences with a lower rate of off-target recognition (Meng et al. 2017). CRISPR/Cas9 variants have been generated to expand the range of genome editing (Hu et al. 2016). Hu et al. (2016) used a modified CRISPR/Cas9 system in editing genomic target sites harboring 5′-NGA-3′ PAMs. A SpCas9 variant, xCas9, has shown its ability to identify a broad class of PAMs including 5′-GAT-3′, 5′-GAA-3′ and 5′-NG-3′. However, there is a constant need for new customized

CRISPR/Cas9 systems that can identify wider ranges of PAM sequences. The efficiency of sequence replacement is another issue. Gain-of-function mutations contribute to the development of different traits in crops. Gene replacement and gene knock-in using homologous recombination (HR) are important for the improvement of agronomic traits. Sequence replacement is still a difficult task since the efficacy of HR is low in plants. Therefore, NHEJ and HR pathways need to be thoroughly studied.

17.2.4 CRISPR/Cpf1 System

With various modifications in the current CRISPR/Cas9 technology, another similar one CRISPR/Cpf1 system has been developed. It has been taken from *Prevotella* and *Francisella 1* bacteria. Cpf1 a type-V CRISPR/Cas endonuclease. It has been used in mammals (Zetsche et al. 2015; Kim et al. 2016). In mammals, four genes were edited by Cpf1 at one go (Zetsche et al. 2015). This system has also been characterized in *O. sativa*, Arabidopsis, soybean, and tobacco (Endo et al. 2016; Tang et al. 2017; Xu et al. 2017). This tool is also based on an endonuclease that is guided by a single RNA. The endonuclease eventually identifies thymidine-rich PAM and produces staggered cuts (Zetsche et al. 2015). Cpf1 is both a DNase and RNAse. First, it chops up target DNA and second, it also processes its own crRNA (Fonfara et al. 2016). In the CRISPR/Cas9 system, Cas9 requires a longer chimeric section of crRNA and tracrRNA. In lieu, the maturation of crRNA by Cpf1 does not involve the aid of tracrRNA. The PAM sequence recognized by Cas9 is 5′-NGG-3′ that is basically G rich sequence and is placed downstream of the target site, while Cpf1 identifies a T-rich PAM sequence situated upstream of the target site. So, the difference lies in their chopping sites as Cpf1 and Cas9 cut DNA at distal and proximal sites in relation to PAM sequence respectively (Xie et al. 2015). Though previous studies have proved that the simultaneous targeting of multiple genes in plants is possible with CRISPR/Cas9 (Xie et al. 2015), the drawback is that a large chunk of sequences contain numerous sgRNA cassettes, which are more labor-intensive to design and generate. Moreover, these may have instability and reduced efficiency after the transformation of explants. In contrast, the CRISPR/Cpf1 system has a greater efficiency of multiplex genome editing using simple short customized direct repeats guided arrays.

17.2.5 Base Editors

CRISPR/Cas9 and its modified form CRISPR/Cpf1 involve NHEJ and HDR mechanisms for DNA damage repair. These methods have a few drawbacks such as lower efficiency of delivery and undesired rearrangements due to DSBs at the targeted loci. As a modified system, the CRISPR/Cas9-based base editor tool has come into play. This technique involves a straightforward and permanent alteration of one base to another without involving a DSB (Komor et al. 2016; Nishida et al. 2016). This technology of base editing comprises a fusion of mutated Cas9 domain with a base-modifying deaminase enzyme that acts on a single-stranded DNA (ssDNA) as a substrate. Similar to CRISPR/Cas9, in the base editing system, a guide RNA binds to the target DNA. In the process of the base pairing of the guide RNA with the target DNA, a DNA bubble having a short segment of ssDNA is created. The deaminase enzyme modifies the targeted base of the ssDNA. Mutated Cas9 with guide RNA is necessary for denaturation of the target DNA duplex (Nishida et al. 2016). As mentioned earlier, wild-type Cas9 contains two nuclease domains RuvC and HNH that act together to cleave dsDNA and induce DSBs. The RuvC domain cuts the strand that pairs with gRNA while the HNH domain generates a nick in the second unpaired complementary strand. The second unpaired ssDNA is a target strand of deaminase enzyme to modify the specific base. Cas9 variants have been created through site-directed mutagenesis of key residues at the catalytic sites within the two nuclease domains. The Cas9 mutant having D10A mutation in the RuvC domain chops the target strand (the strand which undergoes base editing) only while the one having H840A mutation in the HNH domain cuts the non-target strand (the strand paired with gRNA). Thus, both types of variants of Cas9 generate single-strand breaks at the site of the target. However, another variant known as the dead Cas9 (dCas9) does not produce nick in any strand of dsDNA as it contains both mutations. In improved base-editor tools, the Cas9-H840A variant is used that creates a nick in the non-edited DNA strand in the bubble. It induces an *in-vivo* DNA repair system of the cell to fix the non-edited DNA strand using the edited one as a source template. DNA damage repair system, operating within the cell, resists base editing. For example, uracil N-glycosylase (UNG) identifies U-G mismatch and causes hydrolysis of the glycosidic bond between deoxyribose sugar and uracil. Base-excision repair restores C-G base pair. To overcome this problem, the uracil DNA glycosylase inhibitor (UGI) protein is fused to base editors. UNG is inhibited by UGI which is a DNA mimic.

Base-editor tools having cytosine deaminase alter C-G base pairs to T-A base pairs (Lu and Zhu 2017). Cytosine deaminase causes deamination of cytosine into uracil. They have gained success in plants. A precise base editor with a cytidine deaminase from rat and a Cas9 variant- Cas9-D10A nickase (nCas9) was utilized to edit *OsCDC48* gene in rice (Huang et al. 2016). This gene regulates cell death and senescence. Later, nCas9, fused with the enzyme cytidine deaminase, was used for a targeted point mutation in *OsPDS* (*Phytoene desaturase*) and *OsSBEIIb* (gene encoding starch branching enzyme) genes (Li et al. 2016). Base-editing of the nitrogen transporter *OsNRT1.1B* (replacement of C with T) was carried out in rice (Hu et al. 2015). It could enhance nitrogen uptake efficiency. Similarly, the base editing of the DELLA protein-encoding gene *OsSLR1* resulted in the reduction of the height of the plant (Asano et al. 2009). The base-editing strategy was also used for providing herbicide tolerance by targeting *OsALS1* (*acetolactate synthase*) gene (Kuang et al. 2020). In rice, base editing involving C to T alteration in the coding sequence of *OsALS* gene resulted in resistance to imazamox which is a potent herbicide (Xu et al. 2019).

Improved base editors were developed for targeted mutations in rice. Ren et al. (2018) generated a base editor, rice base editor (rBE5), using a cytidine deaminase from human, UGI, and a tailored Cas9n. Cytosine deaminase from humans introduces point mutations at GC and AC target regions. A variant of this human cytidine deaminase was used for proper expression in rice. The efficiency of rBE5 was tested in rice protoplast. This optimized system functioned well regarding the substitution of different bases in rice. Overall, this tool demonstrated highly efficient base editing in terms of both gain and loss of function mutations.

17.3 Genome Editing for Biofortification of Rice

Fortification is the process of increment of specific content such as vitamins and essential elements in a regular diet to improve the nutritional quality of food. It is beneficial for public health. Biofortification is the course through which nutritive values of crops are enriched by agricultural practices, plant breeding or biotechnology. The difference between biofortification and fortification is that the former aims to add nutritional values in crops during regular plant growth and development while the latter aims at this using manual ways during the processing of the crops. Biofortification can provide nourishment to the human population where conventional fortification is difficult to be imposed or fails. Several projects have been implemented on biofortification of various nutrients such as iron, zinc, carotenoid and amino-acid contents in different crops like maize, rice, sorghum, wheat, beans, legumes, sweet potato, etc. (Garg et al. 2018). It has been tried to satisfy the "hidden hunger" for quality nutrients through biofortification. In the next sections of the chapter, recent efforts regarding the biofortification of rice through genome-editing tools have been described. These tools have been applied to generate superior varieties of rice with higher nutritional and commercial values. A comprehensive detail of the manipulation of various rice genes using genome-editing tools has been provided in Table 17.1. Improvement of rice yield is also important for food security, especially for the poor population of rice-consuming countries. So, a part dealing with genome editing for yield-related traits of rice has also been included in this chapter.

17.3.1 Increase in Carbohydrate Content

In rice, the number of published reports on genome-editing tools is limited to date but there is a trend of a steady increase in the number of such publications. One of the key contents of rice grain is carbohydrate, which has been the focus of genome editing. Resistant starch and amylose contents have been significantly improved by modification of starch-branching enzymes or SBEs. *OsSBEIIb* showed high expression in rice grains hence it was targeted for mutation by CRISPR/Cas9 tool (Sun et al. 2017). After a successful Agrobacterium-mediated transformation of the Kitaake variety of rice, mutants contained changes in the exact locus with no noted off-targets. Transgene-free mutants were generated after segregation. Resistant starch and amylose contents increased up to 9% and 25%, respectively in *ossbeII* mutants. In a recent study, *OsISA1* gene encoding isoamylase-type debranching enzyme 1 was edited in rice (Shufen et al. 2019). This enzyme is involved in amylopectin synthesis. Seed size was found to be affected in homozygous mutant lines. The mutants showed shrinking in the seed endosperm and reduction in grain weight. The amounts of amylose, amylopectin, and total starch were also considerably reduced in the generated mutants. In contrast, starch-gel consistency and sugar content increased by four fold in comparison to the wild-type plants (Shufen et al. 2019). This study showed *OsISA1* affects starch biosynthesis and development of the endosperm. Mutants of *OsISA1* gene can be used for the production of natural food stabilizer and thickener owing to the low viscosity and high water solubility of phytoglycogen.

17.3.2 Increase in Glutinous Content

Amylose provides an added value to rice grains regarding its quality for cooking and eating. Enhancement of amylose content of rice grains increases the market value; conversely lowering the content creates glutinous rice that is also valuable (Zhang 2007). In this context, a single dominant gene *Waxy* was identified which regulates the amylose

Table 17.1 Application of genome-editing tools for improvement of nutritional quality and yield-related traits of rice.

Genome editing tool	Target gene	Trait	Rice cultivar/variety	References
CRISPR/Cas9	*OsSBEIIb*	High amylose content	Kitaake	Sun et al. (2017)
CRISPR/Cas9	*OsISA1*	Starch gel consistency	Zhonghua 11	Shufen et al. (2019)
CRISPR/Cas9	*OsWaxy*	Increased glutinous content	Wuyunjing, Xiushui	Zhang et al. (2018)
CRISPR/Cas9	*OsFAD2-1*	Increased oleic acid content	Nipponbare	Abe et al. (2018)
CRISPR/Cas9	*OsNRAMP5*	Low Cd accumulation	Huazhan, Longke638S	Tang et al. (2017)
CRISPR/Cas9	Intergenic region was targeted	Increase in carotenoid content	Kitaake	Dong et al. (2020)
CRISPR/Cas9	*OsF3′H, OsDFR, and OsLDOX*	Change in seed color and anthocyanin content	Heugseonchal, Sinmyungheugchal	Jung et al. (2019)
CRISPR/Cas9	*OsRc* allele	Increase in proanthocyanidins and anthocyanins content	Xiushui134, Shuhui143, ZhiNongS	Zhu et al. (2019)
CRISPR/Cas9	*OsMATL*	Development of haploid rice lines	IR58025B	Yao et al. (2018)
Multiplex CRISPR/Cas9	*OsPYL1, OsPYL4, andOsPYL* 6	Enhancement of yield and growth	Nipponbare	Miao et al. (2018)
CRISPR/Cas9	*OsGn1a, OsGS3, OsDEP1, OsIPA1*	Enhancement in size and number of grains, erect and dense panicles	Zhonghua 11	Li et al. (2016)
Multiplex CRISPR/Cas9	*OsGW2, OsGW5, OsTGW6*	Increased grain weight	LH422	Xu et al. (2016)
CRISPR/Cas9	*OsLAZY1*	Tiller-spreading trait	Kitaake	Miao et al. (2013)
CRISPR/Cas9	*OsFWL4*	Increase in tiller number and grain yield	Zhonghua 11	Gao et al. (2020)

CRISPR/Cas9	*OsTGMS5*	Induction of male sterility	GAZ, Zhenshan97B, HuanongB, TianfengB, YixiangB, ReB, HuahuiB, ZhongzheB, Yuejingsimiao, Yuenongsimiao, Wushansimiao	Zhou et al. (2016)
CRISPR/Cas9	*OsCSA*	Induction of male sterility	Japonica rice 9522, Kongyu131, Jiaoyou5B	Li et al. (2016)
Multiplex CRISPR/Cas9	*OsHd2*, *OsHd4, and OsHd5*	Early heading stage	Longdao16, Longdao18, Daohuaxiang2, Songjing19, Dongnong430	Li et al. (2017)
TALENs	*OsLOX3*	Improvement in quality of seed during storage	Nipponbare	Ma et al. (2015)
TALENs	*OsBADH2*	Fragrant rice	Nipponbare	Shan et al. (2015)

content in rice grains (Zhang et al. 2018). A loss-of-function mutant was generated via the CRISPR/Cas9 technique by mutating the *Waxy* gene (Zhang et al. 2018). In this study, the Agrobacterium-mediated transformation of two elite japonica varieties (Wuyunjing and Xiushui) was carried out with a high mutagenesis rate of 82–86% in the first set of transformants, which subsequently generated 4–15% of the homozygous mutant lines. Off-target mutations were not found in either of the cases. The components of the genome-editing cassette were also removed through segregation from their progenies. The amylose content significantly reduced in mutants in comparison to the wild-type plants. It, in turn, resulted in an increase in glutinosity in mutants. No difference was observed between *waxy* mutants and wild-type plants regarding various yield-related traits. So, the genome-editing tool was effectively used to change the rice into a glutinous one with no effect on other agronomic characters.

17.3.3 Increase in Oleic Acid Content

Rice-bran oil is a marketable derivative of rice, especially in Asia. Bran-oil constituents have health-endorsing features. One such worth mentioning content is a fatty acid called oleic acid. Oleic acid is used to cure and control numerous diseases (Abe et al. 2018). An appropriate change in the fatty acid biosynthetic pathway can bring about enhancement of the amount of oleic acid. Oleic acid is converted into linoleic acid by fatty acid desaturase 2 (FAD2) enzyme. The genome of *O. sativa* harbors four isoforms of *OsFAD2*, of which *OsFAD2-1* shows high expression in rice grains. The knockout of *OsFAD2-1* gene by CRISPR/Cas9 resulted in a drastic increase in oleic acid content in the Nipponbare variety (Abe et al. 2018). In the *osfad2-1* homozygous mutants, the oleic-acid content increased by two fold compared to the wild-type plants. There was no detection of linoleic acid content.

17.3.4 Decrease in Heavy Metal Content

Cadmium (Cd), a heavy metal, is highly lethal for health. It has been observed that a constant intake of heavy metal paves the way to numerous health-related issues (Clemens et al. 2013). Rice grains with high Cd accumulation pose a risk to people who eat it as part of their daily staple diet. Indica rice has a higher rate of Cd retention than japonica rice. Therefore, the development of elite Indica varieties having low Cd content is crucial. Tang et al. (2017) targeted the metal transporter gene *OsNRAMP5* by CRISPR/Cas9 technique and generated low Cd accumulating Indica rice lines. Field trials showed that homozygous knockout lines exhibited less Cd retention (0.05 mg/kg) during cultivation in Cd-treated paddies in comparison to the wild ones which had 0.33–2.90 mg/kg of Cd. No impairment was observed in other traits in mutant plants compared with wild-type ones.

17.3.5 Increase in Carotenoid Content and Pigmentation in Seeds

Five genes namely *OsDSM2*, *OsCCD4a*, *OsCCD4b*, *OsCYP97A4*, and *OsCCD7* were mutated through CRISPR/Cas9 system in rice (Yang et al. 2017). These genes encode enzymes that are involved in the catabolism of carotenoid. This was aimed to increase the accumulation of β-carotene in seed endosperm but, unfortunately, no success was achieved in increasing carotene content in the rice grains. However, recently, rice plants producing carotenoid-rich seeds have been successfully generated using the CRISPR/Cas9 tool (Dong et al. 2020). These rice plants have been reported to be off-target mutations-free as well as marker-free plants without any morphological defect and grain-yield penalty. Interestingly, in this report, no endogenous rice gene was mutated but, rather, a CRISPR/Cas9 tool was used to make a DSB at a predetermined intergenic region (known as plant genomic safe harbor) to insert a carotenoid biosynthesis cassette (Dong et al. 2020).

Targeted gene-editing has also been employed for the alteration of seed color and anthocyanin content. Three genes encoding flavanone 3′-hydroxylase, dihydrofavonol 4-reductase and leucoanthocyanidin dioxygenase were edited individually in a black-rice variety using the CRISPR/Cas9 tool (Jung et al. 2019). These genes are related to the anthocyanin biosynthesis pathway. In many rice consuming countries, red-rice grain producing varieties are cultivated. Red grains provide nutritional benefits. They contain a high amount of proanthocyanidins and anthocyanins. Wild-rice species (*Oryza rufipogon* L.) also produce red grains. The red coloration is regulated by two complementary genes, *Rc* and *Rd* that encode a transcription factor and a dihydroflavonol-4-reductase (DFR) protein respectively. During rice domestication, red color was lost due to frame-shift mutation in the *Rc* gene. Zhu et al. (2019) used CRISPR/Cas9 tool to revert frame-shift mutation and restore the functional in-frame *Rc* allele. They effectively transformed three elite white-rice varieties into red ones (Zhu et al. 2019).

17.4 Genome Editing for Improvement of Agronomic Traits in Rice

Genome-editing technology has also been used to bring about many changes in the agronomic traits of rice (Table 17.1). With the advancement of this technology, the success rate has increased. Mutation in *OsMATL* gene using CRISPR/Cas9 is a noteworthy one that induced the formation of haploid seeds in rice. This gene encodes a pollen-specific phospholipase. *OsMATL* knockout mutation decreased seed set but also resulted in 2–6% haploid induction rate (Yao et al. 2018). This work is a breakthrough for rice breeding. The application of genome editing technology for the improvement of yield and related traits has been discussed in the following sections.

17.4.1 Increase in Rice Harvest Properties

Yield is controlled by an interplay of different hormones and multiple genes. Using the approach of multiplex CRISPR/Cas9, editing of *pyrabactin resistance* (PYL) *genes* (*OsPYL1–OsPYL6*, *OsPYL12*, *OsPYL7–OsPYL11*, and *OsPYL13*) in rice increased its productivity (Miao et al. 2018). Enzymes encoded by these genes are involved in the regulation of abscisic acid receptors. Editing of *OsPYL1–OsPYL6* and *OsPYL12* genes also favored rice growth and development. Among them, a triple mutant *ospyl1/4/6* exhibited maximum productivity and growth with no harm to the seed dormancy period or other agronomic traits (Miao et al. 2018). In light of the harvest index of rice, the most important traits are grains per panicle, panicles per plant, and weight of rice grains. Using the CRISPR/Cas9 method, genes such as *OsGn1a, OsDEP1, OsGS3*, and *OsIPA1* were modified (Li et al. 2016).

Mutant rice plants showed superior panicles, an increase in grain number and grain size. Phenotypes of semi-dwarfism and grain with a long awn were also seen in mutants *of* *OsDEP1* and *OsGS3* genes respectively. Xu et al. (2016) reported the successful simultaneous targeting of three genes using the multiplex genome-editing system. These three genes were *grain weight 2* (*OsGW2*), *OsGW5*, and *thousand-grain weight6* (TGW6). These are negative regulators of grain weight. They reported an increment of 29.3% in grain weight per thousand seeds in triple-mutant rice plants. Using the CRISPR/Cas9 tool, two vacuolar invertases namely OsINV2 and OsINV3 have been identified to control grain-size (Deng et al. 2020). Tillering of rice plants is another important aspect in farming. It was observed that CRISPR/Cas9-mediated mutation in the *OsLAZY1* gene, induced a tiller-spreading phenotype in rice (Miao et al. 2013). Gao et al. (2020) also reported an increase in tiller numbers and grain yield by targeted mutation of *FW2.2 Like* gene (*OsFWL4*) using the CRISPR/Cas9 System. Mutation in this gene also resulted in a wider flag leaf due to an increase in cell number.

17.4.2 Induction of Male Sterility and Shortening of Flowering Period

The process of rice breeding involves the use of a male sterile line for cross-breeding. CRISPR/Cas9 technology has gained a profound interest in the generation of thermo-sensitive genic male sterile (TGMS) lines. This tool was applied to cause precise mutation in *OsTGMS5* gene. A total of 11 novel TGMS rice lines belonging to Indica and Japonica subspecies were readily developed within a single year (Zhou et al. 2016). In japonica rice, photo-sensitive genic male sterile (PGMS) lines were developed by targeting "*Carbon Starved Anther*" (OsCSA) gene using the CRISPR/Cas9 tool. These PGMS lines showed male-sterile trait only under a short-day environment (Li et al. 2016). Heading time is another relevant agricultural property for good rice yield. Three genes namely *OsHd2*, *OsHd4*, and *OsHd5* were edited simultaneously using the multiplex CRISPR/Cas9 system (Li et al. 2017). These genes act as flowering suppressors and negatively regulates the heading date. Early-maturing rice mutants were developed.

17.4.3 Improvement of Traits for Economic Purposes

Maintaining rice-grain quality and shelf life during storage is important in terms of the economy. Lipoxygenases (LOXs) are responsible for reducing rice-grain quality during storage. These enzymes catalyze the process of dioxygenation of polyunsaturated fatty acids to form hydrogen peroxide. The viability of rice seeds is also negatively impacted by *OsLOX1, OsLOX2*, and *OsLOX3* genes. A TALENs-based tool was used to induce an effective mutation in the *OsLOX3* gene (Ma et al. 2015). The *oslox3* knockout lines showed promising seed durability.

Another important trait of the rice grain is its fragrance after cooking. The demand for fragrant rice is very high in the global market. Betaine aldehyde dehydrogenase (BADH2) enzyme hinders generation of 2-acetyl-1-pyrroline (2AP) that significantly contributes to the fragrance of rice. *OsBADH2* gene was effectively mutated by TALEN. It resulted in an increase of 35–75% in 2AP content (Shan et al. 2015).

17.5 Conclusion and Future Aspects

Genome-editing tools have gained ground owing to their ability to generate targeted mutations in the genome to incorporate desired traits in crops within a relatively short period of time. Changes in DNA sequences take place with precision and accuracy using this approach. Editing of the genomic landscape of crop plants, based on epigenetic modifications, also hold great potential in crop improvement because these modifications can be transferred to the next generations without any alteration in the genomic sequence. Recently, CRISPR/dCas9 tool was fused with DNA methyltransferase 3a (DNMT3a) to induce methylation of DNA in mammalian cells (Liu et al. 2016). There is also high scope of ringing epigenetic changes using these tools in rice.

Nevertheless, there are still obstacles in the area of genome editing. The most important challenge is to make the products of this technology available to farmers. Most of the experiments are conducted on a laboratory scale. They are mere proof of the larger concept of genome editing, hence it gives rise to questions and uncertainties pertaining to the performance of the genome-edited crops in the natural environment. Therefore, such studies and validation of proof of the concept should be carried out through field trials. Another dearth of the fact which this technology is facing is biosafety regulations. In the world scenario, several bodies are advocating acceptance of genome-edited crops. These communities include scientists, policymakers, academicians, stakeholders, etc. Proper phenotyping of genome-edited crops, the DNA-repair mechanism involved in the process of genome editing, and the regulatory scenario regarding their release are some of the important issues of concern. Till now, plants with edited DNA fall under the category of genetically modified organisms (GMOs). However, the US Department of Agriculture has exempted many genome-edited crops from GMO regulations (Waltz 2016). Contrarily, the European court of justice has delivered a recent judgment to categorize and regulate genome-edited organisms as GMOs (Callaway 2018). Recently, in India, the Department of Biotechnology prepared a draft document of guidelines and regulatory framework for genome-edited organisms. In the draft, it has been proposed that the process or product of genetic-editing technology is to be regulated as per the existing regulatory framework for GMOs. This technology is to be covered under the rules 1989 ("Rules for the manufacture, use, import, export, and storage of hazardous microorganisms, genetically engineered organisms or cells, 1989" notified under the Environment Protection Act 1986). These rules are implemented through six competent authorities. Nonetheless, this technology can generate rice plants with better traits. It can usher forward a revolution in the field of improvement of the food crops which not only includes rice but also other ones.

Acknowledgment

PS would like to acknowledge the University of Rajasthan, Jaipur for providing basic infrastructure facilities. PS would also like to thank the University Grants Commission, India for Start-up Grant for financial support. PS is also grateful to Departmental Research Support (DRS Phase-II), Department of Botany, University of Rajasthan for providing necessary research facilities.

Conflicts of Interest

The authors declare that they have no conflicts of interest.

References

Abe, K., Araki, E., Suzuki, Y. et al. (2018). Production of high oleic/low linoleic rice by genome editing. *Plant Physiol. Biochem.* 131: 58–62.

Anders, C., Bargsten, K., and Jinek, M. (2016). Structural plasticity of PAM recognition by engineered variants of the RNA-guided endonuclease Cas9. *Mol. Cell* 61: 895–902.

Andersson, M., Turesson, H., Nicolia, A. et al. (2017). Efficient targeted multiallelic mutagenesis in tetraploid potato (Solanum tuberosum) by transient CRISPR-Cas9 expression in protoplasts. *Plant Cell Rep.* 36: 117–128.

Asano, K., Hirano, K., Ueguchi-Tanaka, M. et al. (2009). Isolation and characterization of dominant dwarf mutants, Slr1-d, in rice. *Mol. Gen. Genomics.* 281: 223–231.

Beumer, K., Bhattacharyya, G., Bibikova, M. et al. (2006). Efficient gene targeting in Drosophila with zinc-finger nucleases. *Genetics* 172: 2391–2403.

Bhaya, D., Davison, M., and Barrangou, R. (2011). CRISPR-Cas Systems in Bacteria and Archaea: versatile small RNAs for adaptive defense and regulation. *Annual Review of Genetics* 45: 273–297.

Callaway, E. (2018). CRISPR plants now subject to tough GM laws in European Union. *Nature* 560: 16.

Clemens, S., Aarts, M.G., Thomine, S., and Verbruggen, N. (2013). Plant science: the key to preventing slow cadmium poisoning. *Trends Plant Sci.* 18: 92–99.

Deng, X., Han, X., Yu, S. et al. (2020). OsINV3 and its homolog, OsINV2, control grain size in Rice. *Int. J. Mol. Sci.* 21.

Dong, O.X., Yu, S., Jain, R. et al. (2020). Marker-free carotenoid-enriched rice generated through targeted gene insertion using CRISPR-Cas9. *Nat. Commun.* 11: 1178.

Doyon, Y., McCammon, J.M., Miller, J.C. et al. (2008). Heritable targeted gene disruption in zebrafish using designed zinc-finger nucleases. *Nat. Biotechnol.* 26: 702–708.

Endo, M., Mikami, M., and Toki, S. (2016). Biallelic gene targeting in Rice. *Plant Physiol.* 170: 667–677.

Fonfara, I., Richter, H., Bratovič, M. et al. (2016). The CRISPR-associated DNA-cleaving enzyme Cpf1 also processes precursor CRISPR RNA. *Nature* 532: 517–521.

Gao, Q., Li, G., Sun, H. et al. (2020). Targeted mutagenesis of the rice FW 2.2-like gene family using the CRISPR/Cas9 system reveals OsFWL4 as a regulator of tiller number and plant yield in Rice. *Int. J. Mol. Sci.* 21: 809.

Garg, M., Sharma, N., Sharma, S. et al. (2018). Biofortified crops generated by breeding, agronomy, and transgenic approaches are improving lives of millions of people around the world. *Front. Nutr.* 5: 12.

Hauschild, J., Petersen, B., Santiago, Y. et al. (2011). Efficient generation of a biallelic knockout in pigs using zinc-finger nucleases. *Proc. Natl. Acad. Sci. U. S. A.* 108: 12013–12017.

Hermans, P.W., van Soolingen, D., Bik, E.M. et al. (1991). Insertion element IS987 from Mycobacterium bovis BCG is located in a hot-spot integration region for insertion elements in Mycobacterium tuberculosis complex strains. *Infect. Immun.* 59: 2695–2705.

Hockemeyer, D., Wang, H., Kiani, S. et al. (2011). Genetic engineering of human pluripotent cells using TALE nucleases. *Nat. Biotechnol.* 29: 731–734.
Hu, B., Wang, W., Ou, S. et al. (2015). Variation in NRT1.1B contributes to nitrate-use divergence between rice subspecies. *Nat. Genet.* 47: 834–838.
Hu, X., Wang, C., Fu, Y. et al. (2016). Expanding the range of CRISPR/Cas9 genome editing in rice. *Mol. Plant* 9: 943–945.
Huang, Q.N., Shi, Y.F., Zhang, X.B. et al. (2016). Single base substitution in OsCDC48 is responsible for premature senescence and death phenotype in rice. *J. Integr. Plant Biol.* 58: 12–28.
Ishino, Y., Shinagawa, H., Makino, K. et al. (1987). Nucleotide sequence of the iap gene, responsible for alkaline phosphatase isozyme conversion in Escherichia coli, and identification of the gene product. *J. Bacteriol.* 169: 5429–5433.
Jiang, W., Bikard, D., Cox, D. et al. (2013). RNA-guided editing of bacterial genomes using CRISPR-Cas systems. *Nat. Biotechnol.* 31: 233–239.
Jiang, W., Zhou, H., Bi, H. et al. (2013). Demonstration of CRISPR/Cas9/sgRNA-mediated targeted gene modification in Arabidopsis, tobacco, sorghum and rice, in Nucleic Acids Research. *Nucleic Acids Research* 41: e188–e.
Jinek, M., Chylinski, K., Fonfara, I. et al. (2012). A programmable dual-RNA–guided DNA endonuclease in adaptive bacterial immunity. *Science* 337: 816–821.
Joung, J.K. and Sander, J.D. (2013). TALENs: a widely applicable technology for targeted genome editing. *Nat. Rev. Mol. Cell Biol.* 14: 49–55.
Jung, Y.J., Lee, H.J., Kim, J.H. et al. (2019). CRISPR/Cas9-targeted mutagenesis of F3′H, DFR and LDOX, genes related to anthocyanin biosynthesis in black rice (Oryza sativa L.). *Plant Biotechnol. Rep.* 13: 521–531.
Kaya, H., Mikami, M., Endo, A. et al. (2016). Highly specific targeted mutagenesis in plants using Staphylococcus aureus Cas9. *Sci. Rep.* 6: 26871.
Kim, Y.G., Cha, J., and Chandrasegaran, S. (1996). Hybrid restriction enzymes: zinc finger fusions to Fok I cleavage domain. *Proc. Natl. Acad. Sci. U. S. A.* 93: 1156–1160.
Kim, D., Kim, J., Hur, J.K. et al. (2016). Genome-wide analysis reveals specificities of Cpf1 endonucleases in human cells. *Nat. Biotechnol.* 34: 863–868.
Kleinstiver, B.P., Tsai, S.Q., Prew, M.S. et al. (2016). Genome-wide specificities of CRISPR-Cas Cpf1 nucleases in human cells. *Nat. Biotechnol.* 34: 869–874.
Komor, A.C., Kim, Y.B., Packer, M.S. et al. (2016). Programmable editing of a target base in genomic DNA without double-stranded DNA cleavage. *Nature* 533: 420–424.
Krappmann, S. (2016). CRISPR-Cas9, the new kid on the block of fungal molecular biology. *Med. Mycol.* 55: 16–23.
Kuang, Y., Li, S., Ren, B. et al. (2020). Base-editing-mediated artificial evolution of OsALS1 in planta to develop novel herbicide-tolerant rice germplasms. *Mol. Plant* 13: 565–572.
Li, J., Meng, X., Zong, Y. et al. (2016). Gene replacements and insertions in rice by intron targeting using CRISPR-Cas9. *Nat Plants* 2: 16139.
Li, M., Li, X., Zhou, Z. et al. (2016). Reassessment of the four yield-related genes Gn1a, DEP1, GS3, and IPA1 in rice using a CRISPR/Cas9 system. *Front. Plant Sci.* 7: 377.
Li, Q., Zhang, D., Chen, M. et al. (2016). Development of japonica photo-sensitive genic male sterile rice lines by editing carbon starved anther using CRISPR/Cas9. *J. Genet. Genomics* 43: 415–419.

Li, X., Zhou, W., Ren, Y. et al. (2017). High-efficiency breeding of early-maturing rice cultivars via CRISPR/Cas9-mediated genome editing. *J. Genet. Genomics* 44: 175–178.

Liang, Z., Chen, K., Li, T. et al. (2017). Efficient DNA-free genome editing of bread wheat using CRISPR/Cas9 ribonucleoprotein complexes. *Nat. Commun.* 8: 14261.

Liu, X.S., Wu, H., Ji, X. et al. (2016). Editing DNA methylation in the mammalian genome. *Cell* 167: 233–247. e17.

Lu, Y. and Zhu, J.K. (2017). Precise editing of a target base in the rice genome using a modified CRISPR/Cas9 system. *Mol. Plant* 10: 523–525.

Ma, L., Zhu, F., Li, Z. et al. (2015). TALEN-based mutagenesis of lipoxygenase LOX3 enhances the storage tolerance of rice (Oryza sativa) seeds. *PLoS One* 10: e0143877.

Meng, X., Yu, H., Zhang, Y. et al. (2017). Construction of a genome-wide mutant library in rice using CRISPR/Cas9. *Mol. Plant* 10: 1238–1241.

Miao, J., Guo, D., Zhang, J. et al. (2013). Targeted mutagenesis in rice using CRISPR-Cas system. *Cell Res.* 23: 1233–1236.

Miao, C., Xiao, L., Hua, K. et al. (2018). Mutations in a subfamily of abscisic acid receptor genes promote rice growth and productivity. *Proc. Natl. Acad. Sci. U. S. A.* 115: 6058–6063.

Michael, T.P. and Jackson, S. (2013). *The First 50 Plant Genomes*, vol. 6. Wiley.

Miller, J.C., Holmes, M.C., Wang, J. et al. (2007). An improved zinc-finger nuclease architecture for highly specific genome editing. *Nat. Biotechnol.* 25: 778–785.

Mishra, R., Joshi, R.K., and Zhao, K. (2018). Genome editing in rice: recent advances, challenges, and future implications. *Front. Plant Sci.* 9: 1361.

Moore, M., Klug, A., and Choo, Y. (2001). Improved DNA binding specificity from polyzinc finger peptides by using strings of two-finger units. *Proc. Natl. Acad. Sci. U. S. A.* 98: 1437–1441.

Nakata, A., Amemura, M., and Makino, K. (1989). Unusual nucleotide arrangement with repeated sequences in the Escherichia coli K-12 chromosome. *J. Bacteriol.* 171: 3553–3556.

Nishida, K., Arazoe, T., Yachie, N. et al. (2016). Targeted nucleotide editing using hybrid prokaryotic and vertebrate adaptive immune systems. *Science* 353.

Osiak, A., Radecke, F., Guhl, E. et al. (2011). Selection-independent generation of gene knockout mouse embryonic stem cells using zinc-finger nucleases. *PLoS One* 6: e28911.

Pyott, D.E., Sheehan, E., and Molnar, A. (2016). Engineering of CRISPR/Cas9-mediated potyvirus resistance in transgene-free Arabidopsis plants. *Molecular Plant Pathology* 17: 1276–1288.

Regina, A., Bird, A., Topping, D. et al. (2006). High-amylose wheat generated by RNA interference improves indices of large-bowel health in rats. *Proc. Natl. Acad. Sci. U. S. A.* 103: 3546–3551.

Ren, B., Yan, F., Kuang, Y. et al. (2018). Improved base editor for efficiently inducing genetic variations in rice with CRISPR/Cas9-guided hyperactive hAID mutant. *Mol. Plant* 11: 623–626.

Ricroch, A., Clairand, P., and Harwood, W. (2017). Use of CRISPR systems in plant genome editing: toward new opportunities in agriculture. *Emerg. Top Life Sci.* 1: 169–182.

Shan, Q., Wang, Y., Li, J. et al. (2013). Targeted genome modification of crop plants using a CRISPR-Cas system. *Nat. Biotechnol.* 31: 686–688.

Shan, Q., Zhang, Y., Chen, K. et al. (2015). Creation of fragrant rice by targeted knockout of the OsBADH2 gene using TALEN technology. *Plant Biotechnol. J.* 13: 791–800.

Shufen, C., Yicong, C., Baobing, F. et al. (2019). Editing of rice isoamylase gene ISA1 provides insights into its function in starch formation. *Rice Sci.* 26: 77–87.

Straimer, J., Lee, M.C., Lee, A.H. et al. (2012). Site-specific genome editing in plasmodium falciparum using engineered zinc-finger nucleases. *Nat. Methods* 9: 993–998.

Sun, Y., Jiao, G., Liu, Z. et al. (2017). Generation of high-amylose rice through CRISPR/Cas9-mediated targeted mutagenesis of starch branching enzymes. *Front. Plant Sci.* 8: 298.

Tang, X., Lowder, L.G., Zhang, T. et al. (2017). A CRISPR–Cpf1 system for efficient genome editing and transcriptional repression in plants. *Nat. Plants* 3: 17018.

Tang, L., Mao, B., Li, Y. et al. (2017). Knockout of OsNramp5 using the CRISPR/Cas9 system produces low Cd-accumulating indica rice without compromising yield. *Sci. Rep.* 7: 14438.

Urnov, F.D., Rebar, E.J., Holmes, M.C. et al. (2010). Genome editing with engineered zinc finger nucleases. *Nat. Rev. Genet.* 11: 636–646.

Waltz, E. (2016). CRISPR-edited crops free to enter market, skip regulation. *Nat. Biotechnol.* 34: 582.

Wang, T., Zhang, H., and Zhu, H. (2019). CRISPR technology is revolutionizing the improvement of tomato and other fruit crops. *Hortic Res.* 6: 77.

Wingender, E., Chen, X., Hehl, R. et al. (2000). TRANSFAC: an integrated system for gene expression regulation. *Nucleic Acids Res.* 28: 316–319.

Xie, K., Minkenberg, B., and Yang, Y. (2015). Boosting CRISPR/Cas9 multiplex editing capability with the endogenous tRNA-processing system. *Proceedings of the National Academy of Sciences of the United States of America* 112: 3570–3575.

Xu, R., Yang, Y., Qin, R. et al. (2016). Rapid improvement of grain weight via highly efficient CRISPR/Cas9-mediated multiplex genome editing in rice. *J. Genet. Genomics* 43: 529–532.

Xu, R., Qin, R., Li, H. et al. (2017). Generation of targeted mutant rice using a CRISPR-Cpf1 system. *Plant Biotechnol. J.* 15: 713–717.

Xu, W., Song, W., Yang, Y. et al. (2019). Multiplex nucleotide editing by high-fidelity Cas9 variants with improved efficiency in rice. *BMC Plant Biol.* 19: 511.

Yang, X., Chen, L., He, J., and Yu, W. (2017). Knocking out of carotenoid catabolic genes in rice fails to boost carotenoid accumulation, but reveals a mutation in strigolactone biosynthesis. *Plant Cell Rep.* 36: 1533–1545.

Yao, L., Zhang, Y., Liu, C. et al. (2018). OsMATL mutation induces haploid seed formation in indica rice. *Nat. Plants* 4: 530–533.

Zetsche, B., Gootenberg, J.S., Abudayyeh, O.O. et al. (2015). Cpf1 is a single RNA-guided endonuclease of a class 2 CRISPR-Cas system. *Cell* 163: 759–771.

Zhang, Q. (2007). Strategies for developing green super rice. *Proc. Natl. Acad. Sci. U. S. A.* 104: 16402–16409.

Zhang, Y., Liang, Z., Zong, Y. et al. (2016). Efficient and transgene-free genome editing in wheat through transient expression of CRISPR/Cas9 DNA or RNA. *Nat. Commun.* 7: 12617.

Zhang, J., Zhang, H., Botella, J.R., and Zhu, J.K. (2018). Generation of new glutinous rice by CRISPR/Cas9-targeted mutagenesis of the waxy gene in elite rice varieties. *J. Integr. Plant Biol.* 60: 369–375.

Zhou, H., He, M., Li, J. et al. (2016). Development of commercial Thermo-sensitive genic male sterile Rice accelerates hybrid Rice breeding using the CRISPR/Cas9-mediated TMS5 editing system. *Sci. Rep.* 6: 37395.

Zhu, Y., Lin, Y., Chen, S. et al. (2019). CRISPR/Cas9-mediated functional recovery of the recessive rc allele to develop red rice. *Plant Biotechnol. J.* 17: 2096–2105.

18

Genome Editing for Improving Abiotic Stress Tolerance in Rice

Shweta Roy[1#], Nirbhay Kumar Kushwaha[2#], Hasthi Ram[3], and Praveen Soni[4]

[1] *Department of Cell and Molecular Biology, Uppsala University, Uppsala, Sweden*
[2] *Department of Plant Biology, Swedish University of Agricultural Sciences, Uppsala, Sweden*
[3] *Department of Agri-Biotechnology, National Agri-Food Biotechnology Institute, Mohali, Punjab, India*
[4] *Department of Botany, University of Rajasthan, Jaipur, Rajasthan, India*
[#] *Equal Contribution*

CHAPTER MENU

Genome Engineering for Crop Improvement, First Edition. Edited by Santosh Kumar Upadhyay.

18.1 Introduction

Rice is the main food source for more than 3.5 billion population worldwide. It provides a staple food to a lower and middle-income population of most of the developing countries (Birla et al. 2017). Asia alone consumes around 90% of the total rice production and its demand is continuously rising. Not only in Asia, but the per capita consumption of rice is also continuously rising in African and Latin American countries. In the year 2011, 450 million tons of rice was consumed globally. It is estimated that global rice consumption will rise to 490 million tonnes in 2020. It is expected to further increase up to 650 million tons by the year 2050 (Muthayya et al. 2014). Therefore, rice yield is required to be increased by over 40% to meet global demand by 2050. According to an estimate, agricultural production should increase by 70–85% to feed the population of more than nine billion people in 2050 (Tilman et al. 2011; Alexandratos 2012).

Various abiotic stresses such as extreme temperatures, salinity, drought, flood, and heavy metals cause severe loss in agricultural production and act as a major hurdle in augmenting yield. Generating high yielding stress-tolerant cultivars are the need of the hour to provide a sustainable solution for this challenge. Over the last few decades, scientists were successful in increasing rice productivity by using conventional breeding and mutational-breeding strategies as well as a relatively advanced method of genetic engineering. Conventional crop-breeding techniques involve introgression of a desired trait from a donor variety in an elite variety. The required trait usually comes from existing natural germplasm available in the form of wild relatives of crops. The whole process requires the availability of broad genetic variation and an extensive process of backcrossing to remove any unwanted linked-traits into the elite variety (Scheben et al. 2017). Though the method has generated many new improved varieties with better agronomic traits, the prerequisite of the existence of desired alleles in nature and a time-consuming procedure involving several backcrosses are major limitations of the conventional breeding method. The limitation of the unavailability of naturally occurring desired genetic variation can be overcome by artificial induction of mutations. Mutagenesis by chemicals (e.g. ethyl methane sulfonate), radiation (gamma irradiation) and biological (T-DNA and transposon insertion) methods cause random mutations throughout the genome and results in artificially induced genetic variations (Sikora et al. 2011). However, a large population must be screened for the selection of the desired mutant phenotype. It also requires selection and backcrossing for several generations to eliminate other probable undesirable mutations. The whole procedure is time-consuming and, at the same time, involves rigorous labor. Limitations of random mutagenesis and conventional breeding were overcome by the transgenic approach. In transgenesis, an exogenous gene encoding for a specific trait is transferred into a transformable genotype and later into elite varieties to improve the desired agronomic trait. However, the expression of a foreign gene and its interaction with native genes and the environment, are complex in nature and must be assessed very carefully. Biosafety issues such as the impact on the environment and human health risk evaluation are required before approval of any genetically modified (GM) crop variety (Prado et al. 2014). That is why, despite the development of many GM crop varieties, only a few were approved for commercial usage. Worldwide acceptance of GM crops is still awaited. Therefore, continuous efforts are needed to invent and develop new techniques that are relatively faster than existing methods and can be accepted worldwide for agronomic purposes.

In the past decade, genome editing has become a ground-breaking revolution in the field of molecular biology for the precise and efficient genome modification at the targeted locus (Chen and Gao 2014). The discovery of site-specific nucleases (SSNs) from many organisms has opened the gates to tremendous possibilities for the development of new improved and bio-safe lines by precise genome editing (Gaj et al. 2013). These SSNs can generate several types of mutations such as deletions, insertions, substitutions, replacement, and integration of a specific DNA at a target site. All SSNs identified to date essentially work through the generation of a double-strand break (DSB) in the target DNA. The DSB is repaired either by a non-homologous end-joining (NHEJ) mechanism or a more precise homology dependent recombination (HDR) method (Wyman and Kanaar 2006; Waterworth et al. 2011). NHEJ is an error-prone repair mechanism that includes direct end joining and often generates insertion or deletion (InDels) of few nucleotides at the site of DSB (Gorbunova and Levy 1997). On the contrary, HDR is an accurate repair mechanism that requires a homologous DNA template for a recombination-based repair mechanism. HDR leads to gene replacement or gene insertion-based mutations at a target site (Waterworth et al. 2011). While transgenic strategy results in foreign gene insertion at a random position in the genome, genome-editing tools generate precise mutations at the target site. Therefore, genome-edited crops involve less concern of unpredictable phenotype and relatively fewer biosafety issues. Those genome-edited crops can directly be released as new varieties after clarifying biosafety regulations or can be used in the plant-breeding process. In this chapter, we have discussed different genome-editing technologies, their advantages and disadvantages, and their application in generating abiotic stress-tolerant rice varieties.

18.2 Recent Developments in Genome Editing Technology

In the past few years, a tremendous amount of development in molecular biology has been achieved which has resulted in the advancement of many new tools for targeted-genome modifications. Essentially, all of them comprise of site-directed DNA binding domains for guidance toward the target sequence and an endonuclease domain for its cleavage (Kim and Kim 2014). The complex produces a site-directed DSB at the target site. As mentioned earlier, these DSBs are repaired either by NHEJ or HDR cellular endogenous repair mechanisms. Indels resulting from NHEJ cause gene disruption, frame shift or premature stop codon, while HDR brings about insertions or substitutions. HDR repair mechanisms function at only S and G2 phases whereas NHEJ can take place throughout the cell cycle (Symington and Gautier 2011). Hence, NHEJ is the main repair mechanism of the cell. Thus, this function of genome-editing tools can be employed to manipulate the DNA sequence by targeted mutation at a specific locus.

There are mainly four genome-editing tools that have been developed to date. These are meganucleases (MgNs), zinc-finger nucleases (ZFNs), transcription activator-like effector nucleases (TALENs) and most popular clustered regularly interspaced palindromic repeats (CRISPR) and cas9 protein complex (CRISPR/cas9).

18.2.1 Meganucleases

MgNs, also known as homing endonucleases are unique and highly active enzymes found in many microbes (Stoddard 2011). They have long recognition sequences that can vary from 12–40 base pairs (Epinat et al. 2003). The most widespread MgNs used for genome editing are I-SceI and I-CreI. Despite their high activity and specificity, the use of MgNs is restricted due to difficulty in engineering their DNA-binding domains for any desired target DNA other than their natural DNA binding sites.

18.2.2 Zinc-Finger Nucleases

ZFNs are fusion proteins containing modified zinc-finger (ZF) domain for target DNA identification and nuclease domain of FokI restriction enzyme for generating a DSB (Bibikova et al. 2002). The DNA-binding domain of ZFNs can be customized to target the gene of interest. Usually, they are designed in an array of three to four fingers to maintain high specificity. Specific amino acids of the ZF domain can be changed through site-directed mutagenesis to achieve the specificity for target DNA. Monomeric ZF array forms dimer at the recognition site which is a prerequisite for FokI endonuclease-mediated cleavage. FokI can cleave target DNA located between two ZF domain-binding sites. It is known as "spacer DNA" which contains a five to seven base pairs long palindromic sequence (Durai et al. 2005; Carroll 2011). There are several online programs available to customize the ZF array for a gene of interest. This genome-editing method is highly specific but technically very challenging in designing and, at the same time, its manipulation is restricted to very few target sequences.

18.2.3 Transcription Activator-like Effectors Nucleases

TALENs are modified endonucleases whose DNA-binding domain is derived from bacterial transcription activator-like effectors (TALE) and DNA-cleavage domain is from FokI restriction enzyme (Christian et al. 2010). TALEs from *Xanthomonas oryzae* bacteria are secreted into the host plant cells through the type-III secretion system. After entering into the host cells, TALEs bind to the specific sequence present in the promoter of target-plant genes and activate their transcription (Boch and Bonas 2010). Their property of binding to a specific DNA sequence is exploited in designing TALENs. Like ZFNs, TALENs also possess multiple repeats of TALEs as DNA-binding domains. TALENs function in dimeric form. One TALEN monomer binds to one strand of the target DNA and the other one to its complementary strand. The spacer DNA present between the two TALEN monomers is about 20 base-pairs long. FokI-cleavage domain binds to the spacer DNA and generates a DSB. Each TALE is constituted of a multiunit array (13–28 repeats) of 34 conserved amino acids except for two of them at positions 12 and 13 (Scholze and Boch 2010; Deng et al. 2014). These two hypervariable amino acids recognize one nucleotide of the target DNA and determine the binding specificity. Hypervariable amino-acid pairs NG, NI, HD, and NN recognize thymine, adenine, cytosine, and guanine/adenine respectively and alteration in these amino acids can achieve preferred target binding for a site-specific DSB (Scholze and Boch 2010). Similar to ZFNs, TALENs are also difficult to customize for a variety of targets and require high technical expertise.

18.2.4 Clustered Regularly Interspaced Short Palindromic Repeats/CRISPR Associated Proteins System

CRISPR/Cas9 system is a breakthrough development of recent years that has revolutionized the genome-editing world. CRISPR is an assembly of short palindromic DNA repeats interspaced by spacer DNA (Wiedenheft et al. 2012). It was first reported in a bacterial system and later Cas genes were identified and speculated to be part of this complex due to the proximity of the Cas genes to the CRISPR locus (Jansen et al. 2002). In studies conducted by a few groups (Bolotin et al. 2005; Mojica et al. 2005), spacer DNA was identified as part of the DNA acquired from invading bacteriophage. Bacteria and archaea use this complex as part of an adaptive immune system for defense against invading phages and plasmids (Barrangou et al. 2007). Two classes of the CRISPR/Cas system are known, class I, which comprises of a group of Cas proteins and class II where only one Cas protein is present for cleavage. The most common Cas protein used in genome editing is class II Cas9 protein from *Streptococcus pyrogenes* (SpCas9) (Jinek et al. 2012; Jinek et al. 2014). The most commonly used customized CRISPR/Cas9 tool has two constituents, a single small-guide RNA (sgRNA) and a Cas9 nuclease protein. sgRNA is tailored for the target gene. It is composed of a fusion of crRNA (CRISPR RNA) and a transactivating crRNA (transcr-RNA) (Jinek et al. 2013). The sgRNA guides the complex to the target site by binding to the complementary nucleotides of the target DNA and Cas9 protein cleaves both strands of DNA adjacent to NGG protospacer-adjacent motif (PAM). Cas9 protein comprises two nuclease domains, HNH and RuVC, to cut the DNA strand complementary to sgRNA and the other strand of the target DNA respectively. CRISPR/Cas9 system is much easier in terms of technical requirements for designing the construct. A single-guide RNA specific for the target DNA is sufficient for recruiting the whole complex at the binding site (Jinek et al. 2012; Mali et al. 2013). The whole procedure is easier in comparison to ZFNs and TALENs where multiple protein arrays need to be engineered to customize the gene editor.

18.3 Challenges of Different Genome-Editing Systems

Genome-editing technology has come a long way since the identification of MgNs to the most recent CRISPR/Cas9 system. However, all of them have their own advantages and limitations. MgNs are highly specific for their target sequences but possess a very limited possibility to redesign them according to the requirement. Whereas, ZFNs and TALENs are improved tools in terms of flexibility for customizing, but technically it is very difficult to engineer them due to the extensive redesigning required in multiple arrays of amino-acid repeats. The whole process is relatively complex and their use is restricted for limited target sequences. The CRISPR/Cas9 system is the latest and efficient addition to genome-editing tools. It is highly effective, precise, versatile and easy to construct (Cong et al. 2013). The complex relies on RNA sequence and its complementarity to the target DNA sequence to recognize it effectively. In the case of this tool, the RNA sequence is easier to engineer than the other genome-editing tools which use protein-DNA interaction for the identification of the target sequence. The engineering of a single-guide RNA is sufficient to recruit the Cas9 protein for the cleavage of target DNA in this genome-editing tool. Though this system is the most effective among genome-editing tools, it has its own shortcomings and requires improvement in some aspects. The biggest limitation of this system is the occurrence of off-targeting (Fu et al. 2013; Rees et al. 2017). The Cas9 nuclease can cut the target DNA even in the presence of a mismatch of a few base pairs between target DNA and sgRNA (Kempton and Qi 2019). This can produce off-targets which sometimes become difficult to identify. Further, knowledge of genes acting as negative regulators and controlling agronomic and stress-related traits is a prerequisite for the genome-editing system. At present, there are only a few candidate genes known for editing. This should be the top priority to decipher the function of new genes regulating significant traits.

18.4 Application of Genome-Editing Technology for the Improvement of Abiotic Stress Tolerance in Rice

Earlier genetic engineering technology was extensively employed to enhance stress tolerance in crops. However, due to the concern about the safety of GM crops, they were not accepted by many countries. In such a scenario, the discovery of genome-editing technology has opened several possibilities in the field of agriculture. Over the last few years, the genome-editing technique has successfully been applied to improve rice potential to cope with various abiotic stresses. This technique has also been used to identify the functions of stress-related genes as well as for the characterization of stress-responsive promoter elements in rice (Table 18.1). The next sections describe the application of genome editing in enhancing tolerance to various abiotic stresses in rice.

18.4.1 Salinity Tolerance

Salt stress is one of the major and serious factor that affect soil quality and agricultural production. Being a glycophyte plant, rice is very sensitive to salt stress. The poor drainage system and the use of brackish water for irrigation cause an increase in soil salinization. A high amount of sodium and chloride ions initially causes stomatal closure, nutritional disorder, and reduction in leaf expansion. Later, it causes chemical imbalance and a decline in growth and development that ultimately results in crop yield loss (Ganie et al. 2019). A lot of effort has been made to understand the molecular mechanism conferring salt tolerance and potential genes have been reported that enhance the salt tolerance in crops. Researchers have also applied genome editing tools to develop salinity tolerant rice plants. In a study, the genome-editing tool was applied to modulate the expression of *OsRR22*, which is a type-b response regulator involved in cytokinin signaling. It was earlier noticed that the loss of function of *OsRR22* significantly increases salinity tolerance (Takagi et al. 2015). Cas9-OsRR22-gRNA was used for targeted mutation. Homozygous non-transgenic mutant lines exhibited improved salinity tolerance at the seedling stage (Zhang et al. 2019b). Importantly, mutant lines showed no significant difference regarding agronomic characters compared to wild-type rice plants. The genome-editing approach has also been used for the characterization of salinity stress-specific promoters. One of the studies illustrated that the mRNA level of *OsRAV2* (*Related to ABI3/VP1*) transcription factor is increased by

Table 18.1 Application of genome-editing technology for generating abiotic stress-tolerant rice.

Trait	Target gene	Gene function	Editing method	Type of mutation	Rice cultivar/ variety	Phenotype of the mutant (Tolerant/ sensitive)	References
Salinity tolerance	*OsRR22*	Type-B response regulator involved in cytokinin signaling	CRISPR/Cas9	Knockout	Japonica cultivar WPB106	Tolerant	Zhang et al. (2019b)
	Promoter of *OsRAV2*	AP2/ERF domain-containing protein	CRISPR/Cas9	Knockout	Japonica cultivar Nipponbare	No mutant was generated	Duan et al. (2016)
	OsFLN2	Fructokinase-like protein 2	CRISPR/Cas9	Knockout	Japonica cultivar Nipponbare	Sensitive	Chen et al. (2019)
	OsOTS1	A protease belonging to SUMO Small Ubiquitin-like Modifier class	CRISPR/Cas9	Knockout	Japonica cultivar Kitaake	Sensitive	Zhang et al.(2019a)
	OsMDH1	Malate dehydrogenase 1	CRISPR/Cas9	Knockout	Japonica cultivar ZH11	Tolerant	Nan et al. (2020)
	OsRGA1, OsRGB1, OsRGG1, OsRGG2, OsGS3, OsGGC2, OsDEP1, OsPXLG1-4	Subunits of heteromeric G-complex and putative extra-large G-proteins	CRISPR/Cas9	Knockout	Japonica cultivar Sasanishiki	Tolerant	Cui et al. (2020)
Drought tolerance	*OsSAPK2*	Osmotic stress/ABA-activated protein kinase 2	CRISPR/Cas9	Knockout	Japonica	Sensitive	Lou et al. (2017)
	OsNAC14	Transcription factor	CRISPR/Cas9	Base deletion	Japonica cultivar Dongjin	Sensitive	Shim et al. (2018)
	OsLOGL5 (LOG-like gene)	An enzyme called LONELY GUY involved in cytokinin biosynthesis	CRISPR/Cas9	Knockout	Japonica variety Zhonghua 11	Tolerant	Wang et al. (2020)
	OsRGA1, OsDEP1, OsPXLG4	Gα (OsRGA1), Gγ (OsDEP1), subunits of heteromeric G-complex and putative extra-large G-protein 4	CRISPR/Cas9	Knockout	Japonica cultivar Sasanishiki	Tolerant	Cui et al. (2020)

Cold tolerance	*OsANN3*	Ca^{2+}-dependent phospholipid binding protein	CRISPR/Cas9	Knockout	Japonica cultivar Taipei 309	Sensitive	Shen et al. (2017)
	OsMYB30	R2R3-type MYB transcriptional repressor	CRISPR/Cas9	Single base mutation	Japonica cultivar Nipponbare	Tolerant	Zeng et al. (2019)
	OsRGA1, OsGS3, OsDEP1, OsPXLG4	Gα (OsRGA1), Gγ (OsGS3, OsDEP1), subunits of heteromeric G-complex and putative extra-large G-protein 4	CRISPR/Cas9	Knockout	Japonica cultivar Sasanishiki	Tolerant	Cui et al. (2020)
Herbicide resistance	*OsALS*	Biosynthesis of branched-chain amino acids	CRISPR/Cas9	Substitution	Japonica cultivar Nipponbare	Resistant to bispyribac sodium herbicide	Butt et al. (2017)
	OsALS1	Biosynthesis of branched-chain amino acids	Cas9n-based base editors	Substitution	Japonica cultivar Kitaake and Nangeng 46	Resistant to bispyribac sodium herbicide	Kuang et al. (2020)
	OsALS	Biosynthesis of branched-chain amino acids	CRISPR/Cas9	Substitution	Japonica cultivar Nipponbare	Resistant to imazamox herbicide	Shimatani et al. (2017)
	OsEPSPS	5-Enolpyruvyl shikimate-3-phosphate synthase	CRISPR/Cas9	Gene replacement and insertions	Japonica cultivar Nipponbare	Resistant to glyphosate herbicide	Li et al. (2016)
	OsACC	Acetyl-coenzyme A carboxylase	CRISPR/Cas9	Base editing	Japonica variety Zhonghua 11	Resistant to haloxyfop-R-methyl herbicide	Li et al. (2018)

(*Continued*)

Table 18.1 (Continued)

Trait	Target gene	Gene function	Editing method	Type of mutation	Rice cultivar/ variety	Phenotype of the mutant (Tolerant/ sensitive)	References
Heavy metal tolerance	*OsARM1*	Arsenite-responsive MYB transcription factor	CRISPR/Cas9	Knockout	Japonica cultivar Nipponbare	Tolerant to arsenic stress	Wang et al. (2017)
	OsNramp5	Cadmium transporter	CRISPR/Cas9	knockout	Indica cultivars Huazhan and Longke 638S	Tolerant (Reduction in cadmium content)	Tang et al. (2017)
	OsNramp5	Cadmium transporter	CRISPR/Cas9	knockout	Japonica varieties Nanjing 46 and Huaidao 5	Tolerant (Reduction in cadmium content)	Yang et al. (2019)
	OsLCT1	Cadmium transporter	CRISPR/Cas9	Knockout	Japonica cultivar Xidao 1	Tolerant (Reduction in cadmium content)	Songmei et al. (2019)
	OsHAK1	Potassium transporter	CRISPR/Cas9	Knockout	Japonica cultivar Nipponbare	Tolerant (Reduction in cesium uptake)	Nieves-Cordones et al. (2017)
Alkaline tolerance	*OsPPa6*	Inorganic pyrophosphatase	CRISPR/Cas9	Base deletion and insertion	Japonica cultivar Kitaake	Sensitive	Wang et al. (2019)

salt treatment. OsRAV2 protein contains an AP2/ERF domain. Genome-editing tool CRISPR/Cas9 was applied for the targeted mutation at a specific region (GT-1 element) of the promoter of *OsRAV2* gene to modulate its expression. GT-1 element was found to be a crucial element for the salt induction of *OsRAV2* promoter (Duan et al. 2016). Chen et al. (2019) found that knockout mutation in *OsFLN2* gene encoding a fructokinase-like protein 2 renders rice plant sensitive to salinity stress due to impaired partitioning of sugar to the roots. Similarly, targeted mutagenesis of *OsOTS (Overly Tolerant to Salt1)* encoding a SUMO (Small Ubiquitin-like Modifier) protease caused salt-sensitive phenotype in the mutant (Zhang et al. 2019a). The CRISPR/Cas9 system was used for genome-wide mutagenesis and a library of rice mutants was generated (Lu et al. 2017). A salt-tolerant mutant was identified after screening the library (Nan et al. 2020). The loss of function mutation was identified in *OsMDH1* gene that encodes chloroplast localized malate dehydrogenase 1 enzyme. OsMDH1 was found to negatively regulate the content of pyridoxine (vitamin B6) under salinity stress. Recently, genes encoding subunits of the heteromeric G-protein complex were mutated using the CRISPR/Cas9 tool (Cui et al. 2020). Rice genome contains one *Gα* (*OsRGA1*), one *Gβ* (*OsRGB1*) and five *Gγ* (*OsRGG1, OsRGG2, OsGS3, OsGGC2 and OsDEP1*) genes. Besides, four genes encoding atypically large proteins (*OsPXLG1-4*, putative extra-large G-proteins) similar to Gα subunit are also found in the rice genome. Mutation in each gene improved salt tolerance in rice. However, mutants showed differences in agronomic traits. Mutants also differed from each other regarding response to drought and chilling stresses.

18.4.2 Drought Tolerance

Drought stress negatively affects plant metabolism, physiological and biochemical functions, growth, development, and productivity. Most crops fail to survive in drought conditions. Therefore, researchers are implementing genome-editing tools to modulate different genes linked with drought stress to improve the tolerance of important crops like rice. One of the targeted genes is SnRK2 (*sucrose non-fermenting-1 related protein kinase 2*), which is activated following drought. It plays a key role in the regulation of abscisic acid signaling and response to hyper-osmotic stress. In rice, osmotic stress/ABA–activated protein kinase 2 (OsSAPK2), a member of the OsSnRK2 family, is considered to be a key mediator of ABA signaling (Lou et al. 2017). The CRISPR/Cas9 with sgRNA specific to the third exon of *OsSAPK2* was applied to develop genome-edited *sapk2* mutants. These mutants manifested ABA-insensitive phenotype and were found to be more susceptible to drought than wild-type rice plants. The CRISPR/Cas9 genome-editing tool has also been modulated to achieve CRISPR activation (CRISPRa) of genes involved in the abscisic acid signaling to ameliorate drought tolerance in plants. Though this type of study was carried out in Arabidopsis, it is noteworthy to mention it here. An inactive form of Cas9 (dead Cas9) was fused with histone acetyltransferase (HAT), a transcription activator, to assist targeted epigenome editing. The abscisic acid-responsive element-binding protein 1 (AREB1) plays a crucial role in drought stress response. Its promoter was targeted through the dCas9HAT CRISPR activation tool to induce *AtAREB1* expression. HAT causes acetylation of histone tails which results in disassembly of nucleosome octamers and unpacking of the wrapped DNA around it. Subsequently,

DNA becomes accessible to transcriptional machinery for the expression of genes. The stable lines of CRISPRa dCas9HAT showed enhanced expression of *AtAREB1*, which was also correlated with the higher chlorophyll content. In addition, these lines showed fast closure of stomatal aperture under water-deficient conditions and exhibited improved drought tolerance (Roca Paixao et al. 2019). This type of epigenome editing can also be performed in rice. The CRISPR-Cas9 has also been implemented to study the function of *OsNAC14* gene in rice (Shim et al. 2018). *OsNAC14* confers drought tolerance through the regulation of genes related to plant defense, DNA repair, strigolactone biosynthesis, and stress response. *OsNAC14* overexpressing plants exhibited better growth with higher filling rate and increased number of panicles under drought stress. However, mutation through CRISPR-Cas9 impeded its function. Cytokinins control several characteristics of plants including growth, yield and stress tolerance. LONELY GUY (LOG) enzyme having phosphoribohydrolase activity converts the precursor nucleotides into active cytokinin forms (Kuroha et al. 2009; Seo et al. 2016). Wang et al. (2020) observed that overexpression of the *OsLOGL5* (*LOG-like* gene) caused a reduction in yield and root growth in rice. They also generated six types of mutants by editing the 3′-end of the coding DNA sequence of *OsLOGL5* gene. The resulting mutations affected the C-terminal part of the enzyme without altering the central catalytic domain. The mutants exhibited normal morphology and enhanced yield under normal, drought and nitrogen-deficient conditions in field trials. Targeted mutations in *OsRGA1*, *OsDEP1*, and *OsPXLG4* genes (encoding Gα, Gγ subunits of heteromeric G-complex and putative extra-large G-protein 4 respectively) resulted in improvement in drought tolerance in rice (Cui et al. 2020).

18.4.3 Cold Tolerance

Each plant grows best in its ecological niche specified with optimum temperature and other environmental factors. When plants are exposed to a condition beyond the border of their optimum conditions, it poses stress that affects their growth and development. The cold stress caused by low temperatures also accounts for huge economic losses in agriculture. Under low temperature (chilling stress above 0 °C), most of the cellular enzymatic activities are reduced or completely abolish, the structure of proteins and nucleic acids, and fluidity of the plasma membrane are adversely affected. Freezing stress (below 0 °C) causes the formation of extracellular ice and a drop in water potential, which triggers the outward movement of intracellular water and results in dehydration and shrinkage of cells (Miura and Furumoto 2013). Gradually, water in intercellular and intracellular space is transformed into ice crystals which damage cellular structures causing freezing injuries. Thus, cold stress-induced acute dehydration severely affects the normal cellular processes, which culminate in poor germination, stunted growth, wilting, and cell death (Ding et al. 2019). Many genes play a vital role in cold stress. One of the genes is annexins, which is associated with membrane phospholipids. Annexins are important for the organization of membrane-associated protein networks and also regulate the signaling pathways involved in several cellular processes. Shen et al. (2017) implemented a genome-editing tool and targeted rice annexin *OsAnn3* by sgRNA specific to the second exon of the gene. Six lines with indel mutations were identified. Out of these mutant lines, they tested homozygous and biallelic mutants for cold tolerance. These lines showed high susceptibility to cold stress suggesting a crucial role of annexin in cold stress. They did not notice any off-target effect, which further confirms the specificity of the CRISPR/Cas9 genome-editing tool in rice. Using the multiplex CRISPR–Cas9 system, Zeng et al. (2019) targeted *OsMYB30* gene encoding a cold-responsive transcription factor along with *OsPIN5b* and *OsGS3* genes related to panicle length and grain size respectively. They produced a triple mutant of japonica rice resulting in high yield and improved cold tolerance. However, they also noticed off-target mutations. CRISPR/Cas9 system-mediated mutations in *OsRGA1, OsGS3, OsDEP1, and OsPXLG4* (G-complex related genes) improved tolerance to chilling stress in rice (Cui et al. 2020). Application of genome editing for improvement of tolerance to high temperate is yet to be reported in rice.

18.4.4 Herbicide Tolerance

Herbicides are used to control weed growth in order to enhance crop productivity. Several herbicide-tolerant crops have been developed in the past using the transgenic approach. Genome-editing CRISPR/Cas9 technology has also been employed to develop non-transgenic herbicide-resistant crops. Acetolactate synthase (ALS) is one of the key enzymes that catalyze the biosynthesis of branched-chain amino acids. It is a potential target of many herbicides. Butt et al. (2017) targeted *OsALS* gene using the CRISPR/Cas9 tool which resulted in W548L substitution. Mutant rice plants were found to be resistant to bispyribac sodium herbicide. Recently, cas9n-based adenine and cytosine base editors were used to create *osals1* mutants with different levels of tolerance to bispyribac sodium herbicide (Kuang et al. 2020). Shimatani et al. (2017) used the target-activation-induced cytidine deaminase (Target-AID) genome-editing strategy, which involved Cas9 fusion with a cytidine deaminase (PmCDA1, from *Petromyzon marinus*). In this strategy, two variants of Cas9, dCas9 without cleavage activity and nCas9 with nickase activity were fused with the base editor to target specific sites through sgRNAs. Following this strategy, they introduced a point mutation at a specific site (C287T) of the *OsALS* gene. Mutants showed resistance to imazamox herbicide. In a different case, CRISPR/Cas9 technology was applied to modify the enolpyruvylshikimate-3-phosphate synthase (EPSPS), which can bind to its substrate pyruvate or the herbicide glyphosate. Using genome-editing technology two residues were substituted (P106S and T102I) in the conserved domain of EPSPS that eventually increased glyphosate resistance in rice (Li et al. 2016). In another study, adenine base editor was used to create C2186R mutation in acetyl-coenzyme A carboxylase (ACC) in Zhonghua11 variety of rice. The mutated rice lines exhibited resistance to haloxyfop-R-methyl herbicide (Li et al. 2018).

18.4.5 Heavy Metal Tolerance

Contamination of paddy fields with heavy metals such as arsenic, cadmium, lead, cesium, etc. causes detrimental effects to rice growth and yield. Moreover, their bioaccumulation in rice grains poses a threat to human health. Arsenic is a carcinogenic element and rice efficiently uptakes and translocates it from root to aerial parts. Wang et al. (2017) identified a transcription factor OsARM1 (Arsenite-Responsive Myb1) that controls the transcription levels of genes encoding arsenic transporters. Knockout mutant lines of this transcription factor were generated using the CRISPR/Cas9 tool. They exhibited improved tolerance to arsenic. Rice also accumulates cadmium. Several cadmium transporters have been identified in rice. OsNramp5 and OsNramp1, members belonging to the natural resistance-associated macrophage protein family, are involved in the uptake of cadmium by roots. OsLCT1, a low-affinity cation transporter, transfers cadmium into rice grains. Tang et al. (2017) knocked out *OsNramp5* gene through the CRISPR/Cas9 tool and developed mutant indica rice lines with low cadmium content. Mutant lines did not differ from wild-type rice regarding agronomic traits. However, when the same gene was mutated in japonica rice using the same tool, growth and grain yield were reduced in mutant lines (Yang et al. 2019). Mutant lines of the *OsLCT1* gene produced rice grains with significantly reduced cadmium content (Songmei et al. 2019). The presence of radioactive cesium in rice raised health concerns in Japan following the nuclear mishap at Fukushima in 2011. OsHAK1, a cesium permeable potassium transporter, was inactivated using the CRISPR-Cas9 tool (Nieves-Cordones et al. 2017). The mutants exhibited significantly decreased cesium uptake.

18.4.6 Alkaline Stress Tolerance

Alkaline soil impairs crop growth and yield. It becomes difficult for plants to extract minerals ions at high soil pH. Alkaline stress causes discoloration of leaves due to chlorophyll loss in rice. CRISPR-Cas9-mediated mutagenesis revealed that *OsPPa6* gene encoding an inorganic pyrophosphatase is essential for alkaline tolerance in rice (Wang et al. 2019). Photosynthetic rate, amount of chlorophyll, sucrose, ATP and inorganic phosphate were reduced in mutants under alkaline soil conditions that delayed their growth.

18.5 Challenges of Genome Editing in Rice

Genome editing offers an immense number of prospects in the development of safe and improved crop varieties. Scientists have recognized the potential of its applications. Hence, they are investing an extensive amount of effort in its advancement for crop improvement. Nevertheless, several challenges still need to be overcome to reach the milestone of the commercialization of genome-edited crops.

18.5.1 Efficient Delivery System for Genome Editing Components

Genome-editing components are delivered into the host plant via different transformation methods (Liang et al. 2017). While in most of the transformation methods, cell wall makes the delivery of components very difficult, in addition, all of them require established tissue-culture methods for regeneration of transformed explants. Many wild relatives of the crops with some superior traits have poor regeneration efficiency, therefore, they are recalcitrant to the transformation (Altpeter et al. 2016). Hence, the current delivery system is limited to some species only. Even in varieties where the transformation by tissue culture is well established, the transfected tissue has to be regenerated into the whole plant which is a time-consuming and laborious task. The transformation of indica rice is much more difficult in comparison to the japonica one as indica rice has poor regeneration potential. Improvement in the existing delivery system can be a stepping stone for the development of a wide number of varieties with improved traits. Manipulation of the host plant's morphogenesis regulator genes *Baby boom* and *WUSCHEL* can improve the regeneration ability of the transformed tissue thus, enhance the delivery efficiency (Lowe et al. 2016). Tissue culture-free delivery into the gametophytic tissue or the meristematic cells has been successfully tried in a few crop species and might be a good strategy in rice for a genotype independent plant transformation (Hamada et al. 2017; Zhao et al. 2017).

18.5.2 Target Specificity of the Genome-Editing Tools

The most interesting and significant feature of genome-editing technology is to achieve custom engineering at a particular site in the genome which is not possible through any other molecular biology technique. However, sometimes these methods of gene manipulation are not precise and produce off-targets in the genome which are difficult to identify and result in unpredicted phenotype (Schaefer et al. 2017; Kempton and Qi 2019; Mao et al. 2019). Though HDR-based gene targeting is more precise, it has low efficiency and requires the construction of a relatively complicated genome-editing construct with the DNA repair template (Steinert et al. 2016). Therefore, it is not as convenient as the NHEJ repair pathway. Recently, Li et al. (2019) reported that RNA transcripts can also act as a repair template for genome editing which is a significant discovery for HDR-based precise genome editing. Reduced levels of potential off-targets can be attained by carefully designing more specific sgRNA. Whole-genome sequencing for the unwanted Cas9 cleavage is also possible due to the significant reduction in the cost of sequencing nowadays.

18.5.3 Better Understanding of the Complex Genetic Network of Agronomic Traits

Advancement in technology has made whole-genome sequencing much cheaper and faster. Yet, the understanding of a complex genetic network responsible for agronomic traits is not fully understood. Inadequate knowledge makes the selection of appropriate target genes for the complex trait improvement difficult. A better understanding of the genetic network, together with the other omics studies, can provide a greater possibility for the selection of the best possible target gene for an agronomic trait (Kujur et al. 2013).

18.5.4 Simultaneous Editing of Multiple Genes for Multiple Trait Improvement

While in conventional breeding methods, alleles controlling many superior traits, e.g. yield, nutritional quality, tolerance toward environmental stress, etc. are stacked in an elite variety (Zeng et al. 2017), simultaneous manipulation of multiple target genes is currently limited through genome editing. Though it can be achieved through multiplex genome-editing tools (Miao et al. 2018; Zeng et al. 2019), there are serious challenges regarding the utility of these tools such as technical difficulties in the assembly of long arrays of gRNAs and the efficiency of the delivery of large complex DNAs cargos, rate of mutation of individual genes, and cost-effectiveness (Thompson et al. 2018; McCarty et al. 2020). In some cases, there are possibilities of off-target mutations, unwanted chromosomal translocations and rearrangements due to DSBs at multiple loci, and cytotoxicity as multiple DSBs induces apoptosis (Thompson et al. 2018; McCarty et al. 2020). Therefore, the co-improvement of many diverse traits is a challenging task by genome-editing methods. The collective use of different approaches like genome editing and molecular breeding or sequential genome editing can enhance the possibilities of gene pyramiding (Demorest et al. 2016).

18.5.5 Government Policies for Genome-Edited Crops

Genome-editing technology is very well accepted by the scientific community as a better alternative to the transgenic method that produces GM crops. However, still, issues regarding its biosafety and environmental hazard are debated in public and legal platforms. Changes incorporated into the genome via genome editing are comparable to the natural mutations occurring randomly in the environment. A strong and well-defined policy from the government is required for its public acceptance and for making the scientific community less skeptical about the future of the genome editing research in crops.

18.6 Conclusion and Future Prospects

Genome-editing technology has proved to be a ground-breaking innovation in the field of engineering the genome and development of new improved varieties. Its ability of precise and targeted editing of the gene of interest is the biggest advantage. Among different genome-editing tools, CRISPR/Cas9 is the most versatile and easy to customize for a vast range of target genes. It has expanded the possibility of genome editing by gene knockout, gene replacement, base editing, etc. However, a better understanding of important multigenic traits, enhanced precision for the targeting of gene sequences, improved efficiency of the delivery of the genome-editing components into cells, and the development of efficient transformation methods for recalcitrant varieties are some challenges that have to be overcome. The combined use of system biology and functional genomics can allow the discovery of new genes governing agronomic traits. This information can help in the selection of appropriate target genes for engineering by various genome-editing tools. In the current era where global food security is the biggest threat, genome editing can provide a solution for the generation of new varieties with better stress tolerance and other traits in an accelerated time frame.

Acknowledgment

PS would like to acknowledge the University of Rajasthan, Jaipur and the University Grants Commission, India for providing basic infrastructure facilities and a start-up grant respectively. PS is also grateful to Departmental Research Support (DRS Phase-II), Department of Botany, University of Rajasthan for providing necessary research facilities. HR would like to acknowledge the DST-INSPIRE Faculty grant (grant no. DST/INSPIRE/04/2016/001118) from the Department of Science and Technology, Government of India.

Conflicts of Interest

The authors declare that they have no conflicts of interest.

References

Alexandratos, N.a.J.B. (2012). World agriculture towards 2030/2050: the 2012 revision. ESA Working paper No. 12-03.

Altpeter, F., Springer, N.M., Bartley, L.E. et al. (2016). Advancing crop transformation in the era of genome editing. *Plant Cell* 28 (7): 1510–1520.

Barrangou, R., Fremaux, C., Deveau, H. et al. (2007). CRISPR provides acquired resistance against viruses in prokaryotes. *Science* 315 (5819): 1709–1712.

Bibikova, M., Golic, M., Golic, K.G., and Carroll, D. (2002). Targeted chromosomal cleavage and mutagenesis in Drosophila using zinc-finger nucleases. *Genetics* 161 (3): 1169–1175.

Birla, D.S., Malik, K., Sainger, M. et al. (2017). Progress and challenges in improving the nutritional quality of rice (Oryza sativa L.). *Crit. Rev. Food Sci. Nutr.* 57 (11): 2455–2481.

Boch, J. and Bonas, U. (2010). Xanthomonas AvrBs3 family-type III effectors: discovery and function. *Annu. Rev. Phytopathol.* 48: 419–436.

Bolotin, A., Quinquis, B., Sorokin, A., and Ehrlich, S.D. (2005). Clustered regularly interspaced short palindrome repeats (CRISPRs) have spacers of extrachromosomal origin. *Microbiology* 151 (Pt 8): 2551–2561.

Butt, H., Eid, A., Ali, A. et al. (2017). Efficient CRISPR/Cas9-mediated genome editing using a chimeric single-guide RNA molecule. *Front. Plant Sci.* 8: 1441.

Carroll, D. (2011). Genome engineering with zinc-finger nucleases. *Genetics* 188 (4): 773–782.

Chen, K. and Gao, C. (2014). Targeted genome modification technologies and their applications in crop improvements. *Plant Cell Rep.* 33 (4): 575–583.

Chen, G., Hu, J., Dong, L. et al. (2019). The tolerance of salinity in rice requires the presence of a functional copy of FLN2. *Biomol. Ther.* 10 (1).

Christian, M., Cermak, T., Doyle, E.L. et al. (2010). Targeting DNA double-strand breaks with TAL effector nucleases. *Genetics* 186 (2): 757–761.

Cong, L., Ran, F.A., Cox, D. et al. (2013). Multiplex genome engineering using CRISPR/Cas systems. *Science* 339 (6121): 819–823.

Cui, Y., Jiang, N., Xu, Z., and Xu, Q. (2020). Heterotrimeric G protein are involved in the regulation of multiple agronomic traits and stress tolerance in rice. *BMC Plant Biol.* 20 (1): 90.

Demorest, Z.L., Coffman, A., Baltes, N.J. et al. (2016). Direct stacking of sequence-specific nuclease-induced mutations to produce high oleic and low linolenic soybean oil. *BMC Plant Biol.* 16 (1): 225.

Deng, D., Yan, C., Wu, J. et al. (2014). Revisiting the TALE repeat. *Protein Cell* 5 (4): 297–306.

Ding, Y., Shi, Y., and Yang, S. (2019). Advances and challenges in uncovering cold tolerance regulatory mechanisms in plants. *New Phytol.* 222 (4): 1690–1704.

Duan, Y.B., Li, J., Qin, R.-Y. et al. (2016). Identification of a regulatory element responsible for salt induction of rice OsRAV2 through ex situ and in situ promoter analysis. *Plant Mol. Biol.* 90 (1-2): 49–62.

Durai, S., Mani, M., Kandavelou, K. et al. (2005). Zinc finger nucleases: custom-designed molecular scissors for genome engineering of plant and mammalian cells. *Nucleic Acids Res.* 33 (18): 5978–5990.

Epinat, J.C., Arnould, S., Chames, P. et al. (2003). A novel engineered meganuclease induces homologous recombination in yeast and mammalian cells. *Nucleic Acids Res.* 31 (11): 2952–2962.

Fu, Y., Foden, J.A., Khayter, C. et al. (2013). High-frequency off-target mutagenesis induced by CRISPR-Cas nucleases in human cells. *Nat. Biotechnol.* 31 (9): 822–826.

Gaj, T., Gersbach, C.A., and Barbas, C.F. 3rd (2013). ZFN, TALEN, and CRISPR/Cas-based methods for genome engineering. *Trends Biotechnol.* 31 (7): 397–405.

Ganie, S.A., Molla, K.A., Henry, R.J. et al. (2019). Advances in understanding salt tolerance in rice. *Theor. Appl. Genet.* 132 (4): 851–870.

Gorbunova, V. and Levy, A.A. (1997). Non-homologous DNA end joining in plant cells is associated with deletions and filler DNA insertions. *Nucleic Acids Res.* 25 (22): 4650–4657.

Hamada, H., Linghu, Q., Nagira, Y. et al. (2017). An in planta biolistic method for stable wheat transformation. *Sci. Rep.* 7 (1): 11443.

Jansen, R., van Embden, J.D.A., Gaastra, W. et al. (2002). Identification of genes that are associated with DNA repeats in prokaryotes. *Mol. Microbiol.* 43 (6): 1565–1575.

Jinek, M., Chylinski, K., Fonfara, I. et al. (2012). A programmable dual-RNA-guided DNA endonuclease in adaptive bacterial immunity. *Science* 337 (6096): 816–821.

Jinek, M., East, A., Cheng, A. et al. (2013). RNA-programmed genome editing in human cells. *elife* 2: e00471.

Jinek, M., Jiang, F., Taylor, D.W. et al. (2014). Structures of Cas9 endonucleases reveal RNA-mediated conformational activation. *Science* 343 (6176): 1247997.

Kempton, H.R. and Qi, L.S. (2019). When genome editing goes off-target. *Science* 364 (6437): 234–236.

Kim, H. and Kim, J.S. (2014). A guide to genome engineering with programmable nucleases. *Nat. Rev. Genet.* 15 (5): 321–334.

Kuang, Y., Li, S., Ren, B. et al. (2020). Base-editing-mediated artificial evolution of OsALS1 in planta to develop novel herbicide-tolerant rice germplasms. *Mol. Plant* 13 (4): 565–572.

Kujur, A., Saxena, M.S., Bajaj, D. et al. (2013). Integrated genomics and molecular breeding approaches for dissecting the complex quantitative traits in crop plants. *J. Biosci.* 38 (5): 971–987.

Kuroha, T., Tokunaga, H., Kojima, M. et al. (2009). Functional analyses of LONELY GUY cytokinin-activating enzymes reveal the importance of the direct activation pathway in Arabidopsis. *Plant Cell* 21 (10): 3152–3169.

Li, J., Meng, X., Zong, Y. et al. (2016). Gene replacements and insertions in rice by intron targeting using CRISPR-Cas9. *Nat. Plants* 2: 16139.

Li, C., Zong, Y., Wang, Y. et al. (2018). Expanded base editing in rice and wheat using a Cas9-adenosine deaminase fusion. *Genome Biol.* 19 (1): 59.

Li, S., Li, J., He, Y. et al. (2019). Precise gene replacement in rice by RNA transcript-templated homologous recombination. *Nat. Biotechnol.* 37 (4): 445–450.

Liang, Z., Chen, K., Li, T. et al. (2017). Efficient DNA-free genome editing of bread wheat using CRISPR/Cas9 ribonucleoprotein complexes. *Nat. Commun.* 8: 14261.

Lou, D., Wang, H., Liang, G. et al. (2017). OsSAPK2 confers abscisic acid sensitivity and tolerance to drought stress in rice. *Front. Plant Sci.* 8: 993.

Lowe, K., Wu, E., Wang, N. et al. (2016). Morphogenic regulators baby boom and wuschel improve monocot transformation. *Plant Cell* 28 (9): 1998–2015.

Lu, Y., Ye, X., Guo, R. et al. (2017). Genome-wide targeted mutagenesis in rice using the CRISPR/Cas9 system. *Mol. Plant* 10 (9): 1242–1245.

Mali, P., Yang, L., Esvelt, K.M. et al. (2013). RNA-guided human genome engineering via Cas9. *Science* 339 (6121): 823–826.

Mao, Y., Botella, J.R., Liu, Y., and Zhu, J.K. (2019). Gene editing in plants: progress and challenges. *Natl. Sci. Rev.* 6 (3): 421–437.

McCarty, N.S., Graham, A.E., Studená, L., and Ledesma-Amaro, R. (2020). Multiplexed CRISPR technologies for gene editing and transcriptional regulation. *Nat. Commun.* 11 (1): 1281.

Miao, C., Xiao, L., Hua, K. et al. (2018). Mutations in a subfamily of abscisic acid receptor genes promote rice growth and productivity. *Proc. Natl. Acad. Sci. U. S. A.* 115 (23): 6058–6063.

Miura, K. and Furumoto, T. (2013). Cold signaling and cold response in plants. *Int. J. Mol. Sci.* 14 (3): 5312–5337.

Mojica, F.J., Díez-Villaseñor, C., García-Martínez, J., and Soria, E. (2005). Intervening sequences of regularly spaced prokaryotic repeats derive from foreign genetic elements. *J. Mol. Evol.* 60 (2): 174–182.

Muthayya, S., Sugimoto, J.D., Montgomery, S., and Maberly, G.F. (2014). An overview of global rice production, supply, trade, and consumption. *Ann. N. Y. Acad. Sci.* 1324: 7–14.

Nan, N., Wang, J., Shi, Y. et al. (2020). Rice plastidial NAD-dependent malate dehydrogenase 1 negatively regulates salt stress response by reducing the vitamin B6 content. *Plant Biotechnol. J.* 18 (1): 172–184.

Nieves-Cordones, M., Mohamed, S., Tanoi, K. et al. (2017). Production of low-Cs(+) rice plants by inactivation of the K(+) transporter OsHAK1 with the CRISPR-Cas system. *Plant J.* 92 (1): 43–56.

Prado, J.R., Segers, G., Voelker, T. et al. (2014). Genetically engineered crops: from idea to product. *Annu. Rev. Plant Biol.* 65: 769–790.

Rees, H.A., Komor, A.C., Yeh, W.-H. et al. (2017). Improving the DNA specificity and applicability of base editing through protein engineering and protein delivery. *Nat. Commun.* 8: 15790.

Roca Paixao, J.F. et al. (2019). Improved drought stress tolerance in Arabidopsis by CRISPR/dCas9 fusion with a histone AcetylTransferase. *Sci. Rep.* 9 (1): 8080.

Schaefer, K.A., Wu, W.-H., Colgan, D.F. et al. (2017). Unexpected mutations after CRISPR-Cas9 editing in vivo. *Nat. Methods* 14 (6): 547–548.

Scheben, A., Wolter, F., Batley, J. et al. (2017). Towards CRISPR/Cas crops – bringing together genomics and genome editing. *New Phytol.* 216 (3): 682–698.

Scholze, H. and Boch, J. (2010). TAL effector-DNA specificity. *Virulence* 1 (5): 428–432.

Seo, H., Kim, S., Sagong, H.-Y. et al. (2016). Structural basis for cytokinin production by LOG from Corynebacterium glutamicum. *Sci. Rep.* 6: 31390.

Shen, C., Que, Z., Xia, Y. et al. (2017). Knock out of the annexin gene OsAnn3 via CRISPR/Cas9-mediated genome editing decreased cold tolerance in rice. *J. Plant Biol.* 60: 539–547.

Shim, J.S., Oh, N., Chung, P.J. et al. (2018). Overexpression of OsNAC14 improves drought tolerance in rice. *Front. Plant Sci.* 9: 310.

Shimatani, Z., Kashojiya, S., Takayama, M. et al. (2017). Targeted base editing in rice and tomato using a CRISPR-Cas9 cytidine deaminase fusion. *Nat. Biotechnol.* 35 (5): 441–443.

Sikora, P., Chawade, A., Larsson, M. et al. (2011). Mutagenesis as a tool in plant genetics, functional genomics, and breeding. *Int. J. Plant Genomics* 2011: 314829.
Songmei, L., Jie, J., Yang, L. et al. (2019). Characterization and evaluation of OsLCT1 and OsNramp5 mutants generated through CRISPR/Cas9-mediated mutagenesis for breeding low Cd rice. *Rice Sci.* 26: 88–97.
Steinert, J., Schiml, S., and Puchta, H. (2016). Homology-based double-strand break-induced genome engineering in plants. *Plant Cell Rep.* 35 (7): 1429–1438.
Stoddard, B.L. (2011). Homing endonucleases: from microbial genetic invaders to reagents for targeted DNA modification. *Structure* 19 (1): 7–15.
Symington, L.S. and Gautier, J. (2011). Double-strand break end resection and repair pathway choice. *Annu. Rev. Genet.* 45: 247–271.
Takagi, H., Tamiru, M., Abe, A. et al. (2015). MutMap accelerates breeding of a salt-tolerant rice cultivar. *Nat. Biotechnol.* 33 (5): 445–449.
Tang, L., Mao, B., Li, Y. et al. (2017). Knockout of OsNramp5 using the CRISPR/Cas9 system produces low cd-accumulating indica rice without compromising yield. *Sci. Rep.* 7 (1): 14438.
Thompson, D.B., Aboulhouda, S., Hysolli, E. et al. (2018). The future of multiplexed eukaryotic genome engineering. *ACS Chem. Biol.* 13 (2): 313–325.
Tilman, D., Balzer, C., Hill, J., and Befort, B.L. (2011). Global food demand and the sustainable intensification of agriculture. *Proc. Natl. Acad. Sci. U. S. A.* 108 (50): 20260–20264.
Wang, F.Z., Chen, M.-X., Yu, L.-J. et al. (2017). OsARM1, an R2R3 MYB transcription factor, is involved in regulation of the response to arsenic stress in rice. *Front. Plant Sci.* 8: 1868.
Wang, B., Xie, G., Liu, Z. et al. (2019). Mutagenesis reveals that the OsPPa6 gene is required for enhancing the alkaline tolerance in rice. *Front. Plant Sci.* 10: 759.
Wang, C., Wang, G., Gao, Y. et al. (2020). A cytokinin-activation enzyme-like gene improves grain yield under various field conditions in rice. *Plant Mol. Biol.* 102 (4-5): 373–388.
Waterworth, W.M., Drury, G.E., Bray, C.M., and West, C.E. (2011). Repairing breaks in the plant genome: the importance of keeping it together. *New Phytol.* 192 (4): 805–822.
Wiedenheft, B., Sternberg, S.H., and Doudna, J.A. (2012). RNA-guided genetic silencing systems in bacteria and archaea. *Nature* 482 (7385): 331–338.
Wyman, C. and Kanaar, R. (2006). DNA double-strand break repair: all's well that ends well. *Annu. Rev. Genet.* 40: 363–383.
Yang, C.H., Zhang, Y., and Huang, C.F. (2019). Reduction in cadmium accumulation in japonica rice grains by CRISPR/Cas9-mediated editing of OsNRAMP5. *J. Integr. Agric.* 1: 688–697.
Zeng, D., Tian, Z., Rao, Y. et al. (2017). Rational design of high-yield and superior-quality rice. *Nat. Plants* 3: 17031.
Zeng, Y., Wen, J., Zhao, W. et al. (2019). Rational improvement of Rice yield and cold tolerance by editing the three genes OsPIN5b, GS3, and OsMYB30 with the CRISPR-Cas9 system. *Front. Plant Sci.* 10: 1663.
Zhang, C., Srivastava, A.K., and Sadanandom, A. (2019a). Targeted mutagenesis of the SUMO protease, Overly Tolerant to Salt1 in rice through CRISPR/Cas9-mediated genome editing reveals a major role of this SUMO protease in salt tolerance. *BioRxiv* 1: 555706.
Zhang, A., Liu, Y., Wang, F. et al. (2019b). Enhanced rice salinity tolerance via CRISPR/Cas9-targeted mutagenesis of the OsRR22 gene. *Mol. Breed.* 39: 47–57.
Zhao, X., Meng, Z., Wang, Y. et al. (2017). Pollen magnetofection for genetic modification with magnetic nanoparticles as gene carriers. *Nat. Plants* 3 (12): 956–964.

19

Role of Genome Engineering for the Development of Resistant Starch-Rich, Allergen-Free and Processing Quality Improved Cereal Crops

Anuradha Singh[1], *Amit Yadav*[2], *Joy K. Roy*[3], *and Kaushal Kumar Bhati*[4]

[1] *French Associates Institute for Agriculture and Biotechnology of Drylands (FAAB), Ben-Gurion University of the Negev, Sede Boqer Campus, Sede Boqer, Israel*
[2] *Zuckerberg Institute for Water Research (ZIWR), Ben-Gurion University of the Negev, Israel*
[3] *National Agri-Food Biotechnology Institute (NABI), Mohali, Punjab, India*
[4] *Louvain Institute of Biomolecular Sciences, Catholic University of Louvain, Louvain-la-Neuve, Belgium*

CHAPTER MENU

19.1 Introduction

Cereals, including maize (*Zea mays* L.), wheat (*Triticum aestivum* L.), rice (*Oryza sativa* L.) and barley (*Hordeum vulgare* L.) constitute approximately 80–85% of energy crops for two-thirds of the world's population. Global cereal production is secured at an all-time high of 2714 million tonnes (MT), while its utilization is forecast for 2709MT in the year 2019/2020 (The Food and Agriculture Organization of the United Nations 2020). Based on production volumes, maize becomes one of the largest produced cereals with over 1.09 billion MT, with production dominated by the United States (with over 388 MT). Rice is the second leading cereal with a production volume of 769.6 MT (515.0 MT, milled basis), contributed by Asian countries; mainly China (192 MT) and India (140 MT). The production of wheat is ranked third (766 million MT), with most of production driven by the European Union (153 MT) and India (102 MT), while barley ranks fourth (142.4 MT), mostly contributed by the Russian Federation (16.7 MT) and Canada (8.3 MT) (Statista 2019).

Cereals are the most important food crop with respect to their production and utilization (Shewry et al. 2009) and considered as a major source of carbohydrates (mainly starch),

Genome Engineering for Crop Improvement, First Edition. Edited by Santosh Kumar Upadhyay.

proteins, lipids, minerals and health-promoting phytochemicals (Shewry 2019). Corn is the main cereal grain as measured by production but ranks third as a staple food, after wheat and rice, which are extensively consumed by humans. Cereals, mainly wheat and sometimes barley, are processed into a wide range of food products such as bread, biscuit, unleavened flat bread (chapatti), noodles, macaroni, pasta, rolls, cakes, puddings, baked bread, crackers, breakfast cereals, and baby/weaning foods, while rice by-products include rice bran, rice milk, edible rice paper and husked or milled rice products. The end-use quality of these products primarily depends upon the varieties and stages of processing, together with the biochemical composition of cereal grains such as starch (Kim et al. 2003) and seed-storage proteins (Shewry and Halford 2002) which, in turn, alter organoleptic properties and digestibility, hence consumer acceptability.

Starch is the major and most renewable storage carbohydrate for food and feed (Geigenberger 2011). It contributes between 50–80% to the dry weight of seed at maturity, depending upon genotype and environmental conditions (Smith et al. 1995). It is a discrete, semi-crystalline granule, composed of two types of water-insoluble polymers of glucose molecules, i.e. amylose (AM) (one-quarter) and amylopectin (AP) (three-quarters) (Tetlow et al. 2004). Modification of starch to achieve increased production or an altered ratio of amylose to amylopectin, imparts specific properties that suit particular end-uses. For instance, starch with increased amylose content is termed as resistant starch (RS) and has been found to have a positive impact on human health (Hallstrom et al. 2011). In addition to starch, seed-storage proteins constitute about 10–15% of the grain dry weight and are an important source of protein in the human diet. Particularly in the case of wheat, proteins have been classified into two categories according to their solubility, i.e. soluble (albumins and globulins) and insoluble gluten (gliadins and glutenins) (Biesiekierski 2017). The soluble proteins are mostly metabolic, while insoluble are seed-storage proteins. These proteins play a determinate role in the processing properties of cereal flours, due to their contribution to viscoelastic properties of dough, which will then retain more gas and be more suited for spongy baked goods and leavened bread products (Rustgi et al. 2019). Nonetheless, several gluten-related disorders such as Celiac Disease (CD) and Wheat Allergy (WA) have also occurred in pre-disposed individuals (Brouns et al. 2019). In all these conditions, the ingestion of gluten determinates chronic inflammation of the small intestinal mucosa that results in villous atrophy, crypt hyperplasia and lymphocyte infiltration, leading to nutrient malabsorption (Rustgi et al. 2019).

In this chapter, we will cover recent advances in generating novel starch composition and modification of seed-storage proteins to achieve allergen-free and the process of quality-improved food products. We will first understand the molecular basis of variation in processing and nutrition-related traits using genetic engineering, mutagenesis and new breeding technologies and summarize the usefulness of processing quality-improved food products with value-added consumer benefits to the market.

19.2 Starch Characteristics

Native starches have diverse properties that have been utilized for various food (54%) and non-food (46%) industries, owing to its renewable, biodegradable, abundant, environmentally friendly, cost-competitive and versatile features. Starch is the main storage carbohydrate in higher plants, found in large amounts in various parts of plants, e.g. pollen, leaves, stems, roots, tubers, bulbs, rhizomes, fruits, flowers and seeds (Pfister et al. 2020). Starch occurs naturally as discrete particles called starch granules, which are relatively dense, insoluble and hydrate slightly in water at room temperature (Hsieh et al. 2020). Characteristics of starch granules differ across plants of distinctive botanical origin in terms of size, distribution, shape and morphology (Tester et al. 2014). Rice-starch granules (2–8 μm) tend to exist in clusters of individual granules which are known as compound granules (Bao and Bergman 2004). However, in the case of wheat, it exhibits a bimodal distribution of starch granules (Lloyd 2020). The large starch granules of wheat are disk-like or lenticular in shape, with an average diameter of 15–35 μm, whereas the small starch granules are roughly spherical or polygonal in shape, ranging from 1 to 15 μm in diameter (Lindeboom et al. 2004).

Another unique attribute in starch is that most starch granules are composed of a mixture of two polymers, amylose and amylopectin, arranged into a three-dimensional, semi-crystalline structure (Pfister et al. 2020). These two polymers are built up of six-carbon sugar D-glucose often referred to as the "building block" of starch. However, the biochemical properties of amylose and amylopectin are considerably different. Amylose (linear) contains mainly α-1,4 linkages with relatively few α-1,6 branch points [molecular weight 10^5gmol^{-1}, degree of polymerization (DP) ~100–1000]. In contrast, amylopectin is highly branched, connected in a tree-like structure containing glucose monomers that are linked with α-1,4 linkages but also contains many α-1,6 branch points (molecular weight 10^6–10^7gmol^{-1}, DP ~1000-10 000). The tree-like architecture of amylopectin forms a crystalline structure, where short chains ($DP < 13$) are oriented within clusters and connected by longer chains ($DP > 22$) (Pfister et al. 2020).

Under normal conditions, 75 and 25% of starch granules are composed of amylopectin and amylose respectively and the ratio between these two glucan polymers varies with the genotype depending on their environmental conditions (Hsieh et al. 2019). "Waxy" (*wx*) starch contains less than approximately10% amylose, while amylose extender (*ae*) has more than 50% amylose (Regina et al. 2010). Besides these, starches may also contain some minor constituents of protein, lipids, moisture and ash, that influence the physical and end-uses' quality of starches (Horstmann et al. 2017). The protein and lipid content of starch varies from 0.05–0.5% and 0.6–1.0%, respectively. Starch granule-associated proteins are covalently bound to the amylose-amylopectin structure, whereas lipids exist as an amylose-lipid inclusion complex (Tester et al. 2014).

19.3 Starch Biosynthesis

In plants, both transient and storage starch are synthesized through the concerted and coordinated action of several starch biosynthetic enzymes in photosynthetically active-leaf chloroplasts and in the seed amyloplast, respectively (Pfister and Zeeman 2016). The various biologically active enzymes and their isoforms such as ADP-glucose pyrophosphorylase (AGPase), starch synthase (SS) including granule bound and soluble, starch-branching enzyme (SBE), debranching enzyme (DBE) and amylase, catalyze specific reactions in all starch synthesizing organelles (Singh et al. 2015; Kumar et al. 2018). Briefly, the first committed step of starch biosynthesis, i.e. conversion of Glc-1-P and ATP to ADP-glucose and

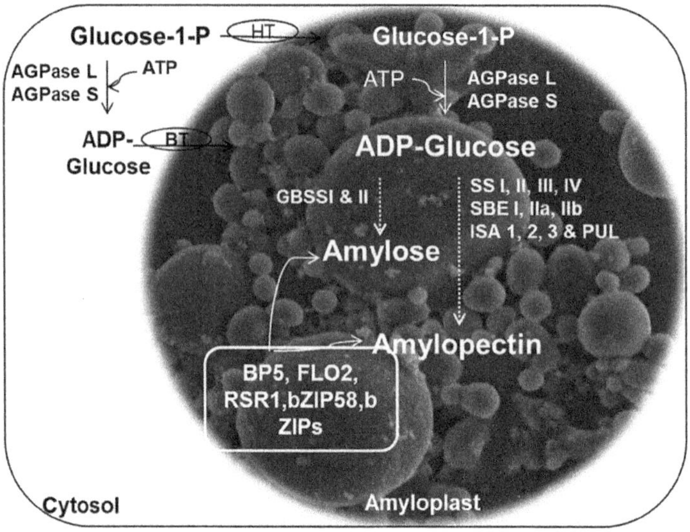

Figure 19.1 Genes involved in starch metabolism in cereal seed. Abbreviation: HT- Hexose Phosphate Transporter; BT – Brittle1; AGPase (L/S) – ADP glucose pyrophosphorylase (large and small subunits); GBSS–Granules Bound Starch Synthase; SS– Soluble Starch Synthase; SBE– Starch Branching Enzyme; ISA– Isoamylases; PUL– Pullulanase.

pyrophosphate (PPi) is catalyzed by AGPase (Huang et al. 2014). Next, the glucose moiety from ADP-glucose is transferred to the non-reducing end of α-glucan receptor of existing chains of amylose and amylopectin, through a new α-1,4-linkage. This step is primarily catalyzed by granule-bound starch synthases (GBSS) in amylose and soluble SS followed by SBE, DBEs in amylopectin (Mishra et al. 2016) (Figure 19.1).

GBSS consists of two isoforms, GBSSI and GBSSII. GBSSI is encoded by the waxy locus in cereals, functioning specifically to elongate amylose (Kumar et al. 2018). Mutants deficient in GBSSI activity or *waxy* have been identified in several plants species, such as maize, wheat, rice and barley with little or no amylose starch in sink tissues, confirming GBSSI is responsible for amylose synthesis (Vrinten and Nakamura 2000).

Unlike GBSS isoforms, SS are partially granule bound or soluble proteins within the starch granules, leading to the extension of the amylopectin molecule. There are at least four SS isoforms viz SSI, SSII, SSIII, and SSIV that have been identified in higher plants (Kumar et al. 2018). Though all isoform contributes to amylopectin synthesis, however, they possess different kinetic properties and seem to extend a specific subset of glucan chains, as suggested by Li et al. (2000).

The branching points in amylopectin, and rarely in amylose polymers, are introduced by SBEs, which hydrolyze α-1,4 glycosidic linkages and re-attach the chain to α-1,6 positions of the growing glucan chain. Multiple isoforms of SBEs (SBEI and SBEII) have been reported in various plant species, such as wheat, maize and rice (Tetlow and Emes 2014). SBEI transfers longer glucan chain than SBEII and the ratio between SBEI and SBEII isoforms are responsible for differences in starch structure during seed development (Burton et al. 1995).

Finally, DBEs play an important role in the development of crystalline amylopectin. There are two types of DBEs. The isoamylases (ISA e. g., ISA1, ISA2, and ISA3) hydrolyze α-1,6 linkages in amylopectin and starch, while pullulanase (PUL) hydrolyzes α-1,6 linkages in amylopectin and pullulan, a fungal polymer of malto-triose (van der Marrel et al. 2002).

Further, starch phosphorylases (Pho1 and Pho2), Disproportionating Enzyme (DPE1 and DPE2), amylases (α- and β-amylases) in conjunction with transporters (Brittle1) and transcription factors such as *OsBP5*, *OsFLO2*, *OsRSR1*, *OsbZIP58*, and *TabZIPs* play essential roles in the regulation of starch biosynthesis (Zhu et al. 2003; Fu and Xue 2010; She et al. 2010; Wang et al. 2013; Kumar et al. 2018).

19.4 Starch Digestibility and Resistant Starch

Starch provides nearly 40–70% of dietary nutrition needs for most of heterotrophic organisms in the world (Lovegrove et al. 2017). The nutrition quality of starch-based food products depends upon the compositional ratio of amylose to amylopectin, processing conditions and their digestibility (Englyst et al. 2018). Based on *in vitro* kinetic assays, i.e. the rate of glucose release and its absorption in the small intestine, starch can be classified as a rapidly digestible starch (RDS, 20 minutes) or slowly digestible starch (SDS, 120 minutes), which is considered as a major determinant of glycemic response (Ahuja et al. 2014). In contrast, the RS is defined as the starch portion which is neither hydrolyzed nor absorbed in the small intestine and has potential to prevent entry of glucose into the blood circulation, hence reduces the risk of type II diabetes (Bird et al. 2007). Fermentation of RS in the large intestine leads to the production of short-chain fatty acids (SCFAs, e.g. acetate, propionate, and butyrate), gases (e.g. H_2, H_2S, CO_2, and CH_4) and organic acids (e.g. lactate, succinate, and pyruvate), which play an important role in modulating immune and inflammatory responses and in preventing colon cancers (Krunic et al. 2018; Fu et al. 2019).

Several extrinsic (starch granule surface characteristics) and intrinsic factors (packing of amorphous and crystalline regions) also influence the digestibility of RS in the small intestine. Based on these factors, RS is usually classified into five categories, which are as follows, RS1, a physically inaccessible starch, RS2, a granular starch, RS3, formed when starches are first cooked and then cooled foods (e.g. potatoes, pasta, rice, etc.), RS4, chemically modified starches, RS5, comprise amylose-lipid complexes that are formed during processing and reforming after cooking or can be created artificially and added to foods (Roman and Martinez 2019).

Generally, high-amylose content in starch is positively linked with RS (Regina et al. 2006). This is due to the fact that the unbranching of amylose limits the access of digestible enzymes, as the two terminal ends are capable of forming single or double helices (Ahuja et al. 2013). A similar relationship was also observed when amylose complexed with lipid resistant to digestion by enzyme (Hu et al. 2018). In contrast, *waxy* starches are composed of highly branched structures of amylopectin, have a high amount of RDS and, as a result β-amylases and maltogenic α-amylase, readily hydrolyze the multiple terminal end glucose units. RS is also produced during cooking, where amylose molecules reassociate rapidly on cooling, resulting in double helices, which resist digestion, whereas amylopectin molecules reassociate slowly and are more susceptible to hydrolysis (Roman and Martinez 2019).

19.5 Genetic Modification in Relation to RS

Recent advancements in biotechnological strategies such as overexpression, targeted mutagenesis and mutation breeding have been employed to increase RS amounts via enhanced levels of amylose. This can be achieved through the alteration of expression in various enzymes involved in either starch elongation (GBSS and SS) or SBEs (Regina et al. 2015; Wang et al. 2017). Modification of these enzymes has a drastic effect on different aspects of starch, such as composition, granule morphology and starch properties.

Linear chains of amylose or starch elongation are primarily catalyzed by GBSSI, encoded by *Wx-A1*, *Wx-B1*, and *Wx-D1*, located on chromosomes 7AS, 4AL, and 7DS, respectively (Kumar et al. 2018). Increasing the expression of GBSS1 leads to high amylose and high RS content. For instance, in the case of rice, introduction of *Wxa* cDNA into null-mutant japonica rice (*wx*) produced 6–11% higher amylose content than that of the original cultivar (Itoh et al. 2003). Overexpression of *Wx-B1* gene in durum wheat transgenic lines did not show any increase of amylose content compared to the untransformed cultivar (Sestili et al. 2012) whereas, in barley, significant changes in expression were found in SSI, SBEI. and SBEIIa genes (Carciofi et al. 2012). Further, unique nucleotide changes within the second and fourth GBSSI introns resulted in variant amylose concentrations in barley (Asare et al. 2012) and site-directed mutagenesis for eight amino acids in *OsGBSSI* affecting starch-binding capacity and modulates amylose content in rice (Liu et al. 2014). Wambugu et al. (2018) reported a genome-based bulk segregate analysis (BSA), identified novel SNP (G/A), which may be responsible for differences in GBSS activity. Additionally, co-expression analysis identified candidate transcription factors (NAC and CCAAT-HAP5), which were co-expressed with various starch biosynthesis genes. Despite the high number of genetic modifications known on GBSSI, there is still a lack of significant increase of amylose content in cereal plants. This might be due to limitations led by the GBSSI protein, the availability of ADP-glucose and the physical space available within the starch granules (Sestili et al. 2012).

Amylose content can also be enhanced by decreasing the amount of amylopectin through suppression of the activity of soluble SS. In rice, retrotransposon Tos17 insertion mediated SSI-deficient mutant lines displayed decreased amylopectin branching (Fujita et al. 2006), whereas SS1 and BEI double mutants exhibited no changes in amylopectin content, branching pattern or starch hydrolysis (Abe et al. 2014). However, silencing of SSI resulted in reduced enzyme activity, with novel starch structure in wheat (McMaugh et al. 2014), with marked increase of amylose content with altered grain-starch composition and amylopectin chain distribution, owing to a pleiotropic effect on GBSS1in rice (Zhao et al. 2019).

In cereals, SSII is the main contributor for intermediate-length branch chains of amylopectin (DP, 12–24) and it exists in two isoforms, SSIIa, found in the amyloplasts of developing endosperm and SSIIb, found in vegetative tissues, with expression patterns varying among tissues (Liu et al. 2012). Loss of function of SSIIa affects amylopectin chain-length distribution and the physico-chemical properties in japonica rice, but no significant changes were found for amylose content (Umemoto et al. 2002; Umemoto et al. 2008), while moderately elevated amylose content (40–50% amylose) in maize (*sugary-2* mutant) (Tziotis et al. 2004; Liu et al. 2012), wheat (Kosar-Hashemi et al. 2007; Luo et al. 2015) and > 50% amylose in barley (*sex6* mutant) (Morell et al. 2003; Bird et al. 2004). Further, introgression of SSIIa null alleles in the durum wheat cultivar "Svevo" caused a more drastic increase in the amylose content of roughly 89% (Lafiandra et al. 2010) and also increased amounts of all major dietary fiber components, including RS (Botticella et al. 2016; Botticella et al. 2018). Additionally, chemical mutagenesis has been employed to produce a new set of partial and complete *SSIIa* null mutant and *SSIIa* triple mutant haplotypes in durum wheat, which was found to cause a hard grain texture, increased RS (amylose content) and protein contents, with overall lower starch content and smaller kernel weight (Hogg et al. 2013; Hogg et al. 2017).

SSIII is the second major contributor for a longer chain of amylopectin (DP, 25–36) and occurs as SSIIIa and SSIIIb, specifically expressed in the developing rice endosperm and leaf, respectively (Ohdan et al. 2005). SSIII mutation in maize (*du1* mutant), resulted in apparent elevated amylose content, decreased proportion of long amylopectin chain with tarnished, glassy and somewhat dull appearance kernels (Wang et al. 1993). Similarly, SSIIIa mutant lines of rice generated retrotransposon Tos17 insertion and N-methyl-N-nitrosourea (MNU) mutagenesis displayed pleiotropic effects on GBSS1, resulting in the changes of structure of amylopectin and amylose (Fujita et al. 2006; Fujita et al. 2007). In contrast, T-DNA insertional mutation for SSIIIa in rice has not been observed to cause this pleiotropic effect on other isoforms, though crystallinity of the starch granules based on their X-ray diffraction patterns decreased (Ryoo et al. 2007). Additionally, double repression of SSIIa/SSIIIa lines in rice displayed a chalky kernel appearance and had increased amylose levels, increased pasting temperatures and decreased viscosities (Zhang et al. 2011).

To further increase RS, suppression of SBEs enzymes has been considered as one of the most prominent existing mechanisms, leading to high amylose content in cereals. Based on primary amino-acid sequence similarity and substrate specificity, three branching enzyme families, e.g. SBEI, SBEIIa, and SBEIIb have been identified and their impacts on starch composition were assessed (Regina et al. 2005). In cereals, spatial expression patterns show that SBEI and SBEIIb are predominantly expressed in the endosperm, while SBEIIa is expressed in all tissues. Antisense inhibition of SBEI had little effect on starch accumulation or had minor effect on structural properties of starch obtained through chemical mutagenesis in rice (Satoh et al. 2003; Xia et al. 2011). Next, antisense-inhibition using hairpin RNA (hp-RNA) of SBEIIa resulted in a noticeable increase in amylose content in wheat and durum wheat (Regina et al. 2006; Regina et al. 2010; Sestili et al. 2010). Chemically induced mutation of SBEIIa in hexaploid and tetraploid durum wheat also resulted in an additional 22% increase in amylose content and a 115% increase in RS content (Hazard et al. 2012; Slade et al. 2012; Mishra et al. 2016). Targeted mutagenesis through CRISPR/Cas9 in SBEI and SBEIIb showed a higher proportion of long chains in debranched amylopectin, with significantly increased AC and RS content (Sun et al. 2017).

Suppression of SBEIIb (amylose-extender, *ae* mutant) by artificial microRNA (amiRNA), hp-RNA and *ssIIIa/sbeIIb* mutant lines in rice, resulted in elevated amylose contents with altered starch chemistry, crystallinity and digestibility (Nishi et al. 2001; Butardo et al. 2011). Similarly, SBEIIb mutations in maize showed significantly longer amylopectin chain length, increased apparent (83.1–85.6%) and absolute amylose-contents (57.4–62.6%) with greater RS content than that of starches from the wild-type (Yao et al. 2004; Shannon et al. 2009).

Additionally, introduction of an unknown number of the high-amylose modifier (HAM) gene into a homozygous mutant of the recessive *ae* allele in maize led to the production of the highest known amylose containing line (>85%), with additive effect on amylose percentages ranging from 68.9 to 88.2% (Wu et al. 2009; Jiang et al. 2010). Likewise, concerted suppression of SBEI, SBEIIa and SBEIIb using a chimeric RNAi hairpin in barley resulted in "amylose-only barley" (Regina et al. 2010; Carciofi et al. 2012). Botticella et al. (2018) generated three single null mutant lines in wheat for GBSSI, SSIIa, and SBEIIa, which displayed no detectable amylose of 0,46, and 79%, respectively.

19.6 Genetic Modification in Relation to Allergen-Free Cereals

An allergy is generally defined as an adverse immune response in genetically predisposed individuals through harmless environmental substances, called allergens and is clinically monitored through the elevated levels of a specific immunoglobulin E (IgE) antibody that play a crucial role during allergic reaction (Singh and Bhalla 2008). In addition to providing dietary energy, certain cereal-derived food products are associated with a range of adverse effects on human health and well-being, notably celiac disease (CD) and WA (Brouns et al. 2019). CD is an autoimmune disorder that affects about 1% of the western world populations, causing mucosal inflammation, small intestinal villous atrophy, chronic diarrhea, abdominal distension, weight loss due to malabsorption and extra intestinal symptoms (Caio et al. 2019). In contrast, only 0.4% of the world population, with a particularly high occurrence in children, is sensitized by WA. Symptoms develop within minutes to hours after ingestion and include itching, swelling, skin rashes and life-threatening anaphylaxis.

The sensitivity of CD and WA are mostly associated with seed-storage proteins such as gluten, hordeins, secalins and avenins from wheat, barley, rye and, oats respectively (Balakireva and Zamyatnin 2016). Based on their differential N-terminal sequence and electrophoresis size and mobility, gluten from wheat is grouped into gliadins and glutenins, which together comprise 70–80% of the total flour protein. Gliadins are a mixture of monomeric polypeptides and are divided into four groups, alpha- (α-), beta- (β-), gamma- (γ-) and omega- (ω-) gliadins, whereas glutenins are polymeric, composed of high-molecular-weight glutenin subunits (HMW-GSs, 90 to 140 KDa) and low-molecular-weight glutenin subunits (LMW-GSs, 30–75 KDa) (Wang et al. 2017). Gluten proteins are rich in proline and glutamine amino acids, which are highly resistant to proteolytic digestion by the gastric, pancreatic and gastrointestinal tract of the human intestine and hence, result in the generation of large peptides (epitopes) (Cebolla et al. 2018).

A variety of epitopes sequence from α, γ and ω gliadins, as well as from the glutenins, have been identified to activate CD. The human leukocyte antigen (HLA) variant DQ2 (DQα1*05, DQβ1*02) and/or DQ8 (DQα1*03, DQβ1*0302) are chiefly responsible for the presentation of α-gliadin-specific CD4+T cell epitopes, resulting in the development of the celiac enteropathy (Molberg et al. 2003). Additionally, these epitopes are better recognized after deamidation of specific glutamine residues into glutamic acid by tissue transglutaminase (tTG) (Arentz-Hansen et al. 2000). The immunological epitopes have also been found in seed-storage protein of other cereals. For example, T-cell epitopes from wheat γ-gliadin as DQ2-γ-I (Sjöström et al. 1998), DQ2-γ-II (Vader et al. 2002), DQ2-γ-III-V (Arentz-Hansen et al. 2002), ω-gliadins as DQ2-ω-I and DQ2-ω-II (Tye-Din et al. 2010), LMW-GS as glutenin-17 and glutenin-156 (Vader et al. 2002), HMW-GSs as DQ8 HMW-glutenin (Sollid et al. 2012), hordeins (Hα2), secalins (Sα9) and avenins (α9A) (Vader et al. 2003; Arentz-Hansen et al. 2004; Tye-Din et al. 2010) have been reported and well characterized. The epitopes present in barley and rye showed similarity to those found in wheat; however the avenins from oat are more distinct and may be considered safe for patients with gluten intolerance/CD (Garsed and Scott 2007).

Lifelong dietary exclusion of gluten and gluten-like proteins is currently the best choice for patients with CD and WA. Adhering to a strict gluten-free diet is demanding, because gluten is a ubiquitous additive in various food and non-food products. Thus, the development of cereal varieties with reduced gluten-toxicity profiles may contribute to the improvement of the diet of CD patients and the reduction of CD incidence (Gil-Humanes et al. 2010). Recently, Juhász et al. (2018) provided novel insights into wheat grain proteins and other allergens using the new International Wheat Genome Sequencing Consortium bread wheat reference genome sequence. RefSeq v1.0. Genes related to CD and WA were primarily expressed in the starchy endosperm and showed wide variation in protein- and transcript-level expression in response to temperature stresses (Juhász et al. 2018).

Conventional breeding of gluten-free wheat is complicated by the presence of three related genomes, containing 20–30 glutenin genes and one hundred gliadin genes, across nine loci. However, being diploid, a novel ultra-low gluten (ULG) barley variety has been developed, in which the hordein (gluten) content was found to be reduced (Tanner et al. 2015). Silencing of α-, γ- and ω-gliadins through RNAi (hairpin construct) in transgenic wheat lead to strongly reduce expression of CD-related gliadin T-cell epitopes, with reasonable baking qualities (Gil-Humanes et al. 2014; Barro et al. 2016). Further, Sánchez-León et al. (2018) using CRISPR/Cas9 technology, precisely targeted the conserved regions of α-gliadin genes in both bread and durum wheat, resulting in reduced amounts of α-gliadins in the seed, providing bread and durum wheat lines with reduced immunoreactivity for gluten-intolerant consumers. Another, low hordein mutant in barley (*lys3a*) induced by EMS mutagenesis identified a missense allele in Barley Prolamin-box Binding Factor (BPBF). Mutation in the wheat homeologs of BPBF, the Wheat Prolamin-box Binding Factors (WPBFs), using the non-transgenic targeting-induced local lesions in genomes (TILLING) method, resulted in decreased gluten and high-Lys wheat (Moehs et al. 2019).

19.7 Genetic Modification in Relation to Improved Processing Quality Cereals

Gluten-containing cereals display adverse immune responses and clinical manifestations, however, they have always has been a major component in people's diet across the world. Gluten (glutenins and gliadins) is defined as a cohesive mass that remains after dough has been made from wheat flour and has the unique viscoelastic properties, which is essential in determining the dough quality of bread and other baked products, while this property is lacking in other cereal proteins (rye, oat, and barley) (Shewry et al. 2009). Variations in the relative amount and composition of glutenins and gliadins have important effects on gluten functionality, dough viscoelasticity and end-use quality (Wang et al. 2017; Rustgi et al. 2019) of cereal products. Numerous studies have defined the molecular basis of bread-making and pasta quality in relation to specific gluten-protein complex (Ortolan and Steel 2017). Comparative transcriptomics (Laudencia-Chingcuanco et al. 2007; Singh et al. 2014), next generation sequencing (Li et al. 2014; Liu et al. 2014) and proteomics analysis (Ma et al. 2014) have been used to improve genome-wide understanding of molecular mechanisms of processing quality gene expression in wheat.

Briefly, HMW-GS are encoded by genes at three loci, *Glu-A1*, *Glu-B1* and *Glu-D1*, located on the long arms of chromosome 1 homoeologous, with each locus containing two tightly linked genes, i.e. the *x*-type and *y*-type (Harberd et al. 1986). The HMW-GS are minor components in terms of quantity (5–10% of total protein), but key factors in the process of bread-making (Cornish et al. 2006). For example, in bread wheat, the HMW-GS *1Dx5 + 1Dy10* encoded by the Glu-D1d locus are associated with good bread-making quality and increased dough strength. However, *1Dx2 + 1Dy12* encoded by Glu-D1aare associated with poor bread-making quality and weak dough. As a result, *Glu-D1d* predominates in high-quality wheat, whereas the allele *Glu-D1a* usually occurs in feed wheat with low bread-making quality (Groeger et al. 1997). The superior quality of the *Glu-D1d* allele is generally attributed to the difference in amino-acid primary structures of *1Dx2* and *1Dx5*. According to Shewry et al. (1995), *1Dx5* has one additional cysteine residue and therefore can form longer polymer chains, resulting in higher elasticity of the dough.

The LMW-GS represent approximately 40% of total wheat gluten fraction and is encoded by genes at the three loci *Glu-A3*, *Glu-B3*, and *Glu-D3*, located on the short arms of chromosomes 1A, 1B, and 1D, respectively (Ciaffi et al. 1999). LMW-GS also has a cysteine residue and inter-molecular di-sulfide bonds and thus are important for determining the rheological properties in dough of bread-making quality. Over-expression of LMW subunits in durum and bread wheat lines determined their functional properties (Tosi et al. 2005).

Gliadins are the major gluten fraction, accounting for ~50% of the total wheat gluten fraction and are encoded by six *Gli* loci mapped to the short arms of homologous groups 1 (*Gli-A1*, *Gli-B1*, and *Gli-D1*) and 6 (*Gli-A2*, *Gli-B2*, and *Gli-D2*) chromosomes and these alleles showed a correlation with dough quality (Barak et al. 2013). A balanced ratio of glutenins to gliadin is needed to achieve optimum viscosity and elasticity in dough (Khatkar et al. 1995; Veraverbeke and Delcour 2002). An increased amount of gliadin in wheat flour reduces peak dough resistance, hence making it suitable for biscuit production (Song and Zheng 2008).

19.8 Conclusions

Starch and proteins from cereals grains play essential roles in the development of various value- added food products. Recent advancement in genetic engineering, biochemical and immunological detection, have enabled the development of cereals with high-amylose leading to high RS and gluten-free properties, with enhanced human nutrition and desired processing quality. High amylose-RS not only provides "dietary fiber" along with low-caloric values, but is also beneficial for pasta-quality characteristics, while dietary intake of cereals with "low-gluten" reduces the risk of CD and WA. The development and commercialization of cereal food items with altered starch and protein composition has just begun and it needs to be intensified as most of the world's populations are dependent upon them.

Acknowledgments

AS would like to acknowledge National Agri-Food Biotechnology Institute (NABI), Mohali for giving fundamental understanding and inspiration to write this chapter. Also, Kreitman School of Advanced Graduate Studies, Ben Gurion University of Negev for Post-doctoral fellowship is thankfully acknowledged.

References

Abe, N., Asai, H., Yago, H. et al. (2014). Relationships between starch synthase I and branching enzyme isozymes determined using double mutant rice lines. *BMC Plant Biology* 14: 80.

Ahuja, G., Jaiswal, S., and Chibbar, R.N. (2013). Starch biosynthesis in relation to resistant starch. In: *Resistant Starch Sources, Applications and Health Benefits* (eds. Y.C. Shi and C.C. Maningat), 1–15. Chichester: Wiley.

Ahuja, G., Jaiswal, S., Hucl, P., and Chibbar, R.N. (2014). Differences in starch granule composition and structure influence in vitro enzymatic hydrolysis of grain meal and extracted starch in two classes of Canadian wheat (*Triticum aestivum* L.). *Cereal Chemistry* 91: 233–239.

Arentz-Hansen, H., Körner, R., Molberg, Q.H. et al. (2000). The intestinal T cell response to α-gliadin in adult celiac disease is focused on a single deamidated glutamine targeted by tissue transglutaminase. *Journal of Experimental Medicine* 191: 603–612.

Arentz-Hansen, H., McAdam, S.N., Molberg, F.B. et al. (2002). Celiac lesion T cells recognize epitopes that cluster in regions of gliadins rich in proline residues. *Gastroenterology* 123: 803–809.

Arentz-Hansen, H., Fleckenstein, B., Molberg, S.H. et al. (2004). The molecular basis for oat intolerance in patients with celiac disease. *PLoS Medicine* 1: e1.

Asare, E.K., Baga, M., Rossnagel, B.G., and andChibbar, R.N. (2012). Polymorphism in the barley granule bound starch synthase 1 (gbss1) gene associated with grain starch variant amylose concentration. *Journal of Agricultural and Food Chemistry* 60: 10082–10092.

Balakireva, A.V. and Zamyatnin, A.A. (2016). Properties of gluten intolerance, gluten structure, evolution, pathogenicity and detoxification capabilities. *Nutrients* 8: 644.

Bao, J.S. and Bergman, C. (2004). The functionality of rice starch. In: *Starch in Food, Structure, Function and Applications* (ed. A.C. Eliasson), 152–184. Cambridge: Woodhead Publishing.

Barak, S., Mudgil, D., and andKhatkar, B.S. (2013). Relationship of gliadin and glutenin proteins with dough rheology, flour pasting and bread making performance of wheat varieties. *LWT- Food Science and Technology* 51: 211–217.

Barro, F., Iehisa, J.C.M., Giménez, M.J. et al. (2016). Targeting of prolamins by RNAi in bread wheat, effectiveness of seven silencing-fragment combinations for obtaining lines devoid of coeliac disease epitopes from highly immunogenic gliadins. *Plant Biotechnology Journal* 14: 986–996.

Biesiekierski, J.R. (2017). What is gluten? *Journal of Gastroenterology and Hepatology* 32: 78–81.

Bird, A.R., Flory, C., Davies, D.A. et al. (2004). A novel barley cultivar (Himalaya 292) with a specific gene mutation in starch synthase IIa raises large bowel starch and short-chain fatty acids in rats. *The Journal of Nutrition* 134: 831–835.

Bird, A.R., Vuaran, M., Brown, I., and Topping, D.L. (2007). Two high-amylose maize starches with different amounts of resistant starch vary in their effects on fermentation, tissue and digesta mass accretion and bacterial populations in the large bowel of pigs. *British Journal of Nutrition* 97: 134–144.

Botticella, E., Sestili, F., Ferrazzano, G. et al. (2016). The impact of the SSIIa null mutations on grain traits and composition in durum wheat. *Breeding Science* 66: 572–579.

Botticella, E., Sestili, F., Sparla, F. et al. (2018). Combining mutations at genes encoding key enzymes involved in starch synthesis affects the amylose content, carbohydrate allocation and hardness in the wheat grain. *Plant Biotechnology Journal* 16: 1723–1734.

Brouns, F., van Rooy, G., Shewry, P. et al. (2019). Adverse reactions to wheat or wheat components. *Comprehensive Reviews in Food Science and Food Safety* 18: 1437–1452.

Burton, R.A., Bewley, J.D., Smith, A.M. et al. (1995). Starch branching enzymes belonging to distinct enzyme families are differentially expressed during pea embryo development. *Plant Journal* 7: 3–15.

Butardo, V.M., Fitzgerald, M.A., Bird, A.R. et al. (2011). Impact of down-regulation of starch branching enzyme IIb in rice by artificial microRNA-and hairpin RNA-mediated RNA silencing. *Journal of Experimental Botany* 62: 4927–4941.

Caio, G., Volta, U., Sapone, A. et al. (2019). Celiac disease, a comprehensive current review. *BMC Medicine* 17: 142.

Carciofi, M., Blennow, A., Nielsen, M.M. et al. (2012). Barley callus, a model system for bioengineering of starch in cereals. *Plant Methods* 8: 36.

Ciaffi, M., Lee, Y.K., Tamas, L. et al. (1999). The low-molecular weight glutenin subunit proteins of primitive wheats. III. The genes from D-genome species. *Theoretical and Applied Genetics* 98: 135–148.

Cornish, G.B., Bekes, F., Eagles, H.A., and Payne, P.I. (2006). Prediction of dough properties for bread wheats. In: *Gliadin and Glutenin,the Unique Balance of Wheat Quality* (eds. C.W. Wrigley, F. Bekes and W. Bushuk), 243–280. St Paul, MN: American Association of Cereal Chemists.

Englyst, K., Goux, A., Meynier, A. et al. (2018). Inter-laboratory validation of the starch digestibility method for determination of rapidly digestible and slowly digestible starch. *Food Chemistry* 245: 1183–1189.

Fu, F.F. and Xue, H.W. (2010). Co-expression analysis identifies rice starch regulator, a rice AP2/EREBP family transcription factor, as a novel rice starch biosynthesis regulator. *Plant Physiology* 154: 927–938.

Fu, X., Liu, Z., Zhu, C. et al. (2019). Nondigestible carbohydrates, butyrate and butyrate-producing bacteria. *Critical Reviews in Food Science and Nutrition* 59: S130–S152.

Fujita, N., Yoshida, M., Asakura, N. et al. (2006). Function and characterization of starch synthase I using mutants in rice. *Plant Physiology* 140: 1070–1084.

Fujita, N., Yoshida, M., Kondo, T. et al. (2007). Characterization of SSIIIa-deficient mutants of rice, the function of SSIIIa and pleiotropic effects by SSIIIa deficiency in the rice endosperm. *Plant Physiology* 144: 2009–2023.

Garsed, K. and Scott, B.B. (2007). Can oats be taken in a gluten-free diet? A systematic review. *Scandinavian Journal of Immunology* 42: 171–178.

Geigenberger, P. (2011). Regulation of starch biosynthesis in response to a fluctuating environment. *Plant Physiology* 155: 1566–1577.

Gil-Humanes, J., Piston, F., Tollefsen, S. et al. (2010). Effective shutdown in the expression of celiac disease-related wheat gliadin T-cell epitopes by RNA interference. *Proceedings of the National Academy of Sciences of the United States of America* 107: 17023–17028.

Gil-Humanes, J., Pistón, F., Altamirano-Fortoul, R. et al. (2014). Reduced-gliadin wheat bread, an alternative to the gluten-free diet for consumers suffering gluten-related pathologies. *PLoS One* 9: e90898.

Groeger, S., Oberforster, M., Werteker, M. et al. (1997). HWM glutenin subunit composition and bread-making quality of Austrian grown wheats. *Cereal Research Communications* 25: 955–962.

Hallstrom, E., Sestili, F., Lafiandra, D. et al. (2011). A novel wheat variety with elevated content of amylose increases resistant starch formation and may beneficially influence glycaemia in healthy subjects. *Food & Nutrition Research* 55 https://doi.org/10.3402/fnr.v55i0.7074.

Harberd, N.P., Bartels, D., and Thompson, R.D. (1986). DNA restriction fragment variation in the gene family encoding high molecular weight (HMW) glutenin subunits of wheat. *Biochemical Genetics* 24: 579–596.

Hazard, B., Zhang, X., Colasuonno, P. et al. (2012). Induced mutations in the starch branching enzyme II (SBEII) genes increase amylose and resistant starch content in durum wheat. *Crop Science* 52: 1754–1766.

Hogg, A.C., Gause, K., Hofer, P. et al. (2013). Creation of a high-amylose durum wheat through mutagenesis of starch synthase II (SSIIa). *Journal of Cereal Science* 57: 377–383.

Hogg, A.C., Martin, J.M., and Giroux, M.J. (2017). Novel ssIIa alleles produce specific seed amylose levels in hexaploid wheat. *Cereal Chemistry* 94: 1008–1015.

Horstmann, S.W., Lynch, K.M., and Arendt, E.K. (2017). Starch characteristics linked to gluten-free products. *MDPI Foods* 6: 29.

Hsieh, C.F., Liu, W., Whaley, J.K., and Shi, Y.C. (2019). Structure, properties and potential applications of waxy tapioca starches – a review. *Trends in Food Science and Technology* 83: 225–234.

Hsieh, C.F., BeMiller, J.N., and Huber, K.C. (2020). Impact of granule hydration on maize and wheat starch chemical reactivity at the granular and molecular levels. *Food Hydrocolloids* 100: 105374.

Hu, P., Fan, X., Lin, L. et al. (2018). Effects of surface proteins and lipids on molecular structure, thermal properties and enzymatic hydrolysis of rice starch. *Food Science and Technology* 38: 84–90.

Huang, B., Hennen-Bierwagen, T.A., and Myers, A.M. (2014). Functions of multiple genes encoding ADP-glucose pyrophosphorylase subunits in maize endosperm, embryo and leaf. *Plant Physiology* 164: 596–611.

Itoh, K., Ozaki, H., Okada, K. et al. (2003). Introduction of wx transgene into rice wx mutants leads to both high- and low-amylose rice. *Plant and Cell Physiology* 44: 473–480.

Jiang, H., Jane, J.L., Acevedo, D. et al. (2010). Variations in starch physicochemical properties from a generation-means analysis study using amylomaize V and VII parents. *Journal of Agricultural and Food Chemistry* 58: 5633–5639.

Juhász, A., Belova, T., Florides, C. et al. (2018). Genome mapping of seed borne allergens and immune-responsive proteins in wheat. *Science Advances* 4: eaar8602.

Khatkar, B.S., Bell, A.E., and Schofield, J.D. (1995). The dynamic rheological properties of glutens and gluten subfractions from wheats of good and poor bread-making quality. *Journal of Cereal Science* 22: 29–44.

Kim, W., Johnson, J.W., Graybosch, R.A., and Gaines, C.S. (2003). Physicochemical properties and end-use quality of wheat starch as a function of waxy protein alleles. *Journal of Cereal Science* 35: 195–204.

Kosar-Hashemi, B., Li, Z., Larroque, O. et al. (2007). Multiple effects of the starch synthase II mutation in developing wheat endosperm. *Functional Plant Biology* 34: 431–438.

Krunic, S.L., Skryhan, K., Mikkelsen, L. et al. (2018). Non-GMO potato lines with an altered starch biosynthesis pathway confer increased-amylose and resistant starch properties. *Starch* 70: 1600310.

Kumar, R., Mukherjee, S., and andAyele, B.T. (2018). Molecular aspects of sucrose transport and its metabolism to starch during seed development in wheat, a comprehensive review. *Biotechnology Advances* 36: 954–967.

Kumar, P., Mishra, A., Sharma, H. et al. (2018). Pivotal role of *bZIPs*in amylose biosynthesis by genome survey and transcriptome analysis in wheat (*Triticum aestivum* L.) mutants. *Scientific Reports* 8: 17240.

Lafiandra, D., Sestili, F., D'Ovidio, R. et al. (2010). Approaches for modification of starch composition in durum wheat. *Cereal Chemistry* 87: 28–34.

Laudencia-Chingcuanco, D.L., Stamova, B.S., You, F.M. et al. (2007). Transcriptional profiling of wheat caryopsis development using cDNA microarrays. *Plant Molecular Biology* 63: 651–668.

Li, Z., Mouille, G., Kosar-Hashemi, B. et al. (2000). The structure and expression of the wheat starch synthase III gene,motifs in the expressed gene define the lincage of the starch synthase III gene family. *Plant Physiology* 123: 613–624.

Li, A., Liu, D., Wu, J. et al. (2014). mRNA and small RNA transcriptomes reveal insights into dynamic homoeolog regulation of allopolyploid heterosis in nascent hexaploid wheat. *Plant Cell* 26: 1878–1900.

Lindeboom, N., Chang, P.R., and Tyler, R.T. (2004). Analytical, biochemical and physicochemical aspects of starch granule size, with emphasis on small granule starches: a review. *Stärke* 56: 89–99.

Liu, F., Romanova, N., Lee, E.A. et al. (2012). Glucan affinity of starch synthase IIa determines binding of starch synthase I and starch-branching enzyme IIb to starch granules. *Biochemical Journal* 448: 373–387.

Liu, D., Wang, W., and Cai, X. (2014). Modulation of amylose content by structure-based modification of OsGBSS1 activity in rice (*Oryza sativa* L.). *Plant Biotechnology Journal* 12: 1297–1307.

Liu, W., Wu, Z., Zhang, Y. et al. (2014). Transcriptome analysis of wheat grain using RNA-Seq. *Frontiers of Agricultural Science and Engineering* 1: 214–222.

Lloyd, J.R. (2020). The a to B of starch granule formation in wheat endosperm. *Journal of Experimental Botany* 71: 1–3.

Lovegrove, A., Edwards, C.H., De Noni, I. et al. (2017). Role of polysaccharides in food, digestion and health. *Critical Reviews in Food Science and Nutrition* 57: 237–253.

Luo, J., Ahmed, R., Kosar-Hashemi, B. et al. (2015). The different effects of starch synthase IIa mutations or variation on endosperm amylose content of barley, wheat and rice are determined by the distribution of starch synthase I and starch branching enzyme IIb between the starch granule and amyloplast stroma. *Theoretical and Applied Genetics* 128: 1407–1419.

Ma, C., Zhou, J., Chen, G. et al. (2014). iTRAQ-based quantitative proteome and phosphoprotein characterization reveals the central metabolism changes involved in wheat grain development. *BMC Genomics* 15: 1029.

van der Marrel, M.J.E.C., van der Veen, B., Uitdehaag, J.C.M. et al. (2002). Properties and applications of starch-converting enzymes of the α-amylase family. *Journal of Biotechnology* 94: 137–155.

McMaugh, S.J., Thistleton, J.L., Anschaw, E. et al. (2014). Suppression of starch synthase I expression affects the granule morphology and granule size and fine structure of starch in wheat endosperm. *Journal of Experimental Botany* 65: 2189–2201.

Mishra, A., Singh, A., Sharma, M. et al. (2016). Development of EMS-induced mutation population for amylose and resistant starch variation in bread wheat (*Triticum aestivum*) and identification of candidate genes responsible for amylose variation. *BMC Plant Biology* 16: 217.

Moehs, C.P., Austil, W.J., Holm, A. et al. (2019). Development of reduced gluten wheat enabled by determination of the genetic basis of the lys3a low hordein barley mutant. *Plant Physiology* 179: 1692–1703.

Molberg, Ø., Flaete, N.S., Jensen, T. et al. (2003). Intestinal T-cell responses to high-molecular-weight glutenins in celiac disease. *Gastroenterology* 125: 337–344.

Morell, M.K., Kosar-Hashemi, B., Cmiel, M. et al. (2003). Barley sex6 mutants lack starch synthase IIa activity and contain a starch with novel properties. *Plant Journal* 34: 173–185.

Cebolla, Á., Moreno, M.L., Coto, L., and Sousa, C. (2018). Gluten immunogenic peptides as standard for the evaluation of potential harmful prolamin content in food and human specimen. *Nutrients* 10: 1927.

Nishi, A., Nakamura, Y., Tanaka, N., and Satoh, H. (2001). Biochemical and genetic analysis of the effects of amylose-extender mutation in rice endosperm. *Plant Physiology* 127: 459–472.

Ohdan, T., Francisco, P.B. Jr., Sawada, T. et al. (2005). Expression profiling of genes involved in starch synthesis in sink and source organs of rice. *Journal of Experimental Botany* 56: 3229–3244.

Ortolan, F. and Steel, C.J. (2017). Protein characteristics that affect the quality of vital wheat gluten to be used in baking. *Comprehensive Reviews in Food Science and Food Safety* 16: 369–381.

Pfister, B. and Zeeman, S.C. (2016). Formation of starch in plant cells. *Cellular and Molecular Life Sciences* 73: 2781–2807.

Pfister, B., Zeeman, S.C., Rugen, M.D. et al. (2020). Theoretical and experimental approaches to understand the biosynthesis of starch granules in a physiological context. *Photosynthesis Research* 145 (1): 55–70. https://link.springer.com/article/10.1007/s11120-019-00704-y.

Regina, A., Kosar-Hashemi, B., Li, Z.Y. et al. (2005). Starch branching enzyme IIb in wheat is expressed at low levels in the endosperm compared to other cereals and encoded at a non-syntenic locus. *Planta* 222: 899–909.

Regina, A., Bird, A., Topping, D. et al. (2006). High-amylose wheat generated by RNA interference improves indices of large-bowel health in rats. *Proceedings of the National Academy of Sciences of the United States of America* 103: 3546–3551.

Regina, A., Kosar-Hashemi, B., Li, Z. et al. (2010). Control of starch branching in barley defined through differential RNAi suppression of starch branching enzyme IIa and IIb. *Journal of Experimental Botany* 61: 1469–1482.

Regina, A., Berbezy, P., Kosar-Hashemi, B. et al. (2015). A genetic strategy generating wheat with very high amylose content. *Plant Biotechnology Journal* 13: 1276–1286.
Roman, L. and Martinez, M.M. (2019). Structural basis of resistant starch (RS) in bread, natural and commercial alternatives. *Food* 8: 267.
Rustgi, S., Shewry, P., Brouns, F. et al. (2019). Wheat seed proteins: factors influencing their content, composition and technological properties and strategies to reduce adverse reactions. *Comprehensive Reviews in Food Science and Food Safety* 18: 1751–1769.
Ryoo, N., Yu, C., Park, C.S. et al. (2007). Knockout of a starch synthase gene OsSSIIIa/Flo5 causes white-core floury endosperm in rice (*Oryza sativa* L.). *Plant Cell Reports* 26: 1083–1095.
Sánchez-León, S., Gil-Humanes, J., Ozuna, C.V. et al. (2018). Low-gluten, nontransgenic wheat engineered with CRISPR/Cas9. *Plant Biotechnology Journal* 16: 902–910.
Satoh, H., Nishi, A., Yamashita, K. et al. (2003). Starch-branching enzyme I-deficient mutation specifically affects the structure and properties of starch in rice endosperm. *Plant Physiology* 133: 1111–1121.
Sestili, F., Janni, M., Doherty, A. et al. (2010). Increasing the amylose content of durum wheat through silencing of the SBEIIa genes. *BMC Plant Biology* 10: 144.
Sestili, F., Botticella, E., Proietti, G. et al. (2012). Amylose content is not affected by overexpression of the Wx-B1 gene in durum wheat. *Plant Breeding* 131: 700–706.
Shannon, J.C., Garwood, D.L., and Boyer, C.D. (2009). Genetics and physiology of starch development. In: *Starch, Chemistry and Technology* (eds. J. BeMiller and R.L. Whistler), 23–82. New York: Academic Press.
She, K.C., Kusano, H., Koizumi, K. et al. (2010). A novel factor FLOURY ENDOSPERM2 is involved in regulation of rice grain size and starch quality. *Plant Cell* 22: 3280–3294.
Shewry, P. (2019). What is gluten-why is it special? *Frontiers in Nutrition* 6: 101.
Shewry, P.R. and Halford, N.G. (2002). Cereal seed storage proteins, structures, properties and role in grain utilization. *Journal of Experimental Botany* 53: 947–958.
Shewry, P.R., Napier, J.A., and Tatham, A.S. (1995). Seed storage proteins, structures and biosynthesis. *Plant Cell* 7: 945–956.
Shewry, P.R., Underwood, C., Wan, Y. et al. (2009). Storage product synthesis and accumulation in developing grains of wheat. *Journal of Cereal Science* 50: 106–112.
Singh, M.B. and Bhalla, P.L. (2008). Genetic engineering for removing food allergens from plants. *Trends in Plant Science* 13: 257–260.
Singh, A., Mantri, S., Sharma, M. et al. (2014). Genome-wide transcriptome study in wheat identified candidate genes related to processing quality, majority of them showing interaction (quality x development) and having temporal and spatial distributions. *BMC Genomics* 15: 29.
Singh, A., Kumar, P., Sharma, M. et al. (2015). Understanding the expression pattern of genes involved in starch biosynthesis during seed development in bread wheat (*Triticum aestivum*). *Molecular Breeding* 35: 184.
Sjöström, H., Lundin, K.E.A., Molberg, K.R. et al. (1998). Identification of a gliadin T-cell epitope in coeliac disease, general importance of gliadin deamidation for intestinal T-cell recognition. *Scandinavian Journal of Immunology* 48: 111–115.
Slade, A.J., McGuire, C., Loeffler, D. et al. (2012). Development of high amylose wheat through TILLING. *BMC Plant Biology* 12: 69.

Smith, A.M., Denyer, K., and Martin, C.R. (1995). What controls the amount and structure of starch in storage organs? *Plant Physiology* 107: 673–677.

Sollid, L.M., Qiao, S.W., Anderson, R.P. et al. (2012). Nomenclature and listing of celiac disease relevant gluten T-cell epitopes restricted by HLA-DQ molecules. *Immunogenetics* 64: 455–460.

Song, Y. and Zheng, Q. (2008). Network formation in glycerol plastisized wheat gluten as viewed by extensional deformation and stress relaxation, final conclusions. *Food Hydrocolloids* 22: 674–681.

Statista (2019). Global leading wheat producing countries 2016/2017-2019/20 http://www.statista.com/statistics/237908/global-top-wheat-producing-countries (accessed 20 July 2020).

Sun, Y., Jiao, G., Liu, Z. et al. (2017). Generation of high-amylose rice through CRISPR/Cas9-mediated targeted mutagenesis of starch branching enzymes. *Frontiers in Plant Science*, 7 8: 298.

Tanner, G.J., Blundell, M.J., Colgrave, M.L., and Howitt, C.A. (2015). Creation of the first ultra-low gluten barley (*Hordeum vulgare* L.) for coeliac and gluten-intolerant populations. *Plant Biotechnology Journal* 14: 1139–1150.

Tester, R.F., Karkalas, J., and Qi, X. (2014). Starch composition, fine structure and architecture (review). *Journal of Cereal Science* 39: 151–165.

Tetlow, I.J. and Emes, M.J. (2014). A review of starch-branching enzymes and their role in amylopectin biosynthesis. *IUBMB Life* 66: 546–558.

Tetlow, I.J., Morell, M.K., and andEmes, M.J. (2004). Recentdevelopments in understanding the regulation of starch metabolism in higher plants. *Journal of Experimental Botany* 55: 2131–2145.

The Food and Agriculture Organization of the United Nations (2020). World Food Situation. http://www.fao.org/worldfoodsituation/csdb/en (accessed 20 July 2020).

Tosi, P., Masci, S., Giovangrossi, A. et al. (2005). Modification of the low molecular weight (LMW) glutenin composition of transgenic durum wheat, effects on glutenin polymer size and gluten functionality. *Molecular Breeding* 16: 113–126.

Tye-Din, J.A., Stewart, J.A., Dromey, J.A. et al. (2010). Comprehensive, quantitative mapping of T cell epitopes in gluten in celiac disease. *Science Translational Medicine* 2: 41ra51.

Tziotis, A., Seetharaman, K., Wong, K.S. et al. (2004). Structural properties of starch fractions isolated from normal and mutant corn genotypes using different methods. *Cereal Chemistry* 81: 611–620.

Umemoto, T., Yano, M., Satoh, H. et al. (2002). Mapping of a gene responsible for the difference in amylopectin structure between japonica-type and indica-type rice varieties. *Theoretical and Applied Genetics* 104: 1–8.

Umemoto, T., Horibata, T., Aoki, N. et al. (2008). Effects of variations in starch synthase on starch properties and eating quality of rice. *Plant Production Science* 11: 472–480.

Vader, W., Kooy, Y., Veelen, P. et al. (2002). The gluten response in children with celiac disease is directed toward multiple gliadin and glutenin peptides. *Gastroenterology* 122: 1729–1737.

Vader, L.W., Stepniak, D.T., Bunnik, E.M. et al. (2003). Characterization of cereal toxicity for celiac disease patients based on protein homology in grains. *Gastroenterology* 125: 1105–1113.

Veraverbeke, W.S. and Delcour, J.A. (2002). Wheat protein composition and properties of wheat glutenin in relation to breadmaking functionality. *Critical Reviews in Food Science and Nutrition* 42: 179–208.

Vrinten, P.L. and Nakamura, T. (2000). Wheat granule-bound starch synthase I and II are encoded by separate genes that are expressed in different tissues. *Plant Physiology* 122: 255–263.

Wambugu, P., Ndjiondjop, M.N., Furtado, A., and Henry, R. (2018). Sequencing of bulks of segregants allows dissection of genetic control of amylose content in rice. *Plant Biotechnology Journal* 16: 100–110.

Wang, Y.J., White, P., Pollack, L., and Jane, J.L. (1993). Characterization of starch structures of 17 maize endosperm mutant genotypes with Oh43 inbred line background. *Cereal Chemistry* 70: 171–179.

Wang, J.C., Xu, H., Zhu, Y. et al. (2013). OsbZIP58, a basic leucine zipper transcription factor, regulates starch biosynthesis in rice endosperm. *Journal of Experimental Botany* 64: 3453–3466.

Wang, J., Hu, P., Chen, Z. et al. (2017). Progress in high-amylose cereal crops through inactivation of starch branching enzymes. *Frontiers in Plant Science* 8: 469.

Wang, Z., Li, Y., Yang, Y. et al. (2017). New insight into the function of wheat glutenin proteins as investigated with two series of genetic mutants. *Scientific Reports* 7: 3428.

Wu, Y., Campbell, M., Yang, Y. et al. (2009). Genetic analysis of high amylose content in maize (*Zea mays* L.) using a triploid endosperm model. *Euphytica* 166: 155–164.

Xia, H., Yandeau-Nelson, M., Thompson, D.B., and andGuiltinan, M.J. (2011). Deficiency of maize starch-branching enzyme i results in altered starch fine structure, decreased digestibility and reduced coleoptile growth during germination. *BMC Plant Biology* 11: 95.

Yao, Y., Thompson, D.B., and andGuiltinan, M.J. (2004). Maize starch-branching enzyme isoforms and amylopectin structure. In the absence of starch-branching enzyme IIb, the further absence of starch-branching enzyme Ia leads to increased branching. *Plant Physiology* 136: 3515–3523.

Zhang, G., Cheng, Z., Zhang, X. et al. (2011). Double repression of soluble starch synthase genes SSIIa and SSIIIa in rice (*Oryza sativa* L.) uncovers interactive effects on the physicochemical properties of starch. *Genome* 54: 448–459.

Zhao, Q., Du, X., Han, Z. et al. (2019). Suppression of starch synthase I (SSI) by RNA interference alters starch biosynthesis and amylopectin chain distribution in rice plants subjected to high temperature. *Crop Journal* 7: 573–586.

Zhu, Y., Cai, X.L., Wang, Z.Y., and Hong, M.M. (2003). An interaction between a MYC protein and an EREBP protein is involved in transcriptional regulation of the rice *Wx* gene. *Journal of Biological Chemistry* 278: 47803–47811.

20

Engineering of Plant Metabolic Pathway for Nutritional Improvement

Recent Advances and Challenges

Sameer Dixit[1#], Akanchha Shukla[1#], Vinayak Singh[2], and Santosh Kumar Upadhyay[3]

[1] *Department of Biology, University of Western Ontario, London, Ontario, Canada*
[2] *Plant Molecular Biology and Genetic Engineering Division, CSIR-National Botanical Research Institute, Lucknow, Uttar Pradesh, India*
[3] *Department of Botany, Panjab University, Chandigarh, India*
 # Equal Contribution

CHAPTER MENU

20.1 Introduction

Agricultural research after the green revolution period has significantly improved crop productivity, which has led to increased food supplies in developed and developing countries. Improved plant breeding approaches and chemicals has enabled self-sufficiency in major staple food crops and food security. However, climatic changes and reduced agricultural biodiversity has negatively affected the nutrient balance in major crops, increasing malnutrition in most of the world population. Nutritional deficiency is a major concern worldwide, affecting around 3 billion people particularly in developing nations (McGuire 2015). In addition, insufficient daily consumption of nutrients in food is also a major concern for human health (Gillespie et al. 2016). Deficiency of vitamin A, zinc and iron are major causes of disabilities and death in Asian and African countries. Combating this nutritional deficiency has been put

Genome Engineering for Crop Improvement, First Edition. Edited by Santosh Kumar Upadhyay.

under high priority by WHO and Consultative Group on International Agricultural Research (Bouis et al. 1999; World Health Organization, WHO 1992).

In an effort to produce healthy food fortified with desired nutrients, biotechnological methods, conventional plant breeding and modern agronomic methods have been developed (Saltzman et al. 2017). However, the long duration of breeding and agronomic practices has limited application in achieving improved nutrition quality. Alternatively, biotechnology, particularly metabolic engineering has aided in efficient development of nutritionally fortified crops (Garg et al. 2018). These nutritionally enhanced crops have helped to combat malnutrition issues in many countries (Garg et al. 2018; Khush et al. 2012). For example, research to enhance growth and fortification of nutritional qualities in rice has aided in attainment of food security and reducing malnutrition in Asian countries, where rice is the major food crop (Zimmermann and Hurrell 2002).

Metabolic engineering employs genetic alteration of metabolic pathways to produce new improved metabolites, which augment nutritional value in crops (Farré et al. 2014). In metabolic engineering, either endogenous metabolic pathways are modified or new metabolic abilities are introduced in crop plants (Fu et al. 2018).

Metabolic pathways are a series of interlinked enzymatic reactions catalyzed by enzymes that form products from substrates via a sequence of intermediates (Zhu et al. 2020). The metabolic pathways may be linear or interlinked employing specific enzymes or families of related enzyme molecules (Farré et al. 2014). The product of these pathways accumulates in plant cells, specific compartments or may be exported to different sites. In addition, intermediates of metabolic pathways are also shuttled at subcellular levels even between different cells, increasing the complexity of these pathways (Caspi et al. 2013). This shuttling of molecules is called a metabolic flux, which can be regulated by metabolites themselves, a phenomenon known as feedback regulation. Feedback inhibition influences enzyme activity in the same pathway. Thus, alterations of these complex pathways to produce the desired metabolic product for crop improvement is a challenging process (Caspi et al. 2013).

In this chapter, we will discuss the latest advances in metabolic engineering approaches for modification of metabolic pathways and improvement of nutrition quality in plants, emphasizing its impact on crop improvement for human nourishment.

20.2 Methods for Metabolic Engineering

In plants, metabolic engineering usually aims to boost the expression or production of desired product which is otherwise produced in small amounts. This can be achieved by directly manipulating the anabolic or catabolic pathways governing production of the desired product (Capell and Christou 2004; Farré et al. 2014). Recently, metabolic engineering has been used to produce metabolites that are not naturally synthesized in the host plant by partial or complete introduction of heterologous metabolic pathway. This approach is known as synthetic metabolic engineering (Hanson and Jez 2018). In general, metabolic engineering involves augmentation, suppression and sequestration strategies (Zhu et al. 2020). Augmentation involves gene manipulation, where the expression of upstream enzymes (most commonly first enzyme of the pathway) is enhanced, and ensuring sufficient supply of precursors for downstream reactions in target pathway. Further, increasing the expression of first or the rate-limiting enzyme avoids feedback inhibition (Capell and Christou 2004).

Suppression strategy involves, inhibition of competing pathways, branch-point pathways and catabolic pathways. This reduces the diversion of intermediates, degradation of target metabolite and improving its yield. Further, sequestration involves formation or increasing the sink compartment for target metabolite storage (Zhu et al. 2020). Initially, single-gene targeting was used in all three approaches. However, plant metabolic pathways are complex and demanding concurrent expression of several genes. Also, the enzymes catalyzing these pathways are often multimeric. Therefore, single-gene manipulation were not able to target multiple metabolites simultaneously and enhance activity of multimeric enzymes (Capell and Christou 2004; Halpin 2005; Naqvi et al. 2010). Therefore, advanced multigene metabolic interventions using expression constructs containing integrated multiple DNA sequences (of multiple transgenes) targeting one or more pathways were developed that proved effective in nutritional augmentation (Zorrilla López et al. 2013).

20.3 Vitamin A

Vitamin A is crucial for human health and its deficiency causes various vision-related abnormalities. Vitamin A deficiency is the major cause of blindness in South Asian countries, and, in severe cases, this deficiency may also lead to morbidity and mortality in children (Martini et al. 2018). Humans are unable to synthesize vitamin A directly, hence consumption of a variety of plant food sources like oranges, spinach, carrot and broccoli, etc. provides provitamin or vitamin precursors (Drouin et al. 2011). These dietary pro-vitamins are used to synthesize vitamin A in the human body (Harrison 2005). There are four kind of plant provitamin A also known as carotenoids. Among these alpha and beta carotenes are found in higher proportions than gamma and beta-cyptoxanthin. Biosynthesis of plant carotenoids occurs in plastids with the help of enzymes encoded by genes from nucleus (Martini et al. 2018, Figure 20.1a).

The recommended dietary intake requirements of vitamin A for women and men is 700 and 900 μg retinol activity equivalents per day, respectively (Trumbo et al. 2001). However, very low amounts of vitamin A are found in crops like rice, millet, sorgum, potato, and other cereals that are major food crops in many countries. Genetic-engineering strategies have been successfully used to augment provitamin A biosynthesis for biofortification of such crops (Table 20.1). The most common strategy to enhance carotenoid biosynthesis involves the overexpression of one or more genes encoding enzymes involved in this pathway (Figure 20.1a). For example, introduction of Phytoene desaturase from *Pantoeaananatis* and Phytoene synthase from *Narcissus pseudonarcissus* in japonica rice cultivar leads to the production of biofortified golden rice, having 1.6ug/g of carotenoid content (Ye et al. 2000). On the other hand, introduction of highly efficient Phytoene synthase from *Zea mays* accumulated 37 μg/g of carotenoids giving rice to golden rice 2 (Paine et al. 2005). Similarly, canola, flax, tomato, potato, wheat, and soybean were also engineered with single or a combination of carotenoid biosynthesis genes to produce the corresponding biofortified cultivars ([Aluru et al. 2008; Ducreux et al. 2005; Fraser et al. 2002; Fujisawa et al. 2008; Schmidt et al. 2015; Wang et al. 2014], Table 20.1). Besides that, another strategy makes use of silencing of gene-encoding lycopene beta cyclase and lycopene epsilon cyclase to equilibrate metabolic flux in favor of high beta-carotene accumulation. This method was used in potato and wheat (Diretto et al. 2007; Van Eck et al. 2007; Zeng et al. 2015). Similarly, con-

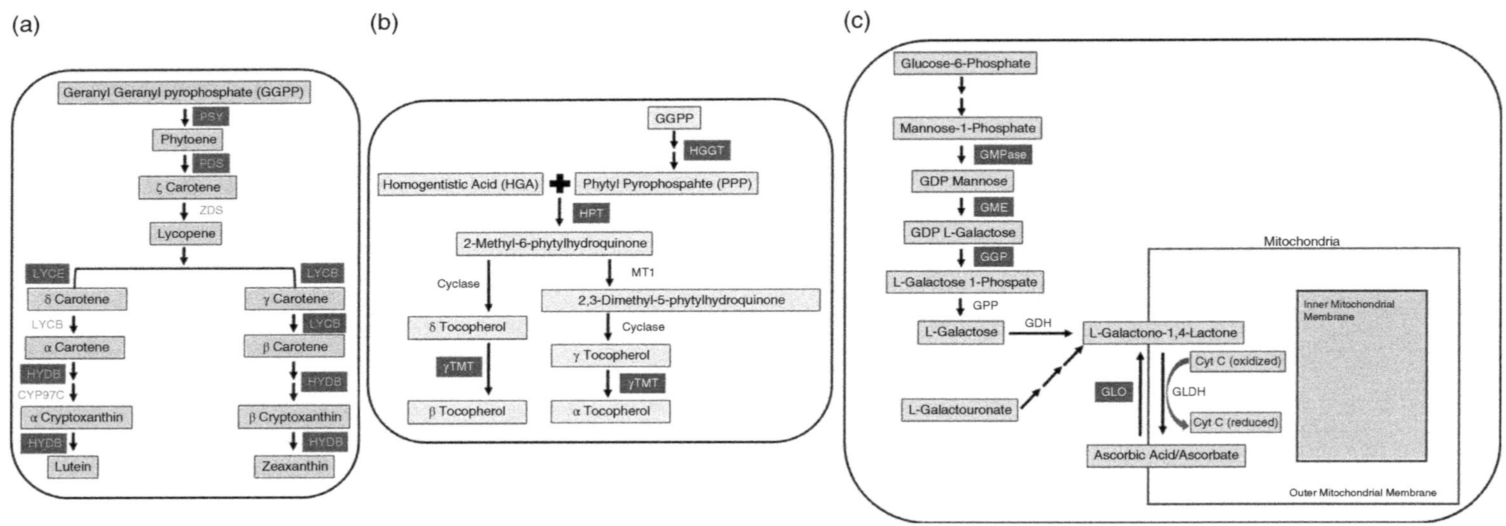

Figure 20.1 Biosynthesis pathways of different vitamins in plants. (a) Provitamin A biosynthesis pathway catalyzed by enzymes: PSY-phytoene synthase, PDS-phytoene desaturase, ZDS- ζ carotene desaturase, LYCE-Lycopene ε cyclase, LYCB-lycopene β cyclase, HYDB-β carotene hydroxylase, CYP97C-carotene ε ring hydroxylase. (b) Vitamin E (tocopherol) biosynthesis pathway catalyzed by enzymes: HGGT-homogentisate geranylgeranyl transferase, HPT-homogentisic acid prenyl transferase, MT1-methyltransferase, γTMT-γ tocopherol methyltransferase. (c) Ascorbic acid/ascorbate (Vitamin C) biosynthesis pathway catalyzed by enzymes: GMPase- GDP mannose pyrophosphorylase, GME-GDP mannose epimerase, GGP-GDP galactose phosphorylase, GPP-galactose 1-phosphate phosphatase, GDH- L-galactose dehydrogenase, GLDH- L-Galactono-1,4-γ-lactone dehydrogenase, GLO- L glucono-1,4-lactone oxidoreductase. The enzymes highlighted in yellow have been metabolically engineered for nutritional augmentation in crop plants.

Table 20.1 Summary of the biofortification for different Vitamins in plants through metabolic engineering.

Vitamin	Gene	Source Organism	Transgenic Plant	Promoter	Expression	Effect	References
Vitamin A	*Phytoene Synthase (psy1)*	Tomato	Tomato	CaMV35S	Over expression	Enhanced Lycopene and Phytoene; Reduction of Gibberellin; Dwarfism	Fray et al. (1995)
	Phytoene Desaturase (crtI)	*Erwinia uredovora*	Tobacco	CaMV35S	Over expression	Enhanced synthesis of β-carotene; Resistance to herbicide norflurazon	Misawa et al. (1993)
	Phytoene Synthase (crtB)	*Erwinia uredovora*	Canola	Napin (BcNA1) promoter (seed specific) from *Brassica rapa*	Over expression	Enhanced synthesis of α and β-carotene; Higher percentage of the oleic acid and a decreased percentage of the linoleic acid linolenic acid.	Shewmaker et al. (1999)
	Phytoene Synthase (psy) + *Phytoene Desaturase (crtl)* + Lycopene β-cyclase (β-Lcy)	Daffodil *Erwinia uredovora* Daffodil	Rice	Endosperm specific, Rice glutelin (Gt1) CaMV35S Endosperm specific, Rice glutelin (Gt1)	Over expression	Enhanced level of total carotenoids and also some extent of lutein and zeaxanthin	Ye et al. (2000) (**Golden Rice**)
	Lycopene β-cyclase (β-Lcy)	*Arabidopsis thaliana*	Tomato	Tomato *pds* promoter	Over expression	Enhanced the level of total carotenoid content as well as β-carotene	Rosati et al. (2000)
	Phytoene Desaturase (crtl)	*Erwinia uredovora*	Tomato	CaMV35S	Over expression	Enhanced synthesis of β-carotene	Römer et al. (2000)
	Phytoene Synthase (crtB)	*Erwinia uredovora*	Tomato	Fruit-specific, Tomato polygalacturonase (PG) promoter	Over expression	Enhanced the level of total carotenoid content as well as β-carotene, phytoene, lycopene, and lutein.	Fraser et al. (2002)

(*Continued*)

Table 20.1 (Continued)

Vitamin	Gene	Source Organism	Transgenic Plant	Promoter	Expression	Effect	References
	Phytoene Synthase (psy) + *Phytoene Desaturase (crtl)* + Lycopene β-cyclase (β-Lcy)	Daffodil *Erwinia uredovora* Daffodil	Rice	Endosperm specific, Rice glutelin (Gt1) CaMV35S CaMV35S	Over expression	Enhanced the level of total carotenoid content as well as β-carotene	Datta et al. (2003)
	Phytoene Synthase (psy) + *Phytoene Desaturase (crtl)*	Daffodil *Erwinia uredovora*	Rice	Endosperm specific, Rice glutelin (Gt1) CaMV35S	Over expression	Enhanced level β-carotene, α-carotene, and β-cryptoxanthin	Hoa et al. (2003)
	Phytoene Synthase (crtB)	*Erwinia uredovora*	Potato	Tuber-specific, Patatin promoter	Over expression	Enhanced the levels of β-carotene and lutein	Ducreux et al. (2005)
	Phytoene Synthase (psy) + *Phytoene Desaturase (crtl)*	Maize *Erwinia uredovora*	Rice	Endosperm specific, Rice glutelin (Gt1) Endosperm specific, Rice glutelin (Gt1)	Over expression	Enhanced the level (higher than golden rice) of carotenoid content as well as β-carotene	Paine et al. (2005) (**Golden Rice 2**)
	Antisense of lycopene epsilon cyclase (LCY-e)	Potato	Potato	Tuber-specific, Patatin promoter	Silencing	Enhanced the level β- β- carotenoid and total carotenoid	Diretto et al. (2006)
	Antisense of β-carotene hydroxylases (CHY1 and CHY2)	Potato	Potato	Tuber-specific, Patatin promoter	Silencing	Enhanced the level of total carotenoid and β-carotene	Diretto et al. (2007)

	*Antisense of β-carotene hydroxylases (*bch*)*	Potato	Potato	Tuber-specific, Granule-bound starch synthase (GBSS) promoter	Silencing	Enhanced the level of total carotenoid, lutein and β-carotene	Van Eck et al. (2007)
	Phytoene Synthase (crtB)	*Erwinia uredovora*	Wheat	Endosperm-specific promoter, 1DxS	Over expression	Enhanced the level of total carotenoid, and β-carotene	Zeng et al. (2015)
	+						
	Antisense of Carotenoid hydroxylase (TaHYD)	*Wheat*		Endosperm-specific promoter, 1DxS	Silencing		
	1-deoxy-D-xylulose 5-phosphate synthase (AtDXS)	*Arabidopsis thaliana*	Rice	Pice RP5 prolamin	Over expression	Enhanced the level (higher than golden rice 2) of carotenoid content as well as β-carotene	Bai et al. (2016)
	+						
	Phytoene synthase-1 (psy1)						
	+						
	Phytoene Desaturase (crtl)	Maize		Wheat low-molecular-weight glutenin			
		Pantoeaananatis		Barley D-hordein			
Vitamin E	*Homogentisic acid geranylgeranyl transferase (HGGT)*	Barley	Corn	CaMV35S	Over expression	Increased in tocotrienol and tocopherol content	Cahoon et al. (2003)
	2-methyl-6-phytylbenzoquinol methyltransferase (vte3 and vte4)	*Arabidopsis thaliana*	Soybean	Napin	Over expression	Increased in α-tocopherol and vitamin E activity in seed	Van Eenennaam et al. (2003)

(Continued)

Table 20.1 (Continued)

Vitamin	Gene	Source Organism	Transgenic Plant	Promoter	Expression	Effect	References
	γ-tocopherol methyltransferase (γ- TMT)	*Arabidopsis thaliana*	Lettuce	CaMV35S	Over expression	Higher conversion of the γ-tocopherol pool to α-tocopherol	Cho et al. (2005)
	Tocopherol cyclase (ZmTC)	Maize	Rapeseed	Seed-specific Napin promoter	Over expression	Increased level of total tocochromanol content	Kumar et al. (2005)
	Tocopherol cyclase (AtTC)	*Arabidopsis thaliana*	Rapeseed	Seed-specific Napin promoter	Over expression	Increased level of total tocochromanol content than ZmTC-transgenics	Kumar et al. (2005)
	Homogentisate prenyltransferases (AtHST)	*Arabidopsis thaliana*	*Arabidopsis thaliana*	CaMV35S	Over expression	Increased levels of prenyllipid, plastoquinone-9 and tocochromanols	Sadre et al. (2006)
	γ-tocopherol methyltransferase (γ- TMT)	*Perilla frutescens*	Soybean	Seed-specific promoter, Vicilin	Over expression	Increased level of the α-tocopherol and β-tocopherol content in T2 seed	Tavva et al. (2007)
	γ-tocopherol methyltransferase (γ- TMT)	*Perilla frutescens*	Perilla	Seed-specific promoter, Vicilin	Over expression	Increased level of the α-tocopherol in seed	Lee et al. (2008)
	Homogentisate phytyl transferase (HPT)	*Malus domestica*	Tomato	CaMV35S	Over expression	Higher levels of α-tocopherol and γ-tocopherol	Seo et al. (2011)
	γ-tocopherol methyltransferase (γ- TMT)	*Arabidopsis thaliana*	Perilla	CaMV35S	Over expression	Increased level of α-tocopherol and decreased level of γ-tocopherol	Ghimire et al. (2011)
	4-hydroxyphenylpruvate dioxygenase (HPPD)	*Arabidopsis thaliana*	Rice	Ubiquitin-1 promoter from maize	Over expression	No change in the total tocopherol level, but a significant shift from γ-tocopherol to α-tocopherol	Farré et al. (2012)
	γ-tocopherol methyltransferase (γ- TMT)	*Arabidopsis thaliana*	Rice	Ubiquitin-1 promoter	Over expression	Decreased level γ- and δ-tocotrienol and increased α isoform	Zhang et al. (2013)

γ-tocopherol methyltransferase (γ- TMT)	*Arabidopsis thaliana*	Rice	Endosperm specific, Rice glutelin (Gt1)	Over expression	Increased in the α-tocotrienol content	Zhang et al. (2013)
Tocopherol cyclase (TC)	*Arabidopsis thaliana*	Transplastomic Tobacco	psbA promoter	Over expression	Increased in total levels of Tocopherol	Yabuta et al. (2013)
γ-tocopherol methyltransferase (γ- TMT)	*Arabidopsis thaliana*	Transplastomic Tobacco	psbA promoter	Over expression	Increased in total Tocopherol levels as well as α-tocotrienol	Yabuta et al. (2013)
Tocopherol cyclase (TC) + *γ-tocopherol methyltransferase (γ- TMT)*	*Arabidopsis thaliana* *Arabidopsis thaliana*	Transplastomic Tobacco	psbA promoter	Over expression	Increased in total Tocopherol levels and α-tocotrienol was a major isoform	Yabuta et al. (2013)
Tocopherol cyclase (TC	*Arabidopsis thaliana*	Transplastomic lettuce	psbA promoter	Over expression	Total Tocopherol levels and vitamin E activity increased	Yabuta et al. (2013)
2-methyl-6-phytyl-1, 4-benzoquinol methyltransferase (MPBQ-MT)	Lettuce	Lettuce	CaMV35S	Over expression	Increased in α- and γ-tocopherol contents and reduction in phylloquinone content	Tang et al. (2016)
γ-tocopherol methyltransferase (γ- TMT)	Maize	*Arabidopsis thaliana*	CaMV35S	Over expression	Increased in α- tocopherol content and α−/γ-tocopherol ratio	Zhang et al. (2020)
γ-tocopherol methyltransferase (γ- TMT)	Maize	Maize	Embryo-preferred promoter Glb1	Over expression	Increased in α- tocopherol content and α−/γ-tocopherol ratio	Zhang et al. (2020)
Tocopherol cyclase (IbTC)	*Ipomoea batatas*	Sweet potato	CaMV35S	Over expression	Increased in α- tocopherol content and abiotic stress tolerance	Kim et al. (2019)

(*Continued*)

Table 20.1 (Continued)

Vitamin	Gene	Source Organism	Transgenic Plant	Promoter	Expression	Effect	References
Vitamin C	*L-gulono-γ-lactone oxidase (L-GAL)*	Rat	Lettuce	CaMV35S	Over expression	Accumulation of seven times more ascorbic acid	Jain and Nessler (2000)
	Dehydroascorbate reductase (DHAR)	Wheat	Maize	Maize ubiquitin (Ub) promoter	Over expression	Increased in foliar and kernel ascorbic acid levels and significantly increased the ascorbate redox	Chen et al. (2003)
	Myo-inositol oxygenase (MIOX4)	*Arabidopsis thaliana*	*Arabidopsis thaliana*	CaMV35S	Over expression	Increased in two to threefold Ascorbate levels	Lorence et al. (2004)
	Phytoene Synthase (psy1)	Maize	Corn	LMW glutenin promoter	Over expression	Increased inβ-carotene, ascorbate and folate	Naqvi et al. (2009)
	+						
	*Phytoene Desaturase (crt*I)	*Erwinia uredovora*		Barley D-hordein promoter			
	+						
	Dehydroascorbate reductase (DHAR)	*Oryza sativa*		Barley D-hordein promoter			
	D-Galacturonic acid reductase (GalUR)	Strawberry	Potato	CaMV35S	Over expression	Increased in ascorbate levels and also enhanced tolerance to abiotic stresses	Hemavathi et al. (2009)
	Dehydroascorbate reductase (DHAR)	Potato	Potato	CaMV35S	Over expression	Increased in ascorbate levels	Qin et al. (2011)
	GDP-Mannose 3′,5′-epimerase (SlGME)	Tomato	Tomato	CaMV35S	Over expression	Increase in total ascorbic acid also enhanced oxidative and abiotic stresses tolerance	Zhang et al. (2011)
	GDP-l-galactose phosphorylase	*Actinidia chinensis*	Tomato	CaMV35S	Over expression	Increase in ascorbate and polyphenolic content in fruit	Bulley et al. (2012)
	GDP-l-galactose phosphorylase	*Actinidia chinensis*	Strawberry	CaMV35S	Over expression	Increase in ascorbate and polyphenolic content in fruit	Bulley et al. (2012)

GDP-l-galactose phosphorylase	*Actinidia chinensis*	Potato	Polyubiquitin promoter	Over expression	Increase in ascorbate in Tuber	Bulley et al. (2012)
GDP-mannose pyrophosphorylase (GMPase)	*Yeast*	Tomato	CaMV35S	Over expression	Increased in ascorbate levels	Cronje et al. (2012)
Arabinono-1,4-lactone oxidase (ALO),	*Arabidopsis thaliana*	Tomato	CaMV35S	Over expression	Increased in ascorbate levels	Cronje et al. (2012)
Myo-inositol oxygenase 2 (MIOX2)	*Arabidopsis thaliana*	Tomato	CaMV35S	Over expression	Increased in ascorbate levels	Cronje et al. (2012)
Antisense of Monodehydroascorbate reductase (MDHAR)	Tomato	Tomato	CaMV35S	Silencing	Increased levels of ascorbate in both fruits and leaves	Gest et al. (2013)
GDP-mannose 3′, 5′-epimerase (GME)	Alfalfa	*Arabidopsis thaliana*	CaMV35S	Over expression	Increased in ascorbate levels and abiotic stresses tolerance	Ma et al. (2014)
Ascorbate peroxidase gene (OsAPX2)	Rice	Alfalfa	CaMV35S	Over expression	Increased in ascorbate levels and salt tolerance	Zhang et al. (2014)
9-cis-epoxycarotenoid dioxygenase (SgNCED1) + *D-arabinono-1,4-lactone oxidase (ALO)*	*Stylosanthes guianensis* Yeast	Stylo	CaMV35S	Over expression	Increased in ascorbate levels and enhanced drought and chilling stresses tolerance	Bao et al. (2016)
GDP-L-galactose phosphorylase (AcGGP)	Kiwifruit	Rice	leaf-specific, Leaf Panicle 2 (LP2) promoter	Over expression	Increased in foliar ascorbate levels and enhanced oxidative and abiotic stresses tolerance	Ali et al. (2019)
Antisense of ascorbate oxidase	Tomato	Tomato	CaMV35S	Silencing	Increased in ascorbic acid content and salinity tolerance	Abdelgawad et al. (2019)

current overexpression of bacterial phytoene synthase (CrtB) and silencing of b-carotene hydroxylase (HYDB) was shown to upregulate beta carotene up to 31-fold in wheat (Zeng et al. 2015). A third strategy was also employed that helped in stabilization at post-translational levels of genes involved in carotenoid biosynthesis. This ensured long-term and continuous accumulation of carotenoids in potato expressing orange (*OR*) gene from cauliflower (Li et al. 2012; Lopez et al. 2008; Lu et al. 2006; Zhou et al. 2015). In addition to manipulating the carotenoid biosynthesis pathways genes, another strategy was employed to ensure supply of isoprenoid precursors for carotenoid synthesis bypassing the rate-limiting step. This method helped in generation of golden rice 2 with introduction of Arabidopsis ORANGE (AtOR) and AtDXS (1-deoxy-D-xylulose-5-phosphate synthase) genes resulting in significantly high carotenoid levels (Bai et al. 2016).

Despite high carotenoid accumulation, the above methods had several drawbacks. For example, the process of introduction of genes in various cultivars involved use of antibiotic resistance genes that raised environmental concerns. Therefore, Hoa et al. (2003) used mannose selection marker to generate biofortified japonica and indica cultivars high in carotenoids. Also, manipulation of phytoene synthase negatively affected gibberellin synthesis causing growth defects like dwarfism (Fray et al. 1995). To counter this, allele pyramiding for biofortification has been widely investigated (Owens et al. 2014). Therefore, it becomes indispensable to investigate other metabolic constraints in carotenoid biosynthesis pathway to ensure successful generation of biofortified cultivars.

20.4 Vitamin E

Vitamin E collectively constitutes a group of hydrophobic compounds with varied antioxidant properties. Among them, alpha tocopherol has the highest biological activity in humans. Plants are the main source of dietary vitamin E, which is absorbed and transported to liver. Vitamin E contributes to antioxidant defense mechanism at cellular levels and has been found effective in prevention of many human diseases (Rizvi et al. 2014). Therefore, fortification of vitamin E in crops aims at either directly increasing alpha tocopherol levels or conversion of other vitamin E group compounds to alpha tocopherol.

The tocopherol biosynthesis pathway in plants involves cyclization, methylation to form gamma tocopherol which is converted to alpha form and transported to plastids (Figure 20.1b). The most common genetic-engineering approach to increase alpha tocopherol levels is by upregulation of genes encoding different enzymes of the biosynthesis pathways ([Cahoon et al. 2003; Cho et al. 2005; Farré et al. 2012; Van Eenennaam et al. 2003; Kumar et al. 2005; Ghimire et al. 2011; Lee et al. 2008; Sadre et al. 2006; Seo et al. 2011; Tang et al. 2016; Tavva et al. 2007; Yabuta et al. 2013; Zhang et al. 2013], Table 20.1). Trials with gamma TMT (gamma tocopherol methyl transferase) introduction in canola and barley HGGT (homogentisate geranylgeranyl transferase) into maize are examples of single-gene manipulation which resulted in high alpha tocopherol accumulation (Cahoon et al. 2003; Van Eenennaam et al. 2003). On the other hand, overexpression of combination of two or more genes like HPT (homogentisate phytyl transferase) and gamma TMT resulted in a 12-fold increase in accumulated alpha tocopherol in canola seeds (Collakova and DellaPenna 2003). Similarly, co-expression of Arabidopsis gene *at-vte3* with *at-vte4* in soybean, resulted in accumulation of more than 90% of alpha tocopherol (Van Eenennaam et al. 2003). However, multigene approach was not found effective in some crops like, maize and potato tubers, where only a 3% increase in alpha tocopherol levels was observed (Crowell et al. 2008; Naqvi et al. 2011). In line with these findings, additional factors affecting vitamin E biosynthesis were investigated. Natural allelic variations like insertions/deletions and single-nucleotide polymorphisms in homogentisate phytyl transferase 4 (*Vte4*) gene was found to significantly regulate alpha tocopherol levels in maize kernels (Li et al. 2012). Further, genome-wide study identified retrotransposon elements upstream to (*Vte3*) affecting vitamin E accumulation in tomato fruits (Quadrana et al. 2014).

20.5 Vitamin C

Vitamin C or ascorbate is a water-soluble compound synthesized in both plants and most animals (Smirnoff 1996). In plants, it serves an antioxidant role and is responsible for alpha tocopherol, zeaxanthin regeneration and pH-mediated regulation of PS II activity (Smirnoff 1996). While in humans, it functions as antioxidant and enzyme co-factor and plays a crucial role in cardiovascular functions, immune defense, iron metabolism, collagen and neurotransmitter synthesis (Carr and Maggini 2017; Hallberg et al. 1989; Martín-Calvo and Martínez-González 2017). Since vitamin Cis stored in the body in limited amounts and excreted through urine, a regular dietary supply is essential for human health (Oreopoulos et al. 1993). Deficiency of vitamin C causes scurvy disease in humans, which can be cured only by the administration of vitamin C (Padayatty and Levine 2016). Vitamin C is majorly found in tomato and potato while very low levels are found in cereal crops. Therefore, metabolic-engineering approaches to increase vitamin C levels in crops becomes indispensable.

Two approaches have mainly been used to biofortify crops with vitamin C. Like other vitamins, overexpression of biosynthesis genes is the main strategy for vitamin C biofortification (Figure 20.1c and Table 20.1), which has been shown to increase ascorbate levels by two- to six-fold in strawberry (Hemavathi et al. 2009), Arabidopsis, potato (Bulley et al. 2012; Lorence et al. 2004) and tomato (Zhang et al. 2011). On the other hand, metabolic engineering to overexpress rat L-glucocno-1,4-lactone oxidoreductase (GLO) increased ascorbate levels by seven-fold in tobacco and lettuce leaves (Jain and Nessler 2000). In addition, increased ascorbate levels were proposed to reduce browning of lettuce leaves (Jain and Nessler 2000). The choice of promoters also impacts the biosynthesis gene overexpression and resultant vitamin c accumulation. For example, use of the desired gene under the control of 35S CaMV promoter resulted in an approximate four-fold increase in ascorbate levels in tobacco leaves than 1.8- to 1.9-fold increase in maize kernels, when maize ubiquitin or shrukun2 promoter was used (Chen et al. 2003).

A second strategy for biofortification, makes use of silencing of ascorbate recycling genes such as monodehydroascorbate reductase (MDHAR) and dehydroascorbate reductase (DHAR). Studies on potato, tomato and maize silenced with either of the two genes increased accumulation of vitamin C, however, it was less in comparison to biofortification achieved during the overexpression approach (Chen et al. 2003; Gest et al. 2013; Naqvi et al. 2009; Qin et al. 2011).

20.6 Vitamin B

Vitamin B are a group of hydrophilic micronutrients comprising of thiamine (B1), riboflavin (B2), niacin (B3), pantothenic acid (B5), pyridoxal and its phosphorylated forms (B6), biotin (B7), folate (B9) and cobalamin (B12). This group of vitamins function as essential co-factors/ co-enzymes for various catabolic and anabolic enzymes required for normal human physiology (Kennedy 2016). Vitamin B subset cannot be synthesized in the human body and therefore must be supplemented through diet (Kennedy 2016). Among all, most of the metabolic-engineering studies have been done to fortify vitamin B6 and folate nutritive levels (Jiang et al. 2017). Folates are synthesized in plants by a highly regulated pathway that takes place in cytosol, plastid and mitochondria (Blancquaert et al. 2010; Storozhenko et al. 2005), while vitamin B6 is synthesized in cytosol from pentose phosphate pathway intermediates in plants (Storozhenko et al. 2007; Tambasco-Studart et al. 2005). Leafy vegetables are rich in folates, however, low levels are also present in lettuce (Nunes et al. 2009), potato (Blancquaert et al. 2013), maize, barley and rice in which the fortification studies have been done (Diaz de la Garza et al. 2007; Naqvi et al. 2009; Ramírez Rivera et al. 2016). Overexpression strategy is mainly studied for folate fortification. For example, increased expression of dihydrofolate synthetase (DHFS) in corn and GTP cyclo hydrolase in lettuce increased two- and eight-fold folate accumulation respectively (Naqvi et al. 2009; Nunes et al. 2009). On the other hand, by crossing pteridine overexpressing traits with para-aminobenzoate producing traits in both tomato and rice resulted in significant 25- to –100-fold upregulation of folate accumulation (Diaz de la Garza et al. 2007; Storozhenko et al. 2007). Moreover, manipulation of folate stability proved beneficial in fortification where, overexpression of folylpolyglutamate synthetase, in combination with GTP cyclo hydrolase, Arabidopsis ADCS and mammalian folate binding proteins, enhanced folate accumulation and stability for up to four months in rice (Blancquaert et al. 2015). However, in some crops like potato, multigene overexpression could only increase two-fold of folate levels (Blancquaert et al. 2013), suggesting more studies need to be done to gain greater insights into folate biosynthesis pathway for its successful biofortification in crops.

Like folate, vitamin B6 fortification also employs overexpression of biosynthesis genes. Initial studies on single-gene overexpression with pyridoxal phosphate synthase (PDX1) and epyridoxal phosphate glutaminase (PDX2) did not achieve significant vitamin B6 accumulation in Arabidopsis (Chen and Xiong 2009; Leuendorf et al. 2010; Raschke et al. 2011). In line with this, an improved strategy was investigated in which, concurrent overexpression of PDX1 and PDX2 coding genes under the control of CaMV promoter in cassava leaves resulted in a nine-fold increase in vitamin B6 levels (Li et al. 2015). The same strategy using PDX1 and PDX2 genes under root patatin promoter achieved more than 15-fold B6 levels in cassava roots (Li et al. 2015). Thus, there is need for better understanding of vitamin biosynthesis pathways for successful nutritional augmentation in crops.

20.7 Amino Acids and Proteins

Amino acids are the building blocks of proteins crucial for normal human physiology. Deficiency of amino acids leads to growth defects, anemia, and defective wound healing, reproductive ailments and, in severe cases, kwashiorkor disease. Moreover, phenylketonuria disorder and maple syrup urine disease are associated with malfunctioning of amino acid metabolism. Essential amino-acid containing diet is crucial during gestation, initial development, lactation and following injury. Most of the amino acids are synthesized *de novo* by human body, except a few, termed as essential amino acids which are supplemented through food (FAO/WHO/UNU 2007). Though all essential amino acids are present in animal-derived food, most of the staple crops lack one or more essential amino acids. For example, cereals have inadequate levels of lysine and threonine while legumes are generally deficient in methionine and cysteine. Therefore, nutritional augmentation strategies have mostly been investigated in these plants.

Aspartate pathway is the most widely investigated, as it synthesizes lysine, methionine and threonine essential amino acids. Aspartate pathway is a highly regulated pathway that occurs in plastids, except the last step where methionine is formed (Galili 2002; Galili and Höfgen 2002). Lysine and threonine act as feedback inhibitors for enzymes like, dihydrodipicolinate synthase (DHPS) and aspartate kinase (AK) involved in this pathway. Therefore, initial metabolic engineering strategies employed overexpression of wild type or mutant *dapA* encoding DHPS and *lysC*-encoding AK genes (Galili 2002; Galili and Höfgen 2002). The mutant version of these genes was unresponsive to lysine and threonine-mediated feedback inhibition. Transgenic rice overexpressing *lysC* from winged bean *(phosphocarpus tetragonoloba)* under regulation of maize ubiquitin promoter, transgenic canola, soybean overexpressing mutant *lysC* and *dapA* genes regulated by seed-specific promoter, transgenic maize overexpressing DHPS gene under embryo-specific promoter and transgenic potato overexpressing bacterial *dapA* achieved increased lysine content (Falco et al. 1995; Mazur et al. 1999; Yue-Feng et al. 2001). In addition, overexpression of mutant *dapA* resulted in significant overproduction of lysine in potato (Sévenier et al. 2002). Increased lysine levels in all plant tissues might cause atypical vegetative growth and flower development, reducing seed yield, hence tissue-specific lysine metabolic engineering can be beneficial (Galili and Höfgen 2002). In addition to lysine synthesis pathway, manipulation of lysine catabolism was also investigated. For example, transgenic tobacco and Arabidopsis with silenced or mutated lysine-ketoglutarate reductase/ saccharopine dehydrogenase (LKR/SDH) genes correlated to significantly high lysine content in seeds, as alpha-amino adipic acid pathway for lysine catabolism was compromised (Karchi et al. 1994, 1995; Zhu and Galili 2003). Similarly, lysine accumulation was 40 times more in LKR/SDH-silenced transgenic corn (Houmard et al. 2007).

A combination of enhanced anabolism and reduced catabolism approach can also be used for lysine fortification. Recently, High Lysine Transgenic lines 1 (HFL1) and 2 (HFL2) were generated, that achieved more than 25-fold lysine contents than wild-type rice (Yang et al. 2017).

Another strategy employs overexpression of storage proteins, which are the main source of essential amino acids in sink tissues. For example, enhanced expression of pea legumin (rich in lysine) increased 4.2% legumin content in rice endosperm, 1.5% in transgenic wheat increasing lysine levels (Sindhu et al. 1997; Stöger et al. 2001). Moreover, expression of modified storage proteins with increased lysine residues like, barley hordothionine protein containing 12 lysine residues (HT12) and the barley high-lysine protein containing eight lysine residues (BHL8) increased lysine content in sorghum by 50% and in corn by four times respectively (Jung and Carl 2000). Expression of another high-lysine storage protein encoding gene sb401, accumulated 54.8% more lysine and 39%increase in total-protein content (Yu et al. 2004). Further expression of animal-derived storage protein alpha lactalbumin, increased 47% lysine content in corn (Bicar et al. 2008). Recently, modified seed-storage protein gene fused with endogenous serine/threonine rich motif coding sequences and regulated by 35S promoter was introduced in rice. Transgenic rice, generated by this approach, accumulated 33.87% lysine, 21.2% threonine, 19.4% total amino acids and 20.25% crude protein more than wild type (Jiang et al. 2016).

Manipulation of another enzyme of aspartate pathway can also be exploited to achieve fortification of methionine and threonine. Increased expression of N-terminally mutated gene encoding feedback insensitive variant of cystathione gamma synthase (CGS) resulted in increased methionine content, which was further enhanced using feedback insensitive AK in Arabidopsis (Hacham et al. 2002, 2006; Hacham et al. 2008). Transgenic potato expressing mutated AK accumulated both methionine and threonine (Yue-Feng et al. 2001). Expression of Brazil nut 2S albumin and sunflower 2S albumin storage proteins increased 33% methionine content in canola and seven-fold methionine content in potato tubers respectively (Chakraborty et al. 2000; Müntz et al. 1998).

Further, overexpression of mutant-rate limiting enzymes can also be employed to fortify other amino acids. For example, overexpression of anthranilate synthase insensitive to tryptophan-mediated inhibition increased tryptophan significantly in rice (400-fold), potato tubers (30-fold), soyabean seeds (20-fold) (Ishimoto et al. 2010; Wakasa et al. 2006; Yamada et al. 2005). Similarly, mutant serine acetyltransferase (SAT) increased cysteine production by 26-fold in lupin seeds accompanied with high glutathione levels (Tabe et al. 2010).

Apart from enhancement of individual amino acids, expression of single proteins, rich in most essential amino acids can also be utilized for fortification. For example, amaranthin protein, AmA1 (from *Amaranthus hypochondriacus*) rich in lysine, tyrosine, threonine, cysteine, methionine, when overexpressed increased total protein content by 32% in corn 45% in potato tubers and 2.5% in wheat seeds (Chakraborty et al. 2000). AmA1 is a non-allergenic protein and correlates with optimum nutritional requirements set by the WHO for humans (Chakraborty et al. 2000).

20.8 Plant Volatiles Compound

Plant volatiles are key contributors to the overall taste and aroma property of fruits (Goff and Klee 2006). The cultivation of many plant species was enhanced due to the presence of specific volatiles in vegetables, fruits, and herbs. Most of the breeding programs were associated to enhance certain traits of the plant such as; yield, size, pigmentation, sugar content, total solids and many more while, at the same time, improving or even preserving volatile production in plant was neglected or ignored. As a result, most current cultivars emit less volatiles than earlier cultivars or their wild relatives (Gutterson 1993). Classical breeding can be an approach to reintroduce or enhance the aroma volatiles in plants. For example, Kamal and their coworkers introduced the distinct aromatic fragrance from *Lycopersicon peruvianum* to the cultivated tomato, *Lycopersicon esculentum* by classical breeding (Kamal et al. 2001). However, this process is time consuming, laborious and also expensive due to the monitoring of volatiles which requires sophisticated instruments like gas chromatography mass spectrometry (GCMS), NMR and others.

A genetic-engineering approach like overexpression of the yeast Δ-9-desaturase gene in tomato driven by constitutive promoter CaMV35S changes the fatty acids and fatty acid-derived flavor compounds (Wang et al. 1996). Speirs and their coworkers also showed that by enhancing the expression of alcohol dehydrogenase gene in tomato changes the balance of some flavor aldehydes and alcohols (Speirs et al. 1998). These transgenic tomatoes have a high level of alcohols which is associated with more intense "ripe fruit" flavor (Speirs et al. 1998). The above reports successfully modulated or enhanced the flavor but were unable to significantly change the aroma of transgenic fruits. In line with this, a major breakthrough was achieved when Lewinsohn and their coworkers successfully enhanced the synthesis of linalool in transgenic tomato (Lewinsohn et al. 2001). In plant, linalool is produced during the hydrolysis of geranyl diphosphate (GPP), an intermediate in carotenoids pathway, by a single-step reaction catalyzed by S-linalool synthase (LIS). Overexpression of LIS from *Clarkia breweri* in tomato under the control of late stages of fruit development specific E8 promoter, transgenic tomato plant accumulates linalool and its oxidation product, 8-hydroxylinalool in the fruit (Lewinsohn et al. 2001). Like linalool, geraniol, an acyclic monoterpene alcohol, also influences the aroma and flavor of tomato. Geraniol is also biosynthesized by geraniol synthase (GES) in single step using GPP as a substrate (Davidovich-Rikanati et al. 2007). Diversion of early plastidial terpenoid pathway by overexpressing GES from *Ocimum basilicum* under the control of polygalacturonase promoter (fruit ripening specific promoter) in tomato (Davidovich-Rikanati et al. 2007). Transgenic tomato accumulated higher amounts of geraniol and has modified the aroma and flavor (Davidovich-Rikanati et al. 2007). Linalool is a tertiary alcohol, so the hydroxyl group of linalool is blocked for further oxidation, whereas geraniol is a primary alcohol and can be further oxidized by non-specific alcohol dehydrogenases. Therefore, accumulation of geraniol has a more significant effect on the aroma and flavor, as 79% untrained taste panelists preferred GES-transgenic over control tomatoes during organoleptic evaluations (Davidovich-Rikanati et al. 2007). The expression of specific-transcription factors such as *Myb, Myc,* globally regulate monoterpene production pathways in plant (Mahmoud and Croteau 2002). Modifying the expression of these transcription factors is also an alternate strategy for pathway engineering. MsMYB, a R2R3-MYB transcription factor specific in Peltate glandular trichomes, negatively regulates monoterpene production, the silencing of this transcription factor increased levels of monoterpenes in basil and tobacco (Reddy et al. 2017).

Methanol is one of the simplest organic products of plant metabolism. It is the second major volatile organic compound after isoprene emitted through leaf surface (Guenther et al. 1995). A major source of methanol production in plant is the de-esterification of cell-wall pectin, a single-step reaction performed by Pectin methylesterase (PME)(Pelloux et al. 2007). The major function of methanol in plant is plant-to-plant communication (Dorokhov et al. 2012) and provides resistance against herbivory (Dixit et al. 2013; Körner et al. 2009). Methanol also stimulates the growth of plants by protecting photosynthetic machinery from photo inhibition (Nonomura and Benson 1992). Recent research suggested that overexpression of PME under the control of wound inducible promoter may increase the growth of the cotton plant and also increase cotton ball size (unpublished data). Identification of new enzymes involved in the biosynthesis of volatiles is significantly progressing. The major reason is the availability of more robust sequencing techniques and bioinformatic tools. However, pathway engineering often showed unpredictable results. The identification of key step/compounds involved in volatile production may significantly improve efficiency.

20.9 Phytic Acid

Phytic acid (PA) is a plant-secondary metabolite which is present in grains and rapidly accumulates in seeds during ripening (Robin and Graham 2004). PA is the primary storage form of phosphorus in plant. It is present in cereals and legumes phosphorus in large amounts as phytin (myo-inositol hexaphosphate), (Deshpande 2002). PA is synthesized in cytoplasm through Glucose-6 Phosphate and stored in vacuole through transporters (Figure 20.2a). Plant seed and grains have been reported to contain 0.5–5% PAs (Loewus 2002). PAs bind proteins causing reduced protein solubility, inhibit enzymes including pepsin, trypsin, alpha–amylase and also reduce absorption of major required minerals to form a complex compound with these minerals such as zinc, calcium, magnesium, and iron (Pusztai et al. 1993).

Maize (*Z. mays*) is a crop which is infested with PA in the aleuron layer of seed. PA, inositol 1, 2, 3, 4, 5, 6-hexakisphosphate, is a secondary metabolite present in maize seeds, which contributes about 75% of phosphorus present in seed. As PA is an anti-nutritional metabolite, it is necessary to reduce the content of PA in maize seed to provide nutritional security.

Several genetic-engineering strategies have been investigated to remove PA genes from seeds and grains. Three genes, ZmIPK1A, ZmIPK, and ZmMRP4involved in PA pathway have undergone construct design of CRISPR CAS and TALENs (Liang et al. 2014).

The designed guide RNA construct for these genes was introduced through protoplast transformation in maize. The efficiencies of designed constructs for T-ZmIPK and T-ZmMRP4 were 39 and 27% respectively. The gene ZmMRP4 involved in the final step of PA accumulation in vacuole, when silenced in maize caused a decrease in the amount of PAs. This transgenic technique is also now being investigated for anti-nutrient removal in other plants.

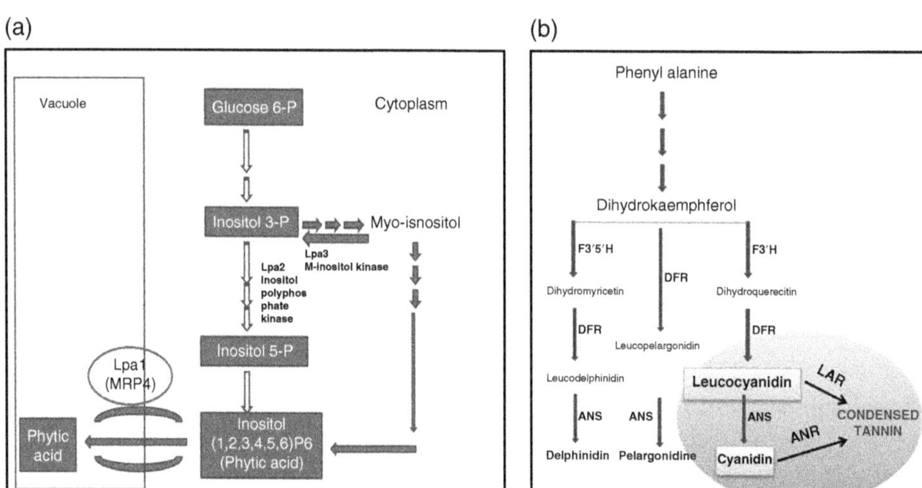

Figure 20.2 (a) Phytic acid synthetic pathway (Zm MRP4: Transporter of phytic acid). (b) Biosynthetic pathway of condensed tannin.

20.10 Condensed Tannin

The chemistry behind the synthesis of condensed tannin is mainly governed in the phenylpropanoid pathway. They are complex organic compounds derived from flavonol, insoluble in water, with a molecular weight of 500 and 3000 Da, and are capable of precipitating proteins (McMahon et al. 2000). These compounds are synthesized through flavonoid biosynthetic pathway in plants (Figure 20.2b) (Kumar and Pandey 2013). They are polymers of two monomers, which are present in isomeric form. These monomers flavan-3-ol (catechin) or flavan-3, 4-diol (epicatechin). Generally, condensed tannin binds to protein, and is regarded as "anti-nutritional compounds," which reduce the protein digestibility. So, gene silencing is a major approach for metabolic engineering in legume and the rape-seed plant.

Anthocyanidin reductase (ANR) or Banylus gene was determined as a potential gene for condensed tannin biosynthesis and accumulation in seed coat. Metabolic engineering of condensed tannin has been performed in *Arabidopsis thaliana* through silencing of ANR gene (Sharma and Dixon 2005). It further opens ways for another plant like oilseed rape (Auger et al. 2009) for production of stable line without condensed tannin. This will help in developing highly fortified protein-rich plant with lowered condensed tannin content.

20.11 Conclusions

Metabolic engineering works as a feasible and sustainable strategy to generate nutritionally enhanced crops (Hefferon 2016; Zhu et al. 2007). It depends on utilizing the genetic pool to introduce and express desired genes that can be endogenous, from other plant/animal species or micro-organisms to improve nutritional quality in crops (Newell-McGloughlin 2008; Pérez-Massot et al. 2013). Transgenic crops generated by metabolic engineering have the potential to combat malnutrition especially among consumers in developing countries (Hirschi 2009). Till now, many crops like legumes, vegetables and cereals have been enriched with vitamins, essential amino acids and minerals by a metabolic-engineering approach (Garg et al. 2018).

Though significant advances have occurred in metabolic engineering, there are still some challenges and constraints with this approach (García-Granados et al. 2019). Insufficient understanding of metabolic pathways and their crucial regulators remains the primary bottleneck of metabolic engineering. One of the reasons being, lack of proper functional annotations of genes which has restricted the determination of gene composition that encodes key metabolic enzymes. Therefore, a collective genomics, transcriptomics, proteomics, and metabolomics analysis would ease understanding of metabolic pathways and their key components (Farré et al. 2014). Continuous metabolite synthesis has a probability to cause atypical cell development and growth in plants. Hence, tissue-specific expression of transgenes will be required to favor biosynthesis and accumulation of desired metabolites in a particular tissue, curtailing the adverse effects on normal growth of plants (Peremarti et al. 2010). Since plant metabolic pathways are complex and regulated by multiple factors and enzymes, conventional molecular biology techniques limit metabolic-engineering development. Thus, recently TGS II multigene expression vector system and gene editing by CRISPR/dCas9 tool has been introduced in metabolic engineering, which have enabled efficient and precise manipulation of metabolic pathways in plants (Li et al. 2017; Zalatan et al. 2015; Zhu et al. 2017). Thus, improved insights of biosynthetic pathways and better developments in metabolic-engineering strategies, will allow precise reconstruction and regulation of complex metabolic pathways, enabling generation of diverse and unique varieties of nutritionally superior crops.

Unfortunately, fortified transgenic crops are not well accepted among populations and poorly adapted by agriculturalists worldwide (Al-Babili and Beyer 2005). The expensive and time-consuming regulatory procedures adopted by different countries for approval and commercialization of transgenic fortified crops makes the situation worse (Inaba and Macer 2004; Watanabe et al. 2005). In addition, agronomic trait assessment, trials on life-forms like humans increases the release time of metabolically engineered cultivars into the market, impacting its success rate. For example, golden rice research details were published in 2000 (Ye et al. 2000), however, it still took 18 years to be approved by the US Food and Drug Administration (US-FDA) for human consumption after three international assessments (Biotechnology Notification File No. BNF 000158 dated2018.05.24). Considerations must also take account of the bioavailability of nutrients after post-harvest processing and cooking of engineered crops (Haas et al. 2005).

Despite these bottlenecks, metabolically engineered crops, fortified with nutrients, have a promising future as these have the potential to eradicate micronutrient malnutrition among diverse populations worldwide.

Acknowledgments

We wish to apologize to all colleagues whose work, because of lack of space, could not be cited. SD is also thankful to MITACS for the MITACS-Accelerate postdoc fellowship.

References

Abdelgawad, K.F., M El-Mogy, M., IA Mohamed, M. et al. (2019). Increasing ascorbic acid content and salinity tolerance of cherry tomato plants by suppressed expression of the ascorbate oxidase gene. *Agronomy* 9: 51.

Al-Babili, S. and Beyer, P. (2005). Golden rice–five years on the road–five years to go? *Trends in Plant Science* 10: 565–573.

Ali, B., Pantha, S., Acharya, R. et al. (2019). Enhanced ascorbate level improves multi-stress tolerance in a widely grown indica rice variety without compromising its agronomic characteristics. *Journal of Plant Physiology* 240: 152998.

Aluru, M., Xu, Y., Guo, R. et al. (2008). Generation of transgenic maize with enhanced provitamin A content. *Journal of Experimental Botany* 59: 3551–3562.

Auger, B., Baron, C., Lucas, M.O. et al. (2009). Brassica orthologs from BANYULS belong to a small multigene family, which is involved in procyanidin accumulation in the seed. *Planta* 230: 1167.

Bai, C., Capell, T., Berman, J. et al. (2016). Bottlenecks in carotenoid biosynthesis and accumulation in rice endosperm are influenced by the precursor–product balance. *Plant Biotechnology Journal* 14: 195–205.

Bao, G., Zhuo, C., Qian, C. et al. (2016). Co-expression of NCED and ALO improves vitamin C level and tolerance to drought and chilling in transgenic tobacco and stylo plants. *Plant Biotechnology Journal* 14: 206–214.

Bicar, E.H., Woodman-Clikeman, W., Sangtong, V. et al. (2008). Transgenic maize endosperm containing a milk protein has improved amino acid balance. *Transgenic Research* 17: 59–71.

Blancquaert, D., Storozhenko, S., Loizeau, K. et al. (2010). Folates and folic acid: from fundamental research toward sustainable health. *Critical Reviews in Plant Sciences* 29: 14–35.

Blancquaert, D., Storozhenko, S., Van Daele, J. et al. (2013). Enhancing pterin and para-aminobenzoate content is not sufficient to successfully biofortify potato tubers and Arabidopsis thaliana plants with folate. *Journal of Experimental Botany* 64: 3899–3909.

Blancquaert, D., Van Daele, J., Strobbe, S. et al. (2015). Improving folate (vitamin B 9) stability in biofortified rice through metabolic engineering. *Nature Biotechnology* 33: 1076–1078.

Bouis, H.E., Graham, R.D., and Welch, R.M. (1999). *The CGIAR Micronutrients Project: Justification, History, Objectives, and Summary of Findings*. International Food Policy Research Institute.

Bulley, S., Wright, M., Rommens, C. et al. (2012). Enhancing ascorbate in fruits and tubers through over-expression of the l-galactose pathway gene GDP-l-galactose phosphorylase. *Plant Biotechnology Journal* 10: 390–397.

Cahoon, E.B., Hall, S.E., Ripp, K.G. et al. (2003). Metabolic redesign of vitamin E biosynthesis in plants for tocotrienol production and increased antioxidant content. *Nature Biotechnology* 21: 1082–1087.

Capell, T. and Christou, P. (2004). Progress in plant metabolic engineering. *Current Opinion in Biotechnology* 15: 148–154.

Carr, A.C. and Maggini, S. (2017). Vitamin C and immune function. *Nutrients* 9: 1211.

Caspi, R., Dreher, K., and Karp, P.D. (2013). The challenge of constructing, classifying, and representing metabolic pathways. *FEMS Microbiology Letters* 345: 85–93.

Chakraborty, S., Chakraborty, N., and Datta, A. (2000). Increased nutritive value of transgenic potato by expressing a nonallergenic seed albumin gene from Amaranthus hypochondriacus. *Proceedings of the National Academy of Sciences* 97: 3724–3729.

Chen, H. and Xiong, L. (2009). Enhancement of vitamin B6 levels in seeds through metabolic engineering. *Plant Biotechnology Journal* 7: 673–681.

Chen, Z., Young, T.E., Ling, J. et al. (2003). Increasing vitamin C content of plants through enhanced ascorbate recycling. *Proceedings of the National Academy of Sciences* 100: 3525–3530.

Cho, E.A., Lee, C.A., Kim, Y.S. et al. (2005). Expression of γ-tocopherol methyltransferase transgene improves tocopherol composition in lettuce (Latuca sativa L.). *Molecules & Cells (Springer Science & Business Media BV)* 19 (1): 16–22.

Collakova, E. and DellaPenna, D. (2003). Homogentisate phytyl transferase activity is limiting for tocopherol biosynthesis in Arabidopsis. *Plant Physiology* 131: 632–642.

Cronje, C., George, G.M., Fernie, A.R. et al. (2012). Manipulation of L-ascorbic acid biosynthesis pathways in Solanum lycopersicum: elevated GDP-mannose pyrophosphorylase activity enhances L-ascorbate levels in red fruit. *Planta* 235: 553–564.

Crowell, E.F., McGrath, J.M., and Douches, D.S. (2008). Accumulation of vitamin E in potato (Solanum tuberosum) tubers. *Transgenic Research* 17: 205–217.

Datta, K., Baisakh, N., Oliva, N. et al. (2003). Bioengineered 'golden' indica rice cultivars with β-carotene metabolism in the endosperm with hygromycin and mannose selection systems. *Plant Biotechnology Journal* 1: 81–90.

Davidovich-Rikanati, R., Sitrit, Y., Tadmor, Y. et al. (2007). Enrichment of tomato flavor by diversion of the early plastidial terpenoid pathway. *Nature Biotechnology* 25: 899–901.

Deshpande, S.S. (2002). *Handbook of Food Toxicology*. CRC Press.

Diaz de la Garza, R., Gregory, J.F., and Hanson, A.D. (2007). Folate biofortification of tomato fruit. *Proceedings of the National Academy of Sciences* 104: 4218–4222.

Diretto, G., Tavazza, R., Welsch, R. et al. (2006). Metabolic engineering of potato tuber carotenoids through tuber-specific silencing of lycopene epsilon cyclase. *BMC Plant Biology* 6: 13.

Diretto, G., Welsch, R., Tavazza, R. et al. (2007). Silencing of beta-carotene hydroxylase increases total carotenoid and beta-carotene levels in potato tubers. *BMC Plant Biology* 7: 11.

Dixit, S., Upadhyay, S.K., Singh, H. et al. (2013). Enhanced methanol production in plants provides broad spectrum insect resistance. *PLoS One* 8: e79664.

Dorokhov, Y.L., Komarova, T.V., Petrunia, I.V. et al. (2012). Airborne signals from a wounded leaf facilitate viral spreading and induce antibacterial resistance in neighboring plants. *PLoS Pathogens* 8 (4): e1002640.

Drouin, G., Godin, J.R., and Pagé, B. (2011). The genetics of vitamin C loss in vertebrates. *Current Genomics* 12: 371–378.

Ducreux, L.J., Morris, W.L., Hedley, P.E. et al. (2005). Metabolic engineering of high carotenoid potato tubers containing enhanced levels of β-carotene and lutein. *Journal of Experimental Botany* 56: 81–89.

Falco, S.C., Guida, T., Locke, M. et al. (1995). Transgenic canola and soybean seeds with increased lysine. *Bio/Technology* 13: 577–582.

Farré, G., Sudhakar, D., Naqvi, S. et al. (2012). Transgenic rice grains expressing a heterologous ρ-hydroxyphenylpyruvate dioxygenase shift tocopherol synthesis from the γ to the α isoform without increasing absolute tocopherol levels. *Transgenic Research* 21: 1093–1097.

Farré, G., Blancquaert, D., Capell, T. et al. (2014). Engineering complex metabolic pathways in plants. *Annual Review of Plant Biology* 65: 187–122.

Fraser, P.D., Romer, S., Shipton, C.A. et al. (2002). Evaluation of transgenic tomato plants expressing an additional phytoene synthase in a fruit-specific manner. *Proceedings of the National Academy of Sciences* 99: 1092–1097.

Fray, R.G., Wallace, A., Fraser, P.D. et al. (1995). Constitutive expression of a fruit phytoene synthase gene in transgenic tomatoes causes dwarfism by redirecting metabolites from the gibberellin pathway. *The Plant Journal* 8: 693–701.

Fu, R., Martin, C., and Zhang, Y. (2018). Next-generation plant metabolic engineering, inspired by an ancient Chinese irrigation system. *Molecular Plant* 11: 47–57.

Fujisawa, M., Watanabe, M., Choi, S.K. et al. (2008). Enrichment of carotenoids in flaxseed (Linumusitatissimum) by metabolic engineering with introduction of bacterial phytoene synthase gene crtB. *Journal of Bioscience and Bioengineering* 105: 636–641.

Galili, G. (2002). New insights into the regulation and functional significance of lysine metabolism in plants. *Annual Review of Plant Biology* 53: 27–43.

Galili, G. and Höfgen, R. (2002). Metabolic engineering of amino acids and storage proteins in plants. *Metabolic Engineering* 4: 3–11.

García-Granados, R., Lerma-Escalera, J.A., and Morones-Ramírez, J.R. (2019). Metabolic engineering and synthetic biology: synergies, future, and challenges. *Frontiers in Bioengineering and Biotechnology* 7: 36.

Garg, M., Sharma, N., Sharma, S. et al. (2018). Biofortified crops generated by breeding, agronomy, and transgenic approaches are improving lives of millions of people around the world. *Frontiers in Nutrition* 5: 12.

Gest, N., Garchery, C., Gautier, H. et al. (2013). Light-dependent regulation of ascorbate in tomato by a monodehydroascorbate reductase localized in peroxisomes and the cytosol. *Plant Biotechnology Journal* 11: 344–354.

Ghimire, B.K., Seong, E.S., Lee, C.O. et al. (2011). Enhancement of alpha-tocopherol content in transgenic Perilla frutescens containing the gamma-TMT gene. *African Journal of Biotechnology* 10: 2430–2439.

Gillespie, S., Hodge, J., Yosef, S., and Pandya-Lorch, R. (eds.) (2016). *Nourishing Millions: Stories of Change in Nutrition*. International Food Policy Research Institute.

Goff, S.A. and Klee, H.J. (2006). Plant volatile compounds: sensory cues for health and nutritional value? *Science* 311: 815–819.

Guenther, A., Hewitt, C.N., Erickson, D. et al. (1995). A global model of natural volatile organic compound emissions. *Journal of Geophysical Research-Atmospheres* 100: 8873–8892.

Gutterson, N.C. (1993). Molecular breeding for color, flavor and fragrance. *Scientia Horticulturae* 55: 141–160.

Haas, J.D., Beard, J.L., Murray-Kolb, L.E. et al. (2005). Iron-biofortified rice improves the iron stores of nonanemic Filipino women. *The Journal of Nutrition* 135: 2823–2830.

Hacham, Y., Avraham, T., and Amir, R. (2002). The N-terminal region of Arabidopsis cystathionine γ-synthase plays an important regulatory role in methionine metabolism. *Plant Physiology* 128: 454–462.

Hacham, Y., Schuster, G., and Amir, R. (2006). An in vivo internal deletion in the N-terminus region of Arabidopsis cystathionine γ-synthase results in CGS expression that is insensitive to methionine. *The Plant Journal* 45: 955–967.

Hacham, Y., Matityahu, I., Schuster, G., and Amir, R. (2008). Overexpression of mutated forms of aspartate kinase and cystathionine γ-synthase in tobacco leaves resulted in the high accumulation of methionine and threonine. *The Plant Journal* 54: 260–271.

Hallberg, L., Brune, M., and Rossander, L. (1989). The role of vitamin C in iron absorption. *International journal for vitamin and nutrition research. Supplement= InternationaleZeitschrift fur Vitamin-und Ernahrungsforschung* 30 (Supplement): 103–108.

Halpin, C. (2005). Gene stacking in transgenic plants–the challenge for 21st century plant biotechnology. *Plant Biotechnology Journal* 3: 141–155.

Hanson, A.D. and Jez, J.M. (2018). Synthetic biology meets plant metabolism. *Plant Science: An International Journal of Experimental Plant Biology* 273: 1.

Harrison, E.H. (2005). Mechanisms of digestion and absorption of dietary vitamin a. *Annual Review of Nutrition* 25: 87–103.

Hefferon, K.L. (2016). Can biofortified crops help attain food security? *Current Molecular Biology Reports* 2: 180–185.

Hemavathi, Upadhyaya, C.P., Young, K.E. et al. (2009). Over-expression of strawberry D-galacturonic acid reductase in potato leads to accumulation of vitamin C with enhanced abiotic stress tolerance. *Plant Science* 177: 659–667.

Hirschi, K.D. (2009). Nutrient biofortification of food crops. *Annual Review of Nutrition* 29: 401–421.

Hoa, T.T.C., Al-Babili, S., Schaub, P. et al. (2003). Golden Indica and Japonica rice lines amenable to deregulation. *Plant Physiology* 133: 161–169.

Houmard, N.M., Mainville, J.L., Bonin, C.P. et al. (2007). High-lysine corn generated by endosperm-specific suppression of lysine catabolism using RNAi. *Plant Biotechnology Journal* 5: 605–614.

Inaba, M. and Macer, D. (2004). Policy, regulation and attitudes towards agricultural biotechnology in Japan. *Journal of International Biotechnology Law* 1: 45–53.

Ishimoto, M., Rahman, S.M., Hanafy, M.S. et al. (2010). Evaluation of amino acid content and nutritional quality of transgenic soybean seeds with high-level tryptophan accumulation. *Molecular Breeding* 25: 313–326.

Jain, A.K. and Nessler, C.L. (2000). Metabolic engineering of an alternative pathway for ascorbic acid biosynthesis in plants. *Molecular Breeding* 6: 73–78.

Jiang, S.Y., Ma, A., Xie, L., and Ramachandran, S. (2016). Improving protein content and quality by over-expressing artificially synthetic fusion proteins with high lysine and threonine constituent in rice plants. *Scientific Reports* 6: 34427.

Jiang, L., Wang, W., Lian, T., and Zhang, C. (2017). Manipulation of metabolic pathways to develop vitamin-enriched crops for human health. *Frontiers in Plant Science* 8: 937.

Jung, R. and Carl, F. (2000).Transgenic corn with an improved amino acid composition. 8th International Symposium on Plant Seeds. Institute of Plant Genetics and Crop Plant Research, Gatersleben.

Kamal, H.M., Takashina, T., Egashira, H. et al. (2001). Introduction of aromatic fragrance into cultivated tomato from the peruvianum complex. *Plant Breeding* 120: 179–181.

Karchi, H., Shaul, O., and Galili, G. (1994). Lysine synthesis and catabolism are coordinately regulated during tobacco seed development. *Proceedings of the National Academy of Sciences* 91: 2577–2581.

Karchi, H., Miron, D., Ben-Yaacov, S., and Galili, G. (1995). The lysine-dependent stimulation of lysine catabolism in tobacco seed requires calcium and protein phosphorylation. *The Plant Cell* 7: 1963–1970.

Kennedy, D.O. (2016). B vitamins and the brain: mechanisms, dose and efficacy – a review. *Nutrients* 8: 68.

Khush, G.S., Lee, S., Cho, J.I., and Jeon, J.S. (2012). Biofortification of crops for reducing malnutrition. *Plant Biotechnology Reports* 6: 195–202.

Kim, S.E., Lee, C.J., Ji, C.Y. et al. (2019). Transgenic sweet potato plants overexpressing tocopherol cyclase display enhanced α-tocopherol content and abiotic stress tolerance. *Plant Physiology and Biochemistry* 144: 436–444.

Körner, E., von Dahl, C.C., Bonaventure, G., and Baldwin, I.T. (2009). Pectin methylesterase Na PME1 contributes to the emission of methanol during insect herbivory and to the elicitation of defence responses in Nicotiana attenuata. *Journal of Experimental Botany* 60: 2631–2640.

Kumar, S. and Pandey, A.K. (2013). Chemistry and biological activities of flavonoids: an overview. *The Scientific World Journal* 2013: 162750.

Kumar, R., Raclaru, M., Schüßeler, T. et al. (2005). Characterisation of plant tocopherol cyclases and their overexpression in transgenic Brassica napus seeds. *FEBS Letters* 579: 1357–1364.

Lee, B.K., Kim, S.L., Kim, K.H. et al. (2008). Seed specific expression of perilla γ-tocopherol methyltransferase gene increases α-tocopherol content in transgenic perilla (Perilla frutescens). *Plant Cell, Tissue and Organ Culture* 92: 47–54.

Leuendorf, J.E., Osorio, S., Szewczyk, A. et al. (2010). Complex assembly and metabolic profiling of Arabidopsis thaliana plants overexpressing vitamin B6 biosynthesis proteins. *Molecular Plant* 3: 890–903.

Lewinsohn, E., Schalechet, F., Wilkinson, J. et al. (2001). Enhanced levels of the aroma and flavor compound S-linalool by metabolic engineering of the terpenoid pathway in tomato fruits. *Plant Physiology* 127: 1256–1265.

Li, L., Yang, Y., Xu, Q. et al. (2012). The Or gene enhances carotenoid accumulation and stability during post-harvest storage of potato tubers. *Molecular Plant* 5: 339–352.

Li, Q., Yang, X., Xu, S. et al. (2012). Genome-wide association studies identified three independent polymorphisms associated with α-tocopherol content in maize kernels. *PLoS One* 7 (5): e36807.

Li, K.T., Moulin, M., Mangel, N. et al. (2015). Increased bioavailable vitamin B 6 in field-grown transgenic cassava for dietary sufficiency. *Nature Biotechnology* 33: 1029–1032.

Li, Z., Zhang, D., Xiong, X. et al. (2017). A potent Cas9-derived gene activator for plant and mammalian cells. *Nature Plants* 3: 930–936.

Liang, Z., Zhang, K., Chen, K., and Gao, C. (2014). Targeted mutagenesis in Zea mays using TALENs and the CRISPR/Cas system. *Journal of Genetics and Genomics* 41: 63–68.

Loewus, F. (2002). Biosynthesis of phytate in food grains and seeds. In: *Food Phytates* (eds. N.R. Reddy and S.K. Sathe), 53–61. Taylor & Francis Group.

Lopez, A.B., Van Eck, J., Conlin, B.J. et al. (2008). Effect of the cauliflower or transgene on carotenoid accumulation and chromoplast formation in transgenic potato tubers. *Journal of Experimental Botany* 59: 213–223.

Lorence, A., Chevone, B.I., Mendes, P., and Nessler, C.L. (2004). Myo-inositol oxygenase offers a possible entry point into plant ascorbate biosynthesis. *Plant Physiology* 134: 1200–1205.

Lu, S., Van Eck, J., Zhou, X. et al. (2006). The cauliflower or gene encodes a DnaJ cysteine-rich domain-containing protein that mediates high levels of β-carotene accumulation. *The Plant Cell* 18: 3594–3605.

Ma, L., Wang, Y., Liu, W., and Liu, Z. (2014). Overexpression of an alfalfa GDP-mannose 3, 5-epimerase gene enhances acid, drought and salt tolerance in transgenic Arabidopsis by increasing ascorbate accumulation. *Biotechnology Letters* 36: 2331–2341.

Mahmoud, S.S. and Croteau, R.B. (2002). Strategies for transgenic manipulation of monoterpene biosynthesis in plants. *Trends in Plant Science* 7: 366–373.

Martín-Calvo, N. and Martínez-González, M.Á. (2017). Vitamin C intake is inversely associated with cardiovascular mortality in a cohort of Spanish graduates: the SUN project. *Nutrients* 9: 954.

Martini, S., Rizzello, A., Corsini, I. et al. (2018). Vitamin A deficiency due to selective eating as a cause of blindness in a high-income setting. *Pediatrics* 141: S439–S444.

Mazur, B., Krebbers, E., and Tingey, S. (1999). Gene discovery and product development for grain quality traits. *Science* 285: 372–375.

McGuireS.(2015). The state of food insecurity in the world. http://www.fao.org/3/a-i4646e.pdf (accessed 20 July 2020).

McMahon, L.R., McAllister, T.A., Berg, B.P. et al. (2000). A review of the effects of forage condensed tannins on ruminal fermentation and bloat in grazing cattle. *Canadian Journal of Plant Science* 80: 469–485.

Misawa, N., Yamano, S., Linden, H. et al. (1993). Functional expression of the Erwinia uredovora carotenoid biosynthesis gene crtl in transgenic plants showing an increase of β-carotene biosynthesis activity and resistance to the bleaching herbicide norflurazon. *The Plant Journal* 4: 833–840.

Müntz, K., Christov, V., Saalbach, G. et al. (1998). Genetic engineering for high methionine grain legumes. *Food/Nahrung* 42: 125–127.
Naqvi, S., Zhu, C., Farre, G. et al. (2009). Transgenic multivitamin corn through biofortification of endosperm with three vitamins representing three distinct metabolic pathways. *Proceedings of the National Academy of Sciences* 106: 7762–7767.
Naqvi, S., Farré, G., Sanahuja, G. et al. (2010). When more is better: multigene engineering in plants. *Trends in Plant Science* 15: 48–56.
Naqvi, S., Farré, G., Zhu, C. et al. (2011). Simultaneous expression of Arabidopsis ρ-hydroxyphenylpyruvate dioxygenase and MPBQ methyltransferase in transgenic corn kernels triples the tocopherol content. *Transgenic Research* 20: 177–181.
Newell-McGloughlin, M. (2008). Nutritionally improved agricultural crops. *Plant Physiology* 147: 939–953.
Nonomura, A.M. and Benson, A.A. (1992). The path of carbon in photosynthesis: improved crop yields with methanol. *Proceedings of the National Academy of Sciences* 89: 9794–9798.
Nunes, A.C., Kalkmann, D.C., and Aragao, F.J. (2009). Folate biofortification of lettuce by expression of a codon optimized chicken GTP cyclohydrolase I gene. *Transgenic Research* 18: 661.
Oreopoulos, D.G., Lindeman, R.D., VanderJagt, D.J. et al. (1993). Renal excretion of ascorbic acid: effect of age and sex. *Journal of the American College of Nutrition* 12: 537–542.
Owens, B.F., Lipka, A.E., Magallanes-Lundback, M. et al. (2014). A foundation for provitamin A biofortification of maize: genome-wide association and genomic prediction models of carotenoid levels. *Genetics* 198: 1699–1716.
Padayatty, S.J. and Levine, M. (2016). Vitamin C: the known and the unknown and goldilocks. *Oral Diseases* 22: 463–493.
Paine, J.A., Shipton, C.A., Chaggar, S. et al. (2005). Improving the nutritional value of Golden Rice through increased pro-vitamin a content. *Nature Biotechnology* 23: 482.
Pelloux, J., Rusterucci, C., and Mellerowicz, E.J. (2007). New insights into pectin methylesterase structure and function. *Trends in Plant Science* 12: 267–277.
Peremarti, A., Twyman, R.M., Gómez-Galera, S. et al. (2010). Promoter diversity in multigene transformation. *Plant Molecular Biology* 73: 363–378.
Pérez-Massot, E., Banakar, R., Gómez-Galera, S. et al. (2013). The contribution of transgenic plants to better health through improved nutrition: opportunities and constraints. *Genes & Nutrition* 8: 29–41.
Pusztai, A., Grant, G., Stewart, J.C. et al. (1993). Nutritional evaluation of RAZ-2, a new Phaseolus vulgaris bean cultivar containing high levels of the natural insecticidal protein arcelin 1. *Journal of Agricultural and Food Chemistry* 41: 436–440.
Qin, A., Shi, Q., and Yu, X. (2011). Ascorbic acid contents in transgenic potato plants overexpressing two dehydroascorbate reductase genes. *Molecular Biology Reports* 38: 1557–1566.
Quadrana, L., Almeida, J., Asis, R. et al. (2014). Natural occurring epialleles determine vitamin E accumulation in tomato fruits. *Nature Communications* 5: 1–11.
Ramírez Rivera, N.G., García-Salinas, C., Aragao, F.J., and Díaz de la Garza, R.I. (2016). Metabolic engineering of folate and its precursors in Mexican common bean (Phaseolus vulgaris L.). *Plant Biotechnology Journal* 14: 2021–2032.
Raschke, M., Boycheva, S., Crèvecoeur, M. et al. (2011). Enhanced levels of vitamin B6 increase aerial organ size and positively affect stress tolerance in Arabidopsis. *The Plant Journal* 66: 414–432.

Reddy, V.A., Wang, Q., Dhar, N. et al. (2017). Spearmint R2R3-MYB transcription factor Ms MYB negatively regulates monoterpene production and suppresses the expression of geranyl diphosphate synthase large subunit (Ms GPPS. LSU). *Plant Biotechnology Journal* 15: 1105–1119.

Rizvi, S., Raza, S.T., Faizal Ahmed, A.A. et al. (2014). The role of vitamin E in human health and some diseases. *Sultan Qaboos University Medical Journal* 14: e157.

Robin, D.G. and Graham, R.D. (2004). Breeding for micronutrients in staple food crops from a human nutrition perspective. *Journal of Experimental Botany* 55: 353–364.

Römer, S., Fraser, P.D., Kiano, J.W. et al. (2000). Elevation of the provitamin A content of transgenic tomato plants. *Nature Biotechnology* 18: 666–669.

Rosati, C., Aquilani, R., Dharmapuri, S. et al. (2000). Metabolic engineering of beta-carotene and lycopene content in tomato fruit. *The Plant Journal* 24: 413–420.

Sadre, R., Gruber, J., and Frentzen, M. (2006). Characterization of homogentisate prenyltransferases involved in plastoquinone-9 and tocochromanol biosynthesis. *FEBS Letters* 580: 5357–5362.

Saltzman, A., Birol, E., Oparinde, A. et al. (2017). Availability, production, and consumption of crops biofortified by plant breeding: current evidence and future potential. *Annals of the New York Academy of Sciences* 1390: 104–114.

Schmidt, M.A., Parrott, W.A., Hildebrand, D.F. et al. (2015). Transgenic soya bean seeds accumulating β-carotene exhibit the collateral enhancements of oleate and protein content traits. *Plant Biotechnology Journal* 13: 590–600.

Seo, Y.S., Kim, S.J., Harn, C.H., and Kim, W.T. (2011). Ectopic expression of apple fruit homogentisate phytyl transferase gene (MdHPT1) increases tocopherol in transgenic tomato (Solanum lycopersicum cv. Micro-Tom) leaves and fruits. *Phytochemistry* 72: 321–329.

Sévenier, R., Van der Meer, I.M., Bino, R., and Koops, A.J. (2002). Increased production of nutriments by genetically engineered crops. *Journal of the American College of Nutrition* 21: 199S–204S.

Sharma, S.B. and Dixon, R.A. (2005). Metabolic engineering of proanthocyanidins by ectopic expression of transcription factors in Arabidopsis thaliana. *The Plant Journal* 44: 62–75.

Shewmaker, C.K., Sheehy, J.A., Daley, M. et al. (1999). Seed-specific overexpression of phytoene synthase: increase in carotenoids and other metabolic effects. *The Plant Journal* 20: 401–412.

Sindhu, A.S., Zheng, Z., and Murai, N. (1997). The pea seed storage protein legumin was synthesized, processed, and accumulated stably in transgenic rice endosperm. *Plant Science* 130: 189–196.

Smirnoff, N. (1996). Botanical briefing: the function and metabolism of ascorbic acid in plants. *Annals of Botany* 78: 661–669.

Speirs, J., Lee, E., Holt, K. et al. (1998). Genetic manipulation of alcohol dehydrogenase levels in ripening tomato fruit affects the balance of some flavor aldehydes and alcohols. *Plant Physiology* 117: 1047–1058.

Stöger, E., Parker, M., Christou, P., and Casey, R. (2001). Pea legumin overexpressed in wheat endosperm assembles into an ordered paracrystalline matrix. *Plant Physiology* 125: 1732–1742.

Storozhenko, S., Ravanel, S., Zhang, G.F. et al. (2005). Folate enhancement in staple crops by metabolic engineering. *Trends in Food Science & Technology* 16: 271–281.

World Health Organization & United Nations University. (2007). Protein and amino acid requirements in human nutrition: report of a joint FAO/WHO/UNU Expert Consultation. World Health Organization, p. 265.

Storozhenko, S., De Brouwer, V., Volckaert, M. et al. (2007). Folate fortification of rice by metabolic engineering. *Nature Biotechnology* 25: 1277–1279.

Tabe, L., Wirtz, M., Molvig, L. et al. (2010). Overexpression of serine acetlytransferase produced large increases in O-acetylserine and free cysteine in developing seeds of a grain legume. *Journal of Experimental Botany* 61: 721–733.

Tambasco-Studart, M., Titiz, O., Raschle, T. et al. (2005). Vitamin B6 biosynthesis in higher plants. *Proceedings of the National Academy of Sciences* 102: 13687–13692.

Tang, Y., Fu, X., Shen, Q., and Tang, K. (2016). Roles of MPBQ-MT in promoting α/γ-tocopherol production and photosynthesis under high light in lettuce. *PLoS One* 11: e0148490.

Tavva, V.S., Kim, Y.H., Kagan, I.A. et al. (2007). Increased α-tocopherol content in soybean seed overexpressing the Perilla frutescens γ-tocopherol methyltransferase gene. *Plant Cell Reports* 26: 61–70.

Trumbo, P., Yates, A.A., Schlicker, S., and Poos, M. (2001). Dietary reference intakes: vitamin A, vitamin K, arsenic, boron, chromium, copper, iodine, iron, manganese, molybdenum, nickel, silicon, vanadium, and zinc. *Journal of the Academy of Nutrition and Dietetics* 101: 294.

Van Eck, J., Conlin, B., Garvin, D.F. et al. (2007). Enhancing beta-carotene content in potato by RNAi-mediated silencing of the beta-carotene hydroxylase gene. *American Journal of Potato Research* 84: 331.

Van Eenennaam, A.L., Lincoln, K., Durrett, T.P. et al. (2003). Engineering vitamin E content: from Arabidopsis mutant to soy oil. *The Plant Cell* 15: 3007–3019.

Wakasa, K., Hasegawa, H., Nemoto, H. et al. (2006). High-level tryptophan accumulation in seeds of transgenic rice and its limited effects on agronomic traits and seed metabolite profile. *Journal of Experimental Botany* 57: 3069–3078.

Wang, C., Chin, C.K., Ho, C.T. et al. (1996). Changes of fatty acids and fatty acid-derived flavor compounds by expressing the yeast Δ-9 desaturase gene in tomato. *Journal of Agricultural and Food Chemistry* 44: 3399–3402.

Wang, C., Zeng, J., Li, Y. et al. (2014). Enrichment of provitamin A content in wheat (Triticumaestivum L.) by introduction of the bacterial carotenoid biosynthetic genes CrtB and CrtI. *Journal of Experimental Botany* 65: 2545–2556.

Watanabe, K.N., Sassa, Y., Suda, E. et al. (2005). Global political, economic, social and technological issues on transgenic crops. *Plant Biotechnology* 22: 515–522.

World Health Organization, WHO (1992). *National Strategies for Overcoming Micronutrient Malnutrition, Document A45/3.* Geneva, Switzerland: WHO.

Yabuta, Y., Tanaka, H., Yoshimura, S. et al. (2013). Improvement of vitamin E quality and quantity in tobacco and lettuce by chloroplast genetic engineering. *Transgenic Research* 22:391–402.

Yamada, T., Tozawa, Y., Hasegawa, H. et al. (2005). Use of a feedback-insensitive α subunit of anthranilate synthase as a selectable marker for transformation of rice and potato. *Molecular Breeding* 14: 363–373.

Yang, Q.Q., Suen, P.K., Zhang, C.Q. et al. (2017). Improved growth performance, food efficiency, and lysine availability in growing rats fed with lysine-biofortified rice. *Scientific Reports* 7: 1–11.

Ye, X., Al-Babili, S., Klöti, A. et al. (2000). Engineering the provitamin A (β-carotene) biosynthetic pathway into (carotenoid-free) rice endosperm. *Science* 287: 303–305.

Yu, J., Peng, P., Zhang, X. et al. (2004). Seed-specific expression of a lysine rich protein sb401 gene significantly increases both lysine and total protein content in maize seeds. *Molecular Breeding* 14: 1–7.

Yue-Feng, G.A.O., Yu-Xiang, J.I.N.G., Shi-Hua, S.H.E.N. et al. (2001). Transfer of lysine-rich protein gene into rice and production of fertile transgenic plants. *Journal of Integrative Plant Biology* 43: 506–511.

Zalatan, J.G., Lee, M.E., Almeida, R. et al. (2015). Engineering complex synthetic transcriptional programs with CRISPR RNA scaffolds. *Cell* 160: 339–350.

Zeng, J., Wang, X., Miao, Y. et al. (2015). Metabolic engineering of wheat provitamin A by simultaneously overexpressing CrtB and silencing carotenoid hydroxylase (TaHYD). *Journal of Agricultural and Food Chemistry* 63: 9083–9092.

Zhang, C., Liu, J., Zhang, Y. et al. (2011). Overexpression of SlGMEs leads to ascorbate accumulation with enhanced oxidative stress, cold, and salt tolerance in tomato. *Plant Cell Reports* 30: 389–398.

Zhang, G.Y., Liu, R.R., Xu, G. et al. (2013). Increased α-tocotrienol content in seeds of transgenic rice overexpressing Arabidopsis γ-tocopherol methyltransferase. *Transgenic Research* 22: 89–99.

Zhang, Q., Cui, M.A., Xin, X.U.E. et al. (2014). Overexpression of a cytosolic ascorbate peroxidase gene, OsAPX2, increases salt tolerance in transgenic alfalfa. *Journal of Integrative Agriculture* 13: 2500–2507.

Zhang, L., Luo, Y., Liu, B. et al. (2020). Overexpression of the maize γ-tocopherol methyltransferase gene (ZmTMT) increases α-tocopherol content in transgenic Arabidopsis and maize seeds. *Transgenic Research* 29: 95–104.

Zhou, X., Welsch, R., Yang, Y. et al. (2015). Arabidopsis OR proteins are the major posttranscriptional regulators of phytoene synthase in controlling carotenoid biosynthesis. *Proceedings of the National Academy of Sciences* 112: 3558–3563.

Zhu, X. and Galili, G. (2003). Increased lysine synthesis coupled with a knockout of its catabolism synergistically boosts lysine content and also transregulates the metabolism of other amino acids in Arabidopsis seeds. *The Plant Cell* 15: 845–853.

Zhu, C., Naqvi, S., Gomez-Galera, S. et al. (2007). Transgenic strategies for the nutritional enhancement of plants. *Trends in Plant Science* 12: 548–555.

Zhu, Q., Yu, S., Zeng, D. et al. (2017). Development of "purple endosperm rice" by engineering anthocyanin biosynthesis in the endosperm with a high-efficiency transgene stacking system. *Molecular Plant* 10: 918–929.

Zhu, Q., Wang, B., Tan, J. et al. (2020). Plant synthetic metabolic engineering for enhancing crop nutritional quality. *Plant Communications* 1: 100017.

Zimmermann, M.B. and Hurrell, R.F. (2002). Improving iron, zinc and vitamin A nutrition through plant biotechnology. *Current Opinion in Biotechnology* 13: 142–145.

ZorrillaLópez, U., MasipVilà, G., Arjó Pont, G. et al. (2013). Engineering metabolic pathways in plants by multigene transformation. *International Journal of Developmental Biology* 57: 565–576.

21

Genome Engineering for Food Security

Sajid Fiaz[1], Sher Aslam Khan[1], Mehmood Ali Noor[2], Afifa Younas[3], Habib Ali[4], Kazim Ali[5], Mahmoud Mohamed Gaballah[6], and Galal Bakr Anis[6]

[1] *Department of Plant Breeding and Genetics, The University of Haripur 22620, Haripur, Khyber Pakhtunkhwa, Pakistan*
[2] *Institute of Crop Sciences, Chinese Academy of Agricultural Sciences, Key Laboratory of Crop Physiology and Ecology, Ministry of Agriculture, Beijing 100081, China*
[3] *Department of Botany, Lahore College for Women University, Lahore, Pakistan*
[4] *Department of Agricultural Engineering, Khawaja Fareed University of Engineering and Information Technology, Rahim Yar Khan, Punjab, Pakistan*
[5] *National Institute for Genomics and Advanced Biotechnology, National Agricultural Research Centrey, Park Road, Islamabad 45500, Pakistan*
[6] *Rice Research and Training Center (RRTC), Rice Research Department, Field Crops Research Institute, Agricultural Research Center, Sakha, Kafr El-Sheikh, Egypt*

CHAPTER MENU

21.1 Introduction

The phenomena of food insecurity and malnutrition are causing serious threats to human lives, especially in developing countries. To stay healthy, it is mandatory to have all essential components of nutrition confirming quality of food. The uncontrolled environmental threats along with unwise utility of non-renewable energy resources are posing a great threat to per capita food availability. Additionally, the global human population is expected to increase by 8.3 billion by 2030, making the situation more serious (United Nations, Department of Economic and Social Affairs, Population Division 2017). To ensure food security, there is dire need to increase the agricultural production by around 50–100% from the available resources (Baulcombe 2010). Plants serve as the primary source for food, feed and energy essential to sustain life on earth. The tremendous upsurge in the human population and global climatic variability over the years requires concrete efforts to breed efficient plants. Plant breeders have utilized natural and induced mutations, heterosis

Genome Engineering for Crop Improvement, First Edition. Edited by Santosh Kumar Upadhyay.

breeding and the genome-engineering approach to address food security (Sedeek et al. 2019). A lot more effort is required to tackle present and emerging food-security challenges. The advent of the green revolution in the mid-twentieth century, the application of agrochemicals, and classical breeding efforts to develop semi-dwarf high yielding varieties helped in maximizing crop production around the globe (Bhargava and Srivastava 2019). At present, researchers have concluded that efforts of classical breeding alone can no longer provide a solution to feed the ever-increasing human population. Therefore, it is now the time to switch on alternative approaches that are much more flexible, reliable and proven to be sustainable for boosting agricultural productivity with least expense of non-renewable natural resources (Chakraborty et al. 2010). To enhance crop production, breeding efforts mainly focused on increasing production from the available unit area. To achieve this objective, breeders gave key importance to yield and yield-related traits (Voss Fels et al. 2019). To avoid crop failure and achieve yield stability, the major focus remains on traits withstanding against stress environment. For this purpose, the germplasm is screened under various abiotic stresses especially heat, cold, salinity, drought and high temperature (Ali et al. 2019). The biotic stresses involve bacterial, viral and fungal diseases which may have the most devastating impact on the global crop production system (Butt et al. 2018). The identification of loci controlling disease resistance and their introduction into the elite germplasm can be utilized as a viable approach. Moreover, it will ultimately reduce the energy requirement of plants to resist against pathogens and the same energy will be converted into high yield. These approaches also hold the ability to develop crops with better nutrition quality essential for human health (Hefferon 2015).

Recent developments being made in the field of agricultural biotechnology have made it possible to develop crops with better yield, nutritional quality and resistance for biotic and abiotic stresses through engineering pathways controlling these traits (Mugode et al. 2014; Ye et al. 2000). These techniques are complementing the drawbacks of classical breeding approaches and are easy to operate as the genomic data of most crops is available publicly. The novel plant breeding techniques (NPBTs) largely rely on genomic data to gain access to genetic information of traits essential for food security (Datta 2013). Genome edit crop have the potential to produce enough under diverse environmental conditions (Ghosh et al. 2011). The area undergoing genome-engineered crops is increasing with every passing year, however, regulatory concerns need the attention of researchers and governing bodies (Dona and Arvanitoyannis 2009). Meanwhile, several countries have adopted policies to cultivate genome-engineered crops and it can be expected the opposition to novel techniques may reduce (Ronald 2011). In this chapter, the possible role of classical and NPBTs for crop improvement programs, the role of genomics in food security, the utilization of genome-editing tools to achieve food security goal will be discussed in detail.

21.2 Plant Breeding for Food Security

21.2.1 Classical Plant Breeding

Crop improvement through traditional breeding approaches has been practiced for quality and production enhancements. The traditional crop improvement programs are largely based on the selection from already available germplasm or mutants with elite attributes

(Cobb et al. 2019). The selection consumes several generations of germplasm to fix desirable traits under severe limitations. The technique of crossing plants with divergent characteristics and selection of suitable recombinant offspring has been adopted as an alternative. The crossing technique is believed to be the foundation of conventional breeding based on the selection of recombinants (Moose and Rita 2008). The recombinants may have one attribute of supreme importance or the most valuable progeny potential to be utilized in future breeding programs. Sometimes, the wide crosses are also performed to introduce gene/genes controlling complex traits, i.e. salt tolerance from tolerant wild relatives to elite cultivars (Pérez-de-Castro et al. 2012). The development of mutants through the application of radiations or chemical mutagenic agents may supplement the objective through the modification of genes at random locations. The resulting mutants need screening and selection latterly, the mutants with desirable phenotypes are backcrossed over multiple generations, sometimes taking several years. The developments in the mid-twentieth century laid the foundation of the green revolution, utility of agrochemicals and cultivars with a potential of high yield developed through traditional breeding methods contributing significantly to achieve food security (Bhargava and Srivastava 2019). However, the traditional breeding methods alone can no longer produce enough to meet the food demands of the ever-growing population. Therefore, it is recommended to promote sustainable agricultural practices for increasing production via conservation of precious non-renewable resources (Singh and Singh 2017). The agricultural biotechnological methods are providing viable alternatives to traditional breeding efforts to cope with worldwide food demands. The modern plant biotechnological approaches have much potential to be tapped into to meet demands for high yield, nutrition, biotic and abiotic resistant crop varieties.

21.2.2 Modern Plant Breeding

The modern breeding program consists of all the allied disciplines making plant breeding program multidisciplinary, versatile and efficient in nature. The availability of whole genome-sequencing data of several crops, i.e. rice, wheat, maize and cotton have revolutionized crop improvement programs. The genomics data have developed a platform for plant breeders to understand the genetic architecture of crops and traits of interest. The availability of genome sequencing of several crops, i.e. rice, sorghum and maize has made the application of DNA markers a reliable approach to identify variation between parents and plant populations (Fiaz et al. 2019). These markers are applied in several ways in the process of breeding commonly for marker-assisted selection (MAS). The MAS commonly applied in hybridization whereas, the markers are also applied for mapping of genes (Sheng et al. 2019). From the last few decades, the utilization of gene-expression techniques has helped to identify the genes essential for the crop improvement programs (Jaganathan et al. 2018). The gene-expression strategies utilize allelic diversity and provide information for the usefulness of specific alleles and can be introgressed into the breeding material. Gene-expression strategies include expressed sequence tags (ESTs), DNA microarray, RNA-seq, HiSeq, and MiSeq techniques based on Illumina and next-generation sequencing (NGS) have provided gene expression for agronomically important genes in maize, wheat, barley and rice essential for food security (Wei et al. 2013). The application of forward genetic approaches to identify naturally existing genetic variations or mutants has helped to generate the germplasm resistance against biotic and abiotic factors. Since, the natural variation or mutation is rare for agronomic traits therefore, plant breeders exploit the induced mutation methods (Viana et al. 2019).

The induced mutation is random in nature and utilizes chemical mutagens and physical agents including gamma rays and fast neutron. These mutagenic agents cause random mutations and a large mutant population is required to be screened to select a desirable mutant plant that is further introgressed for any crop improvement program (Wei et al. 2016). The insertion of jumping genes into the exons disrupts the protein formation and ultimately gives rise to mutation. Similarly, the transfer DNA (T-DNA) has also been used to develop insertion libraries. These engineered T-DNAs are used to carry marker, antibodies and herbicide-resistant genes (Tax and Vernon 2001). To make confirmation, the co-segregation analysis is conducted for the confirmation of inserted allele with novel phenotype. Recently, plant scientists are widely utilizing reverse genetic approaches to understand the functional aspects of genes controlling traits of interest. RNA interference (RNAi) is much-utilized technology for gene silencing; mutagenesis accomplished by a vector which causes double-stranded RNA, ultimately degrading the RNA produced by the gene under investigation (Hammond et al. 2001). Similar to RNAi, virus-induced gene silencing (VIGS), a non-transgenic technique, is employed to produce non-heritable transient phenotypes (Dommes et al. 2019) whereas, RNAi-induced mutations are heritable in nature. The transgenic approaches give rise to the development of genetically modified (GM) crops with the ability to produce higher yields, lower pesticide use, poverty reduction, and improved nutrition (Nicolia et al. 2014). These GM crops are adopted by emerging economies based on the fact that these crops do not pose a risk to human health and the environment. However, some countries in Africa and Asia are reluctant to adopt GM because of erroneously perceived risks and fears of losing export markets to Europe (Zaidi et al. 2019). Before discussing NPBTs, or genome editing, it is important to discuss the role of genomics in ensuring food security.

21.2.3 Genomics for Food Security

Genomics has a role to play in maximizing the utility, diversity and yield of resources, as well as in contributing to sustain food security in the future. Ensuring agricultural production meets the goal of feeding a world experiencing continued human population growth and increasingly severe effects from climate change is an urgent challenge (Barman et al. 2019). To feed the ever-growing population of the world is a scientific, logistical and humanitarian obligation involving a holistic approach, evolving from farmers and plant breeders and impacted by governmental policies and regulations. Global food security is greatly influenced by geographical locations, cultural values in combination with ever-evolving needs based on environmental impact (Fiaz et al. 2016). Though there are several substantial barriers requiring many efforts by all stakeholders to enhance the efficiency and production from available resources. The exploitation of agricultural genomics, being a powerful tool, has proved its potential previously and much can be benefited in the future. The public genome-sequencing forums, genotyping, genome-wide association and genomic-prediction technologies and their application have revolutionized the crop improvement programs (Thottathil et al. 2016). The genomic developments had helped to develop cultivars with premium qualities, i.e. yield, quality, biotic and abiotic stress resistance. However, the greater threat to crops are the ever-changing climatic conditions which impact food security greatly. Genomics have the ability to serve the purpose through the collection of wide sequences and classification of gene banks containing large data from across the species based on their taxonomy and center of origin. These gene banks and wide collections, also known as genetic resources, can further be utilized for crop improvement programs to withstand against changing climatic conditions but need to overcome the logistical challenges (Wambugu et al. 2018). The development of improved cultivars with the potential to maximize yield is the priority for both plant breeders and policymakers to achieve the agenda of zero hunger. The genomics can serve as the reservoir to devise the novel strategies especially the utilization of genome engineering/editing which is more practical and sustainable on the scientific front (Schreiber et al. 2018). These approaches have a better ability to meet future challenges associated with food security to make the world hunger-free.

21.2.4 Genome Engineering for Food Security

Modern agricultural practices emphasize optimized utilization of soil, water, high tech-machinery and facilities, premium seed quality and postharvest management. The implementation of all these components require a significant amount of capital which is difficult to manage on small-scale farming (Virk et al. 2020). Therefore, it becomes immensely important to develop crops through breeding to reduce the production cost. There are several factors impacting crop production including a reduction in arable land, climate change and biotic and abiotic factors (Noor et al. 2019). To withstand against all these barriers, it is absolutely vital to utilize NPBTs especially with utilization of genome-editing technology (Zhang et al. 2018). The research and development in phenotyping and "omics"-based approaches have helped in the identification of a various number of major genes controlling important traits. The identified genes require further validation and functional analysis to understand their mode of action and nature; an interesting scientific front for researchers. Functional gene studies are labor-intensive, laborious and time-consuming, therefore, many of the important and worthy genes remain unnoticed. Advancement in genome-engineering approaches has provided a better alternative to study genetic architecture controlling contrasting phenotypes (Alqudaha et al. 2020). Recent advances in genome editing allow the alteration of endogenous genes to improve traits in crops without transferring transgenes across species boundaries. In recent years, clustered regularly interspaced short palindromic repeats (CRISPR)-associated system (Cas) has emerged as a viable approach to edit crops' genomes important for food security (Fiaz et al. 2019). Previously, transgenic approaches were utilized in which foreign DNA was introduced to develop GM crops which further passed through heavy regulations. The CRISPR-edited crops do not come under the umbrella of GM due to the absence of foreign DNA resulting in lowering the costs of regulatory processes (Wang et al. 2019), better quality seed and affordable seed to farmers, especially in developing countries (Figure 21.1).

The knockout of loci controlling grain yield, i.e. *GS3*, *DEP1*, *GS5*, *GW2*, *Gnla*, and *TGW6*, helped to understand the role of these genes in genetic architecture of grain-yield related traits. The mutants of these genes give the desired, impressive phenotypes with enhanced

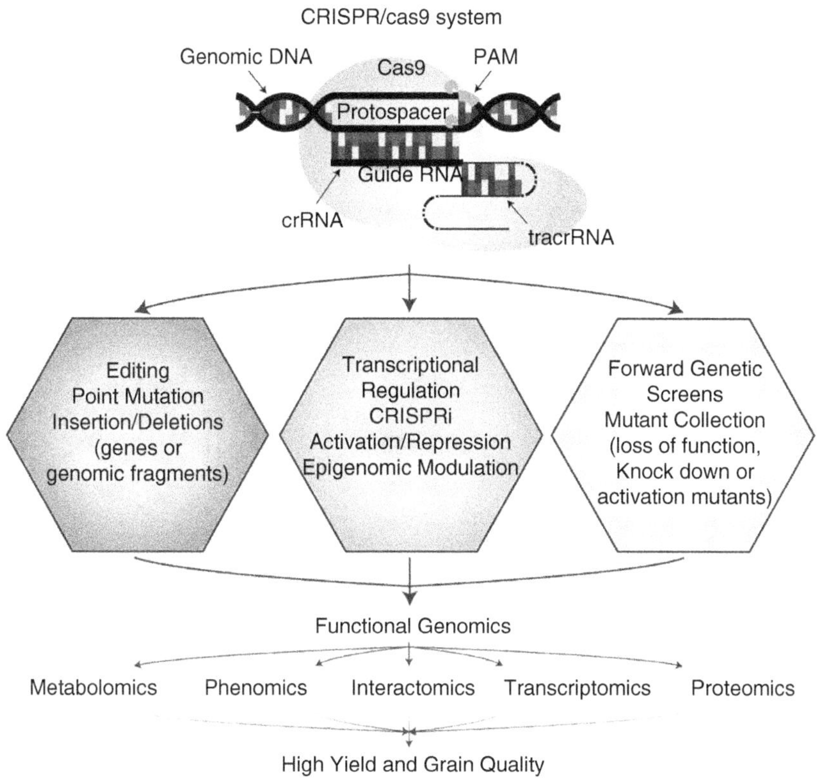

Figure 21.1 A schematic representation of CRISPR/Cas9 system to edit genes controlling traits ensuring high yield and quality. *Source:* Modified from Fiaz et al. (2019).

grain number, dense erect panicles and larger grain size (Li et al. 2016). The application of CRISPR has reduced the time required for hybrid-seed development (Barman et al. 2019). Recently, a novel male fertility related gene in Wheat *Ms1* has been identified and the biallelic frameshift mutation resulted into complete male sterility that can be utilized for commercial hybrid-seed production (Okada et al. 2019). Further, the knockout of *ZmTMS5* and *OsTMS5* in maize and rice, respectively resulted in thermosensitive lines. These lines contain a fertile female part and are successfully utilized for hybrid-seed production (Li et al. 2017). Moreover, a transgene-free maize male sterile line was developed through the targeted mutagenesis of *MS8* gene and proved a valuable source for hybrid development (Chen et al. 2018). The segregation of high value traits such as hybrid vigor or heterosis, has proved to be a major challenge for researchers and farmers. Khanday et al. (2019) triple knockout of the genes *BBM1*, *BBM2*, and *BBM3* caused embryo arrest and abortion, which are fully rescued by male-transmitted *BBM1*. These findings suggest that the requirement for fertilization in embryogenesis is mediated by male-genome transmission of pluripotency factors. When genome editing to substitute the mitosis for meiosis (*MiMe*) phenotype (d'Erfurth et al. 2009; Mieulet et al. 2016) is combined with the expression of *BBM1* in the egg cell, clonal progeny can be obtained that retains genome-wide parental heterozygosity. The synthetic asexual-propagation trait is heritable through multiple generations of clones.

Moreover, multiplex editing of three (*REC8*, *PAIR1*, and *OSD1*) key meiotic genes in hybrid rice leads to the production of clonal diploid gametes and tetraploid seeds. Next, editing of the *MTL* (Kelliher et al. 2017; Li et al. 2017) gene involved in fertilization results in the induction of haploid seeds in hybrid rice. By simultaneous editing of these four endogenous genes in hybrid rice, using the CRISPR/Cas9 system, plants able to propagate clonally through seeds were obtained. The quadruple AOP (Apomictic Offspring Producer) mutants obtained through the knockout of *OsSPO11-1*, *OsREC8*, *OsOSD1*, and *OsMATL* produced the MiMe phenotype. The mutant has the ability to produce apomictic plants (Ferreira et al. 2018). However, thus far, the number of viable clone seeds with their original hybrid genetics intact are much lower than expected and this area needs further research. Similarly, in wheat, *TaMLO* knock out confirmed resistance to powdery mildew disease caused by *Blumeria graminis* f. sp. *Tritici* (Btg) (Wang et al. 2014). Through multiplex system genes, *TaGW2* (a negative regulator of grain traits), *TaLpx-1* (lipoxygenase, which provides resistance to *Fusarium graminearum*) and *TaMLO* (loss of function, confers resistance to powdery mildew resistance) were genome edited in hexaploid wheat. The resulting plants show promising phenotype and proved CRISPR system can be efficiently utilized in polyploidy crops (Wang et al. 2018). In maize, AUXIN REGULATED GENE INVOLVED IN ORGAN SIZE (ARGOS) gene family was studied through the CRISPR method and created new allelic variants of *ARGOS8*. The allelic variant *ARGOS8-v1* and *ARGOS8-v2* were further utilized for the hybrid production at field and multi-location trials. Similarly, CRISPR/Cas9 method has shown excellent promise for introducing resistance against virus-induced diseases mediated by begomoviruses. Begomoviruses are transmitted through white flies and impact greatly in terms of yield losses in several agriculturally important crops including cotton, tomato and cassava (Uniyal et al. 2019). Moreover, the pathogen-derived resistance (PDR) is the most widely adopted method in which the viral sequence is inserted in plant cells for the development of virus resistance. Yin et al. (2019) utilized a Cas9 system for targeting Cotton Leaf Curl Multan virus (*CLCuMuV*) genome. The targeted genome conferred complete resistance to virus infection. In another study, Mubarik et al. (2019) demonstrated CRISPR/Cas9-targeted mutagenesis to suppress cotton leaf curl disease *CLCuD*) disease symptoms instigated by begomoviruses. The indel mutation through CRISPR in *CLCuV* genome showed 65–75% reduction in virus titer. Therefore, this technique could be utilized as a sustainable method for the development of resistant crops against diseases mediated by DNA viruses. The disease-resistant crops can ultimately produce more yield contributing their due share to attain food security. In short, the genome-editing method has demonstrated great potential to be utilized in a crop improvement program to ensure food security.

21.3 Conclusion

Global food security is a big challenge having multifaceted social and economic implications requiring global coordinated efforts. Achieving global food security will require a framework based on the lessons learned from the past. Innovation is essential, and thus an environment facilitating innovation is also essential. To fully exploit the potentials of NPBTs, a multipronged approach is needed, taking into consideration all components involved in technology development, dissemination, adoption, and social acceptance. NPBTs should not be misunderstood as a panacea. Many other technologies and approaches are also needed, including improvements in plant-breeding efforts, genomics, postharvest management, market infrastructure, and social services. However, genome editing is predicted to be a powerful addition in the fight against hunger and poverty. The global community should seize this opportunity by developing conducive regulatory frameworks and support mechanisms. Moreover, rapid developments in genome-editing technologies will further decrease the cost and time for crop optimization based on government support to establish and update regulatory policies. To conclude, ethical standards and food security concerns tend to be regional specific, the genome edit crops require scientific oversights and potential risk assessment keeping consumers and farmers' needs a priority.

Conflict of Interest

The authors have no competing interest to be declared.

Acknowledgement

The authors are thankful to Mr. Shakeel Ahmad from China National Rice Research Institute, Hangzhou, P.R. China for providing illustration 21.1.

References

Ali, U., Zhong, M., Shar, T. et al. (2019). The influence of Ph on cadmium accumulation in seedlings of Rice (*Oryza sativa* L.). *Journal of Plant Growth Regulaiton*: 1–11. https://doi.org/10.1007/s00344-019-10034-x.

Alqudaha, A.M., Sallamb, A., Baenzigerc, P.S., and Börner, A. (2020). GWAS: fast-forwarding gene identification and characterization in temperate cereals: lessons from barley-a review. *Journal of Advance Research* 22: 119–135.

Barman, H.N., Sheng, Z., Fiaz, S. et al. (2019). Generation of a new thermo-sensitive genic male sterile rice line by targeted mutagenesis of TMS5 gene through CRISPR/Cas9 system. *BMC Plant Biology* 19: 109.

Baulcombe, D. (2010). Reaping benefits of crop research. *Science* 327: 761–710.

Srivastava, S. and Bhargava, A. (2019). Plant breeding. In: *Participatory Plant Breeding: Concept and Applications*, 29–102. Singapore: Springer https://doi.org/10.1007/978-981-13-7119-6_2.

Butt, H., Jamil, M., Wang, J.Y. et al. (2018). Engineering plant architecture via CRISPR/Cas9-mediated alteration of strigolactone biosynthesis. *BMC Plant Biology* 18: 174.

Chakraborty, S., Chakraborty, N., Agrawal, L. et al. (2010). Next generation protein rich potato by expressing a seed protein gene AmA1 as a result of proteome rebalancing in transgenic tuber. *Proceeding of National Academy of Sciences* 41: 17533–17538.

Chen, R., Xu, Q., Liu, Y. et al. (2018). Generation of transgene-free maize male sterile lines using the CRISPR/Cas9 system. *Frontiers in Plant Sciences* 9: 1180.

Cobb, J.N., Juma, R.U., Biswas, P.S. et al. (2019). Enhancing the rate of genetic gain in public-sector plant breeding programs: lessons from the breeder's equation. *Theoretical and Applied Genetics* 132: 627–645.

Datta, A. (2013). Genetic engineering for improving quality and productivity of crops. *Agriculture and Food Security* 2: 15. https://doi.org/10.1186/2048-7010-2-15.

d'Erfurth, I., Jolivet, S., Froger, N. et al. (2009). Turning meiosis into mitosis. *PLoS Biology* 7: 1000124.

Dommes, A.B., Gross, T., Herbert, D.B. et al. (2019). Virus-induced gene silencing: empowering genetics in non-model organisms. *Journal of Experimental Botany* 70: 757–770.

Dona, A. and Arvanitoyannis, I.S. (2009). Health risks of genetically modified foods. *Critical Review Food Science and Nutrition* 49: 164–175. https://doi.org/10.1080/10408390701855993.

Ferreira, L.G., de Alencar, D.D.M., Irsigler, A.S.T. et al. (2018). GID1 expression is associated with ovule development of sexual and apomictic plants. *Plant Cell Report* 37: 293–306.

Fiaz, S., Noor, M.A., and Aldosri, F.O. (2016). Achieving food security in the Kingdom of Saudi Arabia through innovation: potential role of agricultural extension. *Journal of Saudi Society of Agricultural Sciences* 17: 365–375.

Fiaz, S., Lv, S., Barman, H.N. et al. (2019). Analysis of genomic regions governing cooking and eating quality traits using a recombinant inbred population in Rice (*Oryza sativa* L.). *International Journal of Agriculture and Biology* 22: 611–619. https://doi.org/10.17957/IJAB/15.1106.

Fiaz, S., Ahmad, S., Riaz, A. et al. (2019). Applications of the CRISPR/Cas9 system for rice grain quality improvement: perspectives and opportunities. *International Journal of Molecular Sciences* 20: 888.

Ghosh, S., Meli, V.K., Kumar, A. et al. (2011). The N-glycan processing enzymes α-mannosidase and β-D-1 N acetyl hexosaminidase are involved in ripening-associated softening in the non-climacteric fruits of capsicum. *Journal of Experimental Botany* 62: 571–582.

Hammond, S.M., Caudy, A.A., and Hannon, G.J. (2001). Post-transcriptional gene silencing by double-stranded RNA. *Nature Reviews Genetics* 2: 110–119.

Hefferon, K.L. (2015). Nutritionally enhanced food crops; progress and perspectives. *International Journal of Molecular Sciences* 16: 3895–3914.

Jaganathan, D., Ramasamy, K., Sellamuthu, G. et al. (2018). CRISPR for crop improvement: an update review. *Frontiers in Plant Sciences* 9: 985.

Kelliher, T., Starr, D., Richbourg, L. et al. (2017). MATRILINEAL, a sperm-specific phospholipase, triggers maize haploid induction. *Nature* 542: 105.

Khanday, I., Skinner, D., Yang, B. et al. (2019). A male-expressed rice embryogenic trigger redirected for asexual propagation through seeds. *Nature* 565: 91–95.

Li, M., Li, X., Zhou, Z. et al. (2016). Reassessment of the four yield-related genes *Gn1a, DEP1, GS3,* and *IPA1* in rice using a CRISPR/Cas9 system. *Frontiers in Plant Sciences* 7: 377.

Li, X., Meng, D., Chen, S. et al. (2017). Single nucleus sequencing reveals spermatid chromosome fragmentation as a possible cause of maize haploid induction. *Nature Communication* 8: 991.

Mieulet, D., Jolivet, S., Rivard, M. et al. (2016). Turning rice meiosis into mitosis. *Cell Research* 26: 1242–1254.

Moose, S.P. and Rita, H. (2008). Molecular plant breeding as the foundation for 21st century crop improvement. *Plant Physiology* 147: 969–977.

Mubarik, M.S., Khan, S.H., Sadia, B., and Ahmad, A. (2019). CRISPR-Cas9 based suppression of cotton leaf curl virus in Nicotiana benthamina. *International Journal of Agriculture and Biology* 22: 517–522.

Mugode, L., Ha, B., Kaunda, A. et al. (2014). Carotenoid retention of biofortified provitamin a maize (*Zea mays* L.) after Zambian traditional methods of milling, cooking and storage. *Journal of Agriculture and Food Chemistry* 62: 6317–6325.

Nicolia, A., Manzo, A., Veronesi, F., and Rosellini, D. (2014). An overview of the last 10 years of genetically engineered crop safety research. *Critical Reviews in Biotechnology* 34: 77–88.

Noor, M.A., Nawaz, M.M., Hassan, M. et al. (2019). Small farmers and sustainable N and P management: implication and potential under changing climate. In: *Carbon and Nitrogen Cycling in Soil*, 1e (eds. R. Datta et al.), 185–219. Springer Nature Singapore Pte Ltd. 2020 https://doi.org/10.1007/978-981-13-7264-3_6185.

Okada, A., Arndell, T., Borisjuk, N. et al. (2019). CRISPR/Cas9-mediated knockout of Ms1 enables the rapid generation of male-sterile hexaploid wheat lines for use in hybrid seed production. *Plant Biotechnology Journal* 17: 1905–1913.

Pérez-de-Castro, A.M., Vilanova, S., Cañizares, J. et al. (2012). Application of genomic tools in plant breeding. *Current Genomics* 13: 179–195.

Ronald, P. (2011). Plant genetics, sustainable agriculture and global food security. *Genetics* 188: 11–20.

Schreiber, M., Stein, N., and Mascher, M. (2018). Genomic approaches for studying crop evolution. *Genome Biology* 19: 140.

Sedeek, K.E.M., Mahas, A., and Mahfouz, M. (2019). Plant genome engineering for targeted improvement of crop traits. *Frontiers in Plant Sciences* 10: 114.

Sheng, Z., Fiaz, S., Li, Q. et al. (2019). Molecular breeding of fragrant early-season hybrid rice using the BADH2 gene. *Pakistan Journal of Botany* 51 (6) https://doi.org/10.30848/PJB2019-6(10).

Singh, R. and Singh, G.S. (2017). Traditional agriculture: a climate-smart approach for sustainable food production. *Energy Ecology and Environment* 2: 296–316. https://doi.org/10.1007/s40974-017-0074-7.

Tax, F.E. and Vernon, D.M. (2001). T-DNA-associated duplication/translocations in Arabidopsis. Implications for mutant analysis and functional genomics. *Plant Physiology* 126: 1527–1538.

Thottathil, G.P., Jayasekaran, K., and Othman, A.S. (2016). Sequencing crop genomes: a gateway to improve tropical agriculture. *Tropical Life Science Research* 27: 93–114.

United Nations, Department of Economic and Social Affairs, Population Division. (2017). The Impact of Population Momentum on Future Population Growth. https://population.un.org/wpp/Publications/Files/PopFacts_2017-4_Population-Momentum.pdf (accessed 20 July 2020).

Uniyal, A.P., Yadav, S.K., and Kumar, V. (2019). The CRISPR-Cas9, genome editing approach: a promising tool for drafting defense strategy against begomoviruses including cotton leaf curl viruses. *J. Plant Biochemistry and Biotechnology* 28: 121–132.

Viana, V.E., Pegoraro, C., Busanello, C., and de Oliveira, A.C. (2019). Mutagenesis in rice: the basis for breeding a new super plant. *Frontiers in Plant Sciences* 10 (1326) https://doi.org/10.3389/fpls.2019.01326.

Virk, A.L., Noor, M.A., Fiaz, S. et al. (2020). Smart farming: an overview. In: *Smart Village Technology, Modeling and Optimization in Science and Technologies* (eds. A.G. Switzerland, S. Patnaik, et al.), 17. Springer Nature https://doi.org/10.1007/978-3-030-37794-6_10191.

Voss Fels, K.P., Stahl, A., and Hickey, L.T. (2019). Q&A: modern crop breeding for future food security. *BMC Biology* 17: 18. https://doi.org/10.1186/s12915-019-0638-4.

Wambugu, P.W., Ndjiondjop, M.N., and Henry, R.J. (2018). Role of genomics in promoting the utilization of plant genetic resources in gene banks. *Briefings in Functional Genomics* 17: 198–206.

Wang, Y., Cheng, X., Shan, Q. et al. (2014). Simultaneous editing of three homoeoalleles in hexaploid bread wheat confers heritable resistance to powdery mildew. *Nature Biotechnology* 32: 947–951.

Wang, W., Pan, Q., He, F. et al. (2018). Transgenerational CRISPR-Cas9 activity facilitates multiplex gene editing in allopolyploid wheat. *CRISPR Journal* 1: 65–74.

Wang, T., Zhang, H., and Zhu, H. (2019). CRISPR technology is revolutionizing the improvement of tomato and other fruit crops. *Horticulture Research* 6: 77.

Wei, L., Xiao, M., Hayward, A., and Donghui, F. (2013). Applications and challenges of next-generation sequencing in Brassica species. *Planta* 238: 1005–1024.

Wei, F.J., Kuang, L.Y., Oung, H.M. et al. (2016). Somaclonal variation does not preclude the use of rice transformants for genetic screening. *Plant Journal* 85: 648–659.

Ye, X., Al-Babili, S., Klöti, A. et al. (2000). Engineering the provitamin A (β-carotene) biosynthetic pathway into (carotenoid-free) rice endosperm. *Science* 287: 303–305.

Yin, K., Han, T., Xie, K. et al. (2019). Engineer complete resistance to cotton leaf curl multan virus by the CRISPR/Cas9 system in Nicotiana benthamiana. *Phytopathology Research* 1: 9.

Zaidi, S.S.A., Vanderschuren, H., Qaim, M. et al. (2019). New plant breeding technologies for food security. *Science* 363: 1390–1391.

Zhang, Y., Massel, K., Godwin, I.D., and Gao, C. (2018). Applications and potential of genome editing in crop improvement. *Genome Biology* 19: 210.

Index